$45.00
DS

D1788427

12/25
STRAND PRICE
$5.00

PHYSICAL CHEMISTRY SOURCE BOOK

THE McGRAW-HILL SCIENCE REFERENCE SERIES

Acoustics Source Book
Communications Source Book
Computer Science Source Book
Fluid Mechanics Source Book
Meteorology Source Book
Nuclear and Particle Physics Source Book
Optics Source Book
Solid-State Physics Source Book
Spectroscopy Source Book

PHYSICAL CHEMISTRY SOURCE BOOK

Sybil P. Parker, *Editor in Chief*

McGRAW-HILL BOOK COMPANY

New York   St. Louis   San Francisco
Auckland   Bogotá   Caracas   Colorado Springs   Hamburg
Lisbon   London   Madrid   Mexico   Milan   Montreal
New Delhi   Oklahoma City   Panama   Paris   San Juan
São Paulo   Singapore   Sydney   Tokyo   Toronto

Cover: Phase equilibrium exemplified by 11 crystals of benzene I coexisting with liquid benzene, at about 680 atmospheres and room temperature. (*National Bureau of Standards*)

This material has appeared previously in the McGRAW-HILL ENCYCLOPEDIA OF SCIENCE AND TECHNOLOGY, 6th Edition, copyright © 1987 by McGraw-Hill, Inc. All rights reserved.

PHYSICAL CHEMISTRY SOURCE BOOK, copyright © 1988 by McGraw-Hill, Inc. All rights reserved. Printed in the United States of America. Except as permitted under the United States Copyright Act of 1976, no part of this publication may be reproduced or distributed in any form or by any means, or stored in a data base or retrieval system, without prior written permission of the publisher.

1 2 3 4 5 6 7 8 9 0   DOC/DOC   8 9 5 4 3 2 1 0 9 8

ISBN 0-07-045504-X

Library of Congress Cataloging in Publication Data:

Physical chemistry source book / Sybil P. Parker, editor in chief.
    p. cm.—(McGraw-Hill science reference series)
    "This material has appeared previously in the McGraw-Hill encyclopedia of science and technology, 6th edition"—CIP t.p. verso.
  Bibliography: p.
  Includes index.
  ISBN 0-07-045504-X
    1. Chemistry, Physical and theoretical—Dictionaries.
I. Parker, Sybil P.   II. McGraw-Hill Book Company.
III. Title: McGraw-Hill encyclopedia of science and technology (6th ed.)   IV. Series.
QD451.P49 1988   87-36629
541.3—dc19

TABLE OF CONTENTS

Introduction	1
Chemical Thermodynamics	7
Chemical Reactions	53
Surface Chemistry	97
Transport Processes	119
Matter: Structure and Properties	135
Electrochemistry	241
Electroanalytical Chemistry	281
Cells and Batteries	297
Optical Phenomena	331
Specialized Fields of Study	359
Contributors	393
Index	399

PHYSICAL CHEMISTRY SOURCE BOOK

INTRODUCTION

PHYSICAL CHEMISTRY deals with the interpretation of chemical phenomena and properties in terms of the underlying physical processes, and with the development of techniques for their investigation. The term chemical physics is often employed to denote a branch of physical chemistry where the emphasis is on the interpretation and analysis of the physical properties of individual molecules and bulk systems, instead of their reactions. Theoretical chemistry is another major branch, where the emphasis is on the calculation of the properties of molecules and systems, and which uses the techniques of quantum mechanics and statistical thermodynamics. For the present purpose it is convenient to regard physical chemistry as dealing with three aspects of matter: its equilibrium properties, structure, and ability to change.

Chemical thermodynamics. The study of matter in a state of equilibrium constitutes the field of chemical thermodynamics. In particular, chemical thermodynamics provides a technique for discussing the response of a system to a change in the external conditions (such as the shift in the boiling and freezing point of either a pure substance or a mixture when the applied pressure is changed, or when the composition of the mixture is modified), and for rationalizing the energy changes that occur in the course of a chemical reaction. The branch of thermodynamics dealing with the latter is called thermochemistry. Chemical thermodynamics also provides a framework for the determination of the maximum amount of work that may be generated by a system undergoing a specified change, and it therefore provides a way of establishing bounds for the efficiencies of a variety of devices, including engines, refrigerators, and electrochemical cells. Thermodynamics is used in chemistry to assess the position of equilibrium of a chemical reaction (that is, how far it will proceed), and to determine what conditions are necessary in order to optimize the yield of a particular product. The branch of chemical thermodynamics dealing with ionic reactions occurring in the presence of electrodes constitutes the field of equilibrium electrochemistry.

Chemical thermodynamics is based on the laws of thermodynamics. In chemistry the most important thermodynamic properties are the enthalpy H and the Gibbs function G (which is also called the free energy). The change of enthalpy may be identified as the heat transferred to a system during a specified change under conditions of constant pressure. Tables of enthalpies of materials have been compiled, and from them it is possible to predict the amount of heat available (or required) for a particular reaction. Not the whole of the heat output of a reaction is available to do work (such as mechanical work, or work to drive other reactions toward a particular desired product). The Gibbs function expresses the maximum amount of work, other than work of expansion, that a specified process may generate under conditions of constant pressure, and as such it may be used to assess whether or not one reaction may be used to drive another in an unnatural direction. For example, the assessment of the changes in the Gibbs function that accompany biochemical reactions may be used to discuss the processes that occur in living cells, where the ingestion of food leads ultimately to growth, mechanical work, and nervous activity.

The fundamental basis of the Gibbs function lies in the tendency of energy to attain a condition of greatest dispersal. This tendency is expressed most generally in terms of the entropy of a system: as the dispersal of energy increases, so too does the entropy. The concept of entropy may be given a sharp and quantitative definition, and entropies of substances have been determined and tabulated. The crucial feature for the present purpose, though, is that whereas the entropy of the universe increases whenever a spontaneous change occurs, the chemist is normally interested in the changes that occur in the system under investigation. The Gibbs function focuses attention on changes in the properties of the system itself, for G automatically takes into account the changes in entropy of the surroundings. In order to determine whether a specified change has a natural tendency to occur under conditions of constant temperature and pressure, it is only necessary to assess whether that change is accompanied by a decrease in the Gibbs function of the system. If a change is accompanied by an increase of Gibbs function, it may still be achieved by coupling the system to another in which a change is occurring such that the overall change in the Gibbs function of the combined system is negative. Since the Gibbs function depends on the composition of the system, it is possible to use it to predict the composition of the system when it has attained equilibrium (such as when some reaction mixture attains some constant composition).

The crucial equation is $\ln K = \Delta G_m^\ominus / RT$, where K is the equilibrium constant of the reaction (in essence a simple function of the concentration of the components at equilibrium), R is the gas constant, T the absolute temperature, and $\Delta G_m^\ominus$ the standard molar Gibbs function for the reaction (the change in the Gibbs function under certain specified, standard conditions). This equation lies at the heart of chemical thermodynamics, and is of great practical importance.

Equilibrium electrochemistry. In equilibrium electrochemistry, attention is focused on systems in which a chemical reaction may release energy by driving electrons through some external circuit. Under certain conditions the potential difference between the two electrodes immersed in the reaction mixture is related to the Gibbs function for the reaction by $E = -\Delta G_m/F$, where F is the Faraday constant, the charge carried by 1 mol of electrons ($F = 96{,}485$ coulombs mol^{-1}). The potential difference under these conditions is denoted E, and is called the electromotive force (emf) of the cell. One object of electrochemistry is to measure the emf of cells, and then to use the tabulated results to discuss the position of equilibria in ionic reactions (via ΔG) and their response to changes of conditions. Practical applications include chemical analysis, the assessment of the power generation and storage capabilities of electrochemical cells and fuel cells, the discussion of tendencies to corrosion, and the analysis of potential differences across biological membranes (such as are responsible for the propagation of nerve impulses).

Quantum chemistry. The principal role of quantum mechanics in chemistry is in the discussion of atomic and molecular structure, and in the interpretation of spectroscopic data. In the branch of physical chemistry known as computational quantum chemistry, interest centers on the numerical solution of the Schrödinger equation in order to obtain wave functions and geometries of molecules. In ab initio calculations the computations are done without any appeal to experimental data, and an attempt is made to predict properties from first principles (that is, from the masses and charges of the electrons and nuclei constituting the molecule). In semiempirical calculations (which are identified by initials such as CNDO and MINDO) some of the computational difficulties are circumvented (but with some loss of reliability) by incorporating experimental data. It is now possible to calculate the shape, electron distribution, and spectroscopic properties of large molecules. One application is referred to as quantum pharmacology, where a pharmacologically active molecule is first screened by calculating and analyzing its electron distribution; to some extent this eliminates lengthy and expensive laboratory screening procedures. Computational quantum chemistry is so developed that it is capable of being used to map the changes in the structures of molecules while they are in the course of reaction, when atoms and groups of atoms are being transferred from one molecule to another.

Investigative techniques. Spectroscopic techniques are used to identify molecules present in a sample and to determine their size, shape, and electron distribution. The techniques fall into four categories: absorption spectroscopy, emission spectroscopy, Raman spectroscopy, and resonance techniques. Other major techniques for the investigation of molecular structure are based on diffraction. These depend on the observation of the direction through which radiation and particles are scattered when they impinge on a sample. The principal example of these techniques is x-ray diffraction, for through it detailed information about the arrangement of atoms in crystals and even in very complex, large, biologically important molecules may be obtained. Also, there are diffraction techniques based on electrons and neutrons.

Other techniques for investigating structure include the electric and magnetic properties of molecules, in particular, the determination of electric polarizabilities and dipole moments (which give information about electron distribution, and are important for the discussion of the dielectric properties), magnetic properties, and the properties based on optical birefringence, such as optical activity and the Faraday effect. In the case of macromolecules and colloids, important structure-analysis techniques include x-ray diffraction, sedimentation rates, and viscosity measurements.

Statistical thermodynamics. Structural properties and thermodynamic properties are brought together by statistical thermodynamics. This major theoretical procedure gives a way of predicting the thermodynamic properties of assemblies of molecules in terms of their individual energy levels. Statistical thermodynamics represents a grand synthesis of the two major aspects of physical chemistry. It provides an understanding of the bulk, macroscopic properties of molecules in terms of the properties of individual molecules, and it also constitutes an important practical technique for calculating otherwise inaccessible properties from readily available spectroscopic data.

Transport processes. The third major branch of physical chemistry is concerned with change: physical change and chemical change. In particular, it is concerned with the rate of change (while thermodynamics is concerned with the possibility of change, and the criteria of spontaneous change). Physical change includes the diffusion of one substance into another, or the migration of ions in an electrolyte solution.

The simplest version of the former are the transport properties of gases. These include thermal conductivity and viscosity. They are treated most simply in terms of the kinetic theory of gases, in which a gas is regarded as a swarm of noninteracting points; modern physical chemistry is concerned with the role of intermolecular forces in determining the transport properties. Ion migration gives rise to questions of the mobility of ions in a variety of solvents, and to the description of diffusion processes in general. The application of thermodynamics to change in general constitutes the field of nonequilibrium thermodynamics.

Chemical kinetics. The major aspect of chemistry is change, and change is therefore also a major aspect of physical chemistry. Chemical change may be studied at a variety of levels. Empirical chemical kinetics is the study of reactions in order to determine how their rates depend on the concentrations of the participants in the reaction and on the conditions, mainly the temperature. Investigation of the time dependence of reactions (a time dependence that can be observed from the order of minutes to picoseconds, now that pulsed, mode-locked lasers have entered chemistry) yields a detailed picture of the sequence of molecular transformations involved in a complex chemical reaction. Each step in the sequence depends on the concentration of the participants and an empirical constant referred to as the rate coefficient. The aim of theoretical chemistry is to account for the value of the rate coefficient and its dependence on the conditions. The earliest broadly successful approach led to the Arrhenius rate law, which is still used as a general guide to the way that the temperature affects the rate of a reaction. The law asserts that the rate coefficient k varies with temperature as $k = A \exp(-E_a/RT)$, where A and E_a are empirical parameters, the latter being known as the activation energy of the reaction. A wide variety of simple reactions conform to this law, and it can be interpreted on the basis that only molecules possessing sufficient energy are able to undergo reaction when they encounter other species; the exponential factor indicates the fraction of collisions that satisfy the energy requirement.

Molecular reaction dynamics. Modern approaches to the prediction and explanation of observed reaction rates have been based either on a statistical view of the reaction (the activated complex theory), where the principles of statistical thermodynamics are applied to a system evolving with time, or a fundamental particle dynamics approach, where the trajectories of molecules are calculated as they undergo reaction, techniques then being used to convert these trajectories to values of the rate coefficients. The latter approach constitutes the field of molecular reaction dynamics. Experimental techniques have also been developed for observing individual molecular encounters and reactions; these are based on molecular beams, where a diffuse beam of one reactant is directed into the path of another, and the pattern of molecular scattering, including the electronic and vibrational states of the products, is interpreted in terms of the forces acting between the reactants during the reactive encounter. These techniques bring investigations closer to an atomistic and molecular interpretation of chemistry than anything that preceded them, but the relation of trajectories and rate coefficients and the relation of gas-phase events to those in solution remain open problems.

Surface chemistry. Chemical kinetics is so important that it is rich in applications and extensions. An important extension is to the reactions that occur on surfaces; these are the processes involved in heterogeneous catalysis. The study of surface chemistry breaks down into the analysis of the steps that lead to the affixation of the species to the surface (that is, the study of adsorption processes), the determination of the structure of the adsorbed species, and finally the reactions, and escape, of the adsorbed species. The entire subject, including the related processes occurring at liquid surfaces, constitutes the field of surface chemistry. A special application of surface chemistry is to the stability of colloidal suspensions of species in fluids, and another is to the processes that occur at the interface between an electrode and the solution in which it is immersed. Electrode reactions are governed largely by the rate of electron transfer between the ions in the solution and the metal electrode. The field of dynamical electrochemistry finds important applications in electrochemical power generation and storage, in corrosion, in electrodeposition, and in electrocatalysis.

P. W. ATKINS

1

CHEMICAL THERMODYNAMICS

Chemical thermodynamics	8
Thermochemistry	17
Chemical equilibrium	21
Phase equilibrium	27
Fused-salt phase equilibria	35
Donnan equilibrium	38
Phase rule	39
Internal energy	41
Enthalpy	41
Entropy	42
Free energy	44
Gibbs function	46
Activity	47
Fugacity	49
pH	50
pK	51

CHEMICAL THERMODYNAMICS
Robert A. Pierotti

The application of thermodynamic principles to systems involving physical and chemical transformations in order to (1) develop quantitative relationships among the identifiable forms of energy and their conjugate variables, (2) establish the criteria for spontaneous change, for equilibrium, and for thermodynamic stability, and (3) provide the macroscopic base for the statistical-mechanical bridge to atomic and molecular properties. The thermodynamic principles applied are the conservation of energy as embodied in the first law of thermodynamics, the principle of internal entropy production as embodied in the second law of thermodynamics, and the principle of absolute entropy and its statistical thermodynamic formulation as embodied in the third law of thermodynamics.

Basic concepts. The basic goal of thermodynamics is to provide a description of a system of interest in order to investigate the nature and extent of changes in the state of that system as it undergoes spontaneous change toward equilibrium and interacts with its surroundings. This goal implicitly carries with it the concept that there are measurable properties of the system which can be used to adequately describe the state of the system and that the system is enclosed by a boundary or wall which separates the system and its surroundings. Properties that define the state of the system can be classified as extensive and intensive properties. Extensive properties are dependent upon the mass of the system, whereas intensive properties are not. Typical extensive properties are the energy, volume, and numbers of moles of each component in the system, while typical intensive properties are temperature, pressure, density, and the mole fractions or concentrations of the components.

Extensive properties can be expressed as functions of other extensive properties, for instance as in Eq. (1), where the volume V of the system is expressed in terms of the internal energy

$$V = V(U,S,\{n_i\}) \tag{1}$$

U, the entropy S, and $\{n_i\}$, the set of numbers of moles of the various components labeled by the index i. A suitable transformation procedure can be used to replace extensive variables by conjugate intensive variables. For example, the volume can be expressed as in Eq. (2a) or (2b). Since

$$V = V'(P,S,\{n_i\}) \qquad (2a) \qquad V = V''(P,T,\{n_i\}) \qquad (2b)$$

temperature T and pressure P are particularly convenient variables to control and measure in chemical systems, the last form is of great utility. All extensive thermodynamic properties X can be rewritten in this form, namely as Eq. (3), and since all such properties are linear homogeneous

$$X = X(T,P,\{n_i\}) \tag{3}$$

functions of the mass, it can be shown that at a given temperature and pressure Eq. (4) holds, where $\overline{X}_i$ is the partial molar value of the extensive property for the ith component, and is given by Eq. (5), where the notation $\{n_j\}'$, means that all of the mole numbers are constant except the

$$X = \sum_i n_i \overline{X}_i \qquad (4) \qquad \overline{X}_i = (\partial X/\partial n_i)_{T,P,\{n_j\}'} \qquad (5)$$

ith one involved in the derivative. $\overline{X}_i$ is itself intensive.

Specification of boundaries. The concept of a boundary enclosing the system and separating it from the surroundings requires specification of the nature of the boundary and of any constraints the boundary places upon the interaction of the system and its surroundings. Boundaries that restrain a system to a particular value of an extensive property are said to be restrictive with respect to that property. A boundary which restrains the system to a given volume V is a fixed wall. A boundary which is restrictive to one component of a system but not to the other components is a semipermeable wall or membrane. A system whose boundaries are restrictive to energy and to mass or moles of components is said to be an isolated system. A system whose boundaries are restrictive only to mass or moles of components is a closed system, whereas an open system has nonrestrictive walls and hence can exchange energy, volume, and mass with its surroundings. Boundaries can be restrictive with respect to specific forms of energy, and two important types are those restrictive to thermal energy but not work (adiabatic walls) and those restrictive to work but not thermal energy (diathermal walls).

Reversible and irreversible processes. Changes in the state of the system can result from processes taking place within the system and from processes involving exchange of mass or energy with the surroundings. After a process is carried out, if it is possible to restore both the system and the surroundings to their original states, the process is said to be reversible; otherwise the process is irreversible. All naturally occurring spontaneous processes are irreversible. The first law defines the internal energy as a state function or property of the state of a system, and restricts the system and its surroundings to those processes which conserve energy. The second law established which of the permissible processes can occur spontaneously.

First law of thermodynamics. The total energy E of a system is the sum of its kinetic energy T, its potential energy V, and its internal energy U, Eq. (6). If a system has constant mass

$$E = T + V + U \tag{6}$$

and its center of mass is moving with uniform velocity in a uniform potential, then changes in the total energy of the system δE are equal to changes in its internal energy δU. Chemical thermodynamics concentrates on the internal energy of the system, but kinetic and potential energy changes of the system as a whole can be important for chemical systems. The principle of conservation of energy requires that the change in the internal energy of a system be the result of energy transfer between the system and its surroundings. The internal energy U is a function of the set of extensive variables associated with the various forms of internal energy. Each form of internal energy is manifest by the product of an extensive variable and its conjugate intensive variable. **Table 1** lists several forms of internal energy, their conjugate pair of variables, and their corresponding work terms.

The internal energy of the system is given by the fundamental equation of state, Eq. (7).

$$U = U(S, V, \{n_i\}, \{X_j\}) \tag{7}$$

Processes which give rise to a change in U are then limited to those for which Eq. (8) holds. Since

$$dU = (\partial U/\partial S)_{V,\{n_i\},\{X_j\}} dS + (\partial U/\partial V)_{S,\{n_i\},\{X_j\}} dV$$
$$+ \sum_i (\partial U/\partial n_i)_{S,V,\{n_i\}',\{X_j\}} dn_i + \sum_j (\partial U/\partial X_j)_{X,V,\{n_i\},\{X_j\}'} dX_j \tag{8}$$

all the extensive state properties are linear homogeneous functions, the coefficients of the differential terms are themselves intensive and correspond to the conjugate variable of the respective extensive variable. Their product is thus the differential work associated with the appropriate form of external energy. Equation (8) then becomes Eq. (9), where μ_i is the chemical potential of the

$$dU = TdS - pdV + \sum_i \mu_i dn_i + \sum_j I_j dX_j \tag{9}$$

ith component and I_j is the conjugate potential for X_j. The internal energy change given by this

Table 1. Internal energy and generalized work

Type of energy	Intensive factor	Extensive factor	Element of work
Mechanical			
Expansion	Pressure (P)	Volume (V)	—PdV
Stretching	Surface tension (γ)	Area (A)	γdA
Extension	Tensile stretch (F)	Length (l)	Fdl
Thermal	Temperature (T)	Entropy (S)	TdS
Chemical	Chemical potential (gm)	Moles (n)	μdn
Electrical	Electric potential (E)	Charge (Q)	EdQ
Gravitational	Gravitational field strength (mg)	Height (h)	mgdh
Polarization			
Electrostatic	Electric field strength (E)	Total electric polarization (P)	EdP
Magnetic	Magnetic field strength (H)	Total magnetic polarization (M)	HdM

expression is dependent only upon the state properties of the system, and hence is independent of the process causing the change. SEE INTERNAL ENERGY.

Heat. Thermal energy exchange or heat (that form of energy transferred as a result of temperature differences between a system and its surroundings) plays a central role in thermodynamics, and is singled out from the other forms of energy or work. This is expressed by Eq. (10), where δq is the differential thermal energy (heat) absorbed by the system from the surroundings

$$dU = \delta q + \delta w \qquad (10)$$

and δw is the differential work performed on the system by the surroundings. Equating Eq. (9) and (10) yields Eq. (11). Since δw is the work performed by the surroundings on the system, Eq. (11) can be rewritten as Eq. (12), where conservation laws for the nonthermal extensive prop-

$$\delta q = TdS - PdV + \sum_i \mu_i dn_i + \sum_j I_j dX_j - \delta w \qquad (11)$$

$$\delta q = TdS - (P - P_s)dV + \sum_i (\mu_i - \mu_{i_s})dn_i + \sum_j (I_j - I_{j_s})dX_j \qquad (12)$$

erties have been utilized and where the subscript s on the intensive variable identifies it as the value for the surroundings. It is convenient to write this equation as Eqs. (13), where $(-\delta a)$ is the

$$\delta q = TdS - \delta a \qquad (13a) \qquad U = TdS + \delta w - \delta a \qquad (13b)$$

sum of the nonthermal differential work terms in Eq. (12). The term δa can be either zero or nonzero. If it is zero, the heat absorbed by the system is equal to TdS. In an adiabatic process δq is zero and $TdS = \delta a$, and hence if δa is nonzero, it must correspond to an internally generated thermal energy. This is frequently referred to as the uncompensated heat of a process, since it does not result from the transfer of heat from the surroundings. The first law or energy conservation principle can provide no further insight concerning either the sign or magnitude of δa. This will remain for the statement of the second law to consider. Of course, given the initial and final states of a system, the first law permits thermochemical calculations pertaining to such changes.

Heat capacity. The heat capacity of a system is of particular importance in such thermochemical calculations. The heat capacity is the amount of thermal energy that can be absorbed by a system for a unit rise in temperature. This is defined by Eq. (14), where C_{process} is the heat

$$\delta q = C_{\text{process}} dT \qquad (14)$$

capacity of a system for a given type of process. Three commonly considered processes for closed systems and their respective heat capacities are the constant-volume process C_v, the constant-pressure process C_p, and the saturated-vapor process C_s. Further understanding is gained by examination of Eqs. (9) and (10). For a closed system in which only mechanical (PV) work is possible, the internal energy change for a constant volume process is given by Eq. (15), and therefore Eqs. (16) hold. In the same closed system, the internal energy change for a constant

$$dU = \delta q = C_v dT = TdS \quad (15) \qquad C_v = (\partial U/\partial T)_V \quad (16a) \qquad \Delta S = \int_{T_1}^{T_2} (C_v/T)dT \quad (16b)$$

pressure process is given by Eq. (17). For convenience, a new state function called the enthalpy

$$dU + PdV = \delta q = C_p dT = TdS \qquad (17)$$

is defined by Eqs. (18), which for a constant-pressure process yields $dH = dU + PdV$, therefore

$$H \equiv U + PV \quad (18a) \qquad dH = dU + PdV + VdP \quad (18b)$$

Eqs. (19) hold. The heat capacity is an extensive property of a system, but it is not a state function

$$C_p = (\partial H/\partial T)_p \quad (19a) \qquad \Delta S = \int_{T_1}^{T_2} (C_p/T)dT \quad (19b)$$

since its value is path- or process-dependent. SEE ENTHALPY.

Second law of thermodynamics. There are many possible and essentially equivalent statements of the second law. It will suffice to state the empirical result that in all spontaneous processes the uncompensated heat δa in Eqs. (13) is always positive. Equation (13a) can be rewritten as Eq. (20), where the term $\delta q/T$ is the contribution to the entropy due to heat exchange

$$dS = \delta q/T + \delta a/T \qquad (20)$$

with the surrounding ($d_e S$), while $\delta a/T$ is the contribution to the entropy produced internally as a result of the interconversion of work terms ($d_i S$). The second law can then be summarized as Eqs. (21), where $d_i S$ greater than zero applies to irreversible processes. When $d_i S = 0$, that is, for a

$$dS = d_e S + d_i S \qquad (21a) \qquad\qquad d_i S \geq 0 \qquad (21b)$$

reversible process, Eq. (22) holds. This is the basic equation for establishing the thermodynamic

$$dS = \delta q_{\text{rev}}/T \qquad (22)$$

temperature scale based upon the theoretical limits of reversible cycles. The requirement that $d_i S > 0$ for spontaneous processes provides the criteria for examining the specific conditions for spontaneous paths, and the criteria for establishing the equilibrium states of a system.

In an isolated system, $dU = 0$, for any process that takes place whether spontaneous or not, whereas $dS = d_i S$ since $d_e S = 0$. Any spontaneous process must therefore increase S until S reaches a maximum value at which point only reversible processes can take place. If a random fluctuation perturbs the equilibrium state, the system will spontaneously return to the equilibrium state. Thus states in the vicinity of the equilibrium state are said to be unstable with respect to such perturbations. SEE ENTROPY.

Helmholtz free energy. While isolated systems are of theoretical value, they do not play an important role in practical chemical systems, and consequently criteria for systems undergoing changes at constant temperature and either constant volume or constant pressure are required. Examination of Eq. (11) indicates that if a state function defined by Eqs. (23) is combined with

$$A \equiv U - TS \qquad (23a) \qquad\qquad dA = dU - TdS - SdT \qquad (23b)$$

Eq. (13b), then Eq. (24) holds, which for constant-temperature processes is Eqs. (25), and where

$$dA = -\delta a + \delta w - SdT \qquad (24)$$

$$dA = -\delta a + \delta w \qquad (25a) \qquad\qquad dA = \delta w_{\text{max}} \qquad (25b)$$
$$\text{Spontaneous process} \qquad\qquad\qquad \text{Reversible process}$$

$-\delta w_{\text{max}}$ is the maximum work that can be performed by the system under constant temperature conditions. If processes take place at constant temperature and constant extensive variables ($V,\{n_i\},\{X_j\}$), then $dA = -\delta a$, and all spontaneous processes are accompanied by a decrease in A and equilibrium is achieved when A is a minimum. The state function A is the Helmholtz free energy, and its characteristic variables are $A(T,V,\{n_i\},\{X_j\})$. SEE FREE ENERGY.

Gibbs free energy. If a state function is defined by Eqs. (26), combining this with Eq.

$$G \equiv U + PV - TS = H - TS \quad (26a) \qquad dG = dU + PdV + VdP - TdS - SdT \quad (26b)$$

(13b) yields Eq. (27), which for a constant temperature and pressure process is Eqs. (28). Here

$$dG = -\delta a + \delta w + PdV + VdP - SdT \qquad (27)$$

$$dG = -\delta a + \delta w + PdV \qquad (28a) \qquad\qquad dG = \delta w_{\text{max}} + PdV = \delta w_{\text{net}} \qquad (28b)$$
$$\text{Spontaneous process} \qquad\qquad\qquad \text{Reversible process}$$

$-\delta w_{\text{net}}$ is the next maximum work a system can perform in excess of expansion work under conditions of constant temperature and pressure. If processes take place at constant temperature and pressure and constant extensive variables ($\{n_i\},\{X_j\}$), then $dG = -\delta a$, and all spontaneous processes are accompanied by a decrease in G and equilibrium is achieved when G is a minimum. The state function G is the Gibbs free energy or free enthalpy, and its characteristic variables are $G(T,P,\{n_i\},\{X_j\})$.

Equations of state. **Table 2** summarizes the fundamental equation of state for the internal energy and entropy, as well as the derived equations of state for practical conditions. A more complete understanding of the relationship of equations involving intensive variables as independent variables is based on the mathematical recognition that intensive variables are Legendre

Table 2. Equations of state and characteristic variables

Basic function	Equation of state	Differential form
Internal energy (U)	$U(S,V,\{n_i\},\{X_i\})$	$dU = TdS - PdV + \sum_i \mu_i dn_i + \sum_j l_j dX_j$
Entropy (S)	$S(U,V,\{n_i\},\{X_i\})$	$dS = (1/T)dU - (P/T)dV + \sum_i (\mu_i/T)dn_i + \sum_j (l_j/T)dX_j$
Enthalpy (H) $(H = U + PV)$	$H(S,P,\{n_i\},\{X_i\})$	$dH = TdS + VdP + \sum_i \mu_i dn_i + \sum_j l_j dX_j$
Helmholtz free energy (A) $(A = U - TS)$	$A(V,T,\{n_i\},\{X_i\})$	$dA = -SdT - PdV + \sum_i \mu_i dn_i + \sum_j l_j dX_j$
Gibbs free energy (G) $(G = U + PV - TS)$	$G(P,T,\{n_i\},\{X_i\})$	$dG = -SdT + VdP + \sum_i \mu_i dn_i + \sum_j l_j dX_j$

transformations of the corresponding conjugate extensive variables. The properties of these transformations are such that derived equations involving intensive variables only incompletely define the state of the system, though they do completely define changes in state.

Affinity and chemical equilibrium. Many chemical systems can be considered closed systems in which a single parameter ξ can be defined as a measure of the extent of the reaction or the degree of advancement of a process. If the reaction proceeds or the process advances spontaneously, entropy must be produced according to the second law and δa must be positive. In terms of the advancement parameter ξ, this uncompensated heat δa can be given by Eq. (29),

$$\delta a = \underline{A} d\xi = T d_i S \qquad (29)$$

where $\underline{A}$ is the affinity of the process or reaction. The affinity is related to internal entropy production by Eq. (30). The condition that the entropy production is zero represents equilibrium, and

$$\underline{A} = T d_i S/d\xi \geq 0 \qquad (30)$$

hence $\underline{A} = 0$ is an equivalent condition for equilibrium in a closed system. For spontaneous processes, since the signs of $\underline{A}$ and $d\xi$ must be the same, for positive $\underline{A}$ the process must advance or go in a forward direction in the usual sense of chemical reactions or physical processes, while for negative $\underline{A}$ the process must proceed in the reverse direction.

Equations (13b) and (26b) can be combined to give Eqs. (31), which indicates that the affinity is itself a state function, Eq. (32).

$$dG = -SdT + VdP - \delta a \qquad (31a)$$
$$= -SdT + VdP - \underline{A}d\xi \qquad (31b)$$
$$\underline{A} = -(\partial G/\partial \xi)_{P,T} = \underline{A}(T,P,\xi) \qquad (32)$$

Consider a closed system in which a chemical reaction can be characterized by the stoichiometry of reaction (33). The stoichiometry requires that at each time element t in the reaction

$$\alpha A + \beta B \rightarrow \gamma C + \delta D \qquad (33)$$

the number of moles of the ith component n_i be given by Eq. (34), where n_i^0 is the number of

$$n_i = n_i^0 + v_i \xi \qquad (34)$$

moles of i in the initial or original state ($t = 0$), v_i is the stoichiometric coefficient for the ith component as given in the balanced equation (the convention is that v_i is positive for products and negative for reactants), and ξ is the degree-of-advancement parameter whose range is normally taken to be zero to unity. In terms of differential changes in advancement Eq. (35) holds.

$$dn_i = v_i d\xi \qquad (35)$$

For closed systems (constant temperature and pressure) in which only thermal, expansion, and chemical work terms are included, Eqs. (36) and (37) hold. The condition for equilibrium is

$$dG = +\sum_i \mu_i\, dn_i = -\delta a \tag{36a}$$

$$= \left(\sum_i v_i\mu_i\right)d\xi = -\underline{A}d\xi \tag{36b}$$

$$\underline{A}(T,P,\xi) = -\sum_i v_i\mu_i \tag{37}$$

$\underline{A} = 0$, and thus for a chemical reaction, equilibrium is achieved when Eq. (38) holds. If electrical

$$\sum_i v_i\mu_i = 0 \tag{38}$$

work is included in Eq. (36), Eqs. (39) and (40) hold. Since $\underline{A} = 0$ is the equilibrium condition,

$$dG = \sum_i \mu_i dn_i + EdQ = -\delta a \tag{39a}$$

$$= \left(\sum_i v_i\mu_i + zFE\right)d\xi = -\underline{A}d\xi \tag{39b}$$

$$\underline{A}(T,P,E,\xi) = -\sum_i v_i\mu_i - zFE \tag{40}$$

equilibrium in an electrochemical system is given by Eq. (41), where z is the number of equiva-

$$\sum_i v_i\mu_i = -zFE \tag{41}$$

lents of charge and F is the Faraday constant (the number of coulombs of charge per equivalent).

More than one reaction can take place in a chemical system, each characterized by a degree-of-advancement parameter, and thus for r independent reactions, Eqs. (42) hold, where the

$$dG = \sum_i \mu_i dn_i = -\delta a \tag{42a}$$

$$= \sum_r \underline{A}_r \delta\xi_r \tag{42b}$$

$$= \sum_r \left(\sum_i v_{ir}\mu_{ir}\delta\xi_r\right) \tag{42c}$$

equilibrium condition is Eq. (43). At equilibrium each of the $\underline{A}_r$ must be zero, but for the sponta-

$$\sum_r \underline{A}_r d\xi_r = 0 \tag{43}$$

neous condition, inequality (44) holds. In a two-reaction system inequality (45) holds, but now $\underline{A}_1$

$$\sum_r \underline{A}_r d\xi_r > 0 \tag{44} \qquad \underline{A}_1 d\xi_1 + \underline{A}_2 d\xi_2 > 0 \tag{45}$$

and $d\xi_1$ do not necessarily have the same sign. If their signs are different, the first reaction can be driven in the nonspontaneous direction by the second reaction. The reactions are then said to be coupled, and this is a common situation in biological systems. SEE CHEMICAL EQUILIBRIUM.

Chemical potential. The affinity of a chemical reaction establishes the spontaneous direction of the reaction, and consequently methods for determining the affinity are important in thermochemical studies. As shown above, the affinity is simply related to the stoichiometric coefficients of the reaction and the chemical potentials of the reactants and products in the reaction. It is necessary therefore to investigate some of the properties of the chemical potential and to develop convenient methods of using it to calculate the affinity.

The chemical potential of a single-phase pure substance can be expressed as $\mu_i = \mu_i(T,P)$, whereas for a component in single-phase solution $\mu_i = \mu_i(T,P,\{x_i\})$, where $\{x_i\}$ is the set of independent mole fractions ($x_i = n_i/\Sigma n_i$). All intensive thermodynamic variables are homogeneous functions of zero degree in mass, and hence Eq. (46) holds, where $\bar{I}_i$ is given by Eq. (47). Equation

$$\sum_i n_i \bar{I}_i = 0 \tag{46}$$

$$\bar{I}_i = (\partial I_i/\partial n_i)_{T,P,\{n_j\}'} \tag{47}$$

(46) is a form of the Gibbs-Duhem relationship. Since the chemical potential is intensive, the Gibbs-Duhem equation for a given solution phase can be written as Eq. (48). This is an important

$$\sum_i x_i \overline{\mu}_i = 0 \tag{48}$$

result in phase equilibria in heterogeneous systems, and places restraints on the number of independent variables in such systems. *SEE PHASE EQUILIBRIUM.*

The chemical potential can be represented in several forms, but for chemical studies the form which expresses that it is a partial molar quantity, Eq. (49), is most useful. Since the order

$$\mu_i = (\partial G/\partial n_i)_{T,P,\{n_j\}'} \tag{49}$$

of differention of exact functions is immaterial, temperature, and pressure derivatives of μ_i are given by Eqs. (50) and (51), where $\overline{S}_i$ and $\overline{V}_i$ are the partial molar entropy and volume of the ith

$$(\partial \mu_i/\partial T)_{P,\{n_j\}} = \partial^2 G/\partial T \partial n_i$$
$$= -(\partial S/\partial n_i)_{T,P,\{n_j\}} = -\overline{S}_i \tag{50}$$

$$(\partial \mu_i/\partial P)_{T,\{n_j\}} = \partial^2 G/\partial P \partial n_i$$
$$= -(\partial V/\partial n_i)_{T,P} = \overline{V}_i \tag{51}$$

component. Two additional useful differential coefficients are given by Eqs. (52) and (53), where

$$[\partial(\mu_i/T)/\partial(1/T)]_{P,\{n_j\}} = \overline{H}_i \tag{52} \qquad (\partial \overline{H}_i/\partial T)_{P,\{n_j\}} = \overline{C}_{P,i} \tag{53}$$

$\overline{H}_i$ and $\overline{C}_{P,i}$ are the partial molar enthalpy and constant-pressure heat capacity.

It is quite apparent that a knowledge of the chemical potentials of pure substances and of substances in solution provides the basis of the thermal properties of the substance, as well as the basis for reaction spontaneity. Usually the chemical potential of a substance is expressed in terms of a standard state and a convenient measure of the deviation from that state. Various functional forms could be used, but it is customary to use Eq. (54), where $\mu_i^{\ominus}$ is the chemical

$$\mu_i(T,P,\{x_i\}) = \mu_i^{\ominus}(T,P,\{x_i\}) + RT \ln a_i^*(^*T,P,\{x_i\}) \tag{54}$$

potential in a designated standard state, R is the gas constant, and a_i^* is the relative activity of the ith substance. The activity is frequently written as Eq. (55), where $f_i^*(T,P,\{x_i\})$ is the activity

$$a_i^* = f_i^* x_i \tag{55}$$

coefficient on the mole fraction basis. Two basic situations arise as the mole fraction approaches its limits of 0 to 1, and it is these limits that determine the functional form of f_i^*. Combining Eq. (54) with (55) gives Eq. (56). If the $\lim_{x_i \to 1} f_i^* = 1$, then Eq. (57), the Raoult law convention, holds

$$\mu_i = \mu_i^* + RT \ln f_i^* + RT \ln x_i \tag{56} \qquad \lim_{x_i \to 1} \mu_i = \mu_i^* = \mu_i^*(T,P) \tag{57}$$

where μ_i^* is the chemical potential of pure i. The other limit, $\lim_{x_i \to 0} f_i = 1$, implies that $\lim_{x_i \to 1} f_i$ equals some finite value, say f_i^∞, and therefore Eq. (58), the Henry law convention, holds, where f_i^∞ is the

$$\lim_{x_i \to 1} \mu_i = \mu_i^{\ominus} + RT \ln f_i^\infty = \mu_i^\infty \tag{58}$$

activity coefficient at mole fraction unity determined by the limiting behavior of the chemical potential in an infinitely dilute solution of i, and μ_i^∞ is the chemical potential i would have if its dilute solution behavior persisted up to mole fraction one. The two situations can be summarized by Eqs. (59), where f_i^* and f_i^∞ are the activity coefficients on the Raoult or Henry law basis, respectively. *SEE ACTIVITY.*

$$\mu_i = \mu_i^*(T,P) + RT \ln f_i^* x_i \qquad \text{Raoult's law} \tag{59a}$$
$$\mu_i = \mu_i^\infty(T,P,x_i) + RT \ln f_i^\infty x_i \qquad \text{Henry law} \tag{59b}$$

The activities discussed above are based upon mole fractions. It is frequently more convenient to use other measures of relative amounts, for example, the molality m, defined as the moles of solute per kilogram of solvent, and the molarity c, defined as the moles of solute per liter of solution. If this is done, the chemical potentials are given by Eqs. (60) or (61), where $\mu_i^{\infty,m}$ and

$$\mu_i = \mu_i^{\infty,m} + RT \ln f_i^m m_i \qquad (60) \qquad \mu_i = \mu_i^{\infty,c} + RT \ln f_i^c c_i \qquad (61)$$

$\mu_i^{\infty,c}$ are infinite dilution based standard state chemical potentials on the molality and molarity convention, and f_i^m and f_i^c are activity coefficients on the same bases. In each case $\lim_{m_i,c_i \to 0} f_i = 1$.

The chemical potential of gases is frequently discussed in terms of its fugacity or pressure, and Eq. (54) is then written as Eq. (62) or (63), where P_i^* is the fugacity of substance i and is equal

$$\mu_i = \mu_i^0(T) + RT \ln P_i^* \qquad (62) \qquad \mu_i = \mu_i^0(T) + RT \ln f_i P_i \qquad (63)$$

to $f_i P_i$, where f_i is the fugacity coefficient and P_i is the partial pressure of i. The fugacity coefficient is defined such that the $\lim_{P \to 0} f_i = 1$. The standard chemical potential $\mu_i^0(T)$ is the value of the chemical potential of a perfect gas, and is a function of temperature only. *See Fugacity*.

The chemical potential of pure liquids or solids is given by Eq. (64), where $P^\ominus$ is the

$$\mu_i(T,P) = \mu_i^*(T,P)$$
$$= \mu_i^\ominus(T,P^\ominus) + \int_{P^\ominus}^{P} \overline{V}_i dP \qquad (64)$$

pressure in a designated standard state. The last term of Eq. (64) is negligible if $(P - P^\ominus)$ is not very large, since $\overline{V}_i$ for condensed phases is relatively small. Generally the standard chemical potential is taken from tables of Gibbs free energies of formation $\Delta G^\ominus$ of pure gases, liquids, or solids or of substances at infinite dilution in particular solvents.

Thermodynamical relationships. The criteria for equilibrium in a chemical system at constant temperature and pressure are given by Eq. (65). If a reaction takes place in a gaseous

$$A = -\sum \nu_i \mu_i = 0 \qquad (65)$$

mixture, substitution of Eq. (62) for the μ_i gives Eq. (66a) or (66b), and Eq. (67) holds, where

$$\sum \nu_i(\mu_i^0(T) + RT \ln P_i^*) = 0 \quad (66a) \qquad \sum \nu_i \mu_i^0(T) = -RT \ln \prod_i (P_i^*)^{\nu_i} \quad (66b)$$

$$\Delta G_{\text{react}}^0(T) = -RT \ln K_{p^*} \qquad (67)$$

$\Delta G_{\text{react}}^0$ is the Gibbs free energy change for a reaction going from a standard state of reactants to a standard state of products in accordance with the stoichiometry of the reaction, and K_{p^*} is the equilibrium constant for the reaction in terms of fugacities. At low pressures $P_i^* = P_i$, and the equilibrium constant for reaction (33) is given by Eq. (68). For a reaction in solution for which the

$$K_p = P_c^\gamma P_d^\delta / P_a^\alpha P_b^\beta \qquad (68)$$

chemical potential is given on a molarity basis, Eqs. (69)–(71) hold, where K_a^c is the equilibrium

$$\sum \nu_i(\mu^{\infty,c} + RT \ln a_i^c) = 0 \quad (69) \qquad \sum \nu_i \mu_i^{\infty,c} = -RT \ln \prod_i (a_i^c)^{\nu_i} \quad (70)$$

$$\Delta G_{\text{react}}^{\infty,c} = -RT \ln K_a^c \qquad (71)$$

constant for the reaction in terms of molarity-based activities. For dilute solutions, $a_i = c_j$, and for reaction (33), Eq. (72) holds. Similar results can be obtained for other concentration-gased activi-

$$K_c = c_c^\gamma c_d^\delta / c_a^\alpha c_b^\beta \qquad (72)$$

ties. The activities can be based on whatever measures of activity are most convenient. Clearly the more general expression of the equilibrium is given by Eq. (73). The standard enthalpy, en-

$$\Delta G_{\text{react}}^\ominus = -RT \ln K_a \qquad (73)$$

tropy, heat capacity, and volume changes for a chemical reaction can be obtained directly from the appropriate temperature derivatives of the standard free energy of the reaction or its equivalent, the equilibrium constant.

The condition for a reaction to take place spontaneously in the direction from reactants to products requires that $\underline{A}$ be positive, and hence Eqs. (74) hold, where Q_a is an expression of the

$$\underline{A} = -\sum_i \nu_i \mu_i > 0 \quad (74a) \qquad \underline{A} = -\left(\sum_i \nu_i \mu_i + RT \ln \prod_i a_i^{\nu_i}\right) > 0 \quad (74b)$$

$$\underline{A} = RT \ln K_a - RT \ln Q_a > 0 \quad (74c) \qquad \underline{A} = RT \ln K_a/Q_a > 0 \text{ or } K_a/Q_a > 1 \quad (74d)$$

form of the equilibrium constant, but involving activities of reactants in their initial state and the activities of the products in their final states. Again, the bases of the activities in Q_a, as in K_a, are arbitrary and selected for convenience.

Third law of thermodynamics. Most understanding of the classical thermodynamics of chemical systems is based upon the first and second laws. As these systems are studied on a molecular rather than on a macroscopic basis, it is apparent that the first and second laws cannot be addressed to this endeavor in a direct manner. Although implicit in the concept of the existence of the fundamental equations of state for U or S is an absolute value of these functions, and therefore an extensive quantity which could be calculated on the basis of molecular properties from quantum mechanics and statistical mechanics, neither the first nor second law considers anything but differences in these state functions. The second law does indeed indicate the existence of an absolute zero for an intensive variable, the temperature, but this is not sufficient to bridge the areas of classical and statistical thermodynamics.

It is found experimentally that for many isothermal processes involving pure phases, Eq. (75) holds. This includes phase transitions between different crystalline modifications, solid-state

$$\lim_{T \to 0} \Delta S = 0 \quad (75)$$

chemical reactions, and even the solid-liquid transition in helium. This, along with Eq. (76), im-

$$S(T,P) - S(0 \text{ K}) = \int_0^T (C_P/T) dT \quad (76)$$

plies that at zero absolute temperature the entropy of pure crystalline phases are equal. If the entropy of pure phases are equal at $T = 0$, it is reasonable to take the value $S(0 \text{ K})$ to be zero. The statement of the third law then is that the entropy of all pure crystalline phases at 0 K is zero. This makes it possible by using Eq. (76), to calculate the absolute or third-law entropy of a substance from experimental measurements of their heat capacities. Comparison of such experimental values with those calculated by statistical thermodynamic methods has provided evidence for the validity of the third law. In some cases thorough investigation of apparent discrepancies from the third law have led to new conclusions concerning the molecular structure of the substances or new information on the energy level system for the molecules. Calculations of the thermodynamic properties for a gas from the spectroscopic properties of molecules is an important result stemming from the third law.

Thermodynamics of irreversible processes. Classical equilibrium thermodynamics is primarily concerned with calculations for reversible processes, and deals with irreversibility in terms of inequalities. In the case of irreversible processes in systems slightly removed from equilibrium, the rate of internal entropy production d_iS/dt is related to the fluxes J_i associated with thermal, concentration, or other differences in intensive parameters or potentials X_i. This entropy production is then given by Eq. (77). The fluxes include heat conduction, diffusion, electric con-

$$d_iS/dt = \sum_i J_i X_i \geq 0 \quad (77)$$

duction, and other direct effects.

In addition, a flux of one type may be coupled to a potential difference of another type. For example, a thermal gradient can result in a mass flux (thermal diffusion), or a concentration gradient in any energy flux. Thermal conductivity, thermoosmosis, and thermoelectric effects are all coupled effects. The fluxes are thus found to be given by Eq. (78), where the L_{ij} are called the

$$J_i = \sum_j L_{ij} X_j \quad (78)$$

phenomenological coefficients. If $L_{ij} \neq 0$, there is a coupling between the flux J_i and the gradient X_j. Microscopic reversibility implies that not far from equilibrium, $L_{ij} = L_{ji}$, and this is known as the Onsager reciprocity relationship. These results, together with the theorem of minimum entropy production, are the basis of investigations of irreversible processes near equilibrium.

Far removed from equilibrium, thermodynamics must be formulated somewhat differently and more cautiously. The interplay of thermodynamic stability and kinetics can give rise to macroscopic structures with both temporal and spatial coherence called dissipative structures. Much theoretical effort has been directed to these studies because of their apparent relevance to biological structures.

Bibliography. I. M. Klotz and R. M. Rosenberg, *Chemical Thermodynamics*, 3d ed., 1972; I. N. Levine, *Physical Chemistry*, 2d ed., 1983; P. A. Rock, *Chemical Thermodynamics*, 1983.

THERMOCHEMISTRY
RANDOLPH C. WILHOIT

A branch of physical chemistry concerned with the absorption or evolution of heat that accompanies chemical reactions. Closely related topics are the latent heat associated with a change in phase (crystal, liquid, gas), the chemical composition of reacting systems at equilibrium, and the electrical potentials of galvanic cells. Thermodynamics provides the link among these phenomena.

A knowledge of such heat effects is important to the chemical engineer for the design and operation of chemical reactors, the determination of the heating values of fuels, the design and operation of refrigerators, the selection of heat storage systems, and the assessment of chemical hazards. Thermochemical information is used by the physiologist and biochemist to study the energetics of living organisms and to determine the calorific values of foods. Thermochemical data give the chemist an insight to the energies of, and interactions among, molecules. *See* CHEMICAL EQUILIBRIUM; PHASE EQUILIBRIUM.

Thermodynamic principles. The first law of thermodynamics expresses the principle of conservation of energy. When a closed system changes from an initial state to a final state, its internal energy, U, changes by the amount shown in Eq. (1), where q is the heat energy trans-

$$U(\text{final}) - U(\text{initial}) = \Delta U = q + W \tag{1}$$

ferred to the system from the outside and w is the work done on the system by external forces. A positive sign of q, w, or ΔU means energy is transferred to the system, while a negative sign means energy is removed. Some authors use the opposite sign convention for w.

Internal energy. Internal energy (U) is a variable of state. This means that its value depends only on the state of the system and not on its previous history. The value of ΔU, the change in internal energy, depends only on the initial and final states. Such states are identified by chemical composition, physical phase, temperature, pressure, and sometimes other relevant variables. The values of q and w, however, depend both on the states and on the way the transformation is brought about. Their sum must always satisfy Eq. (1). The first law governs any system and any combination of states. If a chemical reaction occurs during the transformation, the initial and final states will have different compositions. Then q is called the heat of reaction.

If the change in states takes place with no work done, then $w = 0$ and $\Delta U = q$. The most common example is a process which takes place irreversibly at constant volume. However, in the laboratory, chemical reactions are usually conducted at constant pressure. In this case, $w = -P\Delta V$, where P is the pressure and ΔV is the change in volume for the process. Under such conditions, the value for ΔU is given by Eq. (2). *See* INTERNAL ENERGY.

$$\Delta U = q - P\Delta V \tag{2}$$

Enthalpy. The property H is called enthalpy. It is defined in general as $H = U + PV$. When Eq. (2) is solved for the term q, the result gives the value for ΔH [Eq. (3)], the change in enthalpy.

$$q = \Delta U + P\Delta V = \Delta H \tag{3}$$

If the change in states is brought about reversibly, then w is algebraically a minimum and q is a maximum. In this case, $q = T\Delta S$, where ΔS is the change of entropy of the system and T is the temperature. Heats of reaction are seldom measured directly under such conditions, however.

If q is positive for the irreversible process (energy transferred to the system) the reaction is called endothermic, and if q is negative (heat given off) the reaction is exothermic.

Change in enthalpy (ΔH) and change in internal energy (ΔU) for a chemical reaction are reported as a certain quantity of energy for the number of moles indicated in the balanced chemical equation. Equation (3) provides the relationship between the two quantities. If all reactants and products are liquids or solids, the difference is negligible (except at very high pressures). If gases are involved, the difference is significant only when the number of moles of product gases differs from the number of moles of reactant gases.

A chemical transformation may take place in a series of steps, each corresponding to a certain reaction. Thus, for example, at 1 bar (10^5 pascals) and 25°C (77°F), the reaction steps shown as (4a-d) take place. (The subscript r is the symbol for chemical reaction.) The net result of all of these steps is shown in reaction (4e).

$$CH_4(gas) \rightarrow C(graphite) + 2H_2(gas) \tag{4a}$$
$$\Delta_r H = 74.52 \qquad \Delta_r U = 72.04$$

$$C(graphite) + \tfrac{1}{2}O_2(gas) \rightarrow CO(gas) \tag{4b}$$
$$\Delta_r H = -110.525 \qquad \Delta_r U = -111.764$$

$$CO(gas) + \tfrac{1}{2}O_2(gas) \rightarrow CO_2(gas) \tag{4c}$$
$$\Delta_r H = -282.984 \qquad \Delta_r U = -281.744$$

$$2H_2(gas) + O_2(gas) \rightarrow 2H_2O(liquid) \tag{4d}$$
$$\Delta_r H = -571.66 \qquad \Delta_r U = -564.22$$

$$CH_4(gas) + 2O_2(gas) \rightarrow CO_2(gas) + 2H_2O(liquid) \tag{4e}$$
$$\Delta_r H = -890.64 \qquad \Delta_r U = -885.69$$

The values of $\Delta_r H$ and $\Delta_r U$ are in kilojoules for the reaction specified. Since both U and H are variables of state, values of ΔU and ΔH are sums of the corresponding values for the intermediate steps for any path which leads from the initial to the final state. This result is called Hess's law, but it is simply a consequence of the first law of thermodynamics.

The change of any property, symbolized by X, associated with a chemical reaction may be equated to a sum of terms, one for each reactant and each product, by Eq. (5), where the $X(i)$ are the corresponding properties of reactants and products, i, and ν_i are the coefficients in the balanced chemical equation (positive for products, negative for reactants). The enthalpy of formation, sometimes called heat of formation, of a compound is the change in enthalpy for a reaction in which the compound is synthesized from its component elements. These quantities are seldom measured directly but are calculated from enthalpies of other reactions through the application of Hess's law. The enthalpy change for any reaction can be calculated from the enthalpies of formation of the reactants and products by the substitution of $\Delta_f H$ for X in the right side of Eq. (5)

$$\Delta_r X = \Sigma \nu_i X(i) \tag{5}$$

[the subscript f is the symbol for formation from elements]. SEE ENTHALPY.

Heats of reaction. Enthalpies of reactions may be obtained from several types of measurement. Three classes are recognized. The first-law heat of reaction is measured directly in a calorimeter. The second-law heat of reaction is calculated from the effect of temperature change on the equilibrium constant by the use of the Van't Hoff equation. The third-law heat of reaction is calculated by $\Delta_r H = \Delta_r G + T\Delta_r S$. $\Delta_r G$ is the change in Gibbs energy for the reaction calculated from an equilibrium constant or electrical cell potential, and $\Delta_r S$ is the corresponding change in entropy calculated from measured heat capacities and the third law of thermodynamics. In principle, all three methods give the same result for any particular reaction. The ease of measurement and the attainable accuracy may vary for different situations.

Enthalpies of reactions are slowly changing functions of temperature. The enthalpies of a reaction at the two temperatures T_1 and T_2 may be related to the difference in enthalpies for each component of the reaction between the two temperatures by Eq. (6). The quantities in the last

$$\Delta_r H(T_2) = \Delta_r H(T_1) + \Delta_r[H(i, T_2) - H(i, T_1)] \tag{6}$$

term of Eq. (6) may be calculated from the corresponding heat capacities by Eq. (7).

$$H(i, T_2) - H(i, T_1) = \int_{T_1}^{T_2} C_p(i)\, dT \tag{7}$$

SEE CHEMICAL THERMODYNAMICS; ENTROPY; FREE ENERGY; GIBBS FUNCTION.

Calorimetric measurements. A calorimeter is an instrument for measuring the heat added to or removed from a process. There are many designs, but the following parts can generally be identified: the vessel in which the process is confined, the thermometer which measures its temperature, and the surrounding environment called the jacket. The heat associated with the process is calculated by Eq. (8), where T is the temperature. The quantity C, the energy equivalent

$$q = C[(T(\text{final}) - T(\text{initial})] - q_{\text{ex}} - w \tag{8}$$

of the calorimeter, is obtained from a separate calibration experiment. The work transferred to the process, w, is generally in the form of an electric current (as supplied to a heater, for example) or as mechanical work (as supplied to a stirrer, for example) and can be calculated from appropriate auxiliary measurements. The quantity q_{ex} is the heat exchanged between the container and its jacket during the experiment. It is calculated from the temperature gradients in the system and the measured thermal conductivities of its parts.

Two principal types of calorimeters are used to measure heats of chemical reactions. In a batch calorimeter, known quantities of reactants are placed in the vessel and the initial temperature is measured. The reaction is allowed to occur and then the final equilibrium temperature is measured. If necessary, the final contents are analyzed to determine the amount of reaction which occurred.

In a flow calorimeter, the reactants are directed to the reaction vessel in two or more steady streams. The reaction takes place quickly and the products emerge in a steady stream. The rate of heat production is calculated from the temperatures, flow velocities, and heat capacitites of the incoming and outgoing streams, and the rates of work production and heat transfer to the jacket. Dividing this result by the rate of reaction gives the heat of reaction.

The combustion of a substance in oxygen is often studied in a specially designed reaction calorimeter. The heats of combustion of liquid or solid samples are usually measured in a batch-type calorimeter. The vessel is a strong steel alloy bomb which is placed in a container of water fitted with appropriate thermometers and stirrers. The sample is placed in the bomb in an atmosphere of oxygen at high pressure (around 30 atm or 3 megapascals). It is then ignited by an electrical fuse. Heats of combustion of gases are usually measured in a type of flow calorimeter called a flame calorimeter. Heats of combustion can be measured to accuracies of 1 part in 10,000. The primary limiting factor in such measurements is the purity of the samples. Heats of combustion in fluorine have also been measured.

Units and symbols. The International Union of Pure and Applied Chemistry (IUPAC) Commission on Thermodynamics has recommended the general symbol $\Delta_\alpha^\beta X$ to represent the change of any property of a system when it changes from an initial state (denoted by α) to a final state (denoted by β). In addition, certain commonly observed processes are given special symbols of the type $\Delta_b X$, where the subscript b represents the process. Other symbols are vap, vaporization of a liquid; sub, sublimation of a solid; fus, fusion of a solid; mix, mixing without reaction; r, chemical reaction; f, formation from elements; and c, combustion. The subscript m placed after the property symbol indicates 1 mole of substance. Additional specifications can be placed within parentheses. For example, $\Delta_{\text{vap}} H_m(H_2O, 298.15\text{ K})$ symbolizes the heat of vaporization of 1 mole of water at 298.15 K, $\Delta_r G(1000\text{ K})$ the change in Gibbs energy for a chemical reaction at 1000 K, and $\Delta_c H_m(C_4H_{10}, \text{g}, 300\text{ K})$ the enthalpy of combustion of 1 mole of butane gas at 300 K.

Thermochemical quantities are usually reported and tabulated for substances in their standard states. The standard state of a solid is the thermodynamically stable crystal, of a liquid the liquid, and of a gas the hypothetical ideal gas, all at unit pressure. For the past century the pressure unit has been the atmosphere. It has been suggested that the bar is more suitable for this role as it is more compatible with the International System (SI) of units. Standard states can be defined for any temperature, but 25°C (298.15 K) has been traditional. In Customary units, this standard state is given at 77°F (536.67°R). The ideal gas is a hypothetical state, but its properties can readily be calculated from the equation of state of the real gas. The internal energy and enthalpy of an ideal gas are independent of its pressure.

The standard state for the solvent in a solution is the pure liquid. The standard states for the solutes are hypothetical ideal solutions at unit concentrations. Concentrations are usually expressed as molalities or mole fractions. Properties of real solutions can be related to those of the hypothetical ideal solutions by appropriate auxiliary data. Standard states for individual ions in solution are defined with the help of the additional conventions that the enthalpy and Gibbs energies of formation of the hydrogen ion are zero. A degree symbol (°) designates a property of a standard state. The standard-state concept promotes compactness and explicitness for the tabulation of data.

In the past, thermochemical quantities usually have been given in units of calories. A calorie is defined as the amount of heat needed to raise the temperature of 1 gram of water 1°C. However, since this depends on the initial temperature of the water, various calories have been defined, for example, the 15° calorie, the 20° calorie, and the mean calorie (average from 0 to 100°C). In addition, a number of dry calories have been defined. Those still used are the thermochemical calorie (exactly 4.184 joules) and the International Steam Table calorie (exactly 4.1868 J).

Thermochemical quantities have also been reported in terms of British thermal units (Btu). This is the amount of heat required to raise the temperature of 1 lb of water 1°F. A proliferation of Btu's similar to that for calories has occurred. The Btu in common use is the International Steam Table Btu (1055.056 J).

The SI rules do not recognize either the calorie or the Btu. The energy unit is the absolute joule (J). Most modern literature uses this unit.

Sources of data. Original reports of measured values of thermochemical quantities are widely scattered among the world's scientific literature. A number of compilations of enthalpy of formation $\Delta_f H°$, Gibbs energy of formation $\Delta_f G°$, absolute entropy $S°$, and heat capacity at constant pressure $C_p°$ of pure compounds at 298.15 K have been published during the past century. Some of them contain data at other temperatures and values for mixtures and ions as well.

A unique example is the International Critical Tables which appeared as a series of seven volumes between 1926 and 1930. The series was the result of an international cooperation among scientists to collect all reliable physical and chemical properties of materials available at that time. Volume V contains thermochemical data. More recent compilations have been made for inorganic compounds. The Landolt-Börnstein Tables, which have undergone a series of revisions since the 1890s, contain extensive thermodynamic data. Other compilations which are regularly updated by supplements and revisions are the JANAF Tables for low-molecular-weight inorganic compounds and the Thermodynamics Research Center publications for organic and some nonmetallic inorganic compounds.

A consequence of Hess's law is that thermochemical values such as heats of formation, combustion, reaction, and phase transition at a fixed temperature are interrelated through a system of linear algebraic equations. Hundreds, or even thousands, of such equations are available for even limited sets of compounds. They usually form an overdetermined set. The compiler has the job of selecting values of heats of formation which best fit the experimental data with consideration of the assigned uncertainties. If data at different temperatures and second- and third-law heats of reactions are included, the system of equations becomes nonlinear. Formerly the selection was made manually by a series of iterations; more recently computer programs have been written to help in the task of data management and equation solving. The whole process must be repeated to incorporate new data.

To promote internal consistency among thermochemical compilations, a division of the Committee on Data for Science and Technology (CODATA) has recommended certain values of key properties. These represent a basic starting point for most other compilations.

Heating values. The heating value (also called calorific value) of a fuel is the heat of combustion (with a positive sign) of a certain quantity of fuel expressed in some units when burned under given conditions. The price for wholesale commodity transfer is based on the heating value, rather than mass or volume. Engineers use heating values to carry out heat balance calculations for furnaces, engines, and chemical processes.

Many specific definitions of heating value have been issued by trade and standards organizations around the world. An organization may give different definitions for solid, liquid, and gaseous fuels. Some of these have found their way into long-term sales contracts and legal systems of many countries. Therefore the term heating value does not have a universally recognized quantitative meaning.

In many English-speaking countries, heating values have been reported in Btu per pound at 15°F. For a gross heating value the water produced by the combustion is assumed to be liquid. For a net heating value it is assumed to be a gas. A precise definition requires additional specifications such as the nature of other products (for example, those formed from nitrogen or sulfur if present), the amount of water in the fuel (dry or wet basis), and whether the fuel is burned in air or oxygen. If the fuel is a gas, the state, real or ideal, must be indicated. If it is in a real state, the pressure must also be specified.

Commercial calorimeters are available for the measurement of heating values, but accurate measurements are difficult to make and require skilled technicians. Many fuels are complicated mixtures whose composition is not completely known. If the composition is known, the heating value can be calculated from the standard-state enthalpies of combustion of the pure components. This may require a large amount of auxiliary data such as heat capacities, heats of mixing, equation of state of pure and mixed systems, and values of unit conversions.

Organizations concerned with definitions of heating values are the U.S. National Bureau of Standards and its counterpart in other countries, American Society for Testing and Materials, Gas Processors Association, International Standards Organization, and Groupe International des Importateurs de Gaz Natural Liquifie (GIIGNAL).

CHEMICAL EQUILIBRIUM
Cecil E. Vanderzee

In a dynamic or kinetic sense, chemical equilibrium is a condition in which a chemical reaction is occurring at equal rates in its forward and reverse directions, so that the concentrations of the reacting substances do not change with time. In a thermodynamic sense, it is the condition in which there is no tendency for the composition of the system to change; no change can occur in the system without the expenditure of some form of work upon it. From the viewpoint of statistical mechanics, the equilibrium state places the system in a condition of maximum freedom (or minimum restraint) compatible with the energy, volume, and composition of the system. The statistical approach has been merged with thermodynamics into a field called statistical thermodynamics; this merger has been of immense value for its intellectual stimulus, as well as for its practical contributions to the study of equilibria. *See* CHEMICAL THERMODYNAMICS.

Of the three viewpoints, the thermodynamic approach is by far the most powerful and fruitful in treating the quantitative relationships between the position of equilibrium and the factors which govern it. Since thermodynamics is concerned with relationships among observable properties, such as temperature, pressure, concentration, heat, and work, the relationships possess general validity, independent of theories of molecular behavior. Because of the simplicity of the concepts involved, this article will utilize that approach.

Chemical potential. Thermodynamics attributes to each chemical substance a property called the chemical potential, which may be thought of as the tendency of the substance to enter into chemical (or physical) change. Although the chemical potential of a substance cannot be directly measured (except on a relative basis), differences in chemical potential are measurable. (The units are those of energy per mole.)

The importance of the chemical potential lies in its relation to the affinity of driving force of a chemical reaction. Consider general reaction (1). Let μ_A be the chemical potential per mole

$$aA + bB \rightleftharpoons gG + hH \tag{1}$$

of substance A, μ_B be the chemical potential per mole of B, and so on. Then, according to one of the fundamental principles of thermodynamics (the second law), the reaction will be spontaneous when the total chemical potential of the reactants is greater than that of the products. Thus, for spontaneous change (naturally occurring processes), notation (2) applies. When equilibrium is

$$[g\mu_G + h\mu_H] - [a\mu_A + b\mu_B] < 0 \tag{2}$$

reached, the total chemical potentials of products and reactants become equal; thus Eq. (3) holds

$$[g\mu_G + h\mu_H] - [a\mu_A + b\mu_B] = 0 \tag{3}$$

at equilibrium. The difference in chemical potentials in (2) and (3) is called the driving force or affinity of the process or reaction; naturally, it is zero when the chemical system is in chemical equilibrium.

For reactions at constant temperature and pressure (the usual restraints in a chemical laboratory), the difference in chemical potentials becomes equal to the free energy change ΔG for the process in Eq. (4). The decrease in free energy represents the maximum net work obtainable

$$\Delta G = [g\mu_G + h\mu_H] - [a\mu_A + b\mu_B] \tag{4}$$

from the process. When no more work is obtainable, the system is at equilibrium. Conversely, if the value of ΔG for a process is positive, some useful work will have to be expended upon the process, or reaction, in order to make it proceed; the process cannot proceed naturally or spontaneously. (The term spontaneously as used here implies only that a process can occur. It does not imply that the reaction will be rapid or instantaneous. Thus, the reaction between hydrogen and oxygen is a spontaneous process in the sense of the term as used here, even though a mixture of hydrogen and oxygen can remain unchanged for years unless ignited or exposed to a catalyst.)

Since by definition a catalyst remains unchanged chemically throughout a reaction, its chemical potential does not appear in (2), (3), and (4). A catalyst, therefore, can contribute nothing to the driving force of a reaction, nor can it, in consequence, alter the position of the chemical equilibrium in a system. *See Catalysis*.

In addition to furnishing a criterion for the equilibrium state of a chemical system, the thermodynamic method goes much further. In many cases, it yields a relation between the change in chemical potentials (or change in free energy) and the equilibrium concentrations of the substances involved in the reaction. To do this, the chemical potential must be expressed as a function of concentration (and other properties of the substance).

The chemical potential μ is usually represented by Eq. (5), where R is the ideal gas con-

$$\mu = \mu^0 + RT \ln x + RT \ln f \tag{5}$$

stant, T is the absolute temperature, $\ln x = \log_e x = 2.3026 \log_{10} x$, x is the concentration of the substance, f is the activity coefficient of the substance, and μ^0 is the chemical potential of the substance in its standard state.

For substances obeying the laws of ideal solutions (or ideal gases), the last term, $RT \ln f$, is zero, since it is a measure of the deviation from ideal behavior caused by intermolecular or interionic forces. An ideal solution would then be a solution for which $RT \ln f$ is zero over the whole concentration range. For real solutions, the ideal or reference state where f is unity is generally chosen as a state of ideal purity (mole fraction = 1) for solids and solvents, and as a state of infinite dispersion (concentration = 0) for gases and solutes. Although the activity coefficient f is regarded as dimensionless, its numerical values will depend upon the particular concentration scale x with which it must be associated.

Although the choice of concentration scales is somewhat a matter of convenience, the following are conventionally used: $x = p$ = partial pressure, for gases; $x = c$ or m = molar or molal concentrations, for solutes in electrolytic solutions; x = mole fraction, for solids and solvents. When the choice is not established by convention, the mole fraction scale is to be preferred.

Activity and standard states. It is often convenient to utilize the product, fx, called the activity of the substance and defined by $a = fx$. The activity may be looked upon as an effective concentration of the substance, measured in the same units as the concentration x with which it is associated. The standard state of the substance is then defined as the state of unit activity (where $a = 1$) and is characterized by the standard chemical potential μ^0. Clearly, the terms μ^0, f, and x are not independent; the choice of the activity scale serves to fix the standard state. For example, for an aqueous solution of hydrochloric acid, the standard state for the solute

(HCl) would be an (hypothetical) ideal 1 molar (or molal) solution, and for the solvent (H_2O) the standard state would be pure water (mole fraction = 1). The reference state would be an infinitely dilute solution; here the activity coefficients would be unity for both solute and solvent. For the vapor of HCl above the solution, the standard state would be the ideal gaseous state at 1 atm (100 kilopascals) partial pressure; the reference state would be a state of zero pressure. (For gases, the term fugacity is used instead of activity.) It should be noted that the reference state is a limiting state which in many cases can be reached only through an extrapolation from observed behavior. *See Fugacity*.

Equilibrium constant. If the general reaction in Eq. (1) occurs at constant temperature T and pressure P when all of the substances involved are in their standard states of unit activity, Eq. (4) would become Eq. (6). The quantity ΔG^0 is known as the standard free energy change for

$$\Delta G^0 = [g\mu_G^0 + h\mu_H^0] - [a\mu_A^0 + b\mu_B^0] \tag{6}$$

the reaction at that temperature and pressure for the chosen standard states. (Standard state properties are commonly designated by a superscript, ΔG^0, μ^0.) Since each of the standard chemical potentials (μ^0) is a unique property determined by the temperature, pressure, standard state, and chemical identity of the substance concerned, the standard free energy change ΔG^0 is a constant (parameter) characteristic of the particular reaction for the chosen temperature, pressure, and standard states.

If, in a reaction, Eq. (1), at constant temperature and pressure, the chemical potential of each substance is expressed in terms of Eq. (5), the free energy change for the reaction, from Eq. (4), becomes Eq. (7) in terms of the activities and the standard free energy change, Eq. (6). Equa-

$$\Delta G = \Delta G^0 + RT \ln \frac{a_G^g a_H^h}{a_A^a a_B^b} \tag{7}$$

tion (7) is often written in the form of Eq. (8), where Q^0 is the ratio of the activities of products to

$$\Delta G = \Delta G^0 + RT \ln Q^0 \tag{8}$$

the activities of reactants, each activity bearing as an exponent the corresponding coefficient in the balanced equation for the reaction. The standard free energy change ΔG^0 serves as a reference point from which the actual free energy change ΔG can be calculated in terms of the activities of the reacting substances.

When the system has come to chemical equilibrium at constant temperature and pressure, $\Delta G = 0$, from Eq. (3). Equation (7) then leads to the very important relation shown in Eq. (9), where the value of K^0 is shown as Eq. (10), and the activities are the equilibrium values. The ratio

$$\Delta G^0 = -RT \ln K^0 \tag{9} \qquad K^0 = \left[\frac{a_G^g a_H^h}{a_A^a a_B^b}\right] \tag{10}$$

of the activities at equilibrium, K^0, is called the equilibrium constant or, more precisely, the thermodynamic equilibrium constant. (The terms K^0 and Q^0 are written with superscripts to emphasize that they represent ratios of activities.) The equilibrium constant is a characteristic property of the reaction system, since it is determined uniquely in terms of the standard free energy change. The term $-\Delta G^0$ represents the maximum net work which the reaction could make available when carried out at constant temperature and pressure with the substances in their standard states. It should be clear from Eq. (6) that the magnitude of ΔG^0 is directly proportional to the amount of material represented in the reaction in Eq. (1). Likewise, Eqs. (7) to (10) denote this same proportionality through the exponents a, b, g, and h in the terms K^0 and Q^0. Naturally, the value of ΔG^0, as well as K^0 and Q^0, will depend upon the particular concentration scales and standard states selected for the system, so it is essential that sufficient information be stated about a system to prevent any ambiguity.

Equations (8) and (10) can be combined in the form of Eq. (11). When Q^0 for a specified set

$$\Delta G = RT \ln \frac{Q^0}{K^0} \tag{11}$$

of conditions is larger than K^0 (so that ΔG is positive), the proposed reaction cannot occur. On the other hand, when Q^0 is less than K^0, the proposed process or reaction can occur. The equilibrium constant thus serves as a measure of the position of chemical equilibrium for a system. For the proposed process for which Q^0 is greater than K^0, the reaction system would be moving away from its equilibrium state (impossible of its own accord!), and for a proposed process for which Q^0 is less than K^0, the process would bring the system closer to its equilibrium state (as in all naturally occurring processes).

Instead of activities, values of concentrations and activity coefficients at equilibrium may be used to express the form of the equilibrium constant K^0 in Eq. (12). This gives the equilibrium

$$K^0 = \left[\frac{x_G^g x_H^h}{x_A^a x_B^b}\right] \cdot \left[\frac{f_G^g f_H^h}{f_A^a f_B^b}\right] \quad (12)$$

constant as a product of two terms, each of the same form as K^0 itself. The first term, involving concentrations, is directly measurable if the system can be analyzed at equilibrium. On the other hand, the activity coefficient term, as seen in Eq. (13), is frequently difficult to evaluate.

$$\Gamma = \frac{f_G^g f_H^h}{f_A^a f_B^b} \quad (13)$$

Intensive studies of activity coefficients made upon a wide variety of chemical systems led to a number of simplifying principles and some useful theoretical treatments of the subject. For gases, the activity coefficients differ only slightly from unity for pressures up to 10 atm (1 megapascal) and can be evaluated from equation-of-state data. For mixtures of nonelectrolytes, the values also appear to be close to unity in many cases. For solutions of electrolytes, the activity coefficients vary greatly with concentration, and in many cases approach unity only below a useful or even meaningful concentration. The theoretical treatments of P. Debye, E. Hückel, and others have systematized the patterns of electrolyte behavior, making possible a reasonable estimate of the activity coefficients in many cases. SEE ACTIVITY; SOLUTION.

In general, the function Γ, Eq. (13), approaches unity as the composition of the system approaches that of the reference state, so in practice most equilibrium constants K^0 are evaluated through some suitable extrapolation procedure involving Eq. (12). See the discussion following Eq. (5).

For many approximate calculations or when data for Γ are scarce, it is common to express the equilibrium constant as the concentration term only; that is, Eq. (14) holds. Unless Γ is a

$$K = \frac{x_G^g x_H^h}{x_A^a x_B^b} \quad (14)$$

rather insensitive function of concentration, the so-called constants obtained in this manner will not be constant at all as the composition is varied, and even though approximately constant, may vary considerably from the true value of K^0. Although the practice of assigning Γ a value of unity will often give adequate results and is frequently the only expedient available, the results should be used with caution.

It is appropriate to point out here that the kinetic concept of chemical equilibrium introduced by C. M. Guldberg and P. Waage (1864) led to the formulation of the equilibrium constant in terms of concentrations. Although the concept is correct in terms of the dynamic picture of opposing reactions occurring at equal speeds, it has not been successful in coping with the problems of activity coefficients and cannot lead to the useful relations, Eqs. (4) and (11), in terms of an energetic criterion for the position of equilibrium. Conversely, the thermodynamic approach yields no relationship between the driving force at the reaction and the rate of approach to equilibrium. SEE CHEMICAL DYNAMICS.

The influence of temperature upon the chemical potentials, and hence upon the equilibrium constant, is given by the Gibbs-Helmholtz equation, Eq. (15). The derivative on the left represents

$$\left[\frac{d \ln K^0}{dT}\right]_P = \frac{\Delta H^0}{RT^2} \quad (15)$$

the slope of the curve obtained when values of $\ln K^0$ for a reaction, obtained at different temperatures but always at the same pressure P, are plotted against temperature. The standard heat of reaction ΔH^0 for the temperature T at which the slope is measured, is the heat effect which could also be observed by carrying out the reaction involving the standard states in a calorimeter at the corresponding temperature and pressure.

For endothermic reactions, which absorb heat (ΔH^0 positive), K^0 increases with increasing temperature. For exothermic reactions, which evolve heat (ΔH^0 negative), K^0 decreases with increasing temperature and the yield of products is reduced. A more useful arrangement of Eq. (15) is shown in Eq. (16). In practice, plots of $\ln K^0$ against $1/T$ are nearly linear for many reactions

$$\left[\frac{d \ln K^0}{d(1/T)}\right]_P = \frac{-\Delta H^0}{R} \tag{16}$$

where the value of ΔH^0 changes slowly with temperature. Hence, over small temperature ranges, Eq. (16) becomes, in integrated form, Eq. (17). This relation is much used for calculating heats of

$$\ln \frac{K_2^0}{K_1^0} = \frac{\Delta H^0}{R}\left[\frac{T_2 - T_1}{T_1 T_2}\right] \tag{17}$$

reaction from two equilibrium measurements or for determining a new equilibrium constant K_2^0 from values of K_1^0 and ΔH^0.

For accurate work, or for extending the calculations over a wide range of temperature, ΔH^0 must be known as a function of temperature before Eq. (15) or (16) can be integrated. When sufficient heat capacity data are available, Kirchhoff's equation, involving the difference in heat capacities between the products and reactants, Eq. (18), may be combined with Eq. (15) to yield

$$\Delta H^0 = \Delta H_0^0 + \alpha T + \frac{\beta T^2}{2} + \frac{\gamma T^3}{3} \tag{18}$$

Eq. (19). In these equations, α, β, and γ are determined from heat capacity data; the constants

$$R \ln K^0 = -\frac{\Delta H_0^0}{T} + \alpha \ln T + \frac{\beta T}{2} + \frac{\gamma T^2}{6} + I \tag{19}$$

ΔH_0^0 and I require knowledge of one value of ΔH^0 and one value of K^0, or values of K^0 at two temperatures. *See* Thermochemistry.

When equilibria are studied under conditions of constant temperature and constant volume, and with use of volume concentrations to fix standard states, the preceding treatment will yield ΔE^0, the internal energy change, which is the calorimetric heat of reaction at constant volume.

Homogeneous equilibria. These involve singlephase systems: gaseous, liquid, and solid solutions. In most cases, solid solutions are so far from ideal that equilibrium constants cannot be evaluated, and such systems are treated in terms of the phase rule. *See* Phase equilibrium.

A typical gas-phase equilibrium is the ammonia synthesis shown in (20). The most natural

$$N_2 + 3H_2 \rightleftharpoons 2NH_3 \tag{20}$$

concentration measures are mole fraction or partial pressure; molar concentrations might be used. The partial pressures p_i are defined in terms of mole fraction N_i and the total pressure P by $p_i = N_i P$; note that partial pressures in general are not directly observable. For low pressures, where activity coefficients are practically unity, Eq. (21) holds. In turn Eq. (21) may be shown as Eq. (22), where Δn is the increase in the number of moles of gases (here $\Delta n = 2$). The mole

$$K_p^0 = \frac{p_{NH_3}^2}{p_{N_2} p_{H_2}^3} = \frac{N_{NH_3}^2}{N_{N_2} N_{H_2}^3} P^{-2} \tag{21} \qquad K_p^0 = K_N^0 P^{\Delta n} \tag{22}$$

fraction equilibrium constant K_N^0 is pressure dependent, but K_p^0 is independent of pressure because of the difference in standard states. Consequently, an increase in total pressure P must lead to an increase in K_N^0 in this case. If the increase in total pressure is due to a decrease in volume of the system, the result will be an increased yield of products (NH_3). An increase in pressure brought about by injection of an inert gas into a constant volume system would not affect the partial pressures of the reacting gases nor the ultimate yield of products. The value of K_N^0, and thus that of ΔG_N^0, will be affected by change in total pressure due to a change in the net work of mixing and unmixing the gases. This reaction, Eq. (20), is exothermic, so best yields will be obtained at lower temperatures. See Eq. (15).

A typical liquid-phase equilibrium is the dissociation of acetic acid in water, reaction (23) and Eq. (24). In this particular case, the activity coefficient ratio Γ is slightly less than unity in

$$HC_2H_3O_2 + H_2O \rightleftharpoons H_3O^+ + C_2H_3O_2^- \quad (23)$$

$$\frac{K^0}{\Gamma} = \frac{m_{H_3O^+} m_{C_2H_3O_2^-}}{m_{HC_2H_3O_2} N_{H_2O}} \quad (24)$$

dilute solutions; the reference state will be the infinitely dilute solution. The mole fraction of water in the applicable concentration range is close to unity, and so numerically plays little part in evaluation of K^0. For this reason, it is commonly omitted in the formulation of the equilibrium constant, and many erroneous explanations exist in the literature. Letting α be the fraction of acetic acid dissociated and m be the total concentration, $K^0/\Gamma = m\alpha^2/(1 - \alpha)$; inspection shows that α increases with dilution.

The solvent appears to be inert, since its chemical potential remains practically unchanged over the useful concentration range. As a result of this apparent inertness of the solvent, it is not possible to determine the extent of hydration of any dissolved species from equilibrium studies. Thus, where the actual ion is H^+, H_3O^+, or $H_9O_4^+$, it is the total stoichiometric concentration that is measured and used in Eq. (24). SEE IONIC EQUILIBRIUM.

Heterogeneous equilibria. These are usually studied at constant pressure, since at least one of the phases will be a solid or liquid. The imposed pressure may be that of an equilibrium gaseous phase, or it may be an externally controlled pressure.

In describing such systems, the nature of each phase must be specified. In the examples to follow, s, l, and g identify solid, liquid, and gaseous phases, respectively. For solutions or mixtures, the composition is needed, in addition to the temperature and pressure, to complete the specification of the system. If not obvious, the identity of the solvent must be given.

In the equilibrium shown as (25), the relationship of Eq. (26) holds. Here $K^0 = p/N$, the

$$H_2O(l) \rightleftharpoons H_2O(g) \quad (25) \qquad \Delta G^0 = \mu_g^0 - \mu_l^0 = -RT \ln \frac{p}{N} \quad (26)$$

ratio of the vapor pressure p to the liquid mole fraction N. For pure water, the equilibrium constant is simply the standard vapor pressure p^0, and the Clausius-Clapeyron equation is just a special case of the Gibbs-Helmholtz equation, Eq. (15). Now when a small amount of solute is added, decreasing the mole fraction of solvent, the vapor pressure p must be lowered to maintain equilibrium (Raoult's law). The effect of the total applied pressure P upon the vapor pressure p of the liquid is given by the Gibbs-Poynting equation, Eq. (27). Here V_l and V_g are the molar volumes of

$$\left[\frac{dp}{dP}\right]_T = \frac{V_l}{V_g} \quad (27)$$

liquid and vapor. The vapor pressure will increase as external pressure is applied (activity increases with pressure). If the external pressure is applied to a solution by a semipermeable membrane, an applied pressure can be found which will restore the vapor pressure (or activity) of the solvent to its standard state value. SEE OSMOSIS.

Solubility equilibria are of wide variety. For a solid, such as barium sulfate, $BaSO_4$, which dissociates as shown in (28), the equilibrium relationship is shown in Eq. (29). When the solid

$$BaSO_4(s) \rightleftharpoons Ba^{2+} + SO_4^{2-} \quad (28) \qquad K^0 = \frac{a_{Ba^{2+}} a_{SO_4}}{a_{BaSO_4}} \quad (29)$$

state is pure, its mole fraction will be unity. If it is extremely finely divided, its activity coefficient will become greater than unity; with this increase in the activity of the solid state, the solubility must increase to maintain equilibrium. On the other hand, inclusion of foreign ions in the crystal lattice (solid solution formation) may lower the activity of the solid state.

When a gas, such as CO_2, is dissolved in a liquid, its equilibrium with the gas phase is shown in (30). Equation (31), Henry's law, represents the equilibrium constant for this situation.

$$CO_2(soln) \rightleftharpoons CO_2(g) \qquad (30) \qquad\qquad K^0 = \frac{p}{N} \qquad (31)$$

When the gas dissociates in the liquid, reaction (32), then Eq. (33) may be utilized. Here m is the

$$(H^+ + Cl^-)(aq) \rightleftharpoons HCl(aq) \rightleftharpoons HCl(g) \qquad (32) \qquad\qquad K^0 = \frac{p}{m^2} \cdot \Gamma \qquad (33)$$

molal concentration. The equilibrium constant must reflect this behavior through the proper exponents. Equation (33) might correctly imply that the molecules were dimerized in the gaseous phase, if it were not known from a study of vapor densities that in this case they are not.

Similarly, when a solute distributes itself between two immiscible phases, the equilibrium constant takes the form of Eq. (34). Here n is ratio of the molecular weight in phase 1 to that in

$$K^0 = \frac{c_1}{c_2^n} \cdot \Gamma \qquad (34)$$

phase 2. The equilibrium concentrations c in Eq. (34) reflect the relative solubilities of the solute in the two phases; since solubilities may vary widely, distribution operations may provide an effective means for concentrating a widely dispersed solute.

When two components form two immiscible phases at equilibrium, each condensed phase will be a saturated solution; complete immiscibility is impossible in principle, since the chemical potential of any component must be the same in both phases. The separation of a liquid system into two liquid phases is a manifestation of the nonideality of the solutions. For practical purposes, however, many solids may be regarded as immiscible because of the stringent requirements associated with formation of the crystal lattice.

In reactions involving condensed and immiscible phases, for example, reaction (35), there

$$Pb(s) + 2AgCl(s) \rightleftharpoons PbCl_2(s) + 2Ag(s) \qquad (35)$$

can be no change in concentration of any phase during the reaction. Then the Q^0_r term in Eq. (10) will be constant, and ΔG can never become zero. Although reaction is possible, there can be no equilibrium until one of the reactants is completely used up. There might, of course, be one condition of temperature and pressure for which ΔG could be zero. Such is the case in transition phenomena, or melting-freezing phenomena. From the phase-rule viewpoint, the system in reaction (35) lacks one degree of freedom if reaction is possible, so one phase must disappear in order to attain equilibrium.

Bibliography. R. A. Alberty, *Physical Chemistry*, 7th ed., 1987; G. M. Barrow, *Physical Chemistry*, 4th ed., 1979; K. G. Denbigh, *The Principles of Chemical Equilibrium*, 4th ed., 1981; I. N. Levine, *Physical Chemistry*, 2d ed., 1983.

PHASE EQUILIBRIUM
ROBERT L. SCOTT

A general field of physical chemistry dealing with the various situations in which two or more phases (or states of aggregation) can coexist in thermodynamic equilibrium with each other, with the nature of the transitions between phases, and with the effects of temperature and pressure upon these equilibria. Many superficial aspects of the subject are largely qualitative, for example, the empirical classification of types of phase diagrams; but the basic problems always are susceptible to quantitative thermodynamic treatment, and in many cases, statistical thermodynamic methods can be applied to simple molecular models.

Thermodynamics requires that when two phases, α and β, are free to exchange heat, mechanical work, and matter (chemical species), the temperature T, the pressure P, and the chemical potential (partial molar free energy) μ_i of each particular component i must be equal in both phases at equilibrium. Algebraically, equilibrium exists when $T_\alpha = T_\beta$, $P_\alpha = P_\beta$, $\mu_{i,\alpha} = \mu_{i,\beta}$, and $\mu_{j,\alpha} = \mu_{j,\beta}$.

These conditions of thermal, mechanical, and material equilibrium need not all be present if the equilibrium between phases is subject to inhibiting restrictions. Thus, for a solution of a nonvolatile solute in equilibrium with the solvent vapor, the condition of equality of solute chemical potentials $\mu_{2,\alpha} = \mu_{2,\beta}$ need not apply, since there can be no solute molecules in the vapor phase. Similarly, in osmotic equilibria, in which solvent molecules can pass through a semipermeable membrane, whereas solute molecules cannot, $\mu_{1,\alpha} = \mu_{1,\beta}$ and $T_{1,\alpha} = T_{2,\beta}$, but the solute chemical potentials μ_2 are unequal, as are the pressures on opposite sides of the membrane. SEE OSMOSIS; SOLUTION.

If a system consists of P phases and C distinguishable components, there are $C + 2$ thermodynamic variables (C chemical potentials μ_j, plus the temperature and pressure) which are interrelated by an equation for each phase. Since there are P independent equations relating the $C + 2$ variables, one needs to fix only $F = C + 2 - P$ variables to define completely the state of the system at equilibrium; the other variables are then beyond control. This relation for the number of degrees of freedom F, or variance, is called the phase rule and was first derived by Willard Gibbs in 1873. It has proved to be a powerful tool in interpreting and classifying types of phase equilibria.

When chemical changes may occur in the system, the number of components C is the number of independent components, that is, the number of components whose amounts can be varied by the experimenter; this is equal to the total number of chemical species present less the number of independent chemical equilibria between them.

An invariant system has no degrees of freedom ($F = 0$), for which the number of phases $P = C + 2$. For a one-component system, such an invariant point is a triple point at which three phases coexist at a single temperature and pressure only; for a two-component system, a quadruple point (four phases) would be invariant. SEE TRIPLE POINT.

In a univariant system ($F = 1$), $P = C + 1$. With a one-component system, one can fix the temperature at which two phases (liquid and gas, for instance) can coexist in equilibrium; then the pressure (here the vapor pressure) is determined and not subject to external control. A univariant system is described by a line in a phase diagram, for example, a plot of vapor pressure versus temperature. The differential equation, Eq. (1), for such a univariant line in a one-compo-

$$\frac{dP}{dT} = \frac{\Delta H}{T \Delta V} \tag{1}$$

nent system was first deduced by B. P. E. Clapeyron in 1834. In the equation ΔH and ΔV are the enthalpy change (heat absorbed) and volume change, respectively, for the transition from one phase to another.

In systems of two or more components, more complicated equations for the univariant line replace the Clapeyron equation, but in special cases they may reduce to the simpler form. For example, in the chemical decomposition of a solid calcium carbonate to solid calcium oxide and gaseous carbon dioxide, reaction (2), the three phase system (two solids and one gas) is univariant

$$CaCO_3\,(s) \rightleftharpoons CaO(s) + CO_2\,(g) \tag{2}$$

(since the number of independent components C is two), the equilibrium pressure is the decomposition pressure of $CaCO_3$, and the ΔH and ΔV of the Clapeyron equation are the enthalpy and volume changes associated with the chemical reaction.

BINARY SYSTEMS

Phase diagrams and binary systems containing two components are easily classified. Typical examples of the important classes (liquid-gas, liquid-liquid, solid-liquid, solid-solid) have been selected for description.

Liquid-gas equilibrium. In a one-component system, liquid and vapor are in equilibrium at the boiling point. For two-component systems, the two-phase situation is bivariant and more complex. A complete temperature-pressure-composition diagram would be three-dimensional, so most phase diagrams are made for either constant pressure or constant temperature. **Figure 1** shows the simplest type of binary liquid-vapor temperature-composition diagram, exemplified by

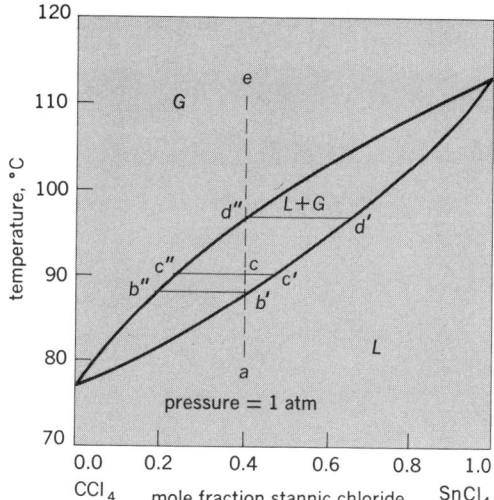

Fig. 1. Binary liquid-vapor temperature-composition diagram for the system carbon tetrachloride + stannic chloride. G = gas (vapor) phase; L = liquid phase; $L + G$ = two coexisting phases. °F = (°C × 1.8) + 32; 1 atm = 10^2 kilopascals.

the system carbon tetrachloride–stannic chloride, which forms essentially ideal solutions. The regions labeled G and L are one-phase regions, gas (vapor) and liquid, respectively; the region labeled $L + G$ is a two-phase region in which liquid and vapor coexist. If the temperature of a liquid mixture of 40 mole percent $SnCl_4$ (mole fraction = 0.40) is increased at a constant pressure of 1 atm, the change in the system can be traced along the straight line $ab'cd''e$. At low temperatures, only one phase, the liquid, is present, but at 87.5°C (190°F) point b', a vapor phase appears. The composition of this vapor phase is given by point b'' (mole fraction $SnCl_4$ = 0.18), and the two conjugate phases are connected on the diagram by the tie line $b''b'$. As the temperature is increased further, more vapor is formed; since the vapor is rich in CCl_4, this component becomes relatively depleted in the liquid phase, and the liquid composition moves along the line $b'c'd'$, while the vapor composition moves along the line $b''c''d''$.

At 90°C (194°F) the overall composition of the two-phase system is represented by point c, but the compositions of vapor and liquid separately are given by the two ends of the tie line, points c'' and c', respectively (mole fractions of 0.22 and 0.47, respectively). The relative amounts of the two phases are given by the lever arm principle of physics. The ratio of the moles of vapor to moles of liquid is given by the ratio of the length cc' to the length $c''c$, here 0.07:0.18, or 28%, in the vapor phase. Further increase in temperature produces more and more vapor until, at 97°C (207°F) the liquid phase (point d', mole fraction 0.62) has become vanishingly small; at higher temperatures, it disappears, and only the vapor phase (point d'', mole fraction = 0.40) remains. Further increase in temperature (along the line $d''e$) is uneventful.

In this simple system, there are no maxima or minima in the liquid and vapor curves; consequently, such liquid mixtures can be separated completely into the two pure components by fractional distillation. Systems with maximum boiling points (acetone + chloroform, **Fig. 2**) or minimum boiling points (ethanol + benzene, **Fig. 3**) cannot be so separated into the pure substances. At the maximum or minimum, the composition of the liquid is identical with that of the vapor with which it is in equilibrium; continued boiling will not alter these compositions, so that these solutions are called constant-boiling mixtures or azeotropes. It should be noted that the solid and liquid lines are smooth curves tangent to each other at such a point; any phase diagram which shows a sharp corner is thermodynamically incorrect.

Maximum boiling mixtures are associated with negative deviations from ideal behavior; this type of deviation usually arises from strong attractions between molecules of the different species (sometimes called compound formation). Minimum boiling mixtures are associated with positive deviations from ideal behavior; this usually arises when the attraction between two unlike molecules (1-2) is weaker than the average of two like pairs (1-1 and 2-2); extreme examples of

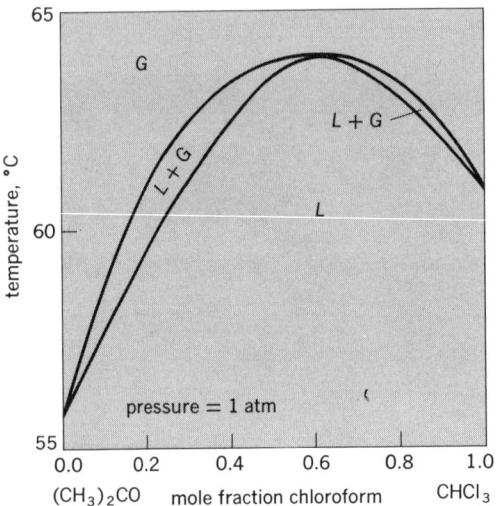

Fig. 2. Temperature-composition diagram for acetone + chloroform, showing maximum boiling point. °F = (°C × 1.8) + 32; 1 atm = 10^2 kilopascals.

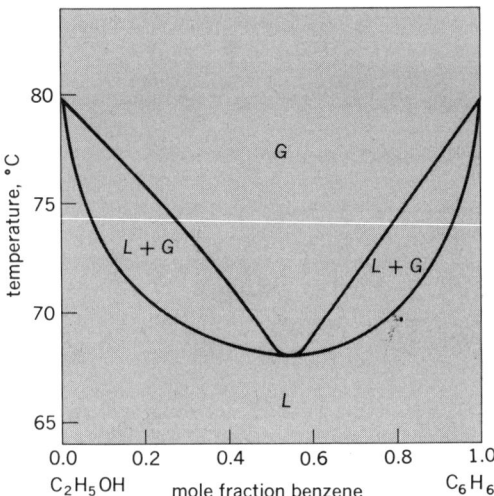

Fig. 3. Temperature-composition diagram for ethanol + benzene, showing minimum boiling point. °F = (°C × 1.8) + 32.

this arise when one component may be described as associated. The simpler type of phase diagram (Fig. 1) occurs when the two components mix nearly ideally or when the boiling points are very different.

Liquid-liquid equilibrium. When two liquids are sufficiently different in their intermolecular forces, they may not mix in all proportions but instead be only partially miscible, part of the phase diagram being occupied by a two-phase region of two immiscible liquid phases. Most liquids (but not all) become more miscible as the temperature increases and are completely miscible at the critical solution temperature (also called the consolute temperature). The system aniline + n-hexane (**Fig. 4**) is a typical example of such liquid-liquid immiscibility and critical phe-

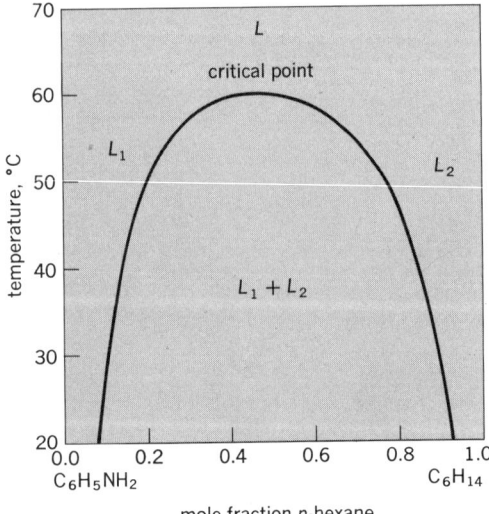

Fig. 4. Liquid-liquid equilibrium for aniline + n-hexane. °F = (°C × 1.8) + 32.

nomena. The liquid-liquid phase boundary and the critical solution temperature are only slightly dependent upon pressure. The size of the two-phase region increases with pressure if (as is usual) the two liquids expand on mixing at constant pressure.

Solid-liquid equilibrium. When two substances are completely miscible with each other and form a complete series of solid solutions, the solid-liquid phase diagrams are entirely analogous to the liquid-vapor diagrams illustrated above. Those with no maximum or minimum, usually associated with nearly ideal liquid and solid solutions, such as methane + krypton, are called type I, according to the Bakhuis Roozeboom classification; those with a maximum melting point, an exceedingly rare type exemplified by d-carvoxime + l-carvoxime, are type II; and those with a minimum melting point (bromobenzene + iodobenzene), type III. SEE SOLID SOLUTION.

In most binary systems, however, extensive solid solutions are impossible because of the incompatibility of the size, shape, and crystal lattices of the two components. In the absence of solid solution formation, the addition of a solute to a liquid solvent invariably depresses the freezing point, that is, the temperature at which, upon cooling, the first trace of solid solvent appears. This depression of the freezing point of a dilute solution is a convenient method (the so-called cryoscopic method) for determining the molecular weight of a solid solute. SEE MOLECULAR WEIGHT.

As the freezing-point curve of a liquid continues to lower temperatures with higher concentrations of the second component, the conventional roles of solute and solvent become reversed, and one speaks of the solubility of a solute rather than of the depression of the freezing point of the solvent; no point of demarcation exists, and the two situations are in fact two aspects of the same phenomenon.

The second component also has a freezing-point–solubility curve marking the temperature at which the solution is in equilibrium with pure solid 2. The point of intersection of the two phase boundaries is the eutectic point which defines a eutectic temperature and a eutectic composition. Below this temperature, the system consists of two solid phases ($S_1 + S_2$ in **Fig. 5a**).

Solid-liquid phase diagrams are conveniently determined by thermal analysis of cooling curves. An initially homogeneous liquid is cooled gradually, and the temperature plotted against time. Figure 5a shows a simple eutectic-type phase diagram, and Fig. 5b sketches typical cooling curves obtained for various compositions (those marked A to F in Fig. 5a). Curves A and F are for the pure components; each shows a single temperature arrest, a horizontal section of the curve

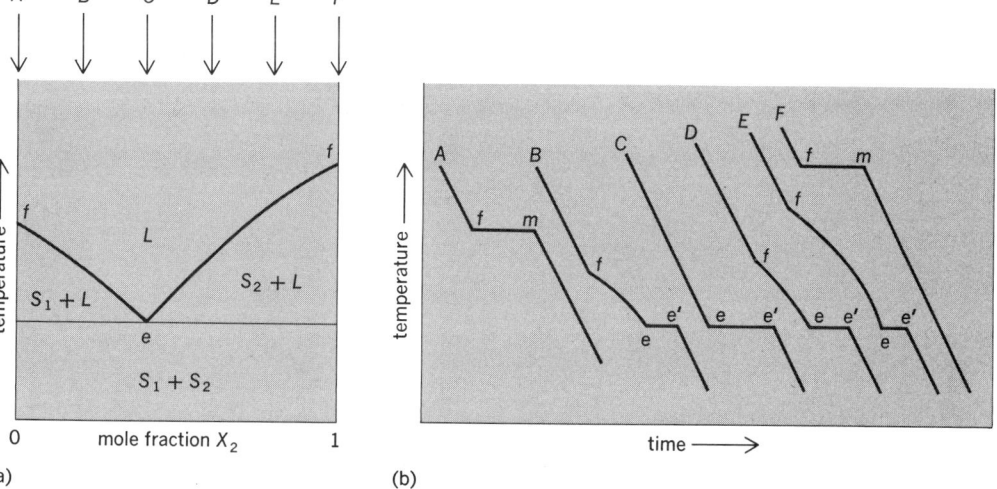

Fig. 5. Solid-liquid equilibrium. (a) Schematic diagram when there are no compounds of solid solutions. Points f are freezing points of pure substance; e is the eutectic point. A, B, C, D, E, and F are the compositions corresponding to the cooling curves of part b. (b) Schematic cooling curves for system shown in part a; f = freezing point, m = melting point, e = beginning of eutectic freezing, e' = end of eutectic freezing.

for the melting point (between f and m), at which the temperature remains constant from the time the first bit of solid is formed until the last bit of liquid disappears.

Curves B and E are for solutions rich in components 1 and 2, respectively. At a temperature below the melting point of the pure substance, the first bit of solvent begins to freeze out; this is indicated by a change in slope (point f). The freezing section of the cooling curve (between points f and e) is not horizontal; as solvent freezes out, the liquid phase becomes richer in solute, and the freezing point is further depressed. When the composition of the liquid phase reaches the eutectic composition, the second component begins to freeze out as well as the first. No further change in liquid composition occurs, so the temperature remains constant until no liquid remains (between points e and e'). The solid which freezes out at the eutectic (the eutectic mixture) appears superficially very different from either pure solid; it is a mixture of very small crystals of each of the two components which have crystallized together. This microcrystalline two-phase mixture is in no sense a compound.

Curve D is a cooling curve for still another composition, and curve C is for the eutectic composition. Alone, curve C is indistinguishable from that for the freezing of a pure substance; only by combining the information from a series of cooling curves can one construct the whole diagram.

When the components form an incomplete series of solid solutions (partial miscibility), the phase diagram can be of the eutectic type (Bakhuis Rooseboom type V) illustrated by the system silver + copper shown in **Fig. 6**. Indeed, since the mutual solubilities are never exactly zero, eutectic diagrams, such as that in Fig. 5a, are in fact merely extreme examples of this more general case.

When a solid phase upon melting transforms into a liquid phase and a solid phase of different composition, one speaks of incongruent melting and a peritectic-type phase diagram. A simple example is the type-IV solid-solution diagram illustrated by the system silver chloride + lithium chloride (**Fig. 7**); similar phenomena occur in systems without any appreciable solid-solution formation.

Many solid-liquid phase diagrams are complicated by the existence of intermediate crystalline phases of different crystal structure. Usually, these intermediate phases are at compositions close to simple mole ratios of the components and have the two kinds of molecules distributed in a regular arrangement; consequently, it is convenient to call these compounds, even if no specific chemical interactions can be demonstrated unequivocally. There is a rich variety of such complex

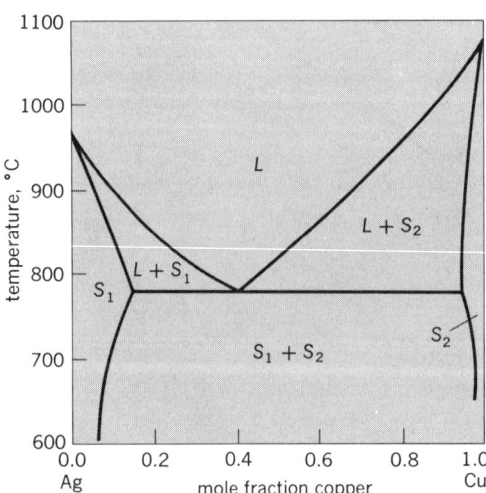

Fig. 6. Solid-liquid equilibrium in the system silver + copper (type-V solid solutions). °F = (°C × 1.8) + 32.

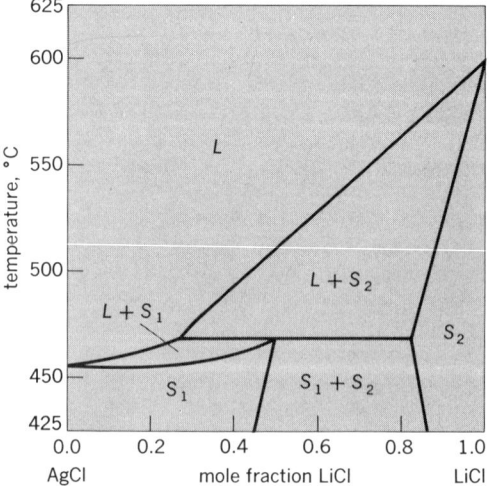

Fig. 7. Solid-liquid equilibrium in the system silver chloride + lithium chloride (type-IV solid solutions). °F = (°C × 1.8) + 32.

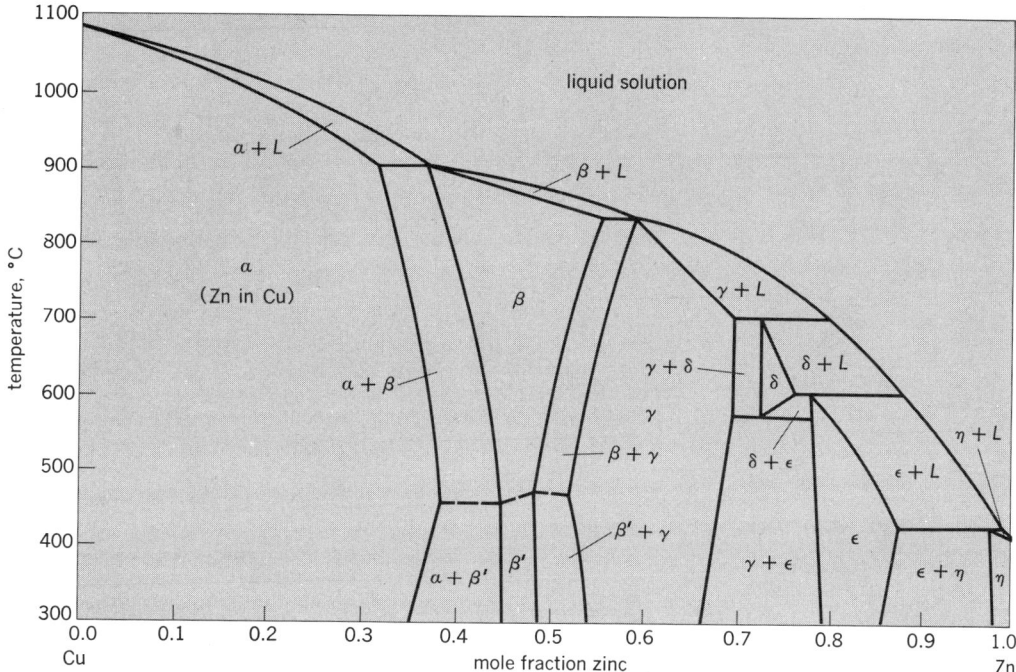

Fig. 8. Temperature-composition diagram for the system copper + zinc (brass). Note the six different solid phases. The broken line separating β from β' denotes a second-order, order-disorder transition. °F = (°C × 1.8) + 32.

phase diagrams, especially among binary systems of ionic salts ($CaCl_2$-KCl) and among binary systems of metals (alloys). **Figure 8** shows the Cu + Zn system (brass), in which a whole series of crystal structures appears: α-brass has the crystal structure of pure copper (face-centered cubic) with an occasional Zn atom in the lattice; β-brass has a body-centered cubic structure with a Cu-Zn ratio of approximately 1:1; γ-brass has a very complex structure related to the formula Cu_5Zn_9; η-brass has the crystal structure of pure Zn (hexagonal, close-packed); the δ and ε phases have still different structures. Note the five peritectic transitions (for example, at 600°C (1112°F) where the ε-solid melts incongruently to the δ-solid and a liquid solution rich in zinc).

Solid-state equilibrium. These are of several types. One can have solid critical solution temperatures (with solid solutions of the same crystal structure) which are analogous to the more familiar liquid-liquid case (the system gold + platinum shown in **Fig. 9**). More common are the transitions between one crystalline form and another, which can occur even in pure substances. One such occurs in the system nitrogen + carbon monoxide (**Fig. 10** shows solid-solid, solid-liquid, and liquid-vapor transitions for this sytem.) SEE TRANSITION POINT.

MULTICOMPONENT SYSTEMS

As one proceeds from binary systems to systems with three or more components, the phase diagrams become more complex. Each component adds another dimension to the representation of the phase equilibria. Thus, for three components, two dimensions are required to represent the phase diagrams for a single temperature and pressure; these are conveniently depicted by a triangular diagram in which each vertex represents a pure component. **Figure 11** shows a diagram of a ternary system in which three liquid phases can coexist. Even when there are three phases, the system is still bivariant. (An example of such a system is water + succinonitrile + diethyl ether.)

A special case of a three-component system is that in which there are two immiscible

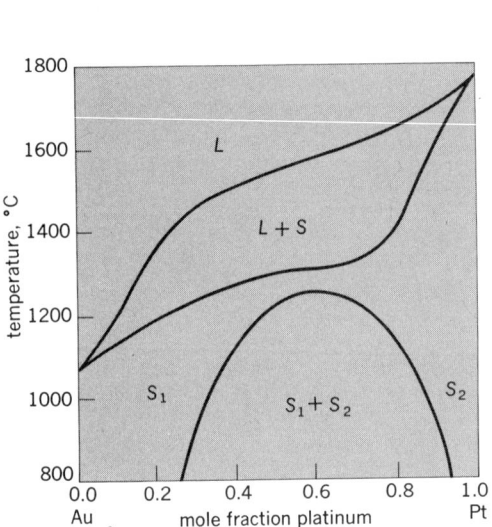

Fig. 9. Solid-liquid and solid-solid equilibria in the system gold + platinum. Critical region of the solid-solid phase boundary nearly touches the solidus curve of the solid-liquid phase boundary. If these touched and coalesced, a type-IV diagram (Fig. 7) would result. °F = (°C × 1.8) + 32.

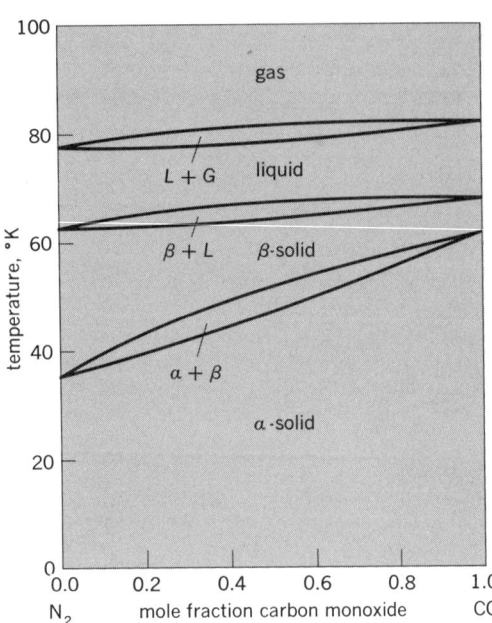

Fig. 10. Phase equilibria for nitrogen + carbon monoxide, being solid-solid, solid-liquid, and liquid-vapor. °F = (K × 1.8) − 459.67.

solvents and a third component, soluble in both, distributed between the two phases. The ratio of the concentrations of the solute in the two solvents is the distribution coefficient; in dilute solutions, this is independent of the concentration, but at higher concentrations, nonideal behavior of the solute can produce systematic variations of the distribution coefficient which ends in the most concentrated solutions with the ratio of the solubilities in the saturated solutions.

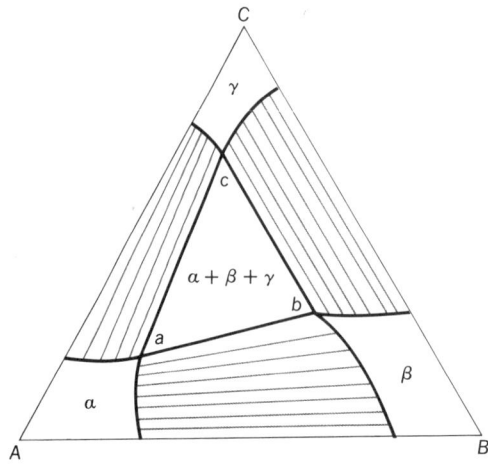

Fig. 11. Schematic diagram of a three-component system at a fixed temperature and pressure. Points A, B, and C represent the pure liquids. The composition corresponding to a point in the diagram is determined by the positions along a line from each vertex to the opposite side; thus point b is 20% A, 50% B, 30% C. Regions α, β, and γ correspond to single phases rich in A, B, and C, respectively; $\alpha + \beta + \gamma$ is a three-phase region; the three saturated solutions have the compositions given by the points a, b, and c. Three two-phase regions, $\alpha + \beta$, $\alpha + \gamma$, and $\beta + \gamma$, are indicated by drawing in the tie lines.

Distribution effects are important in separating similar materials. A small difference in distribution coefficients is amplified by multistage equilibria, such as those used in countercurrent extraction and partition chromatography.

The examples used to illustrate the various types of phase equilibria are not supposed to suggest that these types are restricted to the particular kind of chemical substances shown. In general, examples of each could have been selected from many kinds of substance such as metals, nonmetallic elements, inorganic salts, and organic nonelectrolytes. SEE CHEMICAL EQUILIBRIUM; CHEMICAL THERMODYNAMICS; FUSED-SALT PHASE EQUILIBRIA; INTERFACE OF PHASES.

Bibliography. G. M. Barrow, *Physical Chemistry*, 4th ed., 1979; R. Ginell, *Association Theory: The Phases of Matter and their Transformations*, 1979; I. N. Levine, *Physical Chemistry*, 2d ed., 1983; W. J. Moore, *Basic Physical Chemistry*, 1983; A. Reisman, *Phase Equilibria*, 1970; F. E. Wetmore and D. J. LeRoy, *Principles of Phase Equilibria*, 1951.

FUSED-SALT PHASE EQUILIBRIA
R. E. THOMA

Conditions in which two or more phases of fused-salt mixtures can coexist in thermodynamic equilibrium. Phase diagrams of these equilibrium conditions summarize basic knowledge about fused salts. Numerous advances in the technologies which are based on high-temperature chemistry have become possible through the increase in knowledge about fused salts. The increasingly significant role of fused salts in industrial processes is evident in the widening application of these materials as heat-transfer media, in extractive metallurgy, in nonaqueous reprocessing of nuclear reactor fuels, and in the development of nuclear reactors which create more fuel than they consume (breeder reactors); moreover, these technologies are all direct outgrowths of research and development with fused salts. SEE PHASE EQUILIBRIUM.

Fused-salt media were of great interest before 1900 for the electrolytic reduction of cations to metals. The bulk of research on fused-salt phase equilibria has stemmed from interest in electrometallurgical processes from that time until the mid-twentieth century. In the 1950s and 1960s the U.S. Atomic Energy Commission's programs showed that molten-salt mixtures may serve successfully as nuclear reactor fuels, fuel solvents, coolant fluids, moderators, nuclear reactor breeder blankets for the preparation of plutonium or uranium-233, high-temperature bearing lubricants, and fuel-reprocessing media.

Fused-salt mixtures find application in technology when the need arises for liquids which are stable at high temperatures. For most applications, suitably low melting temperatures and low vapor pressures are primary considerations. To some extent these requirements are conflicting, because salts which are useful in obtaining low freezing temperatures often tend to have appreciable covalent character and therefore to exhibit unfavorably high vapor pressures.

As a special class of liquids, one which is composed entirely of positively and negatively charged ions undiluted by weak-electrolyte supporting media, fused salts are used in many different types of research. For example, advances in solution theory, thermodynamics, and crystal chemistry have come about through studies of fused-salt systems.

A close connection between fused-salt phase diagrams and geochemistry exists and stems from the model principle developed by V. M. Goldschmidt, who noted that isomorphic structures are assumed by ions of the same proportionate size and stoichiometric relations but of different charge. Thus the fluorides of beryllium, calcium, and magnesium, for example, are structural models for SiO_2, TiO_2, and ZrO_2. The fluoride structures are referred to as weakened models because of the smaller electrostatic forces resulting from smaller ionic charges; they have been useful for comparisons with oxide and silicate systems. According to Goldschmidt's interpretation, saltlike materials were derived from components such as H_2O, CO_2, SO_3, Cl_2, and F_2, which were volatilized from molten magmas as they crystallized. Crystallization equilibria in fused-salt systems therefore provide a convenient way to study the mechanisms occurring in the formation of igneous rocks. Such investigations were once pursued vigorously, producing the basic techniques which have been employed routinely in phase studies, essentially since the enunciation of the phase rule by J. W. Gibbs. SEE PHASE RULE.

Methods of investigation. For many years scientists in the United States and elsewhere expended a great deal of effort in attempts to predict the phase relationships of uninvestigated systems; however, the necessary thermochemical data which were required, such as heat capacities of the constituents, the variation of this property with temperature, the heats of fusion, and the free energies of mixing, were not available. It soon became evident that the acquisition of these data was at least as arduous a task as the determination of the actual equilibrium behavior, and that quantitatively the phase relationships predicted from the thermochemical data were too imprecise to make the method attractive. As a result, it has not yet become routine to estimate with practical accuracy the behavior of untested systems. Accordingly, the determination of equilibrium phase diagrams has remained principally in the region of experimental science. To be sure, numerous correlations are available which can assist the experimentalist in minimizing tasks such as determining the interrelationships of crystal structure, thermodynamic quantities, and the predictable degree of ideality of some ionic systems. Such information makes it possible for phase studies to be conducted with greater efficiency and rapidity, and therefore tends to maintain the field as experimental. The largely experimental character of fused-salt phase equilibria is even more a consequence of the development of experimental techniques which permit the delineation of phase relationships in fused-salt systems far more accurately and rapidly than was previously possible.

Innovations in the original means of study have brought about the application of a wide variety of experimental techniques to studies of heterogeneous phase equilibria. These include differential thermal analysis, visual methods for direct observations of melting and freezing behavior at high temperatures, filtration of partially molten mixtures, high-temperature centrifugation, hot-stage microscopy, and high-temperature x-ray diffraction. Electromotive-force measurements and vapor-pressure determinations have also been used with success in the study of heterogeneous equilibria in salt systems.

One of the most important developments among the experimental devices used for solid-state phase studies was the automatic x-ray diffractometer for single crystal structure studies. With automatic diffraction equipment, the average period for determination of new structures is about one-tenth that required previously. With the assistance of computerized equipment, it is commonplace to measure automatically the intensities associated with several thousand reflections from crystal planes and to deduce the structures with significantly greater accuracy than was obtained previously. This has a special function in phase studies: By concurrently determining the equilibrium relations in a system and the structure of the crystalline solids which are found in the system, the stoichiometry, compositional variability, coordination chemistry, and so on, are established unequivocally.

Types of equilibria. In use, fused salts are generally employed under conditions where temperature and composition are varied extensively. Drastic changes in equilibrium phase relationships are thus commonplace. In order to understand how these salts behave chemically, it is necessary to gain a rudimentary knowledge of heterogeneous equilibria. Pragmatically, this approach need not entail a complete nor even extensive study of the systems involved; it will be useful as a framework which prescribes the conditions that can be employed in use and in related investigations with fused salts. Characterization of salt phase equilibria is simple and straightforward with binary and ternary systems, but may become increasingly intricate and of decreasing utility as the effect of additional components must be considered. Considerations here are therefore limited to condensed equilibria in binary and ternary systems.

Binary systems. Consider first the binary system A-B. As drawn (**Fig. 1**) the phase diagram indicates that the components A and B react to form the intermediate compounds A_2B and AB. Limited miscibility in the solid state occurs when mixtures of A and A_2B are present at equilibrium at temperatures between T_1 and T_2, and T_3 and T_4. The compound AB melts incongruently to A_2B and liquid. Crystals of AB are dimorphic; at temperatures above T_5 they exist in a crystalline state which transforms, on cooling below temperature T_5, to a less symmetric (lower-energy form) crystalline state. Liquids formed from mixtures of AB and B are immiscible at compositions such as C_1 and C_2, which separate into immiscible liquids of composition described by horizontal tie lines in the two-liquid region shown in Fig. 1.

Ternary systems. The temperature-composition relations in condensed ternary systems are generally represented by a polythermal projection of the liquidus surfaces. This is a projection

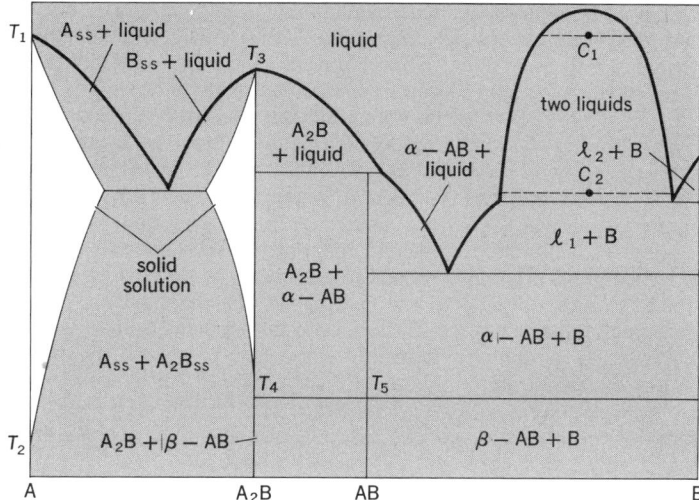

Fig. 1. Diagram of typical phase equilibrium relationships in condensed binary systems of molten salts. Here ℓ is liquid and ss is solid solution.

parallel to the temperature axis on the triangular composition plane; it shows, therefore, the various parts of the liquidus surface or surfaces as viewed from the direction of high temperature. Such diagrams indicate the compositions of liquids saturated with one or more solids. At the invariant reaction points, liquids are in equilibrium with three solids. The direction of temperature change is shown by arrows on boundary curves, and liquidus temperatures are shown by isothermal contours.

The hypothetical ternary system shown in **Fig. 2** displays behavior which is typcially described by three-component phase diagrams. Assume that invariant equilibrium points in the phase diagram A-B-C will ultimately be located as shown in Fig. 2 and that there are no intermediate compounds formed from components B and C; a single intermediate compound AC_2, melting congruently, occurs in the system A-C; the congruently melting compound AB_2 and the incongruently melting compound A_2B are formed from components A and B. A careful examination of several crystalline mixtures cooled from the liquid state can be expected to furnish much information about equilibria in the system A-B-C. Some examples are given below.

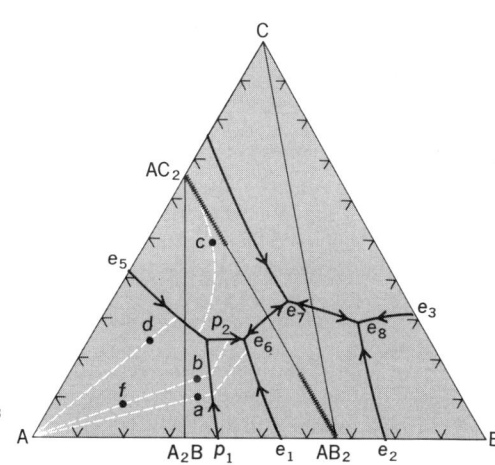

Fig. 2. Typical phase equilibrium relationships in condensed ternary systems of molten salts.

1. All crystallized specimens having compositions within the triangle AB_2-B-C contain only the pure solid phases AB_2, B, and C. These results indicate that the compounds AB_2, B, and C form a subsystem, that AB_2 probably melts congruently, and that the section AB_2-C is a quasibinary system. Under these circumstances, a single invariant point, the eutectic e_8, can occur in AB_2-B-C. It may be possible to estimate its approximate composition by examination of the crystallized mixtures. Morphological data, obtained from petrographic or metallographic methods, may reveal the domains of the primary phases AB_2, B, and C, and the composition of the eutectic.

2. Assume that AC_2 and AB_2 form extensive but limited solid solutions exhibiting maximum solubility at the solidus, as shown in Fig. 2. In general, lattice constants of AB_2 and AC_2 will change as a function of solute concentration. Therefore, x-ray data obtained from crystallized specimens having compositions intermediate between AB_2 and AC_2 can be examined for lattice-spacing changes and correlated employing the parametric method. Some exsolution (precipitation of crystals from the solid solution) is to be expected as temperatures are lowered. The exsolution curve often exhibits considerable slope and precludes the possibility of obtaining data which indicate concentrations of maximum solubility of AB_2 and AC_2 except by static methods.

3. Occurrence of solid solutions of AB_2 and AC_2 indicates that inferences regarding the subsystem AB_2-AC_2-C can be made from phase analyses of mixtures having compositions in triangle AB_2-B-C in addition to those in AB_2-AC_2-C. Some mixtures in triangle AB_2-AC_2-C which crystallize primary-phase AC_2 solid solutions will solidify before the liquid fraction attains the composition e_7 and may not contain detectable concentrations of AB_2 or C. Mixtures which crystallize primary-phase AB_2 solid solutions will behave similarly, producing possibly undetectable amounts of AC_2 or C. The identity of subsystem AB_2-AC_2-C may thus be equivocal without the conclusion based on other data that AB_2-C is a quasibinary.

4. Much information about phase behavior with the triangles A_2B-AC_2-AB_2 and A-A_2B-AC_2 can be obtained from x-ray analyses of specimens crystallizing within triangle A_2B-AC_2-AB_2. The solid-solution formation along AB_2-AC_2 and the melting relations of A_2B will cause mixtures with A_2B-AB_2-AC to undergo a variety of crystallization reactions. Depending on the character and extent of the solid solution and the melting-freezing behavior of A_2B, x-ray analysis may furnish very informative phase data. Crystallization in polycomponent systems often results in the formation of final products which coexist in an unstable equilibrium. Identification of the crystalline phases in the final crystallized mixture provides useful inferences as to the nature of the equilibrium phase diagram.

Bibliography. J. Braunstein et al., *Advances in Molten Salt Chemistry*, vols. 1–3, 1971–1975; L. O. Case, in I. M. Kolthoff (ed.), *Treatise on Analytical Chemistry*, pt. 1: *Theory and Instrumentation*, 1961; L. Eyring (ed.), *Progress in the Science and Technology of the Rare Earths*, vol. 2, 1966; D. Inman and D. G. Lovering (eds.), *Ionic Liquids*, 1981; G. J. Janz, *Molten Salts Handbook*, 1967; E. M. Levin, C. R. Robbins, and H. F. McMurdie, *Phase Diagrams for Ceramists*, 1964, Supplements 1969 and 1975; J. Lumsden, *Thermodynamics of Molten Salt Mixtures*, 1966; J. Plambeck, *Fused Salt Systems*, 1976.

DONNAN EQUILIBRIUM
GEORGE S. MILL AND W. O. MILLIGAN

The particular equilibrium set up when two coexisting phases are subject to the restriction that one or more of the ionic components cannot pass from one phase into the other. This equilibrium was first recognized by F. G. Donnan in 1911. Commonly, the restriction is caused by a membrane which is permeable to the solvent and small ions but impermeable to colloidal ions or charged particles of colloidal size. The presence of a membrane is not essential, since the restriction of movement on the charged colloid can be provided by a centrifugal or gravitational field or by gel coherence.

An immediate consequence of such a restriction in a system is the uneven distribution of diffusible ions at equilibrium. This is apparent in the following example. Let the initial state of a system be a solution of the ions Na^+ and R^- of concentration c_1 separated from a solution of sodium chloride of concentration c_2 by a membrane freely permeable to all but the R^- ions:

Na$^+$	R$^-$	Na$^+$	Cl$^-$
c_1	c_1	c_2	c_2

Then at equilibrium, a certain concentration of chloride ions, x, will have diffused through the membrane, accompanied by the same number of sodium ions in order to preserve electrical neutrality on both sides of the membrane, and the final equilibrium will be:

Na$^+$	R$^-$	Cl$^-$	Na$^+$	Cl$^-$
$(c_1 + x)$		x	$(c_2 - x)$	$(c_2 - x)$
	(1)			(2)

It can be shown thermodynamically that at equilibrium the product of the concentrations, or more strictly, the activities of the sodium and chloride ions, will be the same on both sides of the membrane. Hence may be formed

$$[Na^+]_1[Cl^-]_1 = [Na^+]_2[Cl^-]_2 \quad \text{or} \quad (c_1 + x)x = (c_2 - x)^2$$

Obviously, the diffusion of the chloride ions (and an equal number of sodium ions) through the membrane has been hindered by the presence of the nondiffusable ion R$^-$. Calculations based on this equation, which are confirmed by experimental measurements, show that sodium chloride is almost completely prevented from diffusing through the membrane if it is present in small concentration relative to the concentration of the nondiffusible ion R$^-$. As the relative concentration of sodium chloride is increased, more of it diffuses through the membrane. Finally, when the salt concentration is very high relative to that of R$^-$, an even distribution of sodium and chloride ions on either side of the membrane is approached. A similar equilibrium is also attained when the diffusible salt has no ion common to the colloidal electrolyte.

An important example of Donnan equilibrium is the dialysis of a solution of a colloidal electrolyte against pure water. Sodium ions from the colloidal R$^-$ · Na$^+$ will diffuse through the membrane and be replaced by an equivalent number of hydrogen ions. This phenomenon is called membrane hydrolysis and is helpful in explaining certain membrane equilibria in biological cells and tissues. Obviously if the water is renewed continuously, complete hydrolysis will ultimately ensue.

Two important consequences arise from the Donnan equilibrium. The first is that the observed osmotic pressure, that is, the difference in hydrostatic pressure on the two sides of the membrane, will always exceed that of R$^-$ except when a large excess of salt is added. An illustration of this effect is the behavior of ionic gels (for example, protein gels) when immersed in water. Ionic groups attached to the structure of the gel cannot diffuse out into the surrounding solution, and osmosis causes swelling of the gel. The gel will behave according to the Donnan equilibrium; the swelling is found to be reduced by addition of salts. The second consequence of the Donnan distribution is that a potential difference E is set up at the membrane. It is given by the equation below; R is the gas constant, T is the absolute temperature, and F is Faraday's constant. This is

$$E = \frac{RT}{F} \ln \frac{[Na^+]_1}{[Na^+]_2} = \frac{RT}{F} \ln \frac{[Cl^-]_2}{[Cl^-]_1}$$

the origin of the difference in potential between a suspension and its intermicellar liquid. In soil chemistry this is known as the suspension effect. SEE COLLOID; DIALYSIS; ION-SELECTIVE MEMBRANES AND ELECTRODES.

Bibliography. F. G. Donnan, The theory of membrane equilibria, *Chem. Rev.*, 1(1):73–90, 1924.

PHASE RULE
STANLEY I. SANDLER

A relationship used to determine the number of state variables F, usually chosen from among temperature, pressure, and species compositions in each phase, which must be specified to fix the thermodynamic state of a system in equilibrium. It was derived by J. Willard Gibbs between

1875 and 1878. The phase rule (in the absence of electric, magnetic, and gravitational phenomena) is given by Eq. (1), where C is the number of chemical species present at equilibrium, P is the

$$F = C - P - M + 2 \tag{1}$$

number of phases, and M is the number of independent chemical reactions. Here the term phase is used to indicate a homogeneous, mechanically separable portion of the system, and the term independent reactions refers to the smallest number of chemical reactions which, upon forming various linear combinations, includes all reactions which occur among the species present. The number of independent state variables F is referred to as the degrees of freedom or variance of the system.

Examples. The system constituted by liquid water and water vapor contains two phases ($P = 2$) and one component ($C = 1$), and there are no chemical reactions ($M = 0$). Therefore this system has $F = 1 - 2 - 0 + 2 = 1$ degree of freedom. This is in accord with the observation that the vapor and liquid forms of water exist in equilibrium only for values of temperature and pressure along the coexistance curve, so that specifying either of these variables fixes the other.

At high temperatures the three reactions shown in (2) occur between sulfur and oxygen in

$$S + O_2 = SO_3 \quad (2a) \qquad SO_2 + 1/2\, O_2 = SO_3 \quad (2b) \qquad S + 3/2\, O_2 = SO_3 \quad (2c)$$

the gas phase. There are only two independent chemical reactions in this single-phase, four-component system, since the last reaction is the sum of the first two. Therefore this system has $F = 4 - 1 - 2 + 2 = -3$ degrees of freedom.

Derivation. The derivation of the phase rule starts with the experimental observation that a single phase of C nonreacting components has $C + 1$ degrees of freedom (for example, once temperature, pressure, and $C - 1$ mole fractions are specified, no other intensive variables of the system can vary). Now consider chemical and phase equilibrium in a more general system in which there are C components and P phases, and in which M independent chemical reactions occur. It might appear, based on the experimental observation above, that such a system should have $P(C + 1)$ degrees of freedom. However, since the system is in equilibrium, the degrees of freedom are reduced as follows:

1. At equilibrium the temperature of each phase must be the same. Thus it is not possible to set the temperature of each of the P phases separately; once the temperature of one phase is specified, the temperature of all phases is fixed. Consequently the requirement that the temperature must be the same in all phases at equilibrium eliminates $P - 1$ degrees of freedom.

2. At equilibrium the pressure must be the same in each phase. This eliminates another $P - 1$ degrees of freedom.

3. At equilibrium the chemical potential (partial molar Gibbs free energy) of each species must be the same in each phase. This restriction, which holds for each of the C components, eliminates an additional $C(P - 1)$ degrees of freedom.

4. For each chemical reaction (3) to be in equilibrium, Eq. (4) must be satisfied, where μ_i

$$aA + bB + \cdots = rR + sS + \cdots \quad (3) \qquad a\mu_A' + b\mu_B + \cdots = r\mu_R + s\mu_S + \cdots \quad (4)$$

is the chemical potential of species i. (If this equation is satisfied in one phase, it is, by the equality of the chemical potentials of a given species in its various phases, satisfied in all phases.) This requirement eliminates another M degrees of freedom.

Therefore, the actual number of degrees of freedom in a multicomponent, multiphase, chemically reacting system is given by Eq. (5), which is the same as Eq. (1).

$$\begin{aligned} F &= P(C + 1) - 2(P - 1) - C(P - 1) - M \\ &= C - P - M + 2 \end{aligned} \tag{5}$$

SEE CHEMICAL EQUILIBRIUM; CHEMICAL THERMODYNAMICS; PHASE EQUILIBRIUM.

INTERNAL ENERGY
PAUL J. BENDER

A characteristic property of the state of a thermodynamic system, introduced in the first law of thermodynamics. For a static, closed system (no bulk motion, no transfer of matter across its boundaries), the change ΔE in internal energy for a process is equal to the heat Q absorbed by the system from its surroundings minus the work w done by the system on its surroundings. Only a change in internal energy can be measured, not its value for any single state. For a given process, the change in internal energy is fixed by the initial and final states and is independent of the path by which the change in state is accomplished.

The internal energy includes the intrinsic energies of the individual molecules of which the system is composed and contributions from the interactions among them. It does not include contributions from the potential energy or kinetic energy of the system as a whole; these changes must be accounted for explicitly in the treatment of flow systems. Because it is more convenient to use an independent variable (the pressure P for the system instead of its volume V), the working equations of practical thermodynamics are usually written in terms of such functions as the enthalpy $H = E + PV$, instead of the internal energy itself. SEE CHEMICAL THERMODYNAMICS; ENTHALPY.

ENTHALPY
HAROLD C. WEBER AND WILLIAM A. STEELE

For any system, that is, the volume of substance under discussion, enthalpy is the sum of the internal energy of the system plus the system's volume multiplied by the pressure exerted by the system on its surroundings. This may be expressed as $U + PV = H$, where U is the system's internal energy, P the pressure of the system, V the system's volume, and H the enthalpy of the system. The sum of $U + PV$ is given the special symbol H primarily as a matter of convenience because this sum appears repeatedly in thermodynamic discussion. Consistent units must, of course, be used in expressing the terms in the above equation. Previously, enthalpy was referred to as total heat or heat content, but these terms are misleading and should be avoided. Enthalpy is, from the viewpoint of mathematics, a point function, as contrasted with heat and work, which are path functions. Point functions depend only on the initial and final states of the system undergoing a change; they are independent of the paths or character of the change. Mathematically, the differential of a point function is a complete or perfect differential.

Because the absolute value of internal energy of even a simple system is usually unknown, recorded values of enthalpy are relative values measured above some convenient but arbitrarily chosen datum. Thus in the steam tables of Keenan and Keyes, the datum is liquid water at 32°F (°C) and under its own vapor pressure. At this state water is assumed to have an enthalpy equal to zero. Under this assumption the internal energy of water in this state is a negative quantity equal to PV. No complication is introduced by this fact, although visualization of negative energies of this kind may be disturbing to some. There is limited utility for absolute enthalpies because only the changes in enthalpy are measurable. It is instructive to examine the utility of the enthalpy function in terms of some simple but important thermodynamic processes.

The first law of thermodynamics is merely a statement of the law of conservation of energy. The first law alone indicates that:

1. For a chemical reaction carried out at constant pressure and temperature with no work performed except that resulting from keeping the internal and external pressure equal to each other as the volume changes, the change in enthalpy of the system (the material taking part in the chemical reaction) is numerically equal to the heat that must be transferred to maintain the above-mentioned conditions. This heat is often loosely referred to as the heat of reaction. More properly, it is the enthalpy change for the reaction.

2. So-called heat balances on heat exchangers, furnaces, and similar industrial equipment that operate under steady flow conditions are really enthalpy balances.

3. The work developed in a steadily running adiabatic engine or turbine is equivalent to the enthalpy change of the fluid passing through the engine.

4. The adiabatic, irreversible, steady flow of a stream of materials through a porous plug or a partially opened valve under circumstances where the change in kinetic energy of flow is negligible (a Joule-Thomson process) results in no change in enthalpy of the flowing stream. Although no change in enthalpy results from this process, there is a loss in the energy available for doing work as a result of the pressure drop across the plug or valve. *See* Entropy.

ENTROPY
William F. Jaep

A function first introduced in classical thermodynamics to provide a quantitative basis for the common observation that naturally occurring processes have a particular direction. Subsequently, in statistical thermodynamics, entropy was shown to be a measure of the number of microstates a system could assume.

Reversible processes. Any system under constant external conditions is observed to change in such a way as to approach a particularly simple final state called an equilibrium state. For example, two bodies initially at different temperatures are connected by a metal wire. Heat flows from the hot to the cold body until the temperatures of both bodies are the same. As another example, a vessel containing a gas is connected through a stopcock to an evacuated vessel. When the stopcock is opened, the gas expands to fill the whole of the available space uniformly. It is common experience that the reverse processes never occur if the systems are left to themselves; that is, heat is never observed to flow from the cold to the hot body, nor will the gas compress itself into one of the vessels. Max Planck classified all elementary processes into three categories: natural, unnatural, and reversible.

Natural processes do occur, and proceed in a direction toward equilibrium. Unnatural processes move away with equilibrium and never occur. If $A \rightarrow B$ is a natural process between states A and B, then $B \rightarrow A$ is an unnatural process. A reversible process is an idealized natural process that passes through a continuous sequence of equilibrium states. Consider the evaporation of a liquid in the presence of its vapor at a pressure P. Let the equilibrium vapor pressure of the liquid be p. If $P < p$, liquid evaporates as a natural process. If $P > p$, evaporation is an unnatural process and will not occur; indeed, the opposite process—condensation—will take place. Finally, if $P = p$, both processes of condensation and evaporation are reversible and can be initiated by a very slight increase or decrease in the external pressure P.

A useful idea is that a reversible process may be exactly reversed by an infinitesimal change in the external conditions. If a hot object is placed adjacent to a much colder object, the heat-flow direction cannot be reversed by small changes in the temperature of either object. In reversible processes, work is accomplished through small pressure differences, and heat transfer occurs through small temperature differences.

Entropy function. The state function entropy S puts the foregoing discussion on a quantitative basis. The function is not derived in this article; but, rather, some of its properties are stated, and its implications are discussed mainly by example. Entropy is related to q, the heat flowing into the system from its surroundings and to T, the absolute temperature of the system. The important properties for this discussion are:

1. $dS > q/T$ for a natural change.
   $dS = q/T$ for a reversible change.

It is necessary to introduce both S and T together. A formal derivation would show T^{-1} as an integrating factor leading to the complete differential dS.

2. The entropy of the system S is made up of the sum of all the parts of the system so that $S = S_1 + S_2 + S_3 \cdots$.

Heat flow. Consider two bodies, α and β, at different temperatures separated by an adiabatic (no heat transfer) wall. If the two bodies are connected by a fine wire that allows a small heat flow q from α to β, then $dS_\alpha = -q/T_\alpha$ and $dS_\beta = q/T_\beta$.

For the whole system, Eq. (1) holds. If $T_\alpha > T_\beta$, $dS > 0$, and heat flows from α to β as a

$$dS = dS_\alpha + dS_\beta = q\left(\frac{1}{T_\beta} - \frac{1}{T_\alpha}\right) \tag{1}$$

natural process. The process could be continued until $T_\alpha = T_\beta$ and $dS = 0$.

Once the constraint of the adiabatic wall is abrogated, the entropy increases to a maximum value, and T_α becomes equal to T_β. This is a special case of the most important notion in thermodynamics; that is, the system will assume that equilibrium state which maximizes the entropy at constant energy, consistent with the constraints. *See Heat transfer.*

Nonconservation of entropy. In his study of the first law of thermodynamics, J. P. Joule caused work to be expended by rubbing metal blocks together in a large mass of water. By this and similar experiments, he established numerical relationships between heat and work. When the experiment was completed, the apparatus remained unchanged except for a slight increase in the water temperature. Work (W) had been converted into heat (Q) with 100% efficiency. Provided the process was carried out slowly, the temperature difference between the blocks and the water would be small, and heat transfer could be considered a reversible process. The entropy increase of the water at its temperature T is $\Delta S = (Q/T) = (W/T)$.

Since everything but the water is unchanged, this equation also represents the total entropy increase. The entropy has been created from the work input, and this process could be continued indefinitely, creating more and more entropy. Unlike energy, entropy is not conserved.

Although the heat transfer is considered to be reversible in order to calculate the entropy increase, the overall process of converting work into heat is irreversible. The frictional process that converts kinetic energy into the heat of the metal blocks is a natural process. In fact, the impossibility of the reverse process is Lord Kelvin's statement of the second law of thermodynamics. Heat cannot be completely converted into work without other changes occurring in the surroundings. For example, a gas in a cylinder can be expanded reversibly by extracting heat from a large constant-temperature bath. All of the heat extracted from the bath is converted into work, but eventually the pressure of the gas system would be reduced to an unusable level. The system has changed, and the process cannot continue indefinitely. If one tries to convert heat into work through a system undergoing a cycle so that the system will return to its initial state, one finds that only a portion of the heat input does work and that the remainder must be rejected to a lower temperature; this is just the process which takes place in a heat engine.

Degradation of energy. Energy is never destroyed. But in the Joule friction experiment and in heat transfer between bodies, as in any natural process, something is lost. In the Joule experiment, the energy expended in work now resides in the water bath. But if this energy is reused, less useful work is obtained than was originally put in. The original energy input has been degraded to a less useful form. The energy transferred from a high-temperature body to a lower-temperature body is also in a less useful form. If another system is used to restore this degraded energy to its original form, it is found that the restoring system has degraded the energy even more than the original system had. Thus, every process occurring in the world results in an overall increase in entropy and a corresponding degradation in energy. R. Clausius stated the first two laws of thermodynamics as: "The energy of the world is constant. The entropy of the world tends toward a maximum."

Increasing entropy and mixing. Once the atomic theory of matter is accepted, the entropy concept can be made much clearer. It is then found through statistical thermodynamics that the increase of entropy toward its maximum value at equilibrium corresponds to the change of the system toward its most probable state consistent with the constraints. The most probable state represents the most mixed or most random state. Mixing must be given a broad interpretation which includes particle or configurational mixing, and spreading of energy over the particles or thermal mixing. Diffusion of one gas into another represents obvious configurational mixing and increased entropy. Irreversible expansion of a gas represents configurational mixing of the molecules over the available space. Heat flow represents spreading of the kinetic energy between the particles. Friction spreads the kinetic energy of the body over the constituent particles. Sometimes the energy-spread entropy increase and the configurational entropy increase are not compatible, and a compromise is struck. A subcooled liquid adiabatically crystallizes to a lower configurational entropy but gains even more entropy through the additional energy levels made available. The same sort of behavior occurs in partially miscible liquids—some configurational entropy is sacrificed in order to gain a large amount of energy-spread entropy.

Absolute entropy. The third law of thermodynamics (Nernst's heat theorem) refers to the vanishing of entropy at zero temperature. In 1912 Planck proposed that the theorem applied to pure crystalline solids. However, the theorem is now known to be applicable to gases and, by all reasonable expectation, is applicable to any system. Thus, any substance at finite temperatures has an absolute entropy, the value of which can be determined from either calorimetric or spectroscopic data. Absolute entropies, together with thermochemical data, are very useful in the calculation of equilibrium compositions of reaction systems.

The statistical viewpoint is that a thermodynamic state at finite temperatures corresponds to many microstates. During an observation the microstates of a system undergo continuous rapid transitions. Since entropy is proportional to the logarithm of the number of available microstates, the Nernst theorem implies that the thermodynamic state at zero temperature corresponds to a single microstate. Thus, at zero temperature, even a ferromagnetic material should exist in a single state, fully magnetized in a direction determined by its inevitable interactions with the environment.

Bibliography. H. B. Callen, *Thermodynamics*, 1960; K. G. Denbigh, *Principles of Chemical Equilibrium*, 4th ed., 1981; J. D. Fast, *Entropy: The Significance of the Concept of Entropy and Its Applications in Science and Technology*, 2d ed., 1968; K. Wark, *Thermodynamics*, 4th ed., 1983.

FREE ENERGY
PAUL J. BENDER

A term in thermodynamics which in different treatments may designate either of two functions defined in terms of the internal energy E or enthalpy H, and the temperature-entropy product TS.

The function $(E - TS)$ is the Helmholtz free energy and is the function ordinarily meant by free energy in European references. The Gibbs free energy is the function $(H - TS)$. For the Lewis and Randall school of American chemical thermodynamics, this is the function meant by the free energy F. To avoid confusion with the symbol F as applied elsewhere to the Helmholtz free energy, the symbol G has been used. A recent development was the introduction of the name free enthalpy, with symbol G, for the Gibbs function.

Theory. For a closed system (no transfer of matter across its boundaries), the work which can be done in a reversible isothermal process is given by the series shown in Eq. (1). For these conditions, $T\Delta S$ represents the heat given up to the surroundings. Should the process be exothermal, $T\Delta S < 0$, then actual work done on the surroundings is less than the decrease in the internal energy of the system. The quantity $(\Delta E - T\Delta S)$ can then be thought of as a change in free energy, that is, as that part of the internal energy change which can be converted into work under the specified conditions. This then is the origin of the name free energy. Such an interpretation of thermodynamic quantities can be misleading, however; for the case in which $T\Delta S$ is positive, Eq. (1) shows that the decrease in "free" energy is greater than the decrease in internal energy. *See* CHEMICAL THERMODYNAMICS.

$$W_{rev} = -\Delta A = -\Delta(E - TS)$$
$$= -(\Delta E - T\Delta S) \quad (1)$$

For constant temperature and pressure in a reversible process the decrease in the Gibbs function G for the system again corresponds to a free-energy change in the above sense, since it is equal to the work which can be done by the closed system other than that associated with its change in volume ΔV under the given constant pressure P. The relations shown in Eq. (2) can be formed since $\Delta H = \Delta E + P\Delta V$.

$$\Delta G = -(\Delta H - T\Delta S) = W_{net}$$
$$= W_{rev} - P\Delta V \quad (2)$$

Each of these free-energy functions is an extensive property of the state of the thermodynamic system. For a specified change in state, both ΔA and ΔG are independent of the path by which the change is accomplished. Only changes in these functions can be measured, not values for a single state.

The thermodynamic criteria for reversibility, irreversibility, and equilibrium for processes in closed systems at constant temperature and pressure are expressed naturally in terms of the function G. For any infinitesimal process at constant temperature and pressure, $-dG \geqq \delta w_{\text{net}}$. If δw_{net} is never negative, that is, if the surroundings do no net work on the system, then the change dG must be negative or zero. For a reversible differential process, $-dG > \delta w_{\text{net}}$; for an irreversible process, $-dG > \delta w_{\text{net}}$. The free energy G thus decreases to a minimum value characteristic of the equilibrium state at the given temperature and pressure. At equilibrium, $dG = 0$ for any differential process taking place, for example, an infinitesimal change in the degree of completion of a chemical reaction. A parallel role is played by the work function A for conditions of constant temperature and volume. Because temperature and pressure constitute more convenient working variables than temperature and volume, it is the Gibbs free energy which is the more commonly used in thermodynamics.

Partial molal quantities. For a particular homogeneous phase in the absence of surface, gravitational, and magnetic forces, the free energy G depends on the numbers of moles of the constituents present, the temperature T, and the pressure P. Let Ω represent the total number of constituents, n_i the number of moles of typical constituent i, and designate by subscript n constant composition, by subscript n_j constancy of the number of moles of all constituents except n_i; then Eq. (3) is formed.

$$dG(T,P,n_1, \ldots, n_\Omega) = \left(\frac{\partial G}{\partial T}\right)_{P,n} dt + \left(\frac{\partial G}{\partial p}\right)_{T,n} + \sum_{i=1}^{\Omega} \left(\frac{\partial G}{\partial n_i}\right)_{T,P,n_j} dn_i \qquad (3)$$

In Eq. (3) the term $(\partial G/\partial n_i)_{T,P,n_j}$ is the chemical potential μ_i of the ith constituent. It is identical to the partial molal free energy $\overline{G}_i$ of Lewis and Randall. It then follows that Eq. (4) holds. *See Solution.*

$$dG = -S\, dT + V\, dP + \sum_{i=1}^{\Omega} \mu_i\, dn_i \qquad (4)$$

Because the chemical potentials at constant T,P are intensive variables whose values are fixed, like that of the density, by relative number of moles of the various constituents present, and are independent of the total mass of the phase, this equation can be integrated for constant T,P and relative composition starting from $n_i = 0$ to obtain Eq. (5). This yields Eq. (6). Consistency

$$G(T,P,n_1, \ldots, n_\Omega) = \sum_{i=1}^{\Omega} n_i \mu_i \qquad (5) \qquad dG = \sum_{i=1}^{\Omega} \mu_i\, dn_i + \sum_{i=1}^{\Omega} n_i\, d\mu_i \qquad (6)$$

with the expression for dG in Eq. (4) requires that Eq. (7) hold. This is the Gibbs-Duhem equation.

$$S\, dT - V\, dP + \sum_{i=1}^{\Omega} n_i\, d\mu_i = 0 \qquad (7)$$

For constant temperature and pressure, this relation imposes a condition on the composition variation of the set of chemical potentials.

Heterogeneous systems. The free energy of a closed, heterogeneous system is the sum of the free energies of its various phases. In the absence of such a constraint as provided by the subdivision of the system by a rigid, semipermeable membrane, the general thermodynamic criterion of equilibrium requires that the temperature and pressure be uniform throughout the system and that the chemical potential of each constituent have a common value for all phases in which it is present. Further, if any of the constituents can be formed from others, the chemical potentials of the reactants and products are related in accordance with the stoichiometry of the reaction equation. Thus, for the reaction in Eq. (8), at equilibrium Eq. (9) can be formed. Expressing each

$$A + 2B \rightleftharpoons 3C + 4D \qquad (8) \qquad\qquad \mu_A + 2\mu_B = 3\mu_C + 4\mu_D \qquad (9)$$

chemical potential μ_i in terms of the standard value $\mu_i°$ and its associated activity term $RT \ln a_i$ results in Eq. (10). In Eq. (10) $\Delta G°$ is called the standard free-energy change for the reaction. Its

$$RT \ln \left(\frac{a_C^3 a_D^4}{a_A a_B^2}\right)_{equil} = -(3\mu_C° + 4\mu_D° - \mu_A° - 2\mu_B°) = -\Delta G° \qquad (10)$$

value depends on the standard states chosen, but for a given temperature and pressure, it is a constant characteristic of the reaction involved. A true equilibrium constant K then results as shown by Eqs. (11) and (12). If the pressure for each standard state is fixed and independent of

$$K = \left(\frac{a_C^3 a_D^4}{a_A a_B^2}\right)_{equil} \qquad (11) \qquad\qquad RT \ln K = -\Delta G° \qquad (12)$$

the pressure of the reaction system, $\Delta G°$ and hence K are functions of temperature only. This is the conventional approach in treating gas-phase equilibria, but not the approach that is used ordinarily for condensed phases. SEE ACTIVITY.

Since the activities can be correlated with partial pressures or concentrations through fugacity coefficients or activity coefficients, this thermodynamic approach eliminates the uncertainties otherwise associated with equilibrium calculations that are based on the law of mass action. SEE CHEMICAL EQUILIBRIUM; FUGACITY.

The prediction of equilibrium constant then requires the calculation of $\Delta G°$ for the reaction. The so-called third-law method involves calculation for the reaction at 25°C (77°F) of the value of $\Delta H°$, the standard heat of reaction, from tabulated standard heat of formation data and of $\Delta S°$ from tabulated third-law entropies. These are combined in the sense of $\Delta G° = \Delta H° - T \Delta S°$ to permit calculation of the equilibrium constant for 25°C (77°F). This in turn is used for evaluation of the integration constant in the integration of the relation in Eq. (13). The integration requires

$$\frac{d \ln K}{dT} = \frac{\Delta H°}{RT^2} \qquad (13)$$

expression of $\Delta H°$ as a function of temperature, which necessitates a knowledge of the heat capacities $C°_{P(i)}$ for the various reactants over the temperature range involved.

Alternatively, if values of the free-energy function $(G° - H°_{298}/T)$ are available, either from experimental measurement or from statistical thermodynamical computations, they can be combined with the standard heat of reaction at 25°C (77°F) to give the desired result, Eq. (14).

$$\Delta G° = \Delta \left(\frac{G° - H°_{298}}{T}\right) + \frac{\Delta H°_{298}}{T} \qquad (14)$$

SEE ENTROPY; THERMOCHEMISTRY.
Bibliography. K. G. Denbigh, *Principles of Chemical Equilibrium*, 4th ed., 1981; E. Fermi, *Thermodynamics*, 1937; K. S. Pitzer and L. Brewer, *Thermodynamics*, rev. ed., 1961; P. A. Rock, *Chemical Thermodynamics*, 1983.

GIBBS FUNCTION
WILLIAM F. JAEP

The Gibbs function G, also known as Gibbs free energy or free enthalpy, is defined in the equation shown, where E is the internal energy, p is the pressure, v is the volume, T is the absolute

$$G = E + pv - TS$$

temperature and S is the entropy. The Gibbs function is most useful in analyzing systems held at constant temperature and pressure. Under these conditions, the change in the Gibbs function ΔG of a system is a measure of the maximum attainable work, not including the work of displacing the environment. For example, ΔG represents the maximum electrical work obtainable from a galvanic cell. When the only work done by the system at constant temperature and pressure is displacing the environment, the equilibrium state is characterized by G having reached its minimum value. Since chemical processes frequently occur at constant temperature and pressure, the Gibbs function is extensively used in chemical engineering for calculating phase equilibrium and reaction equilibrium. SEE CHEMICAL THERMODYNAMICS; FREE ENERGY.

ACTIVITY
PAUL J. BENDER

A function introduced by G. N. Lewis to aid in the thermodynamic treatment of real systems. Like the fugacity, the activity makes possible the correlation of changes in the chemical potential with changes in experimentally measurable quantities, such as concentrations or partial pressures, through relations formally equivalent to those holding for ideal systems. The activity concept retains its usefulness, however, for such cases as condensed phases of low volatility, for which fugacity determinations are impractical. SEE CHEMICAL THERMODYNAMICS; FREE ENERGY; FUGACITY.

The activity a_i of a constituent i for a given state is defined as Eq. (1), where f_i is the

$$a_i = f_i/f_i^0 \tag{1}$$

fugacity for the given state and f_i^0 the fugacity of the constituent for a reference state, or standard state, at the same temperature. This definition then requires that Eq. (2) hold, where μ_i and μ_i^0

$$RT \ln a_i = \mu_i - \mu_i^0 \tag{2}$$

are the values of the chemical potential for the given state and standard state, respectively; R is the gas constant and T the absolute temperature. Because an activity is a relative quantity, a numerical value is meaningless unless the standard state involved is known. The standard state used may in principle be selected arbitrarily, but in practice, certain conventional choices are ordinarily made because a maximum of convenience can be obtained thereby.

Theory. The activity is dependent on the temperature, pressure, and composition of the system. Its temperature coefficient is determined by the relative partial molal enthalpy, Eq. (3),

$$\left(\frac{\partial \ln a_i}{\partial T}\right)_{P,\text{comp}} = -\frac{\overline{H}_i - \overline{H}_i^0}{RT^2} \tag{3}$$

where $\overline{H}_i$ and $\overline{H}_i^0$ represent the relative partial molal enthalpy for the constituent i in the given state and the standard state, respectively. If the pressure for the standard state is fixed at the given temperature and independent of the pressure for the system, Eq. (4) holds. If the standard-

$$\left(\frac{\partial \ln a_i}{\partial P}\right)_{T,\text{comp}} = \frac{\overline{V}_i}{RT} \tag{4}$$

state pressure is always made equal to that for the system, as in the conventional treatment of condensed solutions, Eq. (5) holds. Here, $\overline{V}_i$ and $\overline{V}_i^0$ represent the partial molar volumes for the

$$\left(\frac{\partial \ln a_i}{\partial P}\right)_{T,\text{comp}} = \frac{\overline{V}_i - \overline{V}_i^0}{RT} \tag{5}$$

constituent in the given state and standard state, respectively.

The changes in the activities of the various components of a solution phase resulting from composition changes at constant temperature and pressure are not all independent, but instead must satisfy the condition shown by Eq. (6), where n_i is the instantaneous number of moles of

$$\sum_i n_i d \ln a_i = 0 \quad (T,P \text{ const}) \tag{6}$$

component i. This relation stems from the Gibbs-Duhem equation. For a binary solution, this permits the calculation of the activity of one component from a knowledge of the activity of the other as a function of composition. SEE SOLUTION.

Standard states. For a constituent of a gaseous mixture, the activity is normally required to be equal numerically to the fugacity for the given state, and thus, in the ideal gas limit, it becomes numerically equal to the partial pressure of the constituent. This result is obtained by acceptance as the standard state of that state for which the chemical potential is equal to the quantity μ_i^* appearing in the definition of the fugacity. This standard state is thus a state of unit fugacity, because the temperature coefficient of μ_i^*/T is determined by the enthalpy of the real gas at very low pressure, instead of, as in a real gas, at unit fugacity.

For a pure liquid or solid, the standard state is ordinarily taken to be the pure condensed phase at the given temperature, at a pressure variously specified as 1 atm (100 kilopascals), the vapor pressure at this temperature, or the pressure for the system under consideration. The first two choices make the standard chemical potential, and hence the contribution of the constituent to the standard free energy change for any reaction in which it takes part, independent of pressure, whereas the activity of the constituent changes with the pressure of the system; for a molar volume at about 6.1 in.^{3} (100 ml), for example, at 77°F (25°C), a change of pressure of 30 atm (3000 kPa) will change the activity by about 10%. The third standard-state convention cited above makes the activity for the pure condensed phase unity at all pressures, but the standard chemical potential changes with the pressure.

In liquid solution systems the activity of a constituent is correlated with its concentration X_i through an activity coefficient ξ_i, Eq. (7). For a given solution the numerical values of μ_i^0, a_i,

$$a_i = \xi_i X_i \qquad (7)$$

and ξ_i depend on the concentration scale used.

Activity coefficients. The standard state in general is fixed by imposition of a requirement on the activity coefficient. For a constituent considered as acting as a solvent, the standard state is taken to be the pure liquid constituent at the temperature and pressure of the solutions. This is equivalent to a requirement that the activity coefficient γ_i on the mole fraction N_i of the constituent approaches unity. For liquid-liquid nonelectrolytic solutions, this convention is applied to all constituents. For the binary solution case the dependence of the two activity coefficients on composition is often adequately expressed by the van Laar equations, Eqs. (8) and (9). Here, the

$$\log \gamma_1 = \frac{A_1}{\left(1 + \dfrac{A_1 N_1}{A_2 N_2}\right)^2} \qquad (8) \qquad \log \gamma_2 = \frac{A_2}{\left(1 + \dfrac{A_2 N_2}{A_1 N_1}\right)^2} \qquad (9)$$

coefficients A_1 and A_2 are functions of temperature and pressure. Other relations proposed for binary and ternary systems have been summarized and discussed by E. coworkers.

For a constituent considered to act as a nondissociating solute, the activity coefficient is required to approach unity along the given concentration scale at infinite dilution in the given solvent medium. This result can be obtained by acceptance of a standard state for which the fugacity is numerically equal to the Henry's law constant for the concentration scale used. Thus, Eq. (10) can be written, and then Eq. (11) follows. This standard-state fugacity is the value pre-

$$\lim_{\infty \text{dilution}} f_i = k_{x(i)} X_i \qquad (10)$$

$$\lim_{\infty \text{dilution}} \left(\xi_i = \frac{a_i}{X_i} = \frac{1}{f_i^0} \frac{f_i}{X_i} \right) = \frac{1}{f_i^0} \lim \frac{f_i}{X_i} = \frac{k_{x(i)}}{f_i^0} = 1 \quad \text{if } f_i^0 = k_{x(i)} \qquad (11)$$

dicted by Henry's law for unit concentration on the scale used. Since Henry's law in general holds only at high dilution, this standard state is again a hypothetical state; the fugacity is equal to $k_{x(i)}$, but the partial molal volume and enthalpy correspond to those of the solute at infinite dilution.

For a dissociating solute the treatment used depends on the degree of dissociation; in any case, the existence of dissociation must be accounted for. For slightly dissociating solutes, where in principle the concentrations of the parent species and the products of dissociation can all be determined, the preceding convention is used for each of these solute species. The standard state being thus fixed for each, the standard free-energy change and hence the equilibrium constant for the dissociation process is fixed at a value which must be determined experimentally.

A solute such as $BaCl_2$ must, however, be considered to be completely dissociated in aqueous solutions, even at finite concentrations. Its chemical potential μ_2 (for example, per gram formula weight $BaCl_2$) must then equal the sum of chemical potentials for the products of dissociation, Eq. (12). Activity coefficients are defined for the dissociation products as solutes in the

$$\mu_2 = \mu_{Ba^{2+}} + 2\mu_{Cl^-} \tag{12}$$

conventional way; the corresponding concentrations are calculated on the assumption of complete dissociation. This assumption will be exact at high dilution; its failure at higher concentrations will yield abnormal values for the activity coefficients which can be interpreted on this basis. Let an activity a_2 be defined for the dissociating solute, as in Eq. (13). If it is required that Eq. (14) apply, then Eq. (15) also applies.

$$\mu_2^0 + RT \ln a_2 = \mu_2$$
$$= \mu_{Ba^{2+}}^0 + RT \ln a_{Ba^{2+}} + 2(\mu_{Cl^-}^0 + RT \ln a_{Cl^-}) \tag{13}$$

$$\mu_2^0 = \mu_{Ba^{2+}}^0 + 2\mu_{Cl^-}^0 \tag{14} \qquad a_2 = a_{Ba^{2+}} a_{Cl^-}^2 \tag{15}$$

Since the solution must be electrically neutral, the concentrations of the ions are not independent, and it is not possible to determine individual ionic activities, only the equivalent of the activity a_2. Hence, a (geometric) mean ionic activity $a_\pm$ and mean ionic activity coefficient $\gamma_\pm$ are defined. Thus, for $BaCl_2$, Eqs. (16) exist.

$$a_\pm = a_2^{1/3} = (a_{Ba^{2+}} a_{Cl^-}^2)^{1/3} \qquad \gamma_\pm = (\gamma_{Ba^{2+}} \gamma_{Cl^-}^2)^{1/3} \tag{16}$$
$$= (\gamma_{Ba^{2+}} \gamma_{Cl^-}^2)^{1/3} (m_{Ba^{2+}} m_{Cl^-}^2)^{1/3}$$

For dilute solutions of strong electrolytes the Debye–Hückel theory predicts the concentration dependence of $\gamma_\pm$. At higher concentrations, values must be obtained experimentally. SEE CHEMICAL EQUILIBRIUM; IONIC EQUILIBRIUM.

Bibliography. K. G. Denbigh, *Principles of Chemical Equilibrium*, 4th ed., 1981; D. A. McQuarrie, *Chemical Thermodynamics*, 1983.

FUGACITY
PAUL J. BENDER

A function introduced by G. N. Lewis to facilitate the application of thermodynamics to real systems. Thus, when fugacities are substituted for partial pressures in the mass action equilibrium constant expression, which applies strictly only to the ideal case, a true equilibrium constant results for real systems as well.

The fugacity f_i of a constituent i of a thermodynamic system is defined by Eq. (1) [where

$$\mu_i = \mu_i^* + RT \ln f_i \tag{1}$$

μ_i is the chemical potential and μ_i^* is a function of temperature only], in combination with the requirement that the fugacity approach the partial pressure as the total pressure of the gas phase approaches zero. At a given temperature, this is possible only for a particular value for μ_i^*, which may be shown to correspond to the chemical potential the constituent would have as the pure gas in the ideal gas state at 1 atm pressure. This definition makes the fugacity identical to the partial pressure in the ideal gas case. For real gases, the ratio of fugacity to partial pressure, called the fugacity coefficient, will be close to unity for moderate temperatures and pressures. At low temperatures and appropriate pressures, it may be as small as 0.2 or less, whereas at high pressures at any temperature it can become very large.

The fugacity concept is not restricted to gaseous systems, however. Because of its relation to the chemical potential, the basic thermodynamic criterion of equilibrium requires that the fugacity of a constituent have the same value at equilibrium for every phase in which it is present. This permits the indirect determination of the fugacity for a condensed phase through the calculation of the value for the equilibrium vapor phase, for which the fugacity may be computed routinely if the dependence of its volume on temperature, pressure, and composition is known.

Thus, results are readily obtained for a pure gas, but because of the more extensive data required, accurate calculations have been made for very few mixtures.

For an ideal solution, the fugacity is given by the mole fraction of the constituent multiplied by the fugacity of the pure constituent at the temperature and pressure of the solution. For liquid solutions, this is the thermodynamic counterpart of Raoult's law, but the relation applies also to ideal gaseous solutions and can serve for the prediction of the properties of real gas mixtures. Where no equation of state data are available for pure gases, their fugacities can be estimated by means of the generalized fugacity coefficient chart. At sufficiently high dilution in liquid solution, the fugacity of a nondissociating solute will become proportional to its concentration, the proportionality constant depending on the concentration scale used; this is the thermodynamic statement of Henry's law.

The dependence of the fugacity on temperature at constant pressure and composition is given by Eq. (2). Here H_i^* is the enthalpy per mole for the constituent in the gas phase at very

$$\left(\frac{\partial \ln f_i}{\partial T}\right)_{P,\,comp} = \frac{H_i^* - \overline{H}_i}{RT^2} \qquad (2)$$

low pressure, and $\overline{H}_i$ is its contribution per mole to the enthalpy of the system for the state under consideration. SEE ACTIVITY; CHEMICAL EQUILIBRIUM; CHEMICAL THERMODYNAMICS; GAS; PHASE EQUILIBRIUM; SOLUTION.

Bibliography. K. S. Pitzer and L. Brewer, *Thermodynamics*, rev. ed., 1961; P. A. Rock, *Chemical Thermodynamics*, 1983.

pH
FRANCIS J. JOHNSTON

A term used to describe the hydrogen-ion activity of a system. It is defined by the expression $pH = -\log a_{H^+}$. Here a_{H^+} is the activity of the hydrogen ion. In dilute solutions, the activity is essentially equal to the concentration, and the pH may be approximately defined as $pH = -\log_{10}[H^+]$, where $[H^+]$ is the concentration of the hydrogen ion in moles per liter. The use of the pH makes negative exponents unnecessary in describing the hydrogen-ion activity. In a system in which the hydrogen-ion activity is 10^{-3} mole/liter, the pH is 3. The **table** shows the pH scale and

The pH scale	pH	$[H_3O^+]$
Battery acid	0	1
	1	10^{-1}
Stomach acid	2	10^{-2}
Lemon juice	3	10^{-3}
	4	10^{-4}
Black coffee	5	10^{-5}
	6	10^{-6}
Pure water	7	10^{-7}
Sea water	8	10^{-8}
Baking soda	9	10^{-9}
Toilet soap	10	10^{-10}
	11	10^{-11}
	12	10^{-12}
Household ammonia	13	10^{-13}
Drain cleaner	14	10^{-14}

approximate pH values of some representative solutions. The lower the pH, the more acidic the system. The term is seldom used to describe solutions in which the hydrogen-ion activity is 1 or greater. The expression defining the pH is closely related to the free energy of the hydrogen ion with respect to a standard reference state. SEE BUFFERS.

pK
THOMAS F. YOUNG

The logarithm (to the base 10) of the reciprocal of the equilibrium constant for a specified reaction under specified conditions (for example, solvent and temperature). pK values are often more convenient to tabulate and use than the equilibrium constants themselves. The value of K for the dissociation of the HSO_4^- ion in aqueous solution at 77°F (25°C) is 0.0102 mole/liter. The logarithm is $0.008_6 - 2 = -1.991_4$. pK is therefore $+1.991_4$. The choice of algebraic sign, although arbitrary, results in positive values for most dissociation constants applicable to aqueous solutions. The concept of pK is especially valuable in the study of solutions. SEE CHEMICAL EQUILIBRIUM; IONIC EQUILIBRIUM; PH.

CHEMICAL REACTIONS

Chemical dynamics	54
Chain reaction	63
Catalysis	64
Homogeneous catalysis	65
Heterogeneous catalysis	69
Phase-transfer catalysis	72
Buffers	75
Inhibitor	77
Ionization	79
Ionic equilibrium	79
Oscillatory reaction	81
Oxidation-reduction	84
Steric effect	89

CHEMICAL DYNAMICS

PHILIP R. BROOKS, WILLIAM H. MILLER, AND EDWARD M. EYRING

W. H. Miller wrote the section Theoretical Methods, and E. M. Eyring the section Relaxation Methods.

That branch of physical chemistry which seeks to explain time-dependent phenomena, such as energy transfer and chemical reaction, in terms of the detailed motion of the nuclei and electrons which constitute the system.

REACTION KINETICS

Although the ultimate state of a chemical system is specified by thermodynamics, the time required to reach that equilibrium state is highly variable. As a consequence, determining the rate of chemical reactions has proved to be important for practical reasons. For example, diamonds are thermodynamically unstable with respect to graphite, but the rate of transformation of diamonds to graphite is negligible. Rate studies have also yielded fundamental information about the details of the nuclear rearrangements which constitute the chemical reaction.

Traditional chemical kinetic investigations of the reaction between species X and Y to form Z and W, reaction (1), sought a rate of the form given in Eq. (2), where $d[Z]/dt$ is the rate of

$$X + Y \rightarrow Z + W \qquad (1) \qquad d[Z]/dt = kf([X], [Y], [Z], [W]) \qquad (2)$$

appearance of product Z, f is some function of concentrations of X, Y, Z, and W which are themselves functions of time, and d is the rate constant. Chemical reactions are incredibly diverse, and often the function f is quite complicated, even for seemingly simple reactions such as that in which hydrogen and bromine combine directly to form hydrogen bromide. These are actually complex reactions which proceed through a sequence of simpler reactions, called elementary reactions. For reaction (3d), the sequence of elementary reactions is a chain mechanism known to involve a series of steps, reactions (3a)–(3c). This sequence of elementary reactions was formerly

$$Br_2 \rightarrow 2Br \qquad (3a) \qquad Br + H_2 \rightarrow HBr + H \qquad (3b)$$

$$H + Br_2 \rightarrow HBr + Br \qquad (3c) \qquad H_2 + Br_2 \rightarrow 2HBr \qquad (3d)$$

known as the reaction mechanism, but in the chemical dynamical sense the word mechanism is reserved to mean the detailed motion of the nuclei during a collision.

Bimolecular processes. An elementary reaction is considered to occur exactly as written. Reaction (3b) is assumed to occur when a bromine atom hits a hydrogen molecule. The products of the collision are a hydrogen bromide molecule and a hydrogen atom. On the other hand, the overall reaction is a sequence of these elementary steps and on a molecular basis does not occur as reaction (3d) is written. With few exceptions, the rate law for an elementary reaction $A + B \rightarrow C + D$ is given by $d[C]/dt = k[A][B]$. The order (sum of the exponents of the concentrations) is two, which is expected if the reaction is bimolecular (requires only species A to collide with species B). The rate constant k for such a reaction depends very strongly on temperature, and is usually expressed as $k = Z_{AB}\rho \exp(-E_a/RT)$. Z_{AB} is the frequency of collision between A and B calculated from molecular diameters and temperature; ρ is an empirically determined steric factor which arises because only collisions with the proper orientation of reagents will be effective; and E_a, the experimentally determined activation energy, apparently reflects the need to overcome repulsive forces before the reagents can get close enough to react.

Unimolecular processes. In some instances, especially for decompositions, $AB \rightarrow A + B$, the elementary reaction step is first-order, Eq. (4), which means that the reaction is

$$d[A]/dt = d[B]/dt = k[AB] \qquad (4)$$

unimolecular. The species AB does not spontaneously dissociate; it must first be given some critical amount of energy, usually through collisions, to form an excited species AB*. It is the species AB* which decomposes unimolecularly.

MOLECULAR DYNAMICS

In principle, it is possible to prepare two reagents in specific quantum states and to determine the quantum-state distribution of the products. In practice, this is much too difficult, and experiments have been limited to preparing one reagent or to determining some aspect of the product distribution. This approach yields data concerning the gross aspects of the dynamics rather than the fine details.

Bimolecular reactions. Molecular-beam and luminescence techniques have played a major role in the development of chemical dynamics. Since these techniques largely complement each other, they are illustrated by discussing the results of a single reaction, the formation of deuterium chloride (DCl), reaction (5).

$$D + Cl_2 \rightarrow DCl + Cl \tag{5}$$

Chemiluminescence. Reaction (5) is 46 kilocalories per mole (192 kilojoules per mole) exoergic. If all of the exoergicity were to go into vibration of the newly formed DCl molecule, enough energy is available so that vibrational levels (v') up to $v' = 9$ could be formed. In a gas-phase reaction at high pressure, much of this energy would be dissipated at the walls of the container, but at very low pressures DCl molecules in excited vibrational states, DCl*, will emit infrared emission prior to undergoing a collision.

An apparatus to study this infrared chemiluminescence is shown in **Fig. 1**. Deuterium atoms are made by dissociating D_2 gas in an electrical discharge, and are injected into the observation cell at the top. The reagents mix and react inside a vessel with walls at 77 K (320°F), which freeze out species hitting the wall. The pressure is kept low to minimize vibrationally deactivating collisions. Infrared (IR) emission from the products is gathered by mirrors at the end, taken out through the sapphire window, and analyzed with an IR spectrometer. By analyzing the spectrum, it is possible to determine which vibration-rotation states are emitting and, as a consequence, which vibration-rotation states are formed in the reaction. **Figure 2** shows the relative distribution of vibrational states from the reaction, and also shows, for comparison, the relative distribution of vibrational states calculated from the Boltzmann equation for hot DCl (6000 K or 10,800°F). Thermal distributions at any other temperatures would still show a monotonic decline. (The population for $v' = 0$ is not determined by the chemiluminescence experiments, because that state does not emit.) The DCl formed in the reaction clearly has different properties than hot DCl: the vibrational population displays an inversion, and this system (the hydrogen isotope) was the active medium for the first chemically pumped laser. SEE CHEMILUMINESCENCE.

Molecular-beam experiments. Molecules can be isolated in molecular beams, and collisions between these isolated molecules can be observed by crossing two tenuous molecular beams in a region of otherwise high vacuum. **Figure 3** shows such an experiment. Gaseous atoms or molecules emerge from the ovens, and collimating slits select the particles which are all going in the same direction. The molecular beams cross at the center of rotation of a large platform

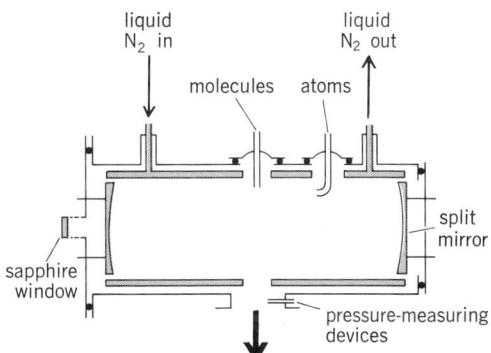

Fig. 1. Reaction vessel for studying infrared chemiluminescence between atoms and molecules at low pressure. (*After D. H. Maylotte, J. C. Polanyi, and K. B. Woodall, J. Chem. Phys., 57:1547–1561, 1972*)

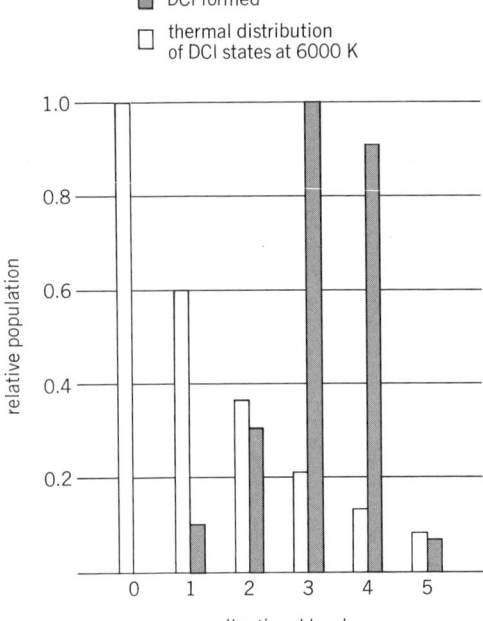

Fig. 2. Relative populations of DCl molecules in different vibrational states formed in the reaction D + Cl_2 → DCl + Cl.

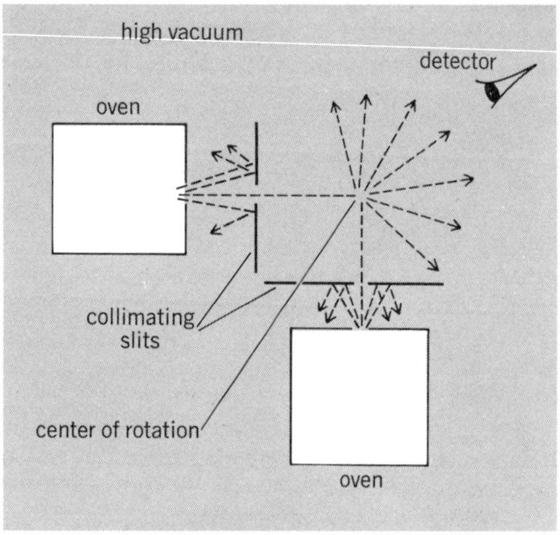

Fig. 3. Schematic of a molecular beam experiment.

which can be rotated under vacuum relative to the two beams. Large vacuum pumps maintain a high vacuum, which ensures that collisions take place only at the intersection of the two beams. Product molecules are ionized by electron bombardment, and detected with a quadrupole mass spectrometer housed within a region of ultrahigh vacuum.

Measurements are made of the scattered product intensity and speed at various scattering angles. For ease of interpretation, these data are transformed into the center-of-mass system in which the two reagents approach each other with equal and opposite momenta. **Figure 4** shows a contour map of the DCl intensity in the center-of-mass coordinate system in which the D atom is incident from the left and the Cl_2 molecule is incident from the right. The product DCl recoils backward (in the direction from which the D came) in a broad but nonetheless anisotropic distribution. The speed of the product is high, and corresponds to about half of the reaction exoergicity appearing in translation recoil of the products, with the balance appearing in vibration and rotation of the DCl consistent with the chemiluminescence results.

The anisotropic product distribution shows that reaction occurs in the time less than a molecular rotation, ~ 1 picosecond. The partitioning of energy roughly equally between vibration and translation suggests that the major amount of energy is released in repulsion between the DCl and Cl. This repulsion is similar to that experienced by a Cl_2 molecule in photodissociation is shown in Fig. 4. Because the deuterium atom is so light, the direction in which the product is expelled is a measure of the orientation of the Cl_2 molecule during reaction. For this reaction the collinear arrangement D-Cl-Cl is preferred.

Other investigations. Molecular-beam machines (and to some extent chemiluminescence machines) have been modified in various ways to explore even finer details of specific reactions. For example, electric and magnetic fields have been used to prepare reagents in various

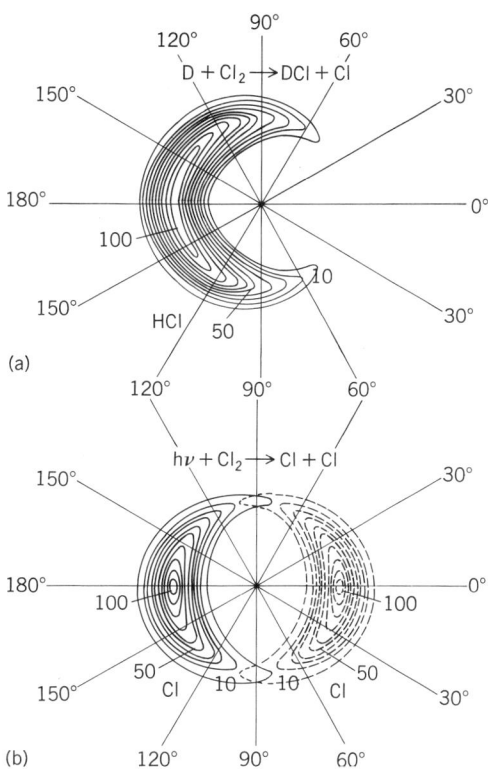

Fig. 4. Contour map of the DCl intensity. (a) Center-of-mass angular distribution of DCl from the reaction D + Cl₂. (b) Angular distribution of Cl atoms from photodissociation of Cl₂ molecules. Broken lines correspond to the second Cl atom. (*After D. R. Herschback, Pure Appl. Chem. 47:61–73, 1976*)

orientations or to determine the magnitude and polarization of the product rotational angular momentum; lasers have been used to prepare reagents in initial vibrational states and to determine the final vibration-rotation state of products by inducing fluorescence from the products; and reactions may be conducted at hyperthermal energies to produce dissociation or even ionization. Much work has been concerned with the interaction of light with isolated molecules.

Unimolecular reactions. Unimolecular reactions of collisionally activated reaction complexes have been studied using both cross-beam techniques and chemiluminescence techniques. The availability of lasers opened new vistas in chemical dynamics, not only because lasers facilitate conventional measurements, as in detecting reaction products by laser-induced fluorescence, but also because they uncovered new phenomena. Absorption of photons by molecules has been known for many years, and it became apparent that under the right circumstances a molecule can absorb not just one photon, but so many as to cause the molecule to dissociate or even to ionize.

These multiphoton dissociation processes can yield different products from thermal reactions, and much confusion surrounded their discovery. The dissociation occurs after a sequence of single-photon absorptions. Absorption of the first photon is a resonant process occurring at a wavelength corresponding to a normal infrared absorption. For polyatomic molecules, a variety of normal vibrations and various combinations and differences exist, and the second photon can be absorbed to give not necessarily the $v = 2$ state of the original normal mode, but possibly some new combination or different state. Sequential photons are absorbed in this fashion to climb the vibrational ladder both vertically and horizontally. The density of available rotation-vibration levels grows exponentially to become a quasicontinuum where succeeding photons are always resonant. Evidence suggests that an isolated molecule absorbs photons until it is sufficiently activated, whereupon it decomposes randomly.

THEORETICAL METHODS

The goal of chemical dynamics is to understand kinetic phenomena from the basic laws of molecular mechanics, and it is thus a field which sees close interplay between experimental and theoretical research.

Energy distribution. An important question regarding the dynamics of chemical reactions has to do with the product energy distribution in exothermic reactions. For example, because the HF molecule is more strongly bound than the H_2 molecule, reaction (6) releases a con-

$$F + H_2 \rightarrow HF + H \tag{6}$$

siderable amount of energy (more than 30 kcal/mol or 126 kJ/mol). The two possible paths for this energy release to follow are into translations, that is, with HF and H speeding away from each other, or into vibrational motion of HF.

In this case it is vibration, and this has rather dramatic consequences; the reaction creates a population inversion among the vibrational energy levels of HF—that is, the higher vibrational levels have more population than the lower levels—and the emission of infrared light from these excited vibrational levels can be made to form a chemical laser. A number of other reactions such as reaction (5) also give a population inversion among the vibrational energy levels, and can thus be used to make lasers.

Most effective energy. The rates of most chemical reactions are increased if they are given more energy. In macroscopic kinetics this corresponds to increasing the temperature, and most reactions are faster at higher temperatures. It seems reasonable, though, that some types of energy will be more effective in accelerating the reaction than others. For example, in reaction (7), which has been studied in a molecular beam, if HCl is vibrationally excited (by using a laser),

$$K + HCl \rightarrow KCl + H \tag{7}$$

this reaction is found to proceed approximately 100 times faster, while the same amount of energy in translational kinetic energy has a smaller effect. Here, therefore, vibrational energy is much more effective than translational energy in accelerating the reaction.

For reaction (6), however, translational energy is more effective than vibration in accelerating the reaction. The general rule of thumb is that vibrational energy is more effective for endothermic reactions (those for which the new molecule is less stable than the original molecule), while translational energy is most effective for exothermic reactions.

Lasers. As seen from above, lasers are also an important supplement to molecular-beam techniques for probing the dynamics of chemical reactions. Because they are light sources with a very narrow wavelength, they are able to excite molecules to specific quantum states (and also to detect what states molecules are in), an example of which is reaction (7). For polyatomic molecules—that is, those with more than two atoms—there is the even more interesting question of how the rate of reaction depends on which vibration is excited.

For example, when the molecule allyl isocyanide, $CH_2=CH-CH_2-NC$, is given sufficient vibrational energy, the isocyanide part (—NC) will rearrange to the cyanide (—CN) configuration. A laser can be used to excite a C-H bond vibrationally. An interesting question is whether the rate of the rearrangement process depends on which C-H bond is excited. Only with a laser is it possible to excite different C-H bonds and begin to answer such questions. This question of mode-specific chemistry, that is, the question of whether excitation of specific modes of a molecule causes specific chemistry to result, has been a subject of great interest. (For the example above, the reaction is fastest if the C-H bond closest to the NC group is excited.) Mode-specific chemistry would allow much greater control over the course of chemical reactions, and it would be possible to accelerate the rate of some reactions (or reactions at one part of a molecule) and not others.

Models and methods. Many different theoretical models and methods have been useful in understanding and analyzing all of the phenomena described above. Probably the single most useful approach has been the calculation of classical trajectories. Assuming that the potential energy function or a reasonable approximation is known for the three atoms in reaction (6), for example, it is possible by use of electronic computers to calculate the classical motion of the three atoms. It is thus an easy matter to give the initial molecule more or less vibrational or translational energy, and then compute the probability of reaction. Similarly, the final molecule and atom can be studied to see where the energy appears, that is, as translation or as vibration.

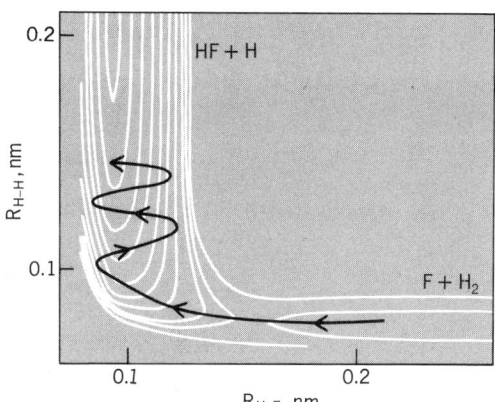

Fig. 5. Contour plot of the potential energy surface for the reaction F + H$_2$ = HF + H, with a typical reactive trajectory indicated.

It is thus a relatively straightforward matter theoretically to answer the questions and to see whether or not mode-specific excitation leads to significantly different chemistry than simply increasing the temperature under bulk conditions.

The most crucial step in carrying out these calculations is obtaining the potential energy surface—the potential energy as a function of the positions of the atoms—for the system. **Figure 5** shows a plot of the contours of the potential energy surface for reaction (6). Even without carrying out classical trajectory calculations, it is possible to deduce some of the dynamical features of this reaction; for example, the motion of the system first surmounts a small potential barrier, and then slides down a steep hill, turning the corner at the bottom of the hill. It is clear that such motion will cause much of the energy released in going down the hill to appear in vibrational motion of HF.

This and other theoretical methods have interacted strongly with experimental research in helping to understand the dynamics of chemical reactions. SEE INORGANIC PHOTOCHEMISTRY; LASER PHOTOCHEMISTRY; PHOTOCHEMISTRY.

RELAXATION METHODS

Considerable use has been made of perturbation techniques to measure rates and determine mechanisms of rapid chemical reactions. These methods provide measurements of chemical reaction rates by displacing equilibria. In situations where the reaction of interest occurs in a system at equilibrium, perturbation techniques called relaxation methods have been found most effective for determining reaction rate constants.

A chemical system at equilibrium is one in which the rate of a forward reaction is exactly balanced by the rate of the corresponding back reaction. Examples are chemical reactions occurring in liquid solutions, such as the familiar equilibrium in pure water, shown in reaction (8). The

$$H_2O \underset{k_b}{\overset{k_f}{\rightleftharpoons}} H^+(aq) + OH^-(aq) \tag{8}$$

molar equilibrium constant at 25°C (77°F) is given by Eq. (9), where bracketed quantities indicate

$$K_{eq} = \frac{[H^+][OH^-]}{[H_2O]} = \frac{10^{-14}}{55.5} = 1.8 \times 10^{-16} \tag{9}$$

molar concentrations. It arises naturally from the equality of forward and backward reaction rates, Eq. (10). Here k_f and k_b are the respective rate constants that depend on temperature but not

$$k_f[H_2O] = k_b[H^+][OH^-] \tag{10}$$

concentrations. Furthermore, the combination of Eqs. (9) and (10) gives rise to Eq. (11). Thus a

$$K_{eq} = k_f/k_b = 1.8 \times 10^{-16} \tag{11}$$

reasonable question might be what the numerical values of k_f in units of s^{-1} and k_b in units of $dm^3\,mol^{-1}s^{-1}$ must be to satisfy Eqs. (9) through (11) in water at room temperature. Stated another way, when a liter of 1 M hydrochloric acid is poured into a liter of 1 M sodium hydroxide (with considerable hazardous sputtering), how rapidly do the hydronium ions, $H^+(aq)$, react with hydroxide ions, $OH^-(aq)$, to produce a warm 0.5 M aqueous solution of sodium chloride? In the early 1950s it was asserted that such a reaction is instantaneous. Turbulent mixing techniques were (and still are) insufficiently fast (mixing time of the order of 1 ms) for this particular reaction to occur outside the mixing chamber. The relaxation techniques were conceived by M. Eigen, who accepted the implied challenge of measuring the rates of seemingly immeasurably fast reactions.

The essence of any of the relaxation methods is the perturbation of a chemical equilibrium (by a small change in temperature, pressure, electric-field intensity, or solvent composition) in so sudden a fashion that the chemical system, in seeking to reachieve equilibrium, is forced by the comparative slowness of the chemical reactions to lag behind the perturbation **(Fig. 6)**.

Temperature jump. Reaction (8) has a nonzero standard enthalpy change, ΔH^0, associated with it, so that a small increase in the temperature of the water requires the concentrations $[H^+]$ and $[OH^-]$ to increase slightly, and $[H_2O]$ to decrease correspondingly, for chemical equilibrium to be restored at the new higher temperature. Thus a small sample cell containing a very pure sample of water may be made one arm of a Wheatstone conductance bridge, and further configured so that a pulse of energy from a microwave source (or infrared laser of appropriate wavelength) is dissipated in the sample liquid. The resulting $\sim 2°C$ ($\sim 3.6°F$) rise in temperature will produce a small increase in conductance that will have an exponential shape and a time constant or relaxation time $\tau \simeq 27$ microseconds; τ is the time required for the signal amplitude to drop to $1/e = 1/2.718$ of its initial value, where e is the base of natural logarithms.

In pure water at 25°C (77°F), $[H^+] = [OH^-] = 10^{-7}\,M$, and for small perturbations, the value for τ is given by Eq. (12), from which it follows that $k_b \simeq 1.8 \times 10^{11}\,dm^3\,mol^{-1}s^{-1}$. This is

$$\tau^{-1} = k_b([H^+] + [OH^-]) + k_f = k_b([H^+] + [OH^-] + K_{eq}) \tag{12}$$

an exceptionally large rate constant for a bimolecular reaction between oppositely charged ions in aqueous solution and is, in fact, larger than that for any other diffusive encounter between ions in water. Eigen and L. DeMaeyer, who first determined this rate constant (using another relaxation method called the electric-field jump method), attributed the great speed of the back reaction of the equilibrium, reaction (8), to the exceptionally rapid motion of a proton through water, accomplished by the successive rotations of a long string of neighboring water molecules (Grotthuss mechanism). Since sample solutions can be heated by a mode-locked laser on a picosecond time scale or by a bunsen burner on a time scale of minutes, the temperature jump (T-jump) relaxation method just described is very versatile. The choice of the particular means of effecting the temperature perturbation is dictated only by the requirement that the temperature rise somewhat more rapidly than the time constant of the chemical reaction to be explored, so that a tedious deconvolution can be avoided. The discharge of a high-voltage (15- to 30-kV) capacitor through the sample liquid containing sufficient inert electrolyte to make it a good electrical conductor is the now classic Joule heating T-jump method used by Eigen and coworkers in their pioneering

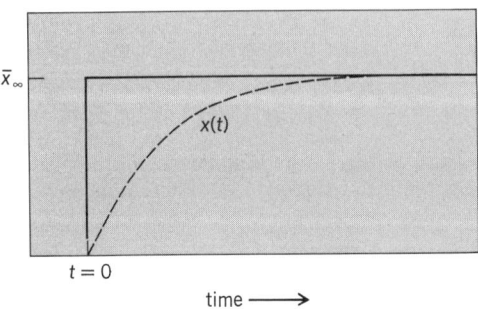

Fig. 6. Relaxational response to a rectangular step function in an external parameter such as temperature or pressure. The broken line represents the time course of the adjustment (relaxation) of the chemical equilibrium to the new temperature or pressure. (*After C. Bernasconi, Relaxation Kinetics, Academic Press, 1976*)

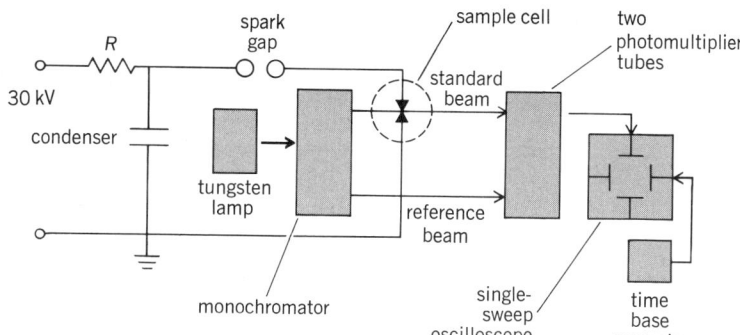

Fig. 7. Schematic of a Joule heating temperature-jump apparatus. (*After H. Eyring and E. M. Eyring, Modern Chemical Kinetics, Reinhold, 1963*)

studies. A schematic of such an apparatus is shown in **Fig. 7**. The 30-kV voltage generator charges the 0.1-microfarad condenser to the voltage at which the spark gap breaks down. The condenser then discharges across the spark gap and through the sample cell, containing an aqueous 0.1 M ionic strength solution, to ground. The sample cell is a ~50 ml (3.05 in.^{3}) Plexiglas cell containing two platinum electrodes spaced 1 cm (0.4 in.) apart and immersed in an aqueous 0.1 M ionic strength solution. The surge of current raises the temperature of the 1-ml (0.061-in.^{3}) volume of solution between the electrodes by 10°C (18°F) in a few microseconds.

Electric-field jump. In a situation, such as reaction (8), in which electrically neutral reactant species dissociate into oppositely charged ions, an especially sensitive tool for measuring rate constants of forward and backward reactions is the electric-field jump (E-jump) technique with conductometric detection. In a strong electric field (of the order of 4×10^6 V m^{-1}), a weak acid in solution is caused to dissociate to a greater degree than it would in the absence of the electric field. For weak electrolytes, such as aqueous acetic acid or ammonia, the effect is the order of 10% or less of the total normal dissociation, even at very high electric-field strengths. However, with a sensitive, high-voltage Wheatstone bridge, the exponential increase with time in the concentration of ions following a precipitous increase in electric-field strength is readily detected. The measured relaxation time (τ) is clearly that corresponding to the high-electric-field environment, but since the rate constants for these reactions differ little in and out of the electric field, no serious problem is posed.

A more serious concern is that the sample solution may have a very high electrical resistance, so that the supposedly square step function in the electric-field strength is not distorted by a significant voltage drop with concomitant heating of the sample liquid. Problems of working with high voltages, balancing capacitive and inductive effects in a very sensitive conductance bridge (now often circumvented by spectrophotometric detection), and the comparative difficulty of evaluating amplitudes of relaxations (as opposed to their readily determined time constants) are all factors that have worked against the wide use of the E-jump technique. There are many more ways of achieving a T-jump than an E-jump, and ΔH^0 values for chemical equilibria are readily available in the thermodynamic literature, whereas the extent to which a chemical equilibrium is displaced by an electric-field increment is rarely already known and is difficult to determine. Thus the commercialization of the T-jump method and the comparative neglect of the E-jump relaxation technique are readily understood.

Notwithstanding these difficulties, the E-jump technique is without peer for the investigation of the kinetics of solvent autoionization or for the exploration of the properties of weak electrolyte solutes in exotic solvents such as acetonitrile or xenon (the latter liquefied under a pressure of ~50 atm or 5 megapascals), so long as the relaxation time to be measured lies in the range 30 nanoseconds $< \tau <$ 100 microseconds.

Ultrasonic absorption. Two other relaxation methods more widely used than the E-jump technique are pressure jump (P-jump) and ultrasonic absorption. Each relies for its effectiveness on a volume change, ΔV^0, occurring in an aqueous sample equilibrium undergoing kinetic investigation. (In a nonaqueous solvent it will frequently be more important that ΔH^0 be large than that

ΔV^0 be so for the equilibrium to be susceptible to study by these two relaxation techniques.) As electrically neutral, weak electrolyte solute species dissociate into ions in aqueous solution, there is an increase in the number of solvent molecules drawn into a highly ordered solvation sheath. The higher the charge density of the ion, the more water molecule dipoles are bound and the greater the change in V^0 as reactants become products. Thus the dissociation of an aqueous neodymium(III) sulfate complex is particularly susceptible to study by one or more of the four or five ultrasonic absorption methods that cover the $f \approx 100$ kHz -1 GHz sound frequency range. Unlike the T-jump and E-jump relaxation methods, which usually employ step function perturbations, the ultrasonic absorption techniques are continuous-wave experiments in which the sample chemical equilibrium absorbs a measurable amount of the sound wave's energy when the frequency of the sound wave (f) and the relaxation time of the chemical equilibrium bear the relation to one another given by Eq. (13).

$$\tau^{-1} = 2\pi f \qquad (13)$$

A particularly easy ultrasonic absorption experiment to understand and perform is the laser Debye-Sears technique. A continuously variable frequency sound wave is introduced by a quartz piezoelectric transducer into a 30-ml (1.83 in.^{3}) sample cell that has entrance and exit windows for a visible laser light beam that passes through the cell at about 90° to the direction of travel of the planar sound wave. The regions of compression and rarefaction in the sound wave act as a diffraction grating for the laser light beam. If a chemical equilibrium in the sample strongly absorbs a particular frequency of sound (f), the definition of the "diffraction grating" will deteriorate and the measured intensity of the first-order diffracted laser light will diminish. The frequency of minimum diffracted light intensity will be that of Eq. (12). **Figure 8** shows a diagram of the apparatus. The piezoelectric (quartz) tranducer cemented to the bottom of a plastic rod that is driven up and down by a stepping motor is controlled by a mini computer. The angle of diffraction of the laser beam by the alternating regions of compression and rarefaction in the liquid (suggested by the horizontal lines) is exaggerated in the diagram.

Ultrasonic absorption techniques have been used in kinetic investigations of quite complicated biophysical systems such as the order-disorder transitions that occur in liquid crystalline phospholipid membranes. While the ultrasonic techniques look through a conveniently broad time window at kinetic processes in solutions, this picture window is difficult to "see through" in the sense that many equilibrium processes in solution can absorb sound energy and the responsible process is not instantly identified by a characteristic absorption of electromagnetic radiation as in a spectrophotometric T-jump or E-jump experiment. A further disadvantage arises from the great breadth of the ultrasonic absorption "peaks" in a plot of normalized sound absorption versus sound frequency. Unless multiple relaxation times in a chemical system are quite widely separated in time, they are difficult to resolve in an ultrasonic absorption spectrum.

Pressure jump. The typical pressure-jump (P-jump) experiment is one in which a liquid sample under about 200 atm (20 MPa) pressure is suddenly brought to atmospheric pressure by the bursting of a metal membrane in the sample cell autoclave. Relaxation times measured spectrophotometrically or conductometrically are thus accessible if $\tau > 100$ microseconds. This technique has proven particularly useful in the elucidation of micellar systems of great interest for catalysis and for petroleum recovery from apparently depleted oil fields.

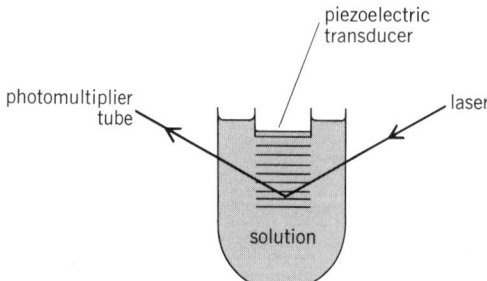

Fig. 8. Schematic of laser Debye-Sears apparatus for measuring ultrasonic absorption ($\sim 15-300$ MHz) in a sample liquid. (*After W. J. Gettins and E. Wyn-Jones, Techniques and Applications of Fast Reactions in Solution, D. Reidel, 1979*)

The continuous- and stopped-flow techniques antedate somewhat the relaxation techniques described above, and have the sometimes important advantage of permitting kinetic measurements in chemical systems far from equilibrium. The stopped-flow experiment is one in which two different liquids in separate syringes are mixed rapidly in a tangential jet mixing chamber and then the rapid flow of mixed reactants is almost immediately brought to a halt in a spectrophotometric, conductometric, or calorimetric observation chamber. Reaction half-lives exceeding 2 milliseconds are easily accessible.

Other relaxation methods. Stopped-flow equipment has been used in concentration-jump and solvent-jump relaxation kinetic studies. An example of an application of the solvent-jump technique to a system insensitive to concentration-jump is a kinetic study of reaction (14)

$$NOCl + n\text{-BuOH} \rightarrow n\text{-BuONO} + HCl \qquad (14)$$

in mixed CCl_4 – acetic acid solvents of varying composition (Bu = butyl). The thermodynamic treatment of the solvent jump is just about the only aspect of the presently known relaxation techniques that was not described in exhaustive detail by the earliest publications of Eigen and DeMaeyer. SEE CHEMICAL THERMODYNAMICS.

Bibliography. C. F. Bernasconi, *Relaxation Kinetics*, 1976; R. B. Bernstein (ed.), *Atom Molecule Collision Theory*, 1979; P. R. Brooks and E. F. Hayes, *State-to-State Chemistry, ACS Symp. Ser. 56*, 1977; M. Eigen, *Nobel Lecture: Immeasurably Fast Reactions*, 1967; H. Eyring et al., *Basic Chemical Kinetics*, 1980; K. Laidler and J. Keith, *Chemical Kinetics*, 1986; W. H. Miller (ed.), *Dynamics of Molecular Collisions*, 1976.

CHAIN REACTION
FARRINGTON DANIELS

A chemical reaction in which many molecules undergo chemical reaction after one molecule becomes activated. In ordinary chemical reactions, every molecule that reacts must first become activated by collision with other rapidly moving molecules. The number of these violent collisions per second is so small that the reaction is slow. Once a chain reaction is started, it is not necessary to wait for more collisions with activated molecules to accelerate the reaction, which now proceeds spontaneously.

Photochemical reactions. A typical chain reaction is the photochemical reaction between hydrogen and chlorine as described by reactions (1).

$$\begin{aligned} Cl_2 + \text{light} &\rightarrow Cl + Cl \\ Cl + H_2 &\rightarrow HCl + H \\ H + Cl_2 &\rightarrow HCl + Cl \\ Cl + H_2 &\rightarrow HCl + H \end{aligned} \qquad (1)$$

The light absorbed by a chlorine molecule dissociates the molecule into chlorine atoms; these in turn react rapidly with hydrogen molecules to give hydrogen chloride and hydrogen atoms. The hydrogen atoms react with chlorine molecules to give hydrogen chloride and chlorine atoms. The chlorine atoms react further with hydrogen and continue the chain until some other reaction uses up the free atoms of chlorine or hydrogen. The chain-stopping reaction may be the reaction between two chlorine atoms to give chlorine molecules, or between two hydrogen atoms to give hydrogen molecules. Again the atoms may collide with the walls of the containing vessel, or they may react with some impurity which is present in the vessel only as a trace.

The length of the chain, that is, the number of molecules reacting per molecule activated, is determined by the relative rates of the competing reactions, namely, the chain-propagating reaction and the chain-stopping reactions. In the chain reaction just described, 10^6 molecules of hydrogen chloride may be formed by the photodissociation of 1 chlorine molecule.

In photochemical chain reactions, the length of the chain can be determined by measuring the number of photons of light absorbed, that is, the number of molecules activated, and dividing by the number of molecules which react chemically. SEE PHOTOCHEMISTRY.

Thermal reactions. In thermal reactions, the length of the chain may sometimes be estimated from a knowledge of the intermediate steps and the kinetics involved. The presence of a chain reaction can often be proved by adding a trace of an inhibitor, such as nitric oxide. If the reaction is slowed down greatly by a very small amount of a substance which reacts with the chain-propagating units, the reaction involves a chain. While the inhibitor is being consumed in this way, the reaction is slow. After an induction period, the inhibitor is consumed and the rapid chain reaction then takes place.

Chain reactions are erratic and are reproduced with difficulty in different laboratories because they depend so much on the presence and concentration of accidental impurities which act as inhibitors.

In many chemical reactions, particularly organic reactions at elevated temperatures, the chains are carried by free radicals which are very reactive fragments of molecules that have unshared electrons, such as $\cdot CH_3$, $\cdot C_2H_5$, $\cdot H$, and $\cdot OH$. The thermal decomposition of propane is a typical free-radical chain which follows reactions (2). One molecule of propane is decomposed into

$$\begin{aligned} C_3H_8 &\rightarrow \cdot CH_3 + \cdot C_2H_5 \\ \cdot CH_3 + C_3H_8 &\rightarrow CH_4 + C_3H_7 \\ \cdot C_3H_7 &\rightarrow \cdot CH_3 + C_2H_4 \\ \cdot CH_3 + C_3H_8 &\rightarrow CH_4 + \cdot C_3H_7 \end{aligned} \qquad (2)$$

free radicals, $\cdot CH_3$ and $\cdot C_2H_5$, which then react with more propane to give the product methane and a free radical, $\cdot C_3H_7$, which decomposes into $\cdot CH_3$ and the product ethylene. The $\cdot CH_3$ reacts with more propane and continues the chain. The chain is terminated by collision of the free radicals with the wall or with each other, in reactions such as (3). Thus it is possible to obtain

$$\cdot CH_3 + \cdot C_3H_7 \rightarrow C_4H_{10} \qquad (3)$$

products of higher molecular weight as well as products of lower molecular weight. The finding of these higher-molecular-weight products supports the theory of free-radical formation and chain reactions.

Certain oxidations in the gas phase are known to be chain reactions. The carbon knock which occurs at times in internal combustion engines is caused by a too-rapid combustion rate caused by chain reactions. This chain reaction is reduced by adding tetraethyllead which acts as an inhibitor.

The polymerization of styrene to give polystyrene and the polymerization of other organic materials to give industrial plastics involve chain reactions. The spoilage of foods, the precipitation of insoluble gums in gasoline, and the deterioration of certain plastics in sunlight involve chain reactions, which can be minimized with inhibitors. *See* CHAIN REACTION; CHEMICAL DYNAMICS; INHIBITOR.

Bibliography. J. H. Espenson, *Chemical Kinetics and Reaction Mechanisms*, 1981; H. Eyring et al., *Basic Chemical Kinetics*, 1980; K. J. Laidler, *Reaction Kinetics*, 2 vols., 1963.

CATALYSIS
ROBERT L. BURWELL, JR.

The phenomenon in which a relatively small amount of foreign material, called a catalyst, augments the rate of a chemical reaction without itself being consumed. A catalyst is material, and not light or heat. It augments a rate. *See* INHIBITOR.

If the reaction $A + B \rightarrow D$ occurs very slowly but is catalyzed by some catalyst, Cat, the addition of Cat must open new channels for the reaction. In a very simple case (1), the two

$$\left. \begin{aligned} A + Cat &\rightarrow ACat \\ ACat + B &\rightarrow D + Cat \end{aligned} \right\} \text{Chain propagation} \qquad (1)$$
$$A + B \rightarrow D \qquad \text{Overall reaction}$$

propagation processes, which are fast compared to the uncatalyzed ractions, $A + B \rightarrow D$, provide the new channel for reaction. The catalyst reacts in the first step, but is regenerated in the second step to commence a new cycle. A catalytic reaction is thus a kind of chain reaction. *See* CHAIN REACTION.

If a reaction is in chemical equilibrium under some fixed conditions, the addition of a catalyst cannot change the position of equilibrium without violating the second law of thermodynamics. Therefore, if a catalyst augments the rate of A + B → D, it must also augment the reverse rate, D → A + B.

Categories. Catalysis is conventionally divided into three categories: homogeneous, heterogeneous, and enzyme. Heterogeneous catalysis plays a dominant role in chemical processes in the petroleum, petrochemical, and chemical industries. Homogeneous catalysis is important in the petrochemical and chemical industries. Enzyme catalysis plays a key role in all metabolic processes and in some industries, such as the fermentation industry.

Homogeneous. In homogeneous catalysis, reactants, products, and catalyst are all present molecularly in one phase, usually liquid. Examples are the hydrogenation of 1-hexene in a hydrocarbon solvent catalyzed by dissolved $[(C_6H_5)_3 P]_3RhH$ [reaction (2)] and the hydrolysis of an ester

$$CH_2=CHCH_2CH_2CH_2CH_3 + H_2 \rightarrow CH_3CH_2CH_2CH_2CH_2CH_3 \qquad (2)$$

$$CH_3COOC_2H_5 + H_2O \xrightarrow{H^+} CH_3COOH + HOC_2H_5 \qquad (3)$$

catalyzed by acid [reaction (3)]. SEE HOMOGENEOUS CATALYSIS.

Heterogeneous. In heterogeneous catalysis, the catalyst is in a separate phase. Usually the reactants and products are in gaseous or liquid phases and the catalyst is a solid. The catalytic reaction occurs on the surface of the solid. Examples are the dehydration and the dehydrogenation of isopropyl alcohol [reactions (4) and (5)]. Reaction (4) can be affected by passing the vapors of

$$CH_3CHOHCH_3 \rightarrow CH_3CH=CH_2 + H_2O \quad \text{Dehydration} \qquad (4)$$

$$CH_3CHOHCH_3 \rightarrow CH_3COCH_3 + H_2 \quad \text{Dehydrogenation} \qquad (5)$$

the alcohol over alumina at about 570°F (300°C), and reaction (5) over copper at 390°F (200°C). SEE HETEROGENEOUS CATALYSIS.

Enzyme. Transformations of matter in living organisms occur by an elaborate sequence of reactions, most of which are catalyzed by biocatalysts called enzymes. Enzymes are proteins and therefore of colloidal dimensions. Although studies of interrelations between homogeneous catalysis and heterogeneous catalysis have been developing, enzyme catalysis remains a rather separate area both in the nature of the catalyst and in the type of reactions catalyzed.

Selectivity. In most cases, a given set of reactants could react in two or more ways, as exemplified by reactions (4) and (5). The degree to which just one of the possible reactions is favored over the other is called selectivity. Selectivity is a key property of a catalyst in any practical application of the catalyst.

Bibliography. J. R. Anderson and M. Boudart (eds.), *Catalysis* vols. 1–7, 1981–1985.

HOMOGENEOUS CATALYSIS
DENIS FORSTER

A process in which a catalyst is in the same phase as the reactant. A homogeneous catalyst is molecularly dispersed (dissolved) in the reactants, which are most commonly in the liquid state. Catalysis of the transformation of organic molecules by acids or bases represents one of the most widespread types of homogeneous catalysis. In addition, the catalysis of organic reactions by metal complexes in solution has grown rapidly in both scientific and industrial importance.

Acid-base catalysis. The two principal areas are specific acid (or base) catalysis and general acid (or base) catalysis. Specific acid catalysis refers to reactions in which only the oxonium ion (H_3O^+) can act as the catalyst. A common example is the hydrolysis of simple acetals, reaction (1). Specific acid catalysis is found to be characteristic of reactions in which there is

$$CH_3-\underset{\underset{OR}{|}}{\overset{\overset{CH_3}{|}}{C}}-OR \underset{}{\overset{H_3O^+}{\rightleftharpoons}} CH_3-\underset{\underset{OR}{|}}{\overset{\overset{CH_3}{|}}{\underset{\oplus}{C}}}-\overset{H}{\underset{}{O}}R \rightleftharpoons CH_3-\underset{\underset{OR}{|}}{\overset{\overset{CH_3}{|}}{C}}\oplus + ROH \underset{}{\overset{H_2O}{\rightleftharpoons}} CH_3-\underset{\underset{OR}{|}}{\overset{\overset{CH_3}{|}}{C}}-OH + H_3O^+ \overset{H_3O^+}{\rightleftharpoons}$$

$$CH_3-\overset{\overset{CH_3}{|}}{C}=O + ROH \qquad (1)$$

rapid, reversible protonation of the substrate before the slow, rate-limiting step.

Reactions which are catalyzed by proton donors in general are considered to be subject to general acid catalysis. General acid catalysis often becomes important only at higher acidity levels. The proton is a convenient and powerful agent for the distortion of the electronic configuration of a substrate in order to facilitate reaction. The mechanism by which this occurs has many variants. For example, a covalent bond may be more easily broken after protonation of one of the bonded atoms; the reaction, $ROH_2^\oplus \rightarrow R^\oplus + H_2O$ is easier than $ROH \rightarrow R^\oplus + OH^\ominus$.

Exactly the same distinction can be made in catalysis by bases as was made above for acids. Thus, in specific base catalysis the reaction rate is proportional to the concentration of $OH^\ominus$.

Metal complexes. In homogeneous catalysis by coordination compounds of transition metals, the catalyst is usually deployed in solution and most commonly exists in a molecularly dispersed form. Thus, all sites are potentially active for catalysis, and in many cases catalysis is observed under much milder reaction conditions than found with heterogeneous catalysis by metals and metal oxides.

The catalysis of the incorporation of carbon monoxide into organic substrates by transition metal complexes is technologically important. The hydroformylation or oxo reaction [reaction (2)]

$$RCH=CH_2 + CO + H_2 \rightarrow RCH_2-CH_2CHO + RCH-CH_3 \atop | \atop CHO \quad (2)$$

in which an olefin is reacted with carbon monoxide and hydrogen to generate a mixture of linear and branched aldehydes, was discovered in 1938. The first catalyst found was dicobalt octacarbonyl, $Co_2(CO)_8$, and this is still used extensively today in commercial operations. The steps involved in this reaction are summarized in reactions (3).

$$Co_2(CO)_8 + H_2 \rightleftharpoons 2HCo(CO)_4 \quad (3a) \qquad HCo(CO)_4 + \text{olefin} \rightleftharpoons HCo(CO)_3(\text{olefin}) + CO \quad (3b)$$

$$HCo(CO)_3(\text{olefin}) \rightleftharpoons RCo(CO)_3 \quad (3c) \qquad RCo(CO)_3 + CO \rightleftharpoons RCo(CO)_4 \quad (3d)$$

$$RCo(CO)_4 + CO \rightleftharpoons (R-CO)Co(CO)_4 \quad (3e) \qquad (RCO)Co(CO)_4 + HCo(CO)_4 \rightarrow RCHO + Co_2(CO)_8 \quad (3f)$$

Some of these steps represent transformations which are common to many sequences found in homogeneous catalysis. Thus, step (3c), the insertion of a coordinated olefin into the metal-hydride bond to generate a metal-alkyl bond, is a frequently encountered method of metal-carbon bond formation. Step (3e), the formation of a metal-acyl bond by alkyl migration to a coordinated carbon monoxide, is a key step in most catalytic (and stoichiometric) syntheses involving the incorporation of carbon monoxide into organic molecules.

While the reaction steps above can be conducted in a stoichiometric manner under very mild reaction conditions, in order for the system to function catalytically at rates which are desirable for industrial processes, the reaction temperature is maintained at greater than 248°F (120°C) and the reaction pressures are usually in excess of 200 atm (20 megapascals).

Other catalyst systems have been discovered which can perform hydroformylation reactions under much milder reaction conditions than cobalt. In particular, rhodium complexes containing triarylphosphine ligands can catalyze hydroformylation reactions at very rapid rates at ~212°F (~100°C) and 30 atm (3 MPa) pressure of synthesis gas (CO + H_2). Another important difference between the rhodium and cobalt catalysts is that the rhodium system can generate a much higher proportion of linear aldehyde product [reaction (2)]. This effect is related to the greater steric crowding around the metal when triarylphosphines are present in the coordination sphere. This is an example of the influence of stereochemistry around the metal on the stereochemical course of the catalytic reaction, and this phenomenon is an important feature of many homogeneously catalyzed reactions. SEE STERIC EFFECT.

Another reaction involving the catalysis of the incorporation of carbon monoxide which has assumed considerable commercial importance is the synthesis of acetic acid from methanol, reaction (4). The reaction is catalyzed by both cobalt and rhodium complexes in the presence of an

$$CH_3OH + CO \rightarrow CH_3CO_2H \quad (4)$$

iodide cocatalyst or promoter. The mechanism of the rhodium-catalyzed reaction is reasonably well understood, as shown in reactions (5). The reaction which generates the metal carbon bond,

$$CH_3OH + HI \rightleftharpoons CH_3I + H_2O \qquad (5a)$$

$$[Rh(CO)_2 I_2]^- + CH_3I \rightleftharpoons [Rh(CO)_2(CH_3)I_3]^- \qquad (5b)$$

$$[Rh(CO)_2(CH_3)I_3]^- \rightarrow [Rh(CO)(COCH_3)I_3]^- \qquad (5c)$$

$$[Rh(CO)(COCH_3)I_3]^- + CO \rightleftharpoons [Rh(CO)_2(COCH_3)I_3]^- \qquad (5d)$$

$$[Rh(CO)_2(COCH_3)I_3] \rightarrow [Rh(CO)_2 I_2]^- + CH_3COI \qquad (5e)$$

$$CH_3COI + H_2O \rightarrow CH_3CO_2H + HI \qquad (5f)$$

that is, step (5b), is rate-determining in the catalytic cycle.

The commercial reactors utilizing rhodium catalysts are operated at temperatures in the range of 300–390°F (150–200°C) and pressures of less than 40 atm (4 MPa). The rate of the reaction is sufficiently rapid that the amount of the very expensive rhodium catalyst required is very small.

A wide range of olefin transformation reactions are catalyzed by transition metal complexes. Some of the more important reactions are isomerization, dimerization, polymerization, and metathesis. Nickel catalysts convert olefins into a mixture of dimers, trimers, and higher oligomers. The rate is particularly rapid with ethylene. The catalytic species are typically generated in place from various nickel complexes by reaction with an alkyl-aluminum compound. Uncharacterized hydridonickel species are postulated to be the active catalysts. The mechanism for dimerization and polymerization can be visualized as a series of sequential olefin insertions into Ni-H and Ni-C bonds followed by β-hydride elimination. The dimerization of propylene and higher olefins can give rise to linear, monobranched and dibranched olefin products. The relative amount of each type of product depends on the ligands coordinated to the nickel.

Migratory insertion of alkenes into alkyl-metal bonds is also represented in the polymerizations catalyzed by a variety of transition metal species. For example, ethylene is polymerized by a catalyst prepared in place by the reduction of $TiCl_4$ with alkyl-aluminum compounds. This is the basis for the commercial Ziegler-Natta process for the preparation of high-density polyethylene, which is practiced throughout the world.

Oxidation. Transition metal complexes act as homogeneous catalysts in many different types of oxidation process. Two main categories of reaction can be recognized, involving either one-electron or two-electron processes.

Autoxidation. The involvement of transition metal complexes in one-electron, radical processes is most evident in the so-called autoxidation reactions whereby hydrocarbons are oxidized to various oxygen-containing compounds by radical chain processes. The general scheme is shown in reactions (6). While metal species can enhance the rate of several of the above steps,

$$\text{Initiation:} \quad \text{Initiator} + RH \rightarrow inH + R \cdot \qquad (6a)$$

$$\text{Propagation:} \quad R \cdot + O_2 \rightarrow RO_2 \cdot \qquad (6b)$$

$$RO_2 \cdot + RH \rightarrow RO_2H + R \cdot \qquad (6c)$$

$$\text{Termination:} \quad R \cdot + RO_2 \cdot \rightarrow RO_2R \qquad (6d)$$

$$2 RO_2 \cdot \rightarrow RO_4R \qquad (6e)$$

$$RO_4R \rightarrow \text{nonradical products} + O_2 \qquad (6f)$$

the most common pathway for catalysis of liquid-phase autoxidations involves the metal-catalyzed decomposition of alkyl hydroperoxides, of which reactions (7) and (8) are examples. Cobalt and

$$RO_2H + Co^{II} \rightarrow RO \cdot + Co^{III}(OH) \qquad (7) \qquad (RO_2)Co^{III} \rightleftharpoons RO_2 \cdot + Co^{II} \qquad (8)$$

manganese salts are particularly effective in promoting autoxidation processes. The oxidation of p-xylene to terephthalic acid (the key monomer involved in the manufacture of polyester) is carried out on a very large scale using a cobalt-bromide catalyst.

Indirect oxidation. Transition metal complexes find utility in the catalysis of various types of indirect, two-electron oxidations. Examples of these indirect processes are the so-called Wacker reaction, in which olefins are oxidized to aldehydes or ketones by palladium (II) compounds, with concomitant reduction of the palladium. The palladium is then reoxidized in a separate reaction by a combination of a copper salt and oxygen. The best-known example of the Wacker reaction is the oxidation of ethylene to acetaldehyde, reaction (9). The reaction is con-

$$CH_2=CH_2 + \tfrac{1}{2} O_2 \rightarrow CH_3CHO \tag{9}$$

ducted in an aqueous medium in the presence of palladium and copper chlorides as the catalyst system. The generally accepted mechanism is shown in reactions (10).

$$PdCl_4^{2-} + C_2H_4 \rightleftharpoons [PdCl_3(C_2H_4)]^- + Cl^- \tag{10a}$$

$$[PdCl_3(C_2H_4)]^- + H_2O \rightleftharpoons [PdCl_2(H_2O)(C_2H_4)] + Cl^- \tag{10b}$$

$$[PdCl_2(H_2O)(C_2H_4)] + H_2O \rightleftharpoons [PdCl_2(OH)(C_2H_4)]^- + H_3O^+ \tag{10c}$$

$$[PdCl_2(OH)(C_2H_4)]^- + H_2O \rightleftharpoons [HOCH_2CH_2PdCl(H_2O)] + Cl^- \tag{10d}$$

$$[HOCH_2CH_2PdCl(H_2O)] \rightarrow CH_3CHO + Pd + HCl + H_2O \tag{10e}$$

$$Pd + 2CuCl_2 \rightarrow PdCl_2 + 2\,CuCl \tag{10f}$$

$$CuCl + \tfrac{1}{2}O_2 + HCl \rightarrow CuCl_2 + \tfrac{1}{2}H_2O \tag{10g}$$

Another important indirect oxidation process is the metal-catalyzed epoxidation of olefins with alkyl hydroperoxides, reaction (11). Various molybdenum, vanadium, and chromium com-

$$\mathrm{C=C} + RO_2H \rightarrow \overset{O}{\underset{}{\mathrm{C-C}}} + ROH \tag{11}$$

plexes act as catalysts for this reaction, by pathways which are still rather poorly understood.

Adiponitrile, $NC(CH_2)_4CN$, is produced as a precursor of hexamethylenediamine, one of the building blocks of Nylon 66. The selective addition of two moles of hydrogen cyanide to butadiene has been developed into a valuable new synthesis of adiponitrile. In one variant of this process, a zero-valent nickel catalyst, $Ni[P(OAryl)_3]_4$, can be used to bring about a series of reactions including HCN addition to butadiene, isomerization of cyanolefins, and hydrocyanation of 4-pentenenitrile, reactions (12)–(14). A key step in reactions (12) and (14) appears to be generation of a

$$CH_2=CH-CH=CH_2 + HCN \rightarrow CH_3-CH=CH-CH_2-CN \tag{12}$$

$$CH_3-CH=CH-CH_2-CN \rightleftharpoons CH_2=CH-CH_2-CH_2CN \tag{13}$$

$$CH_2=CH-CH_2-CH_2CN + HCN \rightarrow NC(CH_2)_4CN \tag{14}$$

nickel hydride species capable of reacting with an olefin. An outline of the mechanism is shown in reactions (15)–(19), where L represents phosphite or phosphine.

$$NiL_4 + HCN \rightleftharpoons HNi(CN)L_3 + L \quad (15) \qquad HNi(CN)L_3 + RCH=CH_2 \rightarrow HNi(CN)(olefin)L_2 + L \quad (16)$$

$$HNi(CN)(olefin)L_2 \rightarrow RCH_2CH_2Ni(CN)L_2 \quad (17) \qquad RCH_2CH_2Ni(CN)L_2 \rightarrow NiL_2 + RCH_2CH_2CN \quad (18)$$

$$NiL_2 + 2L \rightarrow NiL_4 \tag{19}$$

Perhaps the most elegant illustration of the selectivity achievable with homogeneous catalysts is found in the asymmetric hydrogenation of unsymmetrical olefins in the presence of rhodium complexes containing optically active phosphine ligands. Through this process, it is possible to prepare a number of optically active α-amino acids from the corresponding unsaturated precursors, reaction (20). (The carbon marked with an asterisk is an asymmetric center.) The energy

$$\underset{\underset{NHCOR_2}{|}}{\overset{\overset{COOH}{|}}{RCH=C}} \xrightarrow[\text{cat}]{H_2} \underset{\underset{NHCOR_2}{|}}{\overset{\overset{COOH}{|}}{RCH_2-C^*-H}} \rightarrow \underset{\underset{NH_2}{|}}{\overset{\overset{COOH}{|}}{RCH_2-C^*-H}} \tag{20}$$

difference between the optical enantiomers is very small, but nevertheless, with suitable optically active phosphine ligands, one isomer can be produced with greater than 90% selectivity. This approach is used commercially in the synthesis of L-dopa, the drug that is used in the treatment of Parkinson's disease.

HETEROGENEOUS CATALYSIS
ROBERT L. BURWELL, JR.

A chemical process in which the catalyst is present in a separate phase. In the usual case, the catalyst is a solid and the reactants and products are in gaseous or liquid phases.

Heterogeneous catalysis proceeds by the formation and subsequent reaction of chemisorbed complexes which can be considered to be surface chemical compounds. In the very simple case where A → B is slow in the absence of catalyst, one might have reaction (1).

$$\left.\begin{array}{l} * + A \to *A \\ A* \to B* \\ B* \to B + * \\ A \to B \end{array}\right\} \text{chain propagating steps} \qquad (1)$$

Reaction A → B is fast if the three preceding steps are fast. Here, * represents a catalytic site on the surface of the catalyst, A* → B* is called a surface reaction, * + A → *A represents the chemisorption of A, and B* → B + * represents desorption of B. *See* ADSORPTION.

With most sets of reactants, more than one reaction will be thermodynamically possible. The degree to which a given catalyst favors one reaction compared with other possible reactions is called the selectivity of the catalyst for reaction (1). Two aspects of a catalyst are of particular importance: its selectivity, and its activity, which can be taken as the rate of conversion of reactants by a given amount of catalyst under specified conditions. Ideally, the rate will be proportional to the amount of catalyst.

History. Heterogeneous catalytic processes were discovered in the 1810s by Sir Humphry Davy and L. J. Thénard. Indeed, the present use of noble metal catalysts to oxidize the hydrocarbon vapors in automobile exhaust goes back to Davy's discovery of the oxidation of hydrocarbon vapors on platinum, and one could say that the Haber process for the synthesis of ammonia goes back to Thénard's observation that ammonia decomposed into its elements when passed over hot iron.

The first plant for the manufacture of sulfuric acid by the contact process, reactions (2),

$$2SO_2 + O_2 \to 2SO_3 \qquad SO_3 + H_2O \to H_2SO_4 \qquad (2)$$

went on-stream in 1875. This was the first important heterogeneous catalytic process to be used in the chemical industry. For many years, the oxidation catalyst was platinum, but it was replaced by vanadium pentoxide (V_2O_5) on silica gel some years ago. However, major development of heterogeneous catalysis started only about 1900. It could be noted that the first industrial process for the manufacture of nitric acid, reaction (3), began in 1906, although in 1838, C. F. Kuhlmann had

$$NH_3 \xrightarrow{O_2} NO \xrightarrow{O_2} NO_2 \xrightarrow{H_2O,\, O_2} HNO_3 \qquad (3)$$

discovered that platinum catalyzed the oxidation of ammonia to nitric oxide. In current usage, the catalyst is Pt-Rh gauze at 1652°F (900°C).

By the 1950s, heterogeneous catalytic processes had come to dominate the petroleum, petrochemical, and chemical industries. Heterogeneous catalysis is a critical feature in energy conservation and interconversion, and is a key feature in the production of synthetic fuels from coal and oil shale.

Reactors. Both industrially and in the laboratory, catalytic reactions are usually effected in one of three kinds of reactors: batch, flow, and gradientless. Examples are shown in the **illustration**. In the flow reactors, there is usually a large change in concentrations of products and reactants between the entrance and exit of the bed. In the gradientless reactor, the changes are kept very small, either by recirculation of 99% of the exit gases from the catalyst bed in *b* back to

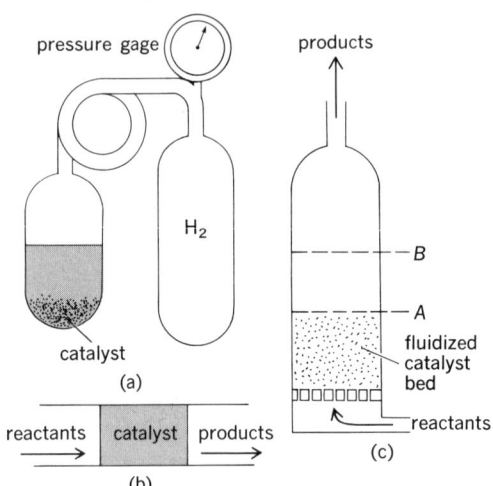

Catalytic reactors. (a) Batch. (b) Continuous fixed-bed. (c) Continuous fluidized reactor.

the entrance to the bed, or by use of a stirred-flow reactor, of which the fluidized reactor in c is one form.

Catalysts. A selection of heterogeneous catalytic processes of scientific or industrial interest is shown in the **table**.

Since catalytic activity will ordinarily be proportional to surface area, most catalysts are used in forms with large specific areas, a_s. The low-area catalyst of reaction (3) is a rather unusual case. Higher-area metal powders are often used for liquid-phase reactions in batch reactors like that shown in a. For example, finely divided nickel, a_s = 25–40 m² g⁻¹, is used for the hydrogenation of unsaturated glycerides in the manufacture of margarine from vegetable oils.

Supported catalysts are widely used. In these, the catalytic ingredient is dispersed in the internal porosity of such supports as silica gel, γ-alumina, and charcoals. These supports have large areas in the internal porosity, 100–1000 m² g⁻¹, and their average pore diameters are 2–20 nanometers. Tiny crystallites of such metals as platinum, palladium, rhodium, and nickel can be formed in the pore structure. Supported oxides and sulfides of transition metals are also used. Supported catalysts have the advantage that the area of the catalytic ingredient can be very large. Since, however, the support granules can have diameters in the 1–5 mm range, large flows of gas produce only moderate pressure drops across a catalyst bed. Supported catalysts are much more resistant to coalescence of the catalytic ingredient than are powders. Further, deposition of carbonaceous residues accompanies many catalytic reactions. In some cases, the catalyst can be regenerated by burning off the residues. Such regeneration would result in drastic losses of area with metal powders.

Supported catalysts are particularly prone to problems with diffusion. Reaction * + A → *A must be preceded by the diffusion of the reactant through the pore structure of the support to a catalytic site. Reaction B* → B + * must be followed by the diffusion of the product out of the support. Heat must also flow in and out of the granule of support. Such matters receive particular attention in the chemical engineering aspects of catalysis.

One important type of catalyst exposes strongly acidic sites in its internal porosity. Such catalysts are used to crack larger molecules of hydrocarbon into smaller ones in petroleum refining. Other catalysts, called dual-functional catalysts, have a hydrogenating catalytic ingredient on an acidic support. These are also of major importance in processing petroleum.

Another type of catalyst consists of an organometallic complex deposited on such supports as silica or alumina. These catalysts have been called heterogenized homogeneous catalysts, and they accompany a recent development in which the nature of surface sites on heterogeneous catalysts has been interpreted in terms of coordination chemistry and homogeneous catalysis.

Most catalysts exhibit coordinatively unsaturated surface sites (*cus*) which are capable of

Some typical heterogeneous catalytic reactions

Catalyst	Reaction	T,°C*
	Hydrogenation	
	$H_2 + C_2H_4 \rightarrow C_2H_6$	-100
	$H_2 + D_2 \rightarrow 2HD$	-180
Pt, Pd, Rh, Ni, as powders or supported on SiO_2, Al_2O_3, or C	$C_3H_8 + D_2 \rightarrow C_3H_7D\text{---}C_3D_8 + HD$	50
	Cyclopropane + $H_2 \rightarrow C_3H_8$	40
	$C_2H_6 + H_2 \rightarrow 2CH_4$	250
Cr$_2$O$_3$ activated to generate coordinatively unsaturated Cr^{3+}	$H_2C_2H_4 \rightarrow C_2H_6$	-100
	$H_2 + D_2 \rightarrow 2HD$	-180
	$D_2 + C_3H_8 \rightarrow C_3H_7D + HD$	200
Fe	$3H_2 + N_2 \rightarrow 2NH_3$	400†
Cu^+/ZnO	$2H_2 + CO \rightarrow CH_3OH$	350†
Pt, Cr_2O_3	Heptane $\rightarrow$ toluene + $4H_2$	450
Pt, Cu	Acetone + $H_2 \rightarrow$ 2-propanol	75
	Polymerization	
Cr^{2+}/SiO$_2$	$C_2H_4 \rightarrow$ linear polyethylene	50
	Olefin metathesis	
Mo^{4+}/Al$_2$O$_3$	$2C_3H_6 \rightarrow C_2H_4 + CH_3CH=CHCH_3$	50
	Oxidation	
	$2H_2 + O_2 \rightarrow 2H_2O$	0–200
Pt, many oxides of transition metals	$2CO + O_2 \rightarrow 2CO_2$	50–200
	$CH_4 + 2O_2 \rightarrow CO_2 + 2H_2O$	200–350
Ag	$2C_2H_4 + O_2 \rightarrow$ ethylene oxide	250
Bismuth molybdate	$C_3H_6 + NH_3 + 3/2 O_2 \rightarrow CH_2=CHCN + 3H_2O$	450
V_2O_5/SiO$_2$	Naphthalene + $O_2 \rightarrow$ phthalic anhydride	350
Fe_3O_4	$H_2O + CO \rightarrow H_2 + CO_2$	450

*The lowest temperature at which significant yields are obtained in a flow reactor. °C = (°F − 32)/1.8.
†Because of the small value of the equilibrium constants at the operating temperatures, these reactions are run at about 100 atm (10 megapascals) in order to get adequate conversions.

reacting with molecules in the gas or liquid phases to form chemisorbed intermediates, as in reaction ∗ + A → ∗A. For example, the atoms at the surface of a crystallite of platinum must be coordinatively unsaturated. There has been interest in applying the studies of surface chemistry and physics on particular crystal surfaces—(111), (110), or (100)—of such metals as platinum to the interpretation of catalytic reactions and chemisorption on metals.

High catalytic activity for a given reaction requires adsorption of the reactants at *cus* sites to be of intermediate strength. If the adsorption is too weak, the reactants are unlikely to be activated; if too strong, A∗ will be unreactive. With good catalysts, then, it is likely that there will be other compounds which adsorb more strongly than the reactants, with partial or complete blockage of the catalytic reaction. Such compounds are said to be poisons. For example, transition metals are poisoned by soft bases like R_2S, R_3P, and CO. Poisoning is common in all types of catalysis.

Mechanism. The kinetics of heterogeneous catalytic reactions are often rather complicated. In general, Rate = (amount of catalyst)$f(T, P_i,$ or $C_i)$, where T is the temperature of the catalyst, P_i is the partial pressure of component i, and C_i is the concentration of component i. The index i covers reactants and products. A reaction having a rate-limiting process. A∗ + B∗ → C∗, will have a rate proportional to $\theta_A \theta_B$ where θ_A is the fraction of surface sites represented by ∗ which have been converted to A∗, and θ_B, that converted to B∗. The rate in terms of the θ_i's (the fraction of the surface covered by A and B) can be related to the P_i's through adsorption isotherms for the components.

Considerable work has been aimed at the determination of the chemical mechanism of particular catalytic reactions. Such work has depended upon kinetics, isotopic tracers, stereochemistry, and structure variation in reactants. A simple example is that which is commonly accepted for the hydrogenations of olefins, the Horiuti-Polanyi mechanism as illustrated for ethylene in reactions (4). It is as yet unclear which formulation for the chemisorption of ethylene is

or

$$H_2C=CH_2 + 2* \rightarrow H_2C-CH_2 \quad \quad \quad (4)$$
$$\quad \quad \quad \quad \quad \quad \quad \quad * \quad \quad *$$

$$H_2C=CH_2 + * \rightarrow H_2C-CH_2$$
$$\quad \quad \quad \quad \quad \quad \quad \quad \quad \downarrow$$
$$\quad \quad \quad \quad \quad \quad \quad \quad \quad *$$

most appropriate. The geometries of the two forms would be nearly the same. The adsorption of ethylene is accompanied by the dissociative chemisorption of hydrogen reaction (5), followed by reactions (6) and (7). Much still remains to be learned about the details of catalytic mechanisms. *See* Homogeneous catalysis.

$$H_2 + 2* \rightarrow 2H* \quad (5) \quad \quad *H_2C-CH_2* + H* \rightarrow *CH_2CH_3 + 2* \quad (6)$$

$$*CH_2CH_3 + H* \rightarrow CH_3CH_3(g) + 2* \quad (7)$$

PHASE-TRANSFER CATALYSIS
R. Alan Jones

A process in which the rate of a reaction occurring in a two-phase organic-water system is enhanced by addition of a compound that helps transfer the water soluble reactant across the interface to the organic phase.

An important factor, which contributes to the slowness of many organic reactions, is the lack of homogeneity of the reaction mixture. This is particularly the case with nucleophilic substitution reactions (1), where RX is an organic reagent and Nu^- is the nucleophilic reagent. The

$$RX + Nu^- \longrightarrow RNu + X^- \quad (1)$$

nucleophilic reagent is frequently an inorganic anion, which is soluble in water in which the organic substrate is insoluble, but is insoluble in the organic phase. The encounter rate between Nu^- and RX is consequently low, as they can only meet at the interface of the heterogeneous system. The water-soluble anion is also frequently highly solvated by water molecules, which stabilize the anion and thus reduce its nucleophilic reactivity. These problems have been overcome in the past by the use of polar aprotic solvents, which will dissolve both the organic and inorganic reagents, or by the use of homogeneous mixed-solvent systems, such as water:ethanol or water:dioxan. Homogeneous reaction systems can also be established by the use of surfactants, and the interfacial area can be increased by rapid agitation or mechanical emulsification. Although these procedures may increase the rate of reaction, they have disadvantages, such as difficulties in the isolation and purification of the products and, on an industrial scale, they can be costly in terms of solvents and energy.

Phase-transfer catalysis involves the transportation of the inorganic anion, Nu^-, from the aqueous phase into the organic phase by the formation of a nonsolvated ion-pair with a cationic phase-transfer catalyst, Q^+. It was originally believed that this process involved the formation of the ion pairs in the aqueous phase, followed by their partition between the aqueous and organic phases. A more current explanation is that with highly lipophilic catalysts, the reactive ion pair $[Q^+Nu^-]$ is formed at the interface between the aqueous and organic phases, followed by rapid transportation into the bulk of the organic phase **(Fig. 1)**. The rate of the reaction is enhanced, as the encounter rate of the nucleophile, Nu^-, with the organic reagent, RX, in the single phase will be significantly higher than at the interface. Moreover, as the anion is transferred without water of solvation, its nucleophilic reactivity can be considerably higher in the organic phase than in the aqueous phase. Rate enhancements of greater than 10^7 have thus been observed.

The efficiency of the catalyst depends upon the ability of the cation, Q^+ to transfer the anion across the interface of the heterogeneous system; that is, the rate should be proportional to the partition coefficient of the ion pair $[Q^+Nu^-]$ and the water solubility of $[Q^+X^-]$ should be higher than that of $[Q^+Nu^-]$. It has been found that bulky quaternary ammonium cations, R_4N^+, such as tetra-*n*-butylammonium and methyltrioctylammonium ions, and phosphonium cations,

Fig. 1. Formation and transportation of the reactive ion pair in phase-transfer catalysis.

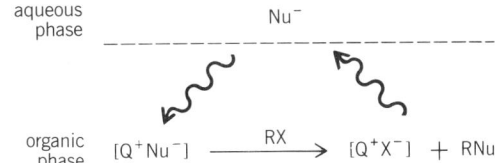

R_4P^+, such as tetra-n-butylphosphonium cations, have a high propensity to form "strong" ion pairs, which have a high solubility in organic solvents, and are excellent phase-transfer catalysts.

As the partition coefficient of the ion pair is a significant factor in the catalytic effect, the choice of the organic phase is important. The most commonly used organic phases are the semi-polar solvents, dichloromethane and 1, 2-dichlorobenzene, and the less polar solvent, toluene.

An alternative phase-transfer catalytic procedure utilizes polyethers, which are capable of complexing alkali metal cations. Ion pairs of the complexed cation and the nucleophilic anion are formed and are transported across the water: organic phase interface in the same manner as the quaternary ammonium and phosphonium salts. Examples of typical polyether catalysts are shown in **Fig. 2**a. They may be acyclic (glymes, I), or monocyclic (crown ethers, II), or polycyclic (cryp-

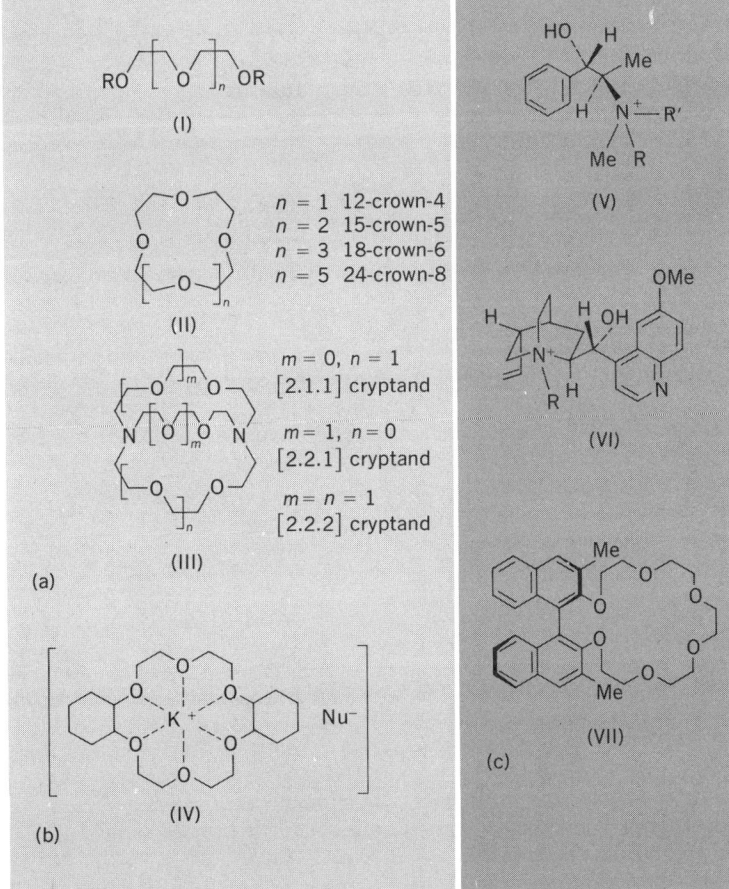

Fig. 2. Structures of some species involved in phase-transfer catalysis. (a) Typical polyether catalysts used in phase-transfer catalyst procedure. (b) Typical complex formed with a polyether catalyst. (c) Chiral catalysts used in asymmetric induction.

tands, III). The most commonly used polyether catalysts are the crowns, such as dicyclohexanono-18-crown-6, which complex readily with potassium and sodium ions to form complexes of the type IV (Fig. 2b).

Applications. Phase-transfer catalysts can be used to enhance reaction rates in nucleophilic substitution reactions, oxidation-reduction reactions, generation of reactive intermediates, and asymmetric induction.

Nucleophilic substitution reactions. By using phase-transfer catalysis, it is possible to enhance the rate of most nucleophilic substitution reactions (1) with a wide range of nucleophilic anions.

Kinetic studies have shown that the nucleophilicity of the nonhydrated halide anions, produced upon dissolution of the quaternary -onium halides in organic solvents, follow the order F > Cl > Br > I. Under these conditions the fluoride ion is a strong base and can be used in preference to the hydroxide ion in base-catalyzed reactions. *See Steric effect.*

Oxidation and reduction reactions. Purple benzene (dichloromethane) and yellow benzene (dichloromethane) are readily obtained by the dissolution of permanganate or dichromate ions, respectively, in the otherwise colorless solvents by using phase-transfer catalysts, and the solutions have been used for a wide range of oxidation reactions in high yields with minimal side reactions. Solid tetraalkylammonium permanganates and dichromates are potentially explosive.

Hydride reductions have been conducted in organic solvents by using tetraalkylammonium borohydrides.

Diborane, B_2H_6, produced from the reaction of tetraalkylammonium borohydride in dichloromethane with simple alkyl halides [reaction (2)] has an activity similar to that of diborane in

$$2[Q^+BH_4^-] + 2CH_3I \longrightarrow 2CH_4 + 2[Q^+I^-] + B_2H_6 \qquad (2)$$

ether and can be used for all the usual reduction and hydroboration reactions. *See Oxidation-reduction.*

Generation of reactive intermediates. Reactive carbanions, which normally require strictly anhydrous reaction conditions for their preparation, can be generated by using the phase-transfer catalytic technique [reaction (3)], where Z = acyl, alkoxycarbonyl, cyano, or nitro, and R = H, alkyl, or aryl.

$$RCHZ_2 \text{(org)} + HO^- \text{(aq)} \xrightarrow{[R'_4N^+X^-]} [RCZ_2^- R'_4N^+] \text{(org)} \xrightarrow{R''X} \begin{array}{c} R \\ | \\ R''-C-Z_2 \\ \text{(org)} \end{array} \qquad (3)$$

Similarly, other systems, which are labile in the presence of water such as RCOCN, can be synthesized by the phase-transfer catalytic procedure.

Tetraalkylammonium hydroxides react with chloroform to yield, initially, the trichloromethyl carbanion, which loses a chloride ion to generate the highly reactive dichlorocarbene [reaction (4)]. It has been proposed that the abstraction of the proton from the chloroform occurs at the

$$CHCl_3 \text{(org)} \xrightarrow{HO^-\text{(aq)}} CCl_3^- \text{(interface)} + H_2O \xrightarrow{[Q^+Cl^-]} [Q^+CCl_3^-] \text{(org)} \qquad (4)$$
$$\downarrow$$
$$[Q^+Cl^-] + \ddot{C}Cl_2$$
$$\text{(org)} \quad \text{(org)}$$

interface and that the role of the catalyst is the transfer of the trichloromethyl carbanion into the bulk of the organic solvent. The dichlorocarbene reacts rapidly and in high product yield with a wide range of organic substrates.

Other dihalogenocarbenes, and vinylidene and vinyl carbenes have been obtained by phase-transfer–catalyzed procedures.

Asymmetric induction. Chiral ammonium catalysts having both a chiral quaternary nitrogen atom and chirality within the carbon skeleton, such as V and VI (Fig. 2c), have been used successfully in the stereochemical control of reactions involving carbonyl-containing compounds.

A critical feature of the chiral catalyst is the presence of a β-hydroxyethyl substituent on the chiral nitrogen atom, which can hydrogen-bond with the carbonyl group of the organic substrate, thereby preferentially presenting one "face" of the substrate for reaction. Reactions in which an enatiomeric excess is >60% have been observed.

Stereochemical control of organic reactions has also been accomplished by using chiral crown ethers, such as VII (Fig. 2c).

Solid-state: liquid-phase reactions and triphase catalysis. The solubilization of solid inorganic salts by quaternary ammonium catalysts and by crown ethers in organic solvents is a simple extension of liquid:liquid phase-transfer catalysis and is widely used.

In a further elaboration of liquid:liquid phase-transfer catalysis, the quaternary ammonium group or the crown ether is attached to a polymeric support. The efficiency of the triphase system is generally lower than that of the conventional phase-transfer process, but the procedure has potential advantages over the two-phase system in that the polymeric catalyst is readily separated from the liquid phases and, on an industrial scale, there is the possibility of continuous phase-transfer–catalyzed processes. SEE CATALYSIS; HOMOGENEOUS CATALYSIS; HETEROGENEOUS CATALYSIS; STEREOCHEMISTRY.

Bibliography. E. V. Dehmlow and S. S. Dehmlow, *Phase-Transfer Catalysis*, 2d ed., 1983; C. M. Starks, *J. Amer. Chem. Soc.*, 93:195–199, 1971; C. M. Starks and C. Liotta, *Phase-Transfer Catalysis: Principles and Techniques*, 1978; W. P. Weber and G. W. Gokel, *Phase-Transfer Catalysis in Organic Synthesis*, 1977.

BUFFERS
A. M. HARTLEY

Solutions selected or prepared to minimize changes in hydrogen ion concentration which would otherwise tend to occur as a result of a chemical reaction. In general, chemical buffers are systems which, once constituted, tend to resist further change due to external influences. Thus it is possible, for example, to make buffers resistant to changes in temperature, pressure, volume, redox potential, or acidity. The commonest buffer in chemical solution systems is the acid-base buffer.

Chemical reactions known or suspected to be dependent on the acidity of the solution, as well as on other variables, are frequently studied by measurements in comixture with an appropriate buffer. For example, it may be desirable to investigate how the rate of a chemical reaction depends upon the hydrogen ion activity (pH). This is accomplished by measurements in several buffer systems, each of which provides a nearly constant, different pH. Alternatively, it may be desirable to measure the effects of other variables on a pH-sensitive system, by stabilizing the pH at a convenient value with a particular buffer.

Effectiveness. Buffer action depends upon the fact that, if two or more reactions coexist in a solution, then the chemical potential of any species is common to all reactions in which it takes part, and may be defined by specification of the chemical potentials of all other species in any one of the reactions. To be effective, a buffer must be able to respond to an increase as well as a decrease of the species to be buffered. In order to do so, it is necessary that the proton transfer step of the buffer be reversible with respect to the species involved, in the reaction to be buffered. In aqueous solution the proton transfer between most acids, their conjugate bases and water, is so rapid and reversible that the dominant direct source of protons for a chemical reaction is H_3O^+, the hydronium ion.

An acid-base buffer reaction in water is defined by reversible reaction (1), and the equilibrium constant K_a shown by Eq. (2).

$$BH^+ + H_2O \rightleftharpoons B + H_3O^+ \quad (1) \qquad K_a = \frac{[H_3O^+][B]}{[BH^+][H_2O]} \quad (2)$$

In Eq. (2) the square brackets designate the activity of the species involved. In normal concentrations of buffer (0.1 mole/liter) the activity of the solvent water is essentially constant and approximately that of pure water (55.5 *M*). Thus the position of the equilibrium may be defined by specifying the activity of any two of the three variable species in reaction (1). Normally this is by means of the equilibrium expression shown as Eq. (3) which, upon converting to a logarithmic form, can be reduced to Eq. (4). Here f is the fraction of the total buffer concentration, (BH^+) +

$$[H_3O^+] = K_a \frac{[BH^+]}{[B]} \qquad (3) \qquad pH = pK_a - \log\frac{(1-f)}{f} - \log\frac{\gamma_{BH+}}{\gamma_B} \qquad (4)$$

(B) existing as B, and γ is the activity coefficient relating activity a to concentration X. This relation is shown by Eq. (5). Thus a buffer pH is approximately defined by the dissociation con-

$$a_x = \gamma_x(X) \qquad (5)$$

stant K_a of the weak acid system and the ratio of acid to conjugate base concentrations. However, the third term in Eq. (4) indicates that the pH is dependent on the change in activity coefficients with concentration. Effects of this dependency may be eliminated in practice by providing a high and essentially invariant ionic environment in the form of an added pH-neutral strong electrolyte such as KNO_3 or NaCl.

Buffer capacity π is defined as the change in added H_3O^+ necessary to produce a given change in pH, $d[H_3O^+]/dpH$. Since the buffer comes to equilibrium with added H_3O^+, $1/\pi$ may also be defined as dpH/df. Inspection of Eq. (4) shows $1/\pi$ to be a minimum when $f = f$; hence a given buffer system has its highest capacity in a solution composed of equal parts BH^+ and B, and the capacity is directly proportional to the concentrations of BH^+ and B. For these reasons buffers are normally used at concentrations 10–100 times higher than the system to be controlled and, if possible, are selected so that the desired pH is approximately equal to pK_a for the buffer system. As a general rule, weak acid systems are not used to stabilize solutions whose pH is more then 2 pH units removed from pK_a, to ensure that the ratio of BH^+ to B will fall in the range 100–0.01.

Water as solvent. Buffers are particularly effective in water, because of the unusual properties of water as a solvent. Its high dielectric constant (80) tends to promote the existence of formally charged ions (ionization). Because it has both an acidic (H) and a basic (O) group, it may form bonds with ionic species leading to an organized sheath of solvent surrounding an ion (solvation). Water also tends to self-ionize to form its own conjugate acid-base system as shown by Eq. (6b), in which K_{ap} is the autoprotolysis constant. The strength of an acid (or base) in solvent

$$2H_2O \rightleftharpoons H_3O^+ + OH^- \qquad (6a) \qquad K_{ap} = [H_3O^+][OH^-] = 10^{-14} \qquad (6b)$$

water cannot be separated from reaction (6a), and the familiar acid (or base) dissociation equilibrium reaction of reaction (1). Strong acids are those for which the K_a of Eq. (2) is very large; weak acids do not completely transfer the proton to water. The strongest acid which may exist in water is H_3O^+; the strongest base is OH^-. Thus, the maximum range of acid level which a solvent can support is governed by its own acid-base properties. In water this range is 14 pH units, or 14 orders of magnitude change in activity of H_3O^+.

The mechanism of buffer action may be regarded as a sequence of the proton transfer steps implied in reaction (1) coupled with reaction (6a). For example, the result of the chemical production or deliberate addition of an acid, HA, is to cause the water autoprotolysis reaction and the buffer acid reaction to respond to the change shown by reactions (7) and (8). Addition of a base would be accommodated by the reverse of (7) and (8). The effect of adding HA depends on the

$$HA + H_2O \rightleftharpoons H_3O^+ + A^- \qquad (7) \qquad H_3O^+ + B \rightarrow BH^+ + H_2O \qquad (8)$$

position of the equilibrium shown in reaction (7); buffer capacity π is usually defined in terms of H_3O^+ added because H_3O^+ is the strongest possible acid in aqueous solution, and would tend to create the maximum possible change in solution pH per mole of added acid. If HA is relatively weak so its degree of dissociation, in reaction (1), is small, its effective H_3O^+ addition may be calculated through Eq. (9), where C_a is the concentration of added HA. A simple calculation using

$$[H_3O^+] \cong \sqrt{K_aC_a} \qquad (9)$$

Eq. (9) shows that a given buffer solution will undergo the same change in pH for the addition of 0.1 mole/liter of a weak acid such as acetic acid ($K_a = 10^{-5}$) as for the addition of 0.01 mole/liter of strong acids such as HCl, $HClO_4$, or HNO_3.

In studies of rates of chemical reactions at constant pH, it is necessary that the proton transfer processes of the buffer acid and base and the solvent be rapid with respect to the primary reaction. The phosphate (HPO_4^{2-}–PO_4^{3-}) and carbonate (HCO_3^-–CO_3^{2-}) systems, among others, sometimes give anomalous effects because this condition may not be obtained. Buffer rate effects are manifested in different reaction rates for a chemical system in two different buffers or otherwise identical ionic strength and nominal (equilibrium) pH. Later evidence seems to suggest that buffers of low-change type, for example, NH_3–NH_4^+, react more rapidly than high-charge types such as HPO_4^{2-}–PO_4^{3-}. *See* IONIC EQUILIBRIUM.

Bibliography. J. J. Cohen and J. P. Kassirer, *Acid-Base*, 1982; H. A. Laitinen, *Chemical Analysis*, 2d ed., 1975; D. D. Perrin and B. Dempsey, *Buffers for pH and Metal Ion Control*, 1979.

INHIBITOR
LEE R. MAHONEY

A substance which is capable of stopping or retarding a chemical reaction. To be technically useful, such compounds must be effective in low concentrations, usually under 1%. The type of reaction which is most easily inhibited is the free-radical chain reaction. The study of inhibitor action is often used as a diagnostic test for free-radical chain character of a reaction. Vinyl polymerization and autoxidation are two important examples of the class. Another reaction type for which inhibitors have been found is corrosion, particularly in aqueous systems. The economic importance of corrosion inhibition can scarcely be overestimated. An understanding of inhibitor action depends on an understanding of the processes which are to be interrupted.

Inhibition of vinyl polymerization. This type of inhibitor action must be considered in terms of the accepted mechanism for the polymerization process, which may be summarized as reaction sequence (1). The symbol P represents a catalyst, often a peroxide, R• is a free radical

$$\left.\begin{array}{l} P \rightarrow 2R\bullet \\ R\bullet + M \rightarrow \sim\sim M\bullet \end{array}\right\} \text{initiation}$$
$$\sim\sim M\bullet + M \rightarrow \sim\sim M\bullet \text{ propagation} \tag{1}$$
$$2\sim\sim M\bullet \rightarrow \text{termination by dimerization or disproportionation}$$

derived from the catalyst, M is a monomer, and ∼∼M• is a growing polymer chain. The polymerization will be stopped or inhibited if some added substance (an inhibitor) reacts more readily than does the monomer with R• to yield a product which will not sustain the polymerization. Every reaction chain is stopped until the inhibitor is consumed. If the added substance (a retarder) is somewhat less reactive, the monomer can compete more successfully for the initiating radicals, so that the result is retardation rather than total inhibition. The difference between inhibition and retardation is one of degree rather than of kind.

Phenolic compounds and quinones. These interact with an initiating radical or a growing polymer chain either by hydrogen atom abstraction or by radical addition to an unsaturated linkage. These interactions are represented by sequence (2).

The phenoxy radicals produced are stabilized by resonance and are not sufficiently reactive to add to the vinyl linkage of another monomer molecule. The usual fate of these radicals is further hydrogen atom loss by reaction with a second polymer radical or by disproportionation with another phenoxy radical to yield a quinone and a hydroquinone, both of which may continue to act as inhibitors. All the reaction possibilities shown have been demonstrated experimentally. Although the efficiencies of phenols and quinones for the interruption of polymerizations vary with the structures of these molecules, they may be classed as inhibitors. Aromatic amines react in a similar manner.

Nitroaromatics. As typified by trinitrobenzene, nitroaromatics function as retarders rather than as inhibitors in most polymerizations. It is necessary, however, to consider the specific reaction involved. Thus polynitroaromatics inhibit the polymerization of vinyl acetate, retard that of styrene, and have no effect on that of methyl methacrylate. No clear-cut mechanism has been established for the interaction of nitro compounds with free radicals.

Monomers. Both monomers and the radicals derived from them differ greatly among themselves in reactivity. Thus, although certain monomers may copolymerize with one another, others may actually function as inhibitors. Styrene and vinyl acetate, for example, both polymerize well when alone. Styrene, however, inhibits the polymerization of vinyl acetate. This occurs because the vinyl acetate radical and the styrene monomer are highly reactive, whereas the styrene radical and the vinyl acetate monomer are not. A small amount of styrene added to vinyl acetate will rapidly react with any vinyl acetate radicals formed when polymerization is initiated. The resulting styrenelike radical will react only very slowly with the vinyl acetate monomer. In the overall process the chain-carrying radical is converted to one which is too unreactive to carry on the chain.

Autoinhibition. This action, sometimes called allylic termination, is exhibited by monomers which contain the highly reactive allylic C—H linkage. Free radicals are capable of hydrogen atom abstraction from copresent molecules as well as of addition to an unsaturated linkage. The ease with which this abstraction reaction is carried out by a given radical is a function of the reactivity of the C—H linkage which is attacked. The reactivity of radicals containing these C—H linkages increases in the order aryl < primary alkyl < secondary alkyl < tertiary alkyl < allyl < benzyl. Because of this high reactivity, a monomer containing an allylic C—H, allyl acetate, for example, functions as its own retarder, as shown by sequence (3). The resonance-stabilized allylic

$$\sim\!\!M\cdot\, + \,CH_2\!\!=\!\!CHCH_2OAc \swarrow \searrow \sim\!\!MCH_2CHCH_2OAc \qquad \sim\!\!MH + CH_2\!\!=\!\!CHCHOAc \tag{3}$$

radical will react with the monomer only very slowly. The predominant further reaction is dimerization. Not only is polymerization slowed in this case, but also the molecular weight of the polymer formed in the reaction is low.

Miscellaneous inhibitors. Oxygen, iodine, and nitric oxide interact rapidly with free radicals to yield stable products. Inclusion of these materials in polymerizing systems thus leads to effective inhibition. It is of particular interest that oxygen will copolymerize, under carefully controlled conditions, with certain monomers, styrene, for example, to yield high-molecular-weight polymeric peroxides. The repeating unit is shown in formula (4). Iodine and nitric oxide have been

$$[-\!\!\operatorname{CH}(C_6H_5)CH_2OO-\!]_n \tag{4}$$

used extensively to detect and in some cases to identify alkyl free radicals in nonchain as well as in chain reactions. This method has proved to be of great value in defining primary processes in photochemical decompositions of aldehydes and ketones. *See* POLYMERIZATION.

Inhibition of corrosion. Metallic corrosion in conducting media is electrochemical in nature. Local electrolytic cells are set up because of the presence of impurities, crystal lattice imperfections, or strains within the metal surface. The result is dissolution of the metal from the anodic regions. Corrosion inhibitors now in use may operate at the anodes or the cathodes or provide physical protection over the entire surface.

Anodic inhibitors. These are mild oxidants which reduce the open-circuit potential difference between local anodes and cathodes and increase the polarization of the former. Sodium chromate and sodium nitrite are most commonly used. The former is used in air conditioners, refrigeration systems, automobile radiators, power plant condensers, and similar equipment. Sodium nitrite finds special use in the protection of petroleum pipelines. It is effective even on rusty, mild steel. An extension of the nitrite type is the use of nitrite salts of secondary amines as vapor-phase inhibitors. The inclusion of a salt such as dicyclohexyl-ammonium nitrite with a packaged steel object provides effective protection against corrosion.

Cathodic inhibitors. Compounds such as calcium bicarbonate and sodium phosphate, in an aqueous medium, deposit on metal surfaces films that provide physical protection against corrosive attack.

Organic inhibitors. These are usually long-chain aliphatic acids and the soaps which are derived from them. Adsorption of these compounds on metal surfaces gives a hydrophobic film which protects the metal from corrosion by many agents. As little as 0.1% of palmitic acid, for example, is sufficient to protect mild steel from attack by nitric acid. SEE CATALYSIS; FREE RADICAL.

IONIZATION
GLENN H. MILLER

The process by which an electron is removed from an atom, molecule, or ion. This process is of basic importance to electrical conduction in gases, liquids, and solids. In the simplest case, ionization may be thought of as a transition between an initial state consisting of a neutral atom and a final state consisting of a positive ion and a free electron. In more complicated cases, a molecule may be converted to a heavy positive ion and a heavy negative ion which are separated.

Ionization may be accomplished by various means. For example, a free electron may collide with a bound atomic electron. If sufficient energy can be exchanged, the atomic electron may be liberated and both electrons separated from the residual positive ion. The incident particle could as well be a positive ion. In this case the reaction may be considerably more complicated, but may again result in a free electron. Another case of considerable importance is the photoelectric effect. Here a photon interacts with a bound electron. If the photon has sufficient energy, the electron may be removed from the atom. The photon is annihilated in the process. Other methods of ionization include thermal processes, chemical reactions, collisions of the second kind, and collisions with neutral molecules or atoms. SEE ELECTRODE POTENTIAL.

IONIC EQUILIBRIUM
THOMAS F. YOUNG

An equilibrium in a chemical reaction in which at least one ionic species is produced, consumed, or changed from one medium to another.

Types of equilibrium. A few examples can illustrate the wide variety of types of ionic equilibrium which are known.

Dissolution of an un-ionized substance. The dissolution of hydrogen chloride (a gas) in water (an ionizing solvent) can be used to illustrate this type. Reactions (1), (2), and (3) all

$$HCl(g) \rightleftharpoons H^+ + Cl^- \quad (1) \qquad HCl(g) + H_2O \rightleftharpoons H_3O^+ + Cl^- \quad (2) \qquad HCl(g) + 4H_2O \rightleftharpoons H_9O_4^+ + Cl^- \quad (3)$$

represent exactly the same equilibrium. Equation (1) ignores the hydration of the proton and is preferred for many purposes when the hydration (or solvation) of the proton is irrelevant to a particular discussion. Equation (2) is written in recognition of the widely held belief that free protons do not exist in aqueous solution. Equation (3) indicates that another three molecules of water are very firmly bound to the H_3O^+ ion (the hydronium ion). There is no implication in Eq. (3), however, that the total number of molecules of water attached to, or weakly affected by, the hydronium ion may not be considerably larger than three.

Not much is known about the solvation of ions, although it has been proved that each chromic ion, Cr^{3+}, in dilute aqueous solution holds at least six water molecules. Only a few other similar data have been clearly established. Hence equations for the other examples cited below are written without regard to solvation, except that the hydrogen ion is usually written in accordance with common practice as H_3O^+.

Dissolution of a crystal in water. The dissociation of solid silver chloride, reaction (4), illustrates this type of equilibrium.

$$AgCl(crystal) \rightleftharpoons Ag^+ + Cl^- \qquad (4)$$

Dissociation of a strong acid. Nitric acid, HNO_3, dissociates as it dissolves in water, as in reaction (5). At 77°F (25°C) about one-half the acid is dissociated in a solution containing 10

$$HNO_3 + H_2O \rightleftharpoons H_3O^+ + NO_3^- \qquad (5)$$

(stoichiometric) moles of nitric acid per liter.

Dissociation of an ion in water. The bisulfate ion, HSO_4^-, dissociates in water, as in reaction (6). About one-half the HSO_4^- is dissociated in an aqueous solution containing about

$$HSO_4^- + H_2O \rightleftharpoons H_3O^+ + SO_4^{2-} \qquad (6)$$

0.011 mole of sulfuric acid per liter at 77°F (25°C).

Dissociation of water itself. In pure water at 77°F (25°C) the concentration of each ion is about 10^{-7} mole/liter, but increases rapidly as temperature is increased. This equilibrium is represented by reaction (7) below.

$$2H_2O \rightleftharpoons H_3O^+ + OH^- \qquad (7)$$

Formation of a complex ion. In water or in a mixture of fused (chloride) salts, complex ions, such as $ZnCl_4^{2-}$, may be formed, as in reaction (8).

$$Zn^{2+} + 4Cl^- \rightleftharpoons ZnCl_4^{2-} \qquad (8)$$

Dissociation of a weak acid. In water acetic acid dissociates to form hydrogen (hydronium) ion and acetate ion, as in reaction (9).

$$CH_3CO_2H + H_2O \rightleftharpoons H_3O^+ + CH_3CO_2^- \qquad (9)$$

Electrochemical reaction. Reaction (10) takes place "almost reversibly" when the equi-

$$1/2 H_2(g) + AgCl(s) + H_2O(l) \rightleftharpoons H_3O^+ + Cl^- + Ag(s) \qquad (10)$$

librium shown exists. A small current is allowed to flow through an electric cell consisting of an aqueous solution of HCl saturated with silver chloride, a hydrogen electrode, and a silver electrode. Saturation is maintained by an excess of solid silver chloride which for convenience is sometimes mixed with the silver or plated, as a coating, on the metal. The electrode is then called a silver–silver chloride electrode.

Many additional types of equilibria could be mentioned, including those reactions occurring entirely in the gaseous phase and those reactions occurring between substances dissolved in two immiscible liquids.

Quantitative relationships. Each reaction obeys an equilibrium equation of the type shown as Eq. (11).

$$\frac{[H^+][Cl^-]}{[HCl(g)]} \frac{f_+ f_-}{\gamma_g} = Q_c Q_f = K \qquad (11)$$

The activity coefficient γ_g can be ignored here because it is very nearly unity. The terms f_+ and f_- are the respective activity coefficients of H^+ and Cl^- but cannot be determined separately. Their product can be determined experimentally and can also be calculated theoretically for very dilute solutions by means of the Debye-Hückel theory of interionic attraction. Because γ_g and $f_+ f_-$ are nearly unity, Eq. (11) demands that the pressure of HCl gas above a dilute aqueous solution be proportional to the square of the concentration of the solute. Q_c is called the concentration quotient and Q_f the quotient of activity coefficients.

Similarly, the dissociation of acetic acid obeys Eq. (12), where f_u is the activity coefficient

$$\frac{[H^+][CH_3CO_2^-]}{[CH_3CO_2H]} \frac{f_+ f_-}{f_u} = Q_c Q_f = K \qquad (12)$$

of the un-ionized acetic acid. Early work on electrolytes revealed that Q_c, the concentration quotient, was constant within the limits of accuracy attainable at the time. Later work revealed that the measured concentration quotient Q_c first increases as concentration is increased from very small values and then decreases sharply. The initial increase is due largely to the electrical forces between the ions which reduce the product f_+f_-. There is some evidence that the subsequent decrease in Q_c and the concomitant rise in Q_f are due to removal of some of the monomeric acetic acid by the formation of dimeric acetic acid, $(CH_3CO_2H)_2$. The fact is, however, that knowledge concerning activity coefficients in solutions other than very dilute ones is not yet understood. Even the experimental methods for the measurement of the molecular species involved in some equilibria were not evolved until recently. SEE ELECTROLYTIC CONDUCTANCE.

Bibliography. G. M. Barrow, *Physical Chemistry*, 4th ed., 1979; J. Bjerrum et al., Stability constants of metal-ion complexes, with solubility products of inorganic substances, *Chem. Soc. London*, pts. 1 and 2, 1957–1958; I. Levine, *Physical Chemistry* 2d ed., 1983; H. Russotti, *The Study of Ionic Equilibria: An Introduction*, 1978.

OSCILLATORY REACTION
RICHARD M. NOYES

A chemical reaction in which some composition variable of a chemical system exhibits regular periodic variations in time or space. It is a basic tenet of chemistry that a closed system moves inexorably toward an unchanging state called chemical equilibrium. That motion can be described by the monotonic increase of entropy if the system is isolated, and by the monotonic decrease of Gibbs free energy if the system is constrained to constant temperature and pressure. Because of this universal restriction on what is possible in chemistry, it may appear bizarre when electrodes in a solution generate the oscillating potentials shown in **Fig. 1**.

The species taking part in a chemical reaction can be classified as reactants, products, or intermediates. The concentrations of reactants decrease. Intermediates are formed by some steps and destroyed by others. If there is only one intermediate, and if its concentration is always much less than the initial concentrations of reactants, this intermediate attains a stable steady state in

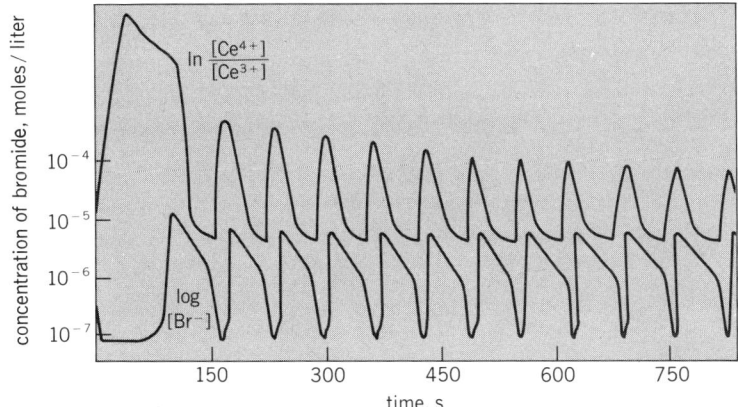

Fig. 1. Oscillatory behavior in a sulfuric acid medium containing potassium bromate, malonic acid, and cerous nitrate. The upper curve is the potential (in unspecified units) of a tungsten electrode and is related to the concentration ratio of cerium(IV)/cerium(III). The lower curve is the potential of an electrode sensitive to bromide ion, and the logarithmic scale at the left relates that potential to the absolute concentration of bromide. (*After R. J. Field, E. Körös, and R. M. Noyes, Oscillations in chemical systems, II. Thorough analysis of temporal oscillations in the bromate-cerium-malonic acid system, J. Amer. Chem. Soc., 94:8649–8664, 1972*)

which the rates of formation and destruction are virtually equal. The kind of oscillation reflected in Fig. 1 requires at least two intermediates which intereact in such a way that the steady state of the total system is unstable to the minor fluctuations present in any collection of molecules. The concentrations of the intermediates may then oscillate regularly, although the oscillations must disappear before the inevitable monotonic approach to equilibrium.

Periodic chemical behavior may be temporal in a uniform system as illustrated in Fig. 1; it may involve simultaneous temporal and spatial variations as in **Fig. 2**; or it may involve spatial periods in a static system as in **Fig. 3**.

Well-authenticated examples of periodic chemical behavior have been known for almost a century, but until the 1970s most chemists either did not know about them or deliberately ignored them. Interest has been developing rapidly, but most examples are only poorly understood. The phenomena are classified here according to types of chemical processes involved. Very different classification schemes may become more appropriate in the future. SEE CHEMICAL EQUILIBRIUM; ELECTRODE POTENTIAL; ENTROPY; GIBBS FUNCTION.

Redox oscillators. The systems whose chemistries are best understood all involve an element that can exist in several different oxidation states. Figure 1 illustrates the so-called Belousov-Zhabotinsky reaction, which was discovered in the Soviet Union in 1951. A strong oxidizing agent (bromate) attacks an organic substrate (such as malonic acid), and the reaction is catalyzed by a metal ion (such as cerium) that can exist in two different oxidation states.

As long as bromide ion (Br) is present, it is oxidized by bromate (BrO_3^-), as in reaction (1).

$$BrO_3^- + 2Br^- + 3H^+ \rightarrow 3HOBr \tag{1}$$

When bromide ion is almost entirely consumed, the cerous ion (Ce^{3+}) is oxidized, as in reaction (2). Reaction (2) is inhibited by Br^-, but when the concentration of bromide has been

$$BrO_3^- + 4Ce^{3+} + 5H^+ \rightarrow HOBr + 4Ce^{4+} + 2H_2O \tag{2}$$

reduced to a critical level, reaction (2) accelerates autocatalytically until bromate is being reduced by Ce^{3+} many times as rapidly as it was by Br^- when reaction (3) was dominant.

The hypobromous acid (HOBr) brominates the organic substrate to form bromomalonic acid (BrMA), as in reaction (3). Reaction (3) creates the bromide ion necessary to shut off fast reaction

$$2Ce^{4+} + BrMA \rightarrow 2Ce^{3+} + Br^- + \text{oxidized organic matter} \tag{3}$$

(2) and throw the system back to dominance by slow reaction (1).

As other redox oscillators become understood, they fit the same pattern of a slow reaction

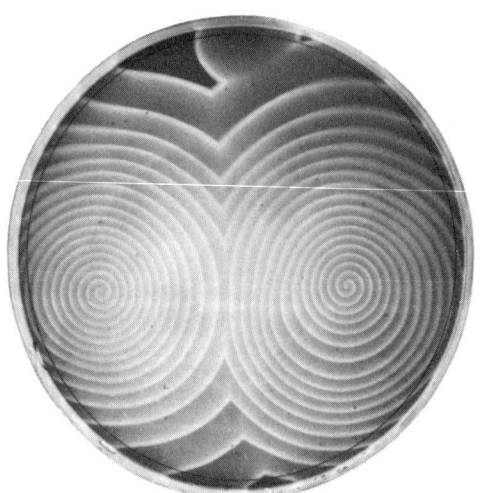

Fig. 2. Rotating spiral bands of oxidation in a thin layer of solution containing the same reagents as in Fig. 1 except that the redox indicator ferrous phenanthroline has been substituted for cerous ion. (*From R. J. Field and R. M. Noyes, Mechanisms of chemical oscillators: Conceptual bases, Acc. Chem. Res., 10:214–221, 1977*).

Fig. 3. Layers of crystallization from an initially uniform magma in the Skaergaard Intrusion in Greenland. (*From A. R. McBirney and R. M. Noyes, Crystallization and layering of the Skaergaard Intrusion, J. Petrol., 20:487–554, 1979*)

destroying a species that inhibits a fast reaction that can be switched on autocatalytically; the fast reaction then generates conditions to produce the inhibitor again.

Until 1982 the known examples involved positive oxidation states of Cl, Br, and I. Elements, like N, S, Cr, and Mn, which have many positive oxidation states may be able to participate in other oscillatory reactions.

The temporal traces in Fig. 1 were obtained with a homogeneous stirred solution. If the solution were unstirred but had a gradient in composition, oscillations in different regions would get out of phase and an apparent wave would traverse the medium much the way flashing lights cause a message to move along a theater marquee.

Figure 2 illustrates a still more complex situation. Each light curve is a region of dominance by reaction (2). A combination of reaction and diffusion triggers an advance outward perpendicular to the wavefront into the dark region dominated by reaction (1). Trigger waves annihilate each other when they meet, and Fig. 2 shows two spirals spinning in opposite directions. This kind of behavior has been suggested to explain the fibrillations when a human heart loses its rhythm and degenerates to uncoordinated local contractions that result in death if the condition is not rapidly reversed. *See* OXIDATION-REDUCTION.

Nucleation oscillators. If a solution contains a population of identical bubbles or crystals, the resulting steady state is unstable to perturbation. Because surface energy is relatively more important in small bubbles, they tend to dissolve while the larger ones grow. In principle, the equilibrium state consists of a single large bubble or crystal. If a chemical reaction produces a gas or a precipitate, periodicities may be observed. A quantitative treatment is complicated because the solution must become supersaturated before nuclei of the gas or solid phase are formed, and the initial nuclei are so small that they grow very slowly at first.

An example of a gas evolution oscillator is the dehydration of formic acid (HCO_2H) in concentrated sulfuric acid. As carbon monoxide escapes, several times a minute the solution foams up like a shocked glass of beer and then subsides.

Figure 3 illustrates a geologic formation on the east coast of Greenland. A large magma chamber cooled very slowly, and the initially uniform material crystallized in a pattern of regular layers.

Thermokinetic oscillators. Many chemical reactions give out heat, and rates are strongly dependent on temperature. If heat is not removed too rapidly from the reactor, reaction rate and temperature may couple to generate oscillations.

The known examples involve highly exothermic reactions like combustion or chlorination of organic compounds. No chemical mechanisms are yet understood in detail. In at least some examples, gradients of temperature and composition in space are important in addition to changes in local rate of reaction.

Reactions on surfaces. Many important industrial reactions take place on the surfaces of catalysts. The occurrence of such a reaction may temporarily alter that surface or its temperature. Such effects sometimes couple to generate oscillations in the rate of reaction. These oscillations may or may not be of value for the reaction being carried out. Specific examples are being studied actively by chemical engineers.

The surfaces of electrodes may also be influenced by processes taking place on them, and periodic changes in voltage or current are well precedented.

Bibliography. R. J. Field and R. M. Noyes, Mechanisms of chemical oscillators, *Acc. Chem. Res.*, 10:214–222, 273–280, 1977; O. Gurel, *Oscillations in Chemical Reactions*, 1983; H. Haken, *Synergetics*, 3d ed., 1983; G. Nicolis and I. Prigogine, *Self-Organization in Nonequilibrium Systems*, 1977; R. M. Noyes, Oscillations in homogeneous systems, *Berichte Bunsen Ges. Phys. Chem.*, 84:295–303, 1980.

OXIDATION-REDUCTION
Henry Taube

An important concept of chemical reactions which is useful in systematizing the chemistry of many substances. Oxidation can be represented as involving a loss of electrons by one molecule and reduction as involving an absorption of electrons by another. Both oxidation and reduction occur simultaneously and in equivalent amounts during any reaction involving either process.

Some important processes which involve oxidation are the rusting of iron or corrosion of metals in general, combustion of hydrocarbons, and the oxidation of carbohydrates (this takes place in a controlled manner in living cells). In each of the foregoing reactions the agent which is reduced is oxygen. Some common important reduction processes are the transformation of carbon dioxide to carbohydrates (this takes place in photosynthesis with water being oxidized), the winning of metals from oxides (carbon is often the reducing agent), electrodeposition of metals (this takes place at the cathode, and an equivalent amount of oxidation takes place at the anode), hydrogenation of fats and of coal, and the introduction of electronegative elements such as oxygen, nitrogen, or halogens into hydrocarbons.

Oxidation number. The oxidation state is a concept which describes some important aspects of the state of combination of the elements. An element in a given substance is characterized by a number, the oxidation number, which specifies whether the element in question is combined with elements which are more electropositive or more electronegative than it is. It further specifies the combining capacity which the element exhibits in a particular combination. A scale of oxidation numbers is defined by assigning to an oxygen atom in a molecule such as SO_4^{2-} the value of $2-$. That for sulfur as $6+$ then follows from the requirement that the sum of the oxidation numbers of all the atoms add up to the net charge on the species. The value of $2-$ for oxygen is not chosen arbitrarily. It recognizes that oxygen is more electronegative than sulfur, and that when it reacts with other elements it seeks to acquire two more electrons, by sharing or outright transfer from the electropositive partner, so as to complete a stable valence shell of eight electrons. For compounds of the halogens an analogous rule is followed, but when a halogen atom is in combination with atoms of a more electropositive element, the oxidation number is taken as $1-$ because only one electron needs to be added to the valence shell to yield a stable octet. The system amounts to a bookkeeping operation on the electrons, so that for this purpose the more electronegative partner is assigned some agreed upon stable electronic configuration, and after taking into account the total charge on the molecule, the net charge left on the electropositive partner is its particular oxidation number. When the combining capacity of an element toward another one is not completely exhausted in a particular combination, as is the case for oxygen in barium peroxide (BaO_2), the electrons shared between atoms of the same kind are evenly divided between them in carrying out the formal decomposition. Thus in the peroxide unit O_2^{2-}, the oxidation number of oxygen is $1-$. This is the net charge left on oxygen in the formal decomposition

$$[:\ddot{O}:\ddot{O}:]^{2-} = 2:\ddot{O}\cdot^{-}$$

The oxidation number by no means gives a complete description of the state of combination of an atom. Specifically, it is not designed to give the actual charge on an atom in a particular compound. Thus it makes no distinction between fluorine in HF, AlF$_3$, or NaF, even though the actual charges residing on the fluorine atoms in these three compounds are different.

The utility of the concept is based in part on just this feature because much of the chemistry of these substances can be understood when it is realized that each of them readily yields F$^-$, as is the case when they dissolve in water. The chemistry of the three substances, in regard to the component fluorine, is concerned with reactions of F$^-$. Although oxidation number is in some respects similar to valence, the two concepts have distinct meanings. In the substance H$_2$, the valence of hydrogen is 1 because each H makes a single bond to another H, but the oxidation number is 0, because the hydrogen is not combined with a different element. *See Valence*.

The systematization of chemistry based on the concept of oxidation number can be illustrated with reference to the chemistry of iodine. The usual oxidation states exhibited by iodine are $1-$, 0, $1+$, $5+$, and $7+$. Examples of substances corresponding to each oxidation state are

$7+$	IO$_4^-$, HIO$_4$, IF$_7$	$5+$	I$_2$O$_5$, IO$_3^-$, HIO$_3$, IF$_5$	$1+$	HIO, IO$^-$, ICl$_2^-$
0	I$_2$			$1-$	I$^-$, HI, NaI

When the oxidation number of an atom in a species is increased, the process is described as oxidation, no matter what reagent produces it; when a decrease in oxidation number takes place, the process is described as reduction, again without regard to the identity of the reducing agent. The term oxidation has been generalized from its historical meaning, implying combination with oxygen, to combination of an element with an element more electronegative than itself.

When classification by oxidation number is adopted, the reactions fall naturally into two classes. In the first class, no change in oxidation number takes place and, in the second, the class of oxidation-reduction reactions, changes in oxidation number do take place. Some examples of the first class are reactions (1)–(4).

$$I_2O_5 + H_2O \rightarrow 2HIO_3 \quad (1) \qquad HIO_3 + OH^- \rightarrow IO_3^- + H_2O \quad (2)$$

$$HOI + H^+ + 2Cl^- \rightarrow H_2O + ICl_2^- \quad (3) \qquad Hg^{2+} + 4I^- \rightarrow HgI_4^{2-} \quad (4)$$

Some samples of the second class are reactions (5)–(8). [In (8) it is implied that electrons

$$Cl_2 + 2I^- \rightarrow 2Cl^- + I_2 \quad (5) \qquad 2Fe^{3+} + 2I^- \rightarrow 2Fe^{2+} + I_2 \quad (6)$$

$$16H^+ + 2MnO_4^- + 10I^- \rightarrow 8H_2O + 2Mn^{2+} + 5I_2 \quad (7) \qquad 2I^- \rightarrow I_2 + 2e^- \quad (8)$$

are being extracted from I$^-$ by an anode in an electrolytic process.]

In reactions of the first class, some center regarded as positive undergoes a change in the nature of the groups associated with it, but provided that the group which replaces the electronegative portion is more electronegative than the center, there is no change in oxidation state. Reaction (3) describes the replacement of OH$^-$ on I$^+$ by Cl$^-$; both Cl and OH are more electronegative than I. In reactions of the second class, changes in oxidation number occur which may or may not be accompanied by changes in the state of association of the centers in question.

Reactions (5), (6), (7), and (8) illustrate the utility of the concept of oxidation number. A variety of reagents as different in their properties as Cl$_2$, Fe^{3+}, MnO$_4^-$, and an anode serve to bring about the change, or oxidation, of I$^-$ to I$_2$. However, their chemical individuality does not affect the state of the product iodine, and no group characteristic of the oxidizing agent is necessarily transferred in the net change. This situation obtains only for reactions in a strongly solvating medium such as water, which provides the groups that associate with the atom being oxidized or reduced. Thus, when the reactions take place in the solid, it is necessary to specify what iodide is being used, whether sodium iodide, NaI, or silver iodide, AgI, for example, and the properties of the reaction would be dependent on the choice.

In representing an element in a particular environment, it is often convenient to specify only the oxidation state, without attempting to identify all of the groups which are attached to the atom in question. Thus, the iron content of a solution made up by dissolving, say, ferric chloride in water will be composed, among others, of the species Fe^{3+}, $FeCl^{2+}$, $FeOH^{2+}$. Collectively, they are correctly, though of course not fully, described by the notation Fe(III). In this kind of usage, the roman numeral represents the oxidation state.

Oxidation-reduction reactions. In an oxidation-reduction reaction, some element decreases in oxidation state and some element increases in oxidation state. The substances containing these elements are defined as the oxidizing agents and reducing agents, and they are said to be reduced and oxidized, respectively. The processes in question can always be represented formally as involving electron absorption by the oxidizing agent and electron donation by the reducing agent. For example, reaction (6) can be regarded as the sum of the two partial processes, or half-reactions, (9) and (10).

$$2I^- \rightarrow I_2 + 2e^- \quad (9) \qquad 2Fe^{3+} + 2e^- \rightarrow 2Fe^{2+} \quad (10)$$

Similarly, reaction (7) consists of the two half-reactions (11) and (12), with half-reaction (11)

$$2I^- \rightarrow I_2 + 2e^- \quad (11) \qquad 16H^+ + 2MnO_4^- + 10e^- \rightarrow 2Mn^{2+} + 8H_2O \quad (12)$$

being taken five times to balance the electron flow from reducing agent to oxidizing agent.

Each half-reaction consists of an oxidation-reduction couple; thus, in half-reaction (12) the reducing agent and oxidizing agent making up the couple are manganous ion, Mn^{2+}, and permanganate ion, MnO_4^-, respectively; in half-reaction (11) the reducing agent is I^- and the oxidizing agent is I_2. The fact that MnO_4^- reacts with I^- to produce I_2 means that MnO_4^- in acid solution is a stronger oxidizing agent than is I_2. Because of the reciprocal relation between the oxidizing agent and reducing agent comprising a couple, this statement is equivalent to saying that I^- is a stronger reducing agent than Mn^{2+} in acid solution. Reducing agents may be ranked in order of tendency to react, and this ranking immediately implies an opposite order of tendency to react for the oxidizing agents which complete the couples. In the list below some common oxidation-reduction couples are ranked in this fashion:

$$
\begin{array}{c}
\text{Strong reducing agent} \\
\uparrow \\
\text{Increasing reducing power} \\
\\
\text{Weak reducing agent}
\end{array}
\qquad
\begin{array}{rcl}
Mg & = & Mg^{2+} + 2e^- \\
Zn & = & Zn^{2+} + 2e^- \\
H_2 & = & 2H^+ + 2e^- \\
Cu & = & Cu^{2+} + 2e^- \\
I^- & = & \tfrac{1}{2}I_2 + e^- \\
Fe^{2+} & = & Fe^{3+} + e^- \\
Br^- & = & \tfrac{1}{2}Br_2 + e^- \\
Cl^- & = & \tfrac{1}{2}Cl_2 + e^- \\
4H_2O + Mn^{2+} & = & MnO_4^- + 8H^+ + 5e^-
\end{array}
\qquad
\begin{array}{c}
\text{Weak oxidizing agent} \\
\\
\text{Increasing oxidizing power} \\
\downarrow \\
\text{Strong oxidizing agent}
\end{array}
$$

A complete list contains the displacement series of the metals. The most powerful reducing agent shown in the list is magnesium, Mg, although this is not the most powerful known. Magnesium is capable of reacting with any oxidizing agent below it in the list to yield Mg^{2+} and to produce the reduced product resulting from the oxidizing agent. Similarly, permanganate ion, MnO_4^-, in acid, the strongest oxidizing agent shown, is capable of reacting with any reducing agent above it in the list. Conversely, an oxidizing agent at the top of the list will not react appreciably with the reducing agent of a couple below it. The list given, containing nine entries, can be used to predict the results of 72 separate experiments (for example, $Mg + Zn^{2+}$ on the one hand and $Mg^{2+} + Zn$ on the other would be counted as separate experiments in arriving at this figure). SEE ELECTROCHEMICAL SERIES; ELECTRONEGATIVITY.

Since the driving force for a reaction depends on concentrations, the concentrations of all reactants and products must be specified, as well as other conditions, in compiling a list such as that given. The order shown obtains for water solutions at 25°C (77°F), approximately 1 M in all soluble reagents and having gases present at approximately 1 atm pressure. A second limitation on the use of this list lies in the fact that it applies only when the expected reaction products are compatible. Although copper is capable in principle of reducing iodine to form I^- and Cu^{2+} at high concentration, these products are not compatible with each other, but they react to form copper (I) iodide, CuI. Allowance for such features can always be made by incorporating the necessary additional half-reactions into the list. Finally, it must be stressed that the list can be used to predict the results of experiments only for systems which reach equilibrium sufficiently rapidly; it does not serve to predict the rate of reaction. To achieve the reduction of Fe^{3+} by H_2 predicted in the list, it would be necessary to use a catalyst in order to realize the reaction in a reasonable time.

The equilibrium information implied by a table of half-reactions can readily be put into quantitative form. Thus, the standard free-energy change for the reaction of 1 equivalent weight of each reducing agent with some common oxidizing agent can be entered opposite each half-reaction. The numerical values of these quantities will be in the same order as are the half-reactions and can be combined algebraically to yield the standard free energy change, and therefore the equilibrium constant, for any reaction which can be written from the table. *See* Chemical equilibrium.

A chemist concerned with reactions of the type under discussion will have a ready vocabulary of facts concerning oxidizing agents and reducing agents, such as their oxidizing or reducing powers, the speed with which they react, and the characteristics which may complicate their application. A typical problem in analytical chemistry is to reduce $Cr_2O_7^{2-}$ to Cr^{3+} in acidic (perchloric acid) solution without introducing elements which are not already present. Metallic reducing agents such as zinc and iron or metal ion reducing agents are immediately eliminated from consideration because the products of oxidation may be difficult to remove from the resulting solution. A solution of hydrogen iodide, HI, would be suitable, except that it would be necessary to take special pains to add it in equivalent amount because excess HI would be difficult to remove (the iodine, I_2, produced by oxidation of I^-, however, is easy to remove by extracting it with a suitable solvent such as carbon tetrachloride). Hydrogen would be ideal (the product of its oxidation, H^-, is already present in solution, and excess reducing agent can easily be removed) except that the rate of reaction would be disappointingly slow. A suitable reducing agent would be hydrogen peroxide, H_2O_2; it reacts rapidly, the product of oxidation is oxygen, which escapes from solution, and excess oxidizing agent is easily destroyed by heating the solution. *See* Electrode potential.

Mechanisms. The data needed to predict the outcome at equilibrium of the reaction of most common oxidizing and reducing agents are known. A list of the kind shown above can be extended, and when it is elaborated with entries carrying the quantitative information, accurate calculations of the equilibrium state for all the reactions implied by the table can be made. By contrast, though the rates of reaction are also of great importance, they are much less completely understood and less completely described. To understand the rates of reaction, it is necessary to consider how the reactions take place. To illustrate one of the problems of mechanism, a reaction is selected which, though not nearly as complicated as some, suffices for present purposes. When an aqueous solution containing Fe^{2+} is added to one containing Br_2, reaction (13) takes place. For

$$2Fe^{2+} + Br_2 = 2Fe^{3+} + 2Br^- \tag{13}$$

the final stable products to be produced in a single step would require that two Fe^{2+} and one Br_2 be brought together in an encounter. This course for the reaction is found to be less probable than one in which a single Fe^{2+} encounters one Br_2. Since in forming stable products the oxidation state of iron increases by one unit while that of a bromine molecule decreases by two (one for each atom), the reaction resulting from the encounter must leave either iron or bromine in an unstable state. Reasonable alternatives for the reactive intermediates produced are represented in reactions (14a) and (15b), together in each case with a sequel reaction which leads to the correct overall stoichiometry.

$$Fe^{2+} + Br_2 = Fe^{3+} + Br^- + Br \tag{14a}$$

$$Br + Fe^{2+} = Fe^{3+} + Br^- \tag{14b}$$

$$Fe^{2+} + Br_2 = Fe(IV) + Br^- \tag{15a}$$

$$Fe(IV) + Fe^{2+} = 2Fe^{3+} \tag{15b}$$

In the present system, the evidence points to the mechanism represented by the first alternative [reactions (14a) and (14b)]. But even after a reaction such as (13) is resolved into the different steps, and reactive intermediates are identified, there remain questions about how the changes in oxidation state are brought about in the individual steps. Thus, when Fe^{2+} reacts with Br_2, do the two reactants make direct contact—this would involve replacement of a water molecule on $Fe(H_2O)_6^{2+}$, the form which Fe^{2+} adopts in water, by Br_2—or does reaction occur by electron transfer from intact $Fe(H_2O)_6^{2-}$ to Br_2? If the latter occurs, does electron transfer take place over large distances from $Fe(H_2O)_6^{2+}$ to Br_2?

These questions are important not only for inorganic systems of the kind illustrated by the steps making up reaction (13) but also for reactions at electrodes, and for oxidation-reduction reactions catalyzed by enzymes in biological systems. These kinds of questions are under investigation and have been partly answered for certain systems.

Two different kinds of mechanisms are recognized for oxidation-reduction reactions. So-called outer-sphere reactions are easier to understand in a fundamental way and will be described first. Reaction (16) introduces a typical such reaction. Here the changes in oxidation state, $4+$ to $3+$ for Ir and $2+$ to $3+$ for Fe, take place without bonds to either atom being broken, and in this particular system there is not even much change in bond distances attending the changes in oxidation state. Electron transfer is explicit in such a system, and the electron transfer act is subject to the Franck-Condon restriction. This imposes a barrier to electron transfer in that it requires that the environments (coordination sphere and solvent) readjust prior to electron transfer so that after readjustment the energy of the system is the same whether the electron is on one metal or the other. The rates of reactions such as (16) can be estimated at least approximately by

$$IrCl_6^{2-} + Fe(CN)_6^{4-} \rightarrow IrCl_6^{3-} + Fe(CN)_6^{3-} \tag{16}$$

calculating the work to bring the partners together, the work of meeting the Franck-Condon restriction, and assuming that electron delocalization when the partners are in contact is adequate. Greater success has been met in attempts at correlating rates and here the equation developed by R. A. Marcus, relating the rate of reaction such as (16) to the standard free energy change and to the rates of the so-called self-exchange reactions [(17) and (18)], is proving to be useful not only

$$Fe^*(CN)_6^{4-} + Fe(CN)_6^{3-} \rightarrow {}^*Fe(CN)_6^{3-} + Fe(CN)_6^{4-} \tag{17}$$

$$^*IrCl_6^{3-} + IrCl_6^{2-} \rightarrow {}^*IrCl_6^{2-} + IrCl_6^{3-} \tag{18}$$

in simple systems but also in understanding electron transfer reactions for large and complex biological molecules.

Much more difficult to systematize and understand is the extensive and very important class of oxidation-reduction reactions in which the changes in oxidation state are linked to bond breaking or bond making. A simple example is provided by reaction (19). Isotopic labeling experiments have shown that in the course of the reaction an oxygen atom originating on ClO_3^- is transferred to the reducing agent. Though the process can formally be represented as involving electron loss by SO_2 and electron gain by ClO_3^-, electron transfer as the means by which the changes in oxidation state are brought about is not at all explicit in a reaction of this kind. These so-called inner-sphere mechanisms operate also for reactions involving metal ions. For example, when $[(NH_3)_5CoCl]^{2+}$ reacts with $Cr(H_2O)_6^{2+}$, the Cr(III) product has the formula $Cr(H_2O)_5Cl^{2+}$, and Cl is transferred from Co(III) to Cr(II) when the former oxidizes the latter. This kind of reaction clearly has much in common with that represented by reaction (19). In the latter system atomic

$$ClO_3^{*-} + SO_2 + H_2O \rightarrow ClO_2^{*-} + SO_3O^{*2-} + 2H^+ \tag{19}$$

oxygen is formally transferred; in the former, atomic chlorine.

A class of inner-sphere reactions of metal-containing molecules which are now recognized as playing an important role in many catalytic processes involves so-called oxidative addition. Reactions of this kind have been known for a long time, but their significance was not appreciated until interest in homogeneous catalytic processes began to develop. A commonplace example of oxidative addition is provided by reaction (20). It will be noted that in reaction (20) both the

$$SnCl_2 + Cl_2 \rightarrow SnCl_4 \quad (20)$$

oxidation number and the coordination number of Sn increase. Oxidative addition for a strong oxidizing agent such as Cl_2 is not surprising, but the reaction type took on new significance when it was discovered that with a suitable metal complex, oxidative addition can be realized also with hydrogen halides, alkyl halides, and even hydrogen. Among the metal complexes which undergo this kind of reaction are Rh(I), Ir(I), or Pt(0) species with 4 or fewer groups attached. In each case, there is the opportunity for an increase in both oxidation and coordination number. A specific example of a molecule which undergoes oxidative addition with H_2 is $[(C_6H_5)_3P]_3RhCl$, which is a useful catalyst for the hydrogenation of alkenes and alkynes. A reaction step in the catalytic sequence is the addition of H_2 to the metal so that the H-H bond is severed; the adduct then reacts with the alkene (or alkyne) transferring two atoms of hydrogen. A substance which will activate H-R bonds (where R is an alkyl radical) in the same way would be desirable. SEE HOMOGENEOUS CATALYSIS.

The fundamental aspects of electron transfer processes in oxidation-reduction reactions have much in common with the electron jump processes in semiconductors. Recognizing this connection is productive both for those interested in chemical effects accompanying electron transfer (that is, in oxidation-reduction processes) and those interested in electron mobility as a subject in its own right.

Bibliography. J. P. Collman and L. S. Hegedus, *Principles and Applications of Organotransition*, 1980; A. Haim, *Account. Chem. Res.*, 8:264, 1975; W. M. Latimer, *Oxidation States of the Elements and Their Potentials in Aqueous Solution*, 2d ed., 1952; R. G. Linck, *Int. Rev. Sci., Inorg. Ser.* [1], 9:303, 1972; H. Taube and E. S. Gould, *Account. Chem. Res.*, 2:321, 1969.

STERIC EFFECT
ERNEST WENKERT

The influence of the spatial configuration of reacting substances upon the rate, nature, and extent of reaction. The sizes and shapes of atoms and molecules, the electrical charge distribution, and the geometry of bond angles influence the courses of chemical reactions.

The steric course of organochemical reactions is greatly dependent on the mode of bond cleavage and formation, the environment of the reaction site, and the nature of the reaction conditions (reagents, reaction time, and temperature). The effect of steric factors is best understood in ionic reactions in solution. The nucleophilic substitution reaction at a saturated carbon atom can serve as an illustration.

Saturated nucleophilic substitution. While the reacting carbon atom is in an electron-deficient state in the transition state (state of highest energy, somewhere between starting material and product) in a nucelophilic substitution process, the reaction can be varied from a two-step, unimolecular ionization to a one-step, bimolecular transformation. The former mode of reaction, a solvolysis or S_N1 process, converts a tetrahedral carbon into a solvated planar carbonium ion intermediate, and hence, leads to racemization (randomization of configuration); for example, reaction (1).

$$\underset{R''}{\overset{R}{\underset{R'}{\diagdown}}}C-X \xrightarrow{-X^{\ominus}} \underset{R'\ R''}{\overset{R}{|}}C^{\oplus} \xrightarrow[-H^{\oplus}]{H_2O} \underset{R'\ R''}{\overset{R}{|}}C-OH + HO-\underset{R''}{\overset{R}{\underset{R'}{\diagup}}}C \quad (1)$$

The second reaction path proceeds by a simultaneous rupture of the old bond and creation of a new one, either by inversion of configuration, a Walden inversion or S_N2 process, for example, reaction (2); or in a few cases by a front-side displacement of part of the substituent already

$$\underset{R''}{\overset{R}{\underset{R'}{\diagdown}}}C-X \xrightarrow{Y^{\ominus}} \left[Y^{\delta\ominus}----\underset{R'\;R''}{\overset{R}{\underset{|}{\overset{|}{C}}}}^{\delta\oplus}----X^{\delta\ominus} \right]^{\ominus} \xrightarrow{-X} Y-\underset{R''}{\overset{R}{\underset{R'}{\diagdown}}}C \qquad (2)$$

present, an $S_N i$ process, and hence, retention of configuration, for example, reaction (3).

$$\text{ROH} \xrightarrow{\text{SOCl}_2} \text{ROSOCl} \longrightarrow \underset{\text{(ion pair)}}{R^{\oplus}\text{OSOCl}^{\ominus}} \xrightarrow{-\text{SO}_2} \underset{\text{(ion pair)}}{R^{\oplus}\text{Cl}^{\ominus}} \longrightarrow \text{RCl} \qquad (3)$$

These substitution reactions are highly solvent-dependent; for example, tert-butyl chloride undergoes solvolysis close to 500,000 times faster in water, a solvent of high ionizing power, than in ethanol, a solvent of low dielectric properties. The nature of R, R', and R'' is one of the factors determining which of the above pathways a compound prefers for its substitution. The larger the size of these three groups, the greater is the tendency to relieve steric strain by extrusion of X, and thus the more the need for an $S_N 1$ process. The smaller the size of the environment, the greater is the accessibility of the reagent from the back side, and thus the more tendency toward an $S_N 2$ process. As a consequence, tertiary compounds undergo racemization readily, whereas primary systems prefer inversion.

Both the rate and the steric course of an ionic displacement may depend often on the ability of groups adjacent to the reaction site to accommodate a positive charge. Substitution of α-halo ethers and allyl or benzyl halides occurs much faster than a similar reaction of unsubstituted halides because of the intermediacy of the stabilized cations or cationlike transition states in notation (4).

$$\text{(4)}$$

In view of the charge distribution over more than one atom in these cases, sometimes the incoming reagent forms a bond at a site different from that of the leaving group. As a consequence, an $S_N 2'$ process, reaction (5), or an $S_N i'$ path, reaction (6), may result. Both processes

$$\underset{C_\beta}{\overset{X}{\underset{|}{\overset{|}{C_\alpha}}}}{\diagdown}\overset{C_\gamma}{=} \xrightarrow{Y^{\ominus}} \left[\underset{C_\beta}{\overset{X^{\delta\ominus}}{\underset{|}{\overset{|}{C^{\delta\oplus}}}}}{\diagdown}\overset{Y^{\delta\ominus}}{\underset{C}{=}} \right] \xrightarrow{-X^{\ominus}} \underset{C_\beta}{\overset{}{\underset{|}{\overset{|}{C_\alpha}}}}{\diagdown}\overset{Y}{\underset{C_\gamma}{=}} \qquad (5)$$

$$\qquad (6)$$

yield retention of configuration; that is, the orientation of the new substituent on its carbon atom is identical with that originally held by the former functional group on a site that is two carbon atoms removed.

In the presence of participating neighboring groups, even solvolyses can lead to retention of configuration. Certain rigidly held homoallyl systems undergo substitution at a site that is three carbons removed from the position of the leaving group, but with retention of configuration, as represented by reaction (7).

$$\text{(7)}$$

Double inversion is responsible for the retained configuration of the products of solvolysis of α-halo acid salts, reaction (8).

$$\text{(8)}$$

Solvolysis of *trans*-β-acetoxy systems in nonaqueous media leads to trans products. In the presence of water, cis compounds are obtained, reaction (9).

$$\text{(9)}$$

Base treatment of *trans*-halohydrins leads to *trans*-vicinal glycols. The intermediate epoxide is isolable. Although ring opening of the latter may yield two different trans products, the diaxial one is formed preferentially in cycloalkane cases, reaction (10).

$$\text{(10)}$$

Rearrangement. Migration of neighboring groups toward the reaction site, resulting in skeletal rearrangements, is a common occurrence. Both the internal displacement of the leaving group by the migrating group and the subsequent external displacement of the migrating group by the solvent or added reagent proceed in a trans sense, that is, by back-side approach. Thus, the overall steric consequence of one migration sequence is retention of configuration.

There are several examples of 1,2-hydride migration, such as reaction (11).

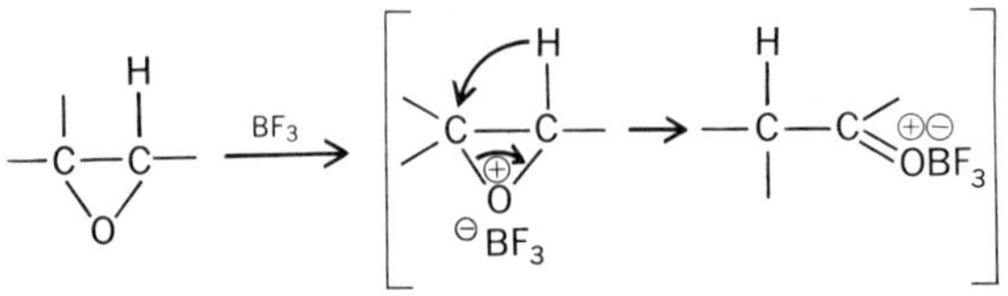

$$\begin{array}{c}\text{-C-C-} \\ \diagdown\text{O}\diagup\end{array} \xrightarrow{BF_3} \left[\begin{array}{c}\text{epoxide-BF}_3 \text{ complex} \end{array} \right] \rightarrow -C-C\begin{array}{c}H\\ \diagdown OBF_3\end{array} \quad (11)$$

Transannular hydride shifts are quite similar in nature, as in reaction (12).

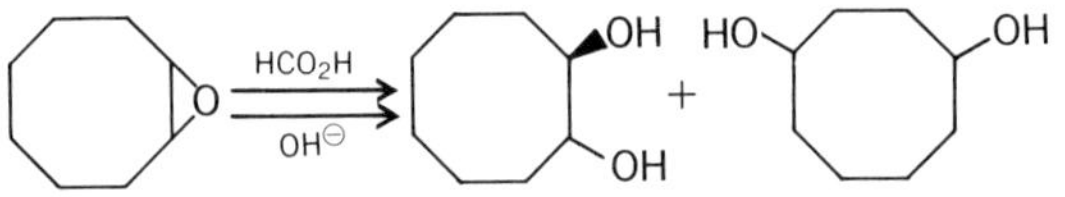

$$\text{cyclooctene oxide} \xrightarrow[OH^\ominus]{HCO_2H} \text{diol products} \quad (12)$$

The 1,2 migration of phenyl groups proceeds by way of fairly stable phenonium ions, reaction (13).

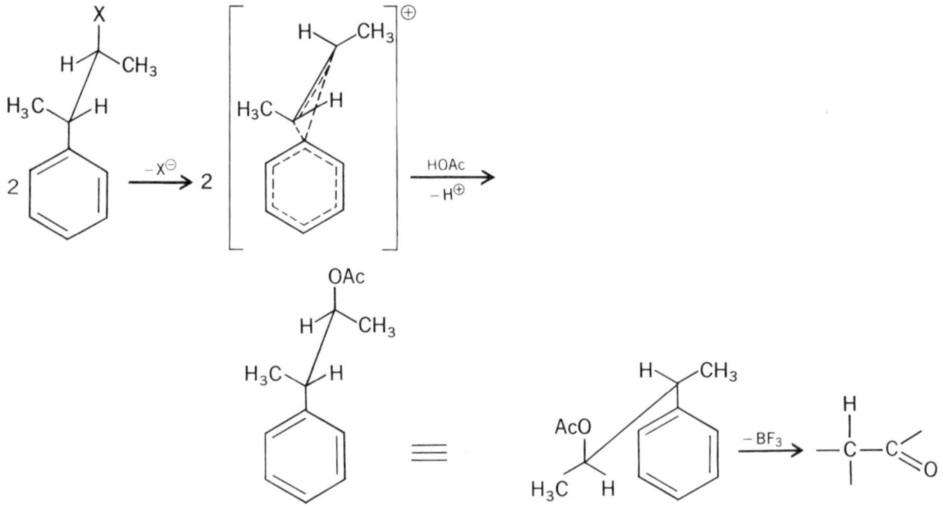

(13)

The Wagner-Meerwein rearrangement of saturated neighboring groups shows directional effects similar to those of the above migrations; for example, the conversion of camphene hydrochloride to bornyl chloride is stereospecific, with retention of configuration, reaction (14).

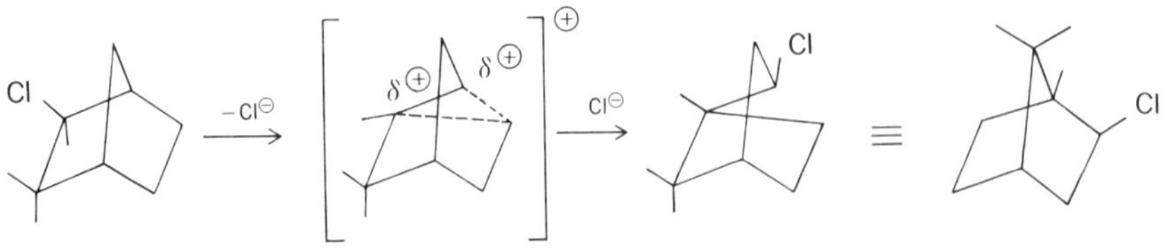

(14)

Neopentyl halides solvolyze to tertiary amyl derivatives, reaction (15).

$$(CH_3)_3CCH_2X \xrightarrow{-X} [(CH_3)\overset{\oplus}{C}CH_2] \rightarrow (CH_3)_2\overset{\oplus}{C}CH_2CH_3 \xrightarrow[H^\oplus]{ROH} (CH_3)_2\underset{OR}{C}CH_2CH_3 \quad (15)$$

Cyclohexane systems with equatorial leaving groups may undergo contraction to five-membered rings, reaction (16).

$$\text{(16)}$$

Organic compounds possessing potential leaving groups at the bridgehead of small bicyclic ring systems undergo substitution processes only sluggishly. In the absence of ready access at the back side of the reaction center, the S_N2 pathway is excluded. The inability of the compounds to form planar carbonium ions precludes a S_N1 route. However, displacements do occur slowly at elevated temperatures, presumably via nonplanar cations.

Unsaturated nucleophilic substitution. Nucleophilic substitution reactions at unsaturated carbon atoms can take place by two possible mechanistic routes, an elimination-addition process and an addition-elimination scheme. The former route is best illustrated by the transformation of aromatic halides into anilines, reaction (17).

$$\text{(17)}$$

A benzyne

The latter is encountered in the interconversion of carboxylic acids and their derivatives, for example, reaction (18). Because the central carbon atom has greater steric requirements in the

$$RCO_2R' \underset{-ROH}{\overset{H_2O}{\rightleftarrows}} [RC(OH)_2OR'] \rightleftarrows RCO_2H \quad (18)$$

reaction intermediate than in the starting material, the reaction velocity is strongly dependent on the size and number of neighboring groups; that is, an increase in the bulkiness of R is reflected in a decrease of the rate of the reaction.

The addition-elimination mechanism is portrayed also by the aromatic nucleophilic substitution reaction, for example, (19). In order to be able to stabilize the reaction intermediate, the all-

$$\text{(19)}$$

important nitro group must be coplanar with the benzene ring. As a consequence, ortho substituents, which may block this steric requirement, retard the reaction rate. Unusual aromatic nucleophilic substitutions have been observed in cases where steric hindrance by ortho substituents has prevented addition to aromatic ketones to occur, for example, reaction (20).

$$\text{(20)}$$

Addition reactions. The steric course of addition reactions at unsaturated sites depends largely on the reagent. Catalytic hydrogenation, a nonhomogeneous process of undetermined mechanism, occurs in a cis manner. In the absence of any steric interference, it leads to thermodynamically stable products. In the presence of steric hindrance, the two new hydrogen atoms are usually introduced on the least hindered side of the unsaturated compounds. However, sometimes some bulky polar groups actually aid, rather than retard, adsorption of the catalyst on their side of the reducing compound, thereby leading to products of opposite configuration.

The oxidation of olefins to vicinal glycols by permanganate salts or osmium tetroxide also proceeds in a cis fashion and also involves the least-hindered side of the reacting substrate. The Diels-Alder reaction behaves similarly, for example, reaction (21). All addition processes, during

$$\text{(21)}$$

which two new bonds are formed more or less simultaneously, yield cis adducts.

Ionic addition reactions of olefins and acetylenes occur in a trans manner. The mode of addition is such as to lead to product via the most stable cations (Markovnikoff addition), for example, reaction (22).

$$R_2C=CH_2 \xrightarrow{HBr} \begin{matrix} [R_2\overset{\oplus}{C}CH_3] \rightarrow R_2CCH_3 \\ | \\ Br \\ \\ [R_2CH\overset{\oplus}{C}H_2] \rightarrow R_2CHCH_2Br \end{matrix} \quad (22)$$

Halogen addition to cyclic olefins leads to trans diaxial dihalides which, on standing, isomerize to the more stable trans diequatorial dihalides, reaction (23).

$$\text{(23)}$$

Ionic addition reactions of carbonyl compounds follow a steric course very similar to those of olefins. However, the reagents are mostly nucleophilic, and some reactions are equilibrium processes, for example, reaction (24). The orientation of attack and the reaction rate are governed

$$\text{(24)}$$

by the environment of the carbonyl group.

Addition reactions of conjugated carbonyl systems can occur through cation as well as anion intermediates, but they uniformly place the nucleophilic part of the reagents on the β-carbon atoms, for example, reactions (25) and (26).

$$\text{(25)}$$

$$-\overset{|}{C}=\overset{|}{C}-\overset{|}{C}=O \xrightarrow{OH^{\ominus}} \left[-\overset{|}{\underset{OH}{C}}-\overset{|}{C}\overset{\delta\ominus}{=\!=\!=}\overset{|}{C}\overset{\delta\ominus}{=\!=\!=}O \right] \xrightarrow[-OH^{\ominus}]{H_2O} -\overset{|}{\underset{OH}{C}}-\overset{|}{\underset{H}{C}}-\overset{|}{C}=O \qquad (26)$$

The reaction of carbonyl compounds as enol or enolate anions with electrophilic reagents can take two different courses. If the process is kinetically controlled, the electrophile, a proton, halonium ion, or others, interacts with the substrate on its least-hindered side. However, if the reaction is thermodynamically controlled, then, independent of mechanism, the most stable product is obtained.

Elimination reactions. Elimination reactions can be carried out by pyrolysis of esters, halides, or amine oxides in the liquid or vapor phase. These eliminations always involve a rupture of vicinal cis bonds.

Alternatively, similar cleavage processes can be made to occur ionically in solution, in which case they proceed in a trans fashion. The direction of elimination depends greatly on the molecularity of the process, as well as on the sizes of the leaving group and the attacking base. The two-step, unimolecular cleavage, an $E1$ process, leads predominantly to the more substituted, hence more stable, olefins, for example, reaction (27).

$$CH_3CH_2\underset{Br}{\overset{|}{C}}(CH_3)_2 \xrightarrow{-Br^{\ominus}} [CH_3CH_2\overset{\oplus}{C}(CH_3)_2] \xrightarrow{-H^{\oplus}} CH_3CH=C(CH_3)_2 \qquad (27)$$

The one-step, bimolecular elimination (an $E2$ process) of neutral compounds yields similar products, for example, reaction (28). However, an $E2$ reaction on positive ions, ammonium or

$$CH_3CH_2\underset{Br}{\overset{|}{C}}(CH_3)_2 \xrightarrow{OH^{\ominus}} \left[\begin{array}{c} H_3C \\ \underset{HO\text{---}H}{\overset{\delta\ominus}{}} CH\overset{}{=\!=\!=}C(CH_3)_2 \end{array} \right] \xrightarrow[-H_2O]{-Br^{\ominus}} CH_3CH=C(CH_3)_2 \qquad (28)$$

sulfonium salts, affords the less substituted olefin in preponderant yield, reaction (29).

$$CH_3CH_2\overset{\oplus}{N}(CH_3)_2CH_2CH_2CH_3 \xrightarrow{OH^{\ominus}} \left[\begin{array}{c} \overset{\oplus}{N}(CH_3)_2CH_2CH_2CH_3 \\ \underset{H}{CH_2}\text{---}CH_2 \end{array} \right] \xrightarrow{-H_2O}$$
$$CH_2=CH_2$$
$$+ \qquad (29)$$
$$CH_3CH_2CH_2N(CH_3)_2$$

Reactions leading to the more stable products are said to follow the Saytzeff rule, whereas those yielding less stable olefins obey the Hofmann rule. Because the transition state in the $E2$ reaction is of lowest energy when all atoms involved in the elimination are in a plane, the fastest rates among cyclic compounds are encountered in the cases which permit a diaxial alignment of vicinal substituents.

Many ionic elimination reactions are known which involve the rupture of more than two bonds, for example, reaction (30).

$$Br\text{---}CH_2\text{---}CH_2\text{---}CH_2\text{---}CH_2\text{---}Br \xrightarrow{:Zn} 2CH_2=CH_2 \qquad (30)$$

The $E2'$ processes, eliminations of two groups of carbon atoms separated by an olefinic linkage, appear to be cis in nature, reaction (31).

$$\text{(reaction 31)} \tag{31}$$

Electrophilic substitution. Steric factors have a fair control over the course of the aromatic electrophilic substitution reaction. In reactions of compounds containing ortho-para directing substituents, the p/o product ratio is usually greater than 1, and increases with the size of the substituent and that of the reacting species. The rate-accelerating participation of electron-donating groups, located ortho or para to the incoming substituent, in stabilizing the transition state is greatly diminished in the presence of bulky ortho neighbors which would prevent the groups from attaining coplanarity with the benzene, for example, reaction (32).

$$\text{(reaction 32)} \tag{32}$$

See Stereochemistry.
Bibliography. A. T. Balaban et al. (eds.), *Steric Fit in Quantitative Structure-Activity Relations*, 1980; G. Natta and M. Farina, *Stereochemistry*, 1973; J. D. Roberts and M. C. Caserio, *Basic Principles of Organic Chemistry*, 2d ed., 1977; F. Vogtle and E. Weber (eds.), *Stereochemistry*, 1984.

SURFACE CHEMISTRY

Surface tension	98
Surface-active agent	99
Molecular adhesion	101
Interface of phases	102
Adsorption	103
Desorption	107
Monomolecular film	111

SURFACE TENSION
Norman H. Nachtrieb

The force acting in the surface of a liquid, tending to minimize the area of the surface. Surface forces, or more generally, interfacial forces, govern such phenomena as the wetting or nonwetting of solids by liquids, the capillary rise of liquids in fine tubes and wicks, and the curvature of free-liquid surfaces. The action of detergents and antifrothing agents, and the flotation separation of minerals depend upon the surface tensions of liquids.

Surface energy. In the body of a liquid, the time-average force exerted on any given molecule by its neighbors is zero. Even though such a molecule may undergo diffusive displacements because of random collisions with other molecules, there exist no directed forces upon it of long duration. It is equally likely to be momentarily displaced in one direction as in any other. In the surface of a liquid, the situation is quite different; beyond the free surface, there exist no molecules to conteract the forces of attraction exerted by molecules in the interior for molecules in the surface. In consequence, molecules in the surface of a liquid experience a net attraction toward the interior of a drop. These centrally directed forces cause the droplet to assume a spherical shape, thereby minimizing both the free energy and surface area.

From the macroscopic point of view, surface tension may be regarded either as a force exerted normally to a unit length in the surface, or as the work which must be expended upon the liquid to increase its area by unity. Accordingly, surface tension is expressed in centimeter-gram-second (cgs) units of dynes/cm or ergs/cm². From the microscopic point of view, the surface tension (or its equivalent, surface energy) is the reversible isothermal work which must be done in bringing molecules from the interior of the liquid to the surface and creating 1 cm² of new surface thereby.

Most liquids have surface tensions of 20–40 dynes/cm at room temperature, but water has the exceptionally high value of 72.75 dynes/cm at 20°C (68°F). Condensed gases such as helium and nitrogen have quite low surface tensions [0.098 dynes/cm at 4.3 K (-452°F) and 6.2 dynes/cm at 90.2 K (-297.2°F) respectively]. Liquid metals have large surface tensions by comparison: mercury, 470 dynes/cm; and liquid copper at 1131°C (2068°F) has a surface tension of 1103 dynes/cm in hydrogen gas. Small but significant differences in the surface tensions of liquids depend upon the composition of the vapor phase.

In the wetting or nonwetting of solids by liquids, the criterion employed is the contact angle between the solid and the liquid (measured through the liquid) (**Fig. 1**). A liquid is said to wet a solid if the contact angle Θ lies between 0 and 90°, and not to wet the solid if the contact angle lies between 90 and 180°. Three interfaces exist when a droplet of liquid contacts a solid, and three corresponding interfacial tensions exist: γ_{SL}, γ_{SV}, and γ_{LV}. The subscripts S, L, and V refer to solid, liquid, and vapor. At equilibrium, a balance of interfacial tensions exists at the line of common contact, which intersects the figures at point 0. For the case of a liquid which wets the solid ($\theta < 90°$), this equilibrium is expressed by relation (1).

$$\gamma_{SV} = \gamma_{SL} + \gamma_{LV} \cos \theta \tag{1}$$

Capillarity. Liquids which wet the walls of fine capillary tubes rise to a height which depends upon the tube radius, the surface tension, the liquid density, and the contact angle. In **Fig. 2**, a liquid of density ρ is shown as having risen to a height H in a capillary whose radius is R. A balance exists between the force exerted by gravity on the mass of liquid raised in the capillary and the opposing force caused by surface tension. The former is $\pi r^2 h \rho g$, whereas the

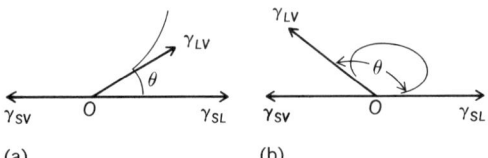

Fig. 1. Contact angle. (a) Liquid wets solid. (b) Liquid does not wet solid.

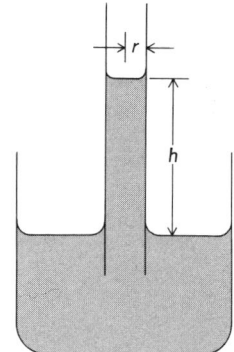

Fig. 2. Rise of liquid in capillary tube.

latter is $2\pi r\gamma$, assuming the contact angle to be zero. It is clear that $h = 2\gamma/r\rho g$, and that the capillary rise varies inversely with the tube radius and the liquid density. Liquids which do not wet the capillary walls are depressed in height according to the same equation.

The shape of the free surface of a liquid in a vessel is only an approximation to a plane. In narrow tubes the meniscus of a liquid is concave upward if the liquid wets the tube and, conversely, convex upward if it does not wet the tube. A pressure difference exists between the concave and convex sides of the surface, the excess pressure on the concave side over the convex side being given by relation (2), where r_1 and r_2 are the principal radii of curvature of the surface.

$$p = \gamma \frac{1}{r_1} + \frac{1}{r_2} \qquad (2)$$

The same equation applies for a bubble of gas within a liquid, with the consequence that the vapor pressure p is larger for small bubbles according to relation (3), where p_0 is the vapor pressure

$$\ln(p/p_0) = (2\gamma/r\rho)(M/RT) \qquad (3)$$

over a liquid surface of infinite radius, R is the gas constant, and M is the molecular weight.

Detergents, soaps, and flotation agents owe their usefulness to their ability to lower the surface tension of water, thereby stabilizing the formation of small bubbles of air. At the same time, the interfacial tension between solid particles and the liquid phase is lowered, so that the particles are more readily wetted and floated after attachment to air bubbles. *See* INTERFACE OF PHASES; SURFACE-ACTIVE AGENT.

Bibliography. R. A. Alberty, *Physical Chemistry* 7th ed., 1987; American Chemical Society, *Contact Angle, Wettability and Adhesion*, 1964.

SURFACE-ACTIVE AGENT
WENDELL H. SLABAUGH

A substance that, even though present in small amounts, exerts a marked effect on the surface behavior of a system. These agents are essentially responsible for producing great changes in the surface energy of liquid or solid surfaces, and their ability to cause these changes is associated with their tendency to migrate to the interface between two phases. Consequently, surface-active agents are of potential interest wherever there are solid-solid, solid-liquid, solid-gas, liquid-liquid, or liquid-gas interfaces, and of particular interest at liquid-gas interfaces at which the surface-active agent is a solute whose presence makes the surface properties of the solution greatly different from those of the solvent. *See* INTERFACE OF PHASES.

Mechanism. Soap, for example, when dissolved in small quantities in water, is responsible for greatly decreasing the surface tension of water, and it is this property of soap that accounts for its ability to act as a detergent. In contrast to soap and other related substances that lower the

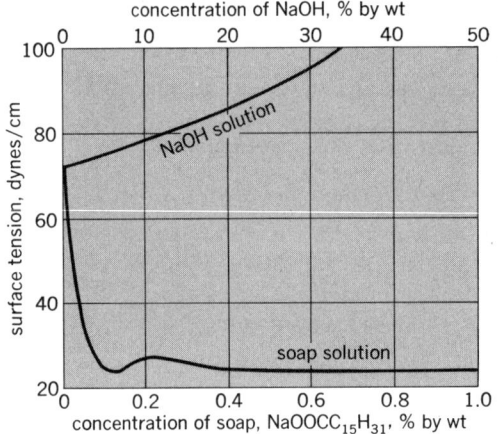

Effect of surface-active agents on water. 1 dyne/cm = 10^{-3} N/m.

surface energy of a liquid, other solutes, such as inorganic salts, acids, and bases may increase the surface tension of a liquid (see **illus**.), but their effect in increasing the surface tension is not nearly so great as the effect of those agents that decrease the surface tension. Occasionally the term surface-inactive solutes is applied to these substances whose presence causes an increase in surface tension.

The importance of surface-active agents is indicated by their strategic necessity in such processes as lubrication, wetting, foaming, emulsification, detergency, water repellence, waterproofing, spreading, and dispersion. In lubrication, for example, the oiliness of a hydrocarbon oil can be improved by the addition of a surface-active agent. In order to achieve lubrication between two solid surfaces, a thin film of liquid must be preserved in the space between the two solid surfaces. The viscosity of this liquid film and the ability of the liquid to wet the solid surfaces determine the resistance of the lubricant system to being squeezed mechanically from the region between the two solid surfaces. Addition of fatty acids, fatty oils, metallic soaps, and various derivatives of aromatic and aliphatic hydrocarbons commonly improves the lubricant qualities of mineral hydrocarbons, and these additives are truly surface-active agents.

The mechanism by which surface-active agents alter the surface energy of a solid or liquid is attributed to the dual nature of the molecules or ions of these substances. Within a single molecule or ion of a surface-active agent, there is a group that is lyophilic toward the dispersing medium or solvent, and at a suitable distance within the same molecule or ion, there is another group that is lyophobic toward the dispersing medium. This ability to embody within the same molecular particle two different groups whose properties are diametrically opposed is sometimes termed amphipathy. For example, the surface activity of sodium oleate, $NaOOCC_{15}H_{31}$, is attributed to the combined effect of the hydrophilic ionic carboxyl salt group at one end of the molecule and the hydrophobic hydrocarbon group that constitutes the remainder of the molecule. In a dilute solution of sodium oleate, the solute migrates to the surface where the hydrophobic parts of the molecules can achieve their lowest energy positions as the result of the solvent's striving to exclude the hydrocarbon group from the solution. Even though the external phase is gaseous, the hydrophobic groups find a sufficiently sympathetic environment at the surface of the liquid. If, on the other hand, the external phase were an oil, the hydrocarbon groups would find an even more sympathetic environment, and in either event, the surface energy of the original solvent would be greatly diminished. SEE MONOMOLECULAR FILM.

Classification. Surface-active agents are usually classified in three groups: anionic, cationic, and nonionic types. Anionic types include carboxylate ions such as occur in sodium oleate. The carboxyl group may be attached directly to the hydrophobic group, or there may be an intermediate ester, amide, or sulfonamide linkage. There are a large number of anionic agents derived from sulfuric and sulfonic acids in which the hydrophobic groups attached to them include aliphatic and aromatic groups that often contain substituents of varying polarity, such as halide, hydroxyl, ether, and ester groups.

Cationic surface-active agents are usually derived from the amino group where, through either primary, secondary, or tertiary amine salts, the hydrophilic character may be achieved by aliphatic and aromatic groups that may be altered by substituents of varying polarity. Other nitrogen compounds, such as quaternary ammonium compounds, guanidine, and thiuronium salts, are included in the cationic class.

The third class of surface-active agents, the nonionic type, are organic substances which contain groups of varying polarity and which render parts of the molecule lyophilic, whereas other parts of the molecule are lyophobic. Examples include polyethylene glycol, polyvinyl alcohol, polyethers, polyesters, and polyhalides. In this class are often included certain colloidal substances such as graphite, powdered metals, metallic oxides, clays, macromolecules, and polymers. SEE SURFACE TENSION.

Bibliography. M. Ash and I. Ash, *Encyclopedia of Surfactants*, vols. 1–3, 1980–1981; D. Attwood and A. T. Florence, *Surfactant Systems: Their Chemistry, Pharmacy,* and *Biology*, 1982; M. J. Rosen, *Surfactants and Interfacial Phenomena*, 1978.

MOLECULAR ADHESION
ROBERT A. WELLER

The tendency of dissimilar solids or liquids to cling together as a result of the interatomic forces which they exert upon each other across their common interface. Some factors affecting molecular adhesion are the physical state of the materials, the composition and the topology of the surfaces in contact, the temperature, and the presence of foreign materials such as adsorbed gases on one or both surfaces.

Mutual forces of attraction, electrical in origin, exist between virtually all atoms and molecules. It is these forces which, under proper conditions of temperature and pressure, compel isolated atoms or molecules to condense into solids and liquids. On average, these forces are approximately equally strong in all directions and result in an atom or molecule achieving its lowest possible energy, that which is thermodynamically favored, when it is surrounded on all sides by other atoms. The presence of a real surface breaks the three-dimensional continuity of the substance, leaving atoms at the surface with an unfulfilled capacity for bonding. It is this capacity for bonding that is responsible for the phenomena of surface tension, adsorption of gases, and adhesion. SEE ADSORPTION; INTERMOLECULAR FORCES; SURFACE TENSION.

In principle, the strength of the adhesive bond between two substances may be quite large—larger, in fact, than the cohesive strength of the weaker material. In practice, however, many factors act to diminish the strength of adhesion. Probably the most significant are incomplete contact of the surfaces due to their microtopography and the presence of adsorbed gases or other surface contaminants. These may diminish the surface's capacity to further bond or simply replace one of the original surfaces with an overlayer which is significantly weaker.

One technological application in which molecular adhesion plays an especially significant role is the deposition by evaporation or sputtering of thin films on substrates of dissimilar composition. A particularly important example is the use of such films to form conducting paths—microscopic wires only a few tens of nanometers thick—in integrated circuits. These films must both adhere well to the underlying substrate and produce contacts with appropriate electrical properties. At present the understanding of adhesion is not sufficiently detailed to be able to predict such properties in advance of experiments.

Enhancement has been observed in the degree of adhesion between evaporated layers and the underlying medium as a result of irradiation by heavy ions with energies of a few MeV. In these experiments, enhanced adhesion has been observed between metal films and metal, semiconductor, and insulating substrates. Materials as dissimilar as gold and Teflon have been joined, and it has also been observed that particle irradiation transforms electrically rectifying interfaces into ohmic ones. Further research in ion-beam enhanced adhesion promises not only new technological advances but also improved understanding of molecular adhesion at the most fundamental level. SEE INTERFACE OF PHASES.

Bibliography. A. W. Adamson, *Physical Chemistry of Surfaces*, 4th ed., 1982; K. W. Allen (ed.), *Adhesion*, vols. 1–8, 1977–1984; R. L. Patrick (ed.), *Treatise on Adhesion and Adhesives*, vols. 1–5, 1967–1981; M. Prutton, *Surface Physics*, 2d ed., 1983; W. C. Wake, *Adhesion and the Formation of Adhesives*, 2d ed., 1982; B. T. Werner et al., Enhanced adhesion from high energy ion irradiation, *Thin Sol. Films*, 104:163–166, 1983.

INTERFACE OF PHASES
WENDELL H. SLABAUGH

The boundary between any two phases. Among the three phases, gas, liquid, and solid, five types of interfaces are possible: gas-liquid, gas-solid, liquid-liquid, liquid-solid, and solid-solid. The abrupt transition from one phase to another at these boundaries, even though subject to the kinetic effects of molecular motion, is statistically a surface only one or two molecules thick.

A unique property of the surfaces of the phases that adjoin at an interface is the surface energy which is the result of unbalanced molecular fields existing at the surfaces of the two phases. Within the bulk of a given phase, the intermolecular forces are uniform because each molecule enjoys a statistically homogeneous field produced by neighboring molecules of the same substance. Molecules in the surface of a phase, however, are bounded on one side by an entirely different environment, with the result that there are intermolecular forces that then tend to pull these surface molecules toward the bulk of the phase. A drop of water, as a result, tends to assume a spherical shape in order to reduce the surface area of the droplet to a minimum.

Surface energy. At an interface, there will be a difference in the tendencies for each phase to attract its own molecules. Consequently, there is always a minimum in the free energy of the surfaces at an interface, the net amount of which is called the interfacial energy. At the water-air interface, for example, the difference in molecular fields in the water and air surfaces accounts for the interfacial energy of 7.2×10^{-6} joule/cm^2 (11.15×10^{-6} cal/in.^{2}) of interfacial surface. The interfacial energy between the two liquids, benzene and water, is 3.5×10^{-6} J/cm^2 (5.4×10^{-6} cal/in.^{2}), and between ethyl ether and mercury is 37.9×10^{-6} J/cm^2 (58.67×10^{-6} cal/in.^{2}). These interfacial energies are also expressed as surface tension in units of dynes per centimeter.

The surface energy at an interface may be altered by the addition of solutes that migrate to the surface and modify the molecular forces there, or the surface energy may be changed by converting the planar interfacial boundary to a curved surface. Both the theoretical and practical implications of this change in surface energy are embodied in the Kelvin equation, Eq. (1), where

$$\ln \frac{P}{P_0} = \frac{2M\gamma}{RT\rho r} \tag{1}$$

P/P_0 is the ratio of the vapor pressure of a liquid droplet with diameter r to the vapor pressure of the pure liquid in bulk, ρ the density, γ the surface energy, and M the molecular weight. Thus, the smaller the droplet the greater the relative vapor pressure, and as a consequence, small droplets of liquid evaporate more rapidly than larger ones. The surface energy of solids is also a function of their size, and the Kelvin equation can be modified to describe the greater solubility of small particles compared to that of larger particles of the same solid. SEE ADSORPTION; SURFACE-ACTIVE AGENT.

Contact angle. At liquid-solid interfaces, where the confluence of the two phases is usually termed wetting, a critical factor called the contact angle is involved. A drop of water placed on a paraffin surface, for example, retains a globular shape, whereas the same drop of water placed on a clean glass surface spreads out into a thin layer. In the first instance, the contact angle is practically 180°, and in the second instance, it is practically 0°. The study of contact angles reveals the interplay of interfacial energies at three boundaries. The **illustration** is a schematic representation of the cross section of a drop of liquid on a solid. There are solid-

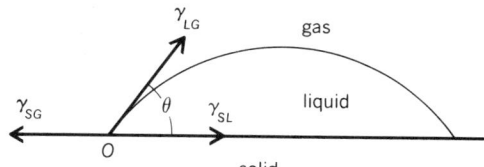

Contact angle θ at interface of three phases.

liquid, solid-gas, and liquid-gas interfaces that meet in a linear zone at O. The forces about O that determine the equilibrium contact angle are related to each other according to Eq. (2), where the

$$\gamma_{SG} = \gamma_{SL} + \gamma_{LG} \cos \theta \qquad (2)$$

γ terms represent free energies at the interfaces and θ is the contact angle. Since only γ_{LG} and θ can be measured readily, the term adhesion tension is defined by Eq. (3). Adhesion tension, which

$$\gamma_{LG} \cos \theta = \gamma_{SG} - \gamma_{SL} = \text{adhesion tension} \qquad (3)$$

is the free energy of setting, is of critical importance in detergency, dispersion of powders and pigments, lubrication, adhesion, and spreading processes.

The measurement of interfacial energies is made directly only upon liquid-gas and liquid-liquid interfaces. In measuring the liquid-gas interfacial energy (surface tension), the methods of capillary rise, drop weight on pendant drop, bubble pressure, sessile drops, Du Nuoy ring, vibrating jets, and ultrasonic action are among those used. There is a small but appreciable temperature effect upon surface tension, and this property is used to determine small differences in the surface tension of a liquid by placing the two ends of a liquid column in a capillary tube whose two ends are at different temperatures. The determination of interfacial energies at other types of interfaces can be inferred only by indirect methods. SEE FREE ENERGY; PHASE EQUILIBRIUM; SURFACE TENSION.

Bibliography. American Chemical Society, *Contact Angle, Wettability, and Adhesion*, Advan. Chem. Ser. 43., 1964; J. T. Davies and E. Rideal, *Interfacial Phenomena*, 2d ed., 1963; M. J. Jaycock and G. D. Parfitt, *Chemistry of Interfaces*, 1981.

ADSORPTION
FREDERICK M. FOWKES

The property of an interface between two immiscible phases (solid, liquid, or vapor) to attract and concentrate components of either phase or both phases as an adsorbed interfacial film. Vapors may adsorb onto solids or liquids, and solutes may adsorb at liquid-vapor, liquid-liquid, liquid-solid, and solid-solid interfaces. Adsorption is a basic thermodynamic property of interfaces, resulting from a discontinuity in intermolecular or interatomic forces. Adsorption is also important in nearly all industrial processes and products.

Some definitions that describe adsorption are as follows: The adsorbent is the solid or liquid which adsorbs. The adsorbate is the solid, liquid, or gas which is adsorbed as molecules, atoms, or ions. Physical adsorption or physisorption is reversible adsorption by weak interactions only; no covalent bonds occur between the adsorbent and the adsorbate; heats of physical adsorption are usually less than 15–20 kilocalories/mole (63–84 kilojoules/mole). Chemical adsorption or chemisorption is adsorption involving stronger interaction between adsorbate and adsorbent usually accompanied by rearrangement of atoms within or between adsorbates; reaction occurs between the surface of the adsorbent and the adsorbate; heats of chemisorption are usually in excess of 20–30 kcal/mole (84–126 kJ/mole).

Applications. Heterogeneous catalysis, in which gas or liquid reactants are specifically adsorbed to a dissimilar phase and chemically altered during their brief retention time, is basic to many industrial processes in the petrochemical, polymer, and chemical industries. SEE HETEROGENEOUS CATALYSIS.

Purification by adsorption is perhaps the oldest known application; examples are wine and beer clarification, color removal in sugar processing, industrial wastewater treatment, and toxic gas adsorption in gas masks.

Adsorption is the basic phenomenon of chromatographic separations, which separate and concentrate components of mixtures according to strength of adsorption onto adsorbents in chromatographic columns.

Adsorption of surface-active substances is the key process in the use of soaps, detergents, emulsifiers, wetting agents, dyes, lubricants, and surface treatments. Other industries which are dependent on adsorption processes include agriculture, mining, petroleum recovery, papermaking, printing, and photography.

Physical adsorption of vapor. Nearly all vapors tend to adsorb onto inorganic solids at temperatures not too much above their boiling point. The intermolecular attractive forces which cause vapors to adsorb (or condense) are generally dominated by the London dispersion forces, an attraction caused by the perturbation of electron orbits by adjacent atoms. These forces, generally proportional to the square of the polarizability per unit volume, are much stronger for most inorganic solids than for water or organic materials, and that is why inorganic solids are the stronger adsorbents.

Another attractive force important in vapor adsorption is the interaction of electron-donor (basic) sites of vapor molecules with electron-acceptor (acidic) sites of adsorbents, or vice versa. These short-range attractions are much stronger than dipole interactions. Silica, an acidic adsorbent, adsorbs basic vapors (water, ammonia, and so forth) much more strongly than acidic vapors (chloroform, CO_2, NO_2, and so forth) regardless of the dipole moments.

The adsorption of water is dominated by hydrogen bonding, an intermolecular acid-base interaction which permits multilayers to adsorb onto acidic or basic adsorbents but prevents adsorption onto neutral surfaces such as graphite or polyethylene, except for the acidic or basic sites provided by impurities on these neutral surfaces. SEE HYDROGEN BOND.

The amount of vapor (or solute) adsorbed per unit mass of adsorbent, σ, is determined as

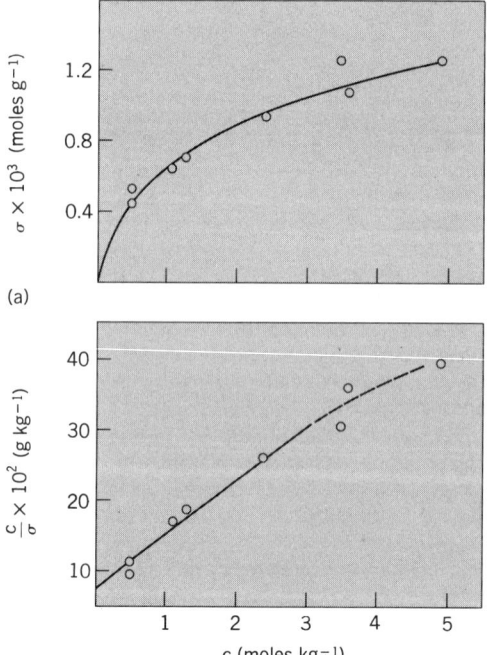

Fig. 1. Adsorption isotherm for benzene on carbon from heptane solution. (a) The data are plotted as σ versus c. (b) The data are plotted according to Eq. (1).

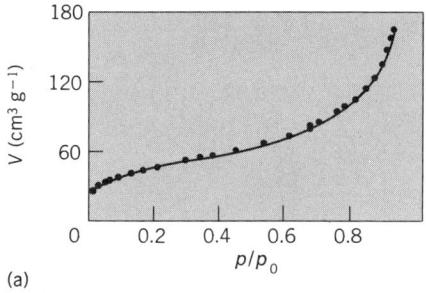
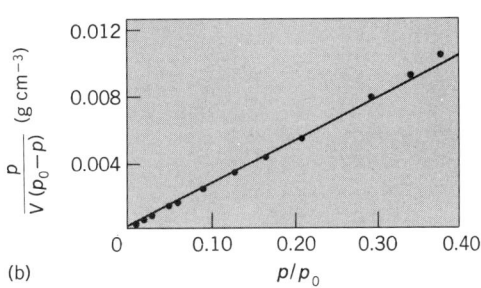

Fig. 2. Nitrogen adsorption on nonporous silica at 77 K (−320°F). (a) Volume per gram (in cubic centimeters at standard temperature and pressure) versus p/p_0. (b) According to the linear form of the BET equation [Eq. (2)].

a function of vapor pressure p (or solute concentration c) at a given temperature: this is the adsorption isotherm. If only a monolayer adsorbs, the results tend to follow the Langmuir adsorption isotherm [Eq. (1)], as shown in **Fig. 1**. In the equation σ_m is the amount of adsorbed vapor

$$\frac{c}{\sigma} = \frac{1}{K\sigma_m} + \frac{c}{\sigma_m} \tag{1}$$

per unit mass of adsorbent necessary for a complete monolayer, and K is the equilibrium constant for adsorption. However, if multilayers adsorb, as is usually observed in the physical adsorption of small molecules, the amount adsorbed is best plotted by the Brunauer-Emmet-Teller (BET) equation [Eq. (2)], as shown in **Fig. 2**. In the equation, c is a constant characteristic of the strength of

$$\frac{p}{\sigma(p_0 - p)} = \frac{1}{\sigma_m c} + \frac{(c-1)}{\sigma_m c}\left(\frac{p}{p_0}\right) \tag{2}$$

adsorption, and p_0 is the saturation pressure, a parameter specifying the pressure for many layers of adsorbate to be present on the surface. Several other adsorption isotherm equations have been developed for physical adsorption, and each has a special usefulness.

The free energy of adsorption of uniform multilayers on an adsorbent is designated by π_e and is called the spreading pressure of an adsorbed film in equilibrium with the vapor pressure p_0 of its liquid. For adsorption of a vapor on a solid, Eq. (3) applies, where γ_S, γ_{SL}, γ_{SV}, and γ_L are

$$\pi_e = \gamma_S - \gamma_{SV} = \gamma_S - (\gamma_{SL} + \gamma_L) \tag{3}$$

surface tensions of the solid, solid-liquid, solid-vapor, and liquid surfaces. The spreading pressure is related to the free energy of interaction per unit area at the adsorbent-liquid interface, $-W_{SL}$, given by Eq. (4). If the adsorption is in uniform multilayers, π_e can be determined from measure-

$$W_{SL} = \gamma_S + \gamma_L - \gamma_{SL} = 2\gamma_L + \pi_e \tag{4}$$

ments of moles of adsorbed vapor per unit area (Γ) versus vapor pressure p, using the Gibbs-Bangham equation, Eq. (5), where R is the gas constant and T is the absolute temperature.

$$\pi_e = RT \int_0^{p_0} \Gamma \, d(\ln p) \tag{5}$$

The free energy of interaction between two phases at an interface depends on the sum of the interaction energies contributed by the different forces operating at the interface, such as dispersion forces (d) [Eq. (6a)], and acid-base interactions (ab) [Eq. (6b)], and for the dispersion

$$W_{SL} = W_{SL}^d + W_{SL}^{ab} + \cdots \quad (6a) \qquad \gamma_L = \gamma_L^d + \gamma_L^{ab} + \cdots \quad (6b)$$

forces the relationship has been established as given in Eq. (7). Thus for the adsorption of saturated hydrocarbons in which $\gamma_L = \gamma_L^d$, the spreading pressure, π_e, is given by Eq. (8). Therefore,

$$W_{SL}{}^d = 2\sqrt{\gamma_S{}^d \gamma_L{}^d} \quad (7) \qquad \pi_e = 2\sqrt{\gamma_S{}^d \gamma_L{}^d} - 2\gamma_L \quad (8)$$

the London attractive forces of an adsorbent can be fully characterized by its value of $\gamma_S{}^d$. Representative values are 98 mJ/m^2 for graphite, 78 mJ/m^2 for silica, and 33.1 mJ/m^2 for polyethylene.

Adsorbents may also be quantitatively characterized for their acidic or basic interactions; the number of acid or basic sites per unit area and their strength can be used to predict such interactions quantitatively.

Chemisorption of vapors. The strong interactions of chemisorption lead to surface compounds with various degrees of covalent bond character. The adsorbed layers are only one molecule thick because covalent bonds exist only between adjacent atoms. Chemisorption occurs on metals and semiconductors and on oxides and sulfides, but is most often observed on transition metals such as silver, nickel, cobalt, platinum, rhodium, and tungsten. Chemisorption is a necessary step in catalysis by these materials.

Because chemisorbed films are tightly bound, high-vacuum techniques such as field-ion microscopy or low-energy electron diffraction (LEED) can be used to determine the two-dimensional architecture of chemisorbed films on some surfaces, and the chemical bonding can be partially elucidated by ultraviolet and x-ray photoelectron spectroscopies (UPS and XPS). Details of the atomic composition of surfaces can be studied by Auger electron spectroscopy (AES), XPS, secondary ion mass spectrometry (SIMS), or ion scattering spectroscopy (ISS).

Adsorption of solutes from aqueous solutions. Soaps and detergents adsorb at the surface of their aqueous solutions and at all interfaces in contact with them. The sorption is driven mainly by rejection of the dissolved soaps and detergents from solution because of changes in water structure induced by these solutes. Adsorption of anionic surfactants such as sodium dodecyl sulfate or sodium dodecylbenzene sulfonate gives surfaces a negative charge, and adsorption of cationic surfactants such as *n*-hexadecyl trimethyl ammonium chloride gives surfaces a positive charge. Such adsorbed films cause similarly charged particles to repel one another, thereby stabilizing emulsions and dispersions and promoting detergency. SEE SURFACE-ACTIVE AGENT.

Inorganic solutes also adsorb from aqueous solutions, but the mechanism may not be the same as for surfactants. Anions such as borates, silicates, phosphates, and polyphosphates adsorb strongly onto fabrics, clays, and metals, conferring negative potentials, an important aspect of detergency.

Water-soluble polymers (such as polyethylene oxides, polyvinyl alcohol, hydrolyzed polyacrylamide, and the naturally occurring polysacharrides and other water-soluble gums) adsorb from aqueous solution onto inorganic surfaces, fabrics, or paper. These adsorbed films can be thick enough to prevent particle-particle adhesion and provide steric stabilization to dispersions. Higher-molecular-weight linear water-soluble polymers can adsorb simultaneously on several particles and thereby flocculate dispersions.

Adsorption from nonaqueous solutions. Oil-soluble surfactants have large hydrocarbon groups and at one end a basic or acidic group which can anchor the molecule to acidic or basic sites on the adsorbent. Strong and rapid adsorption is observed when a strong acid-base interaction occurs at the interface. On the other hand, very little or no adsorption occurs unless there is some acid-base interaction.

Oil-soluble polymers (such as polyalkylmethacrylates, or block copolymers with polybutadiene or polyisobutylene blocks) adsorb onto inorganic surfaces, mainly by acid-base interaction. The olefin groups of rubber are weak bases, but their large number per polymer molecule causes strong adsorption onto acidic surfaces such as acidic carbon blacks and clays; this strong adsorption is the cause of the reinforcement of rubber by such inorganic particles. Such acid-base interactions in polymer adsorption also are the governing factor in the adhesion to metals of paints and adhesives, including those which are polymerized in places on surfaces.

Kinetics. In many cases of adsorption, the rates depend mainly on the rate of arrival of molecules at surfaces. However, not every molecule sticks upon collision, and sticking coefficients vary from unity to less than 10^{-9}, depending on temperature, coverage, and the particular interface.

In the physical adsorption of gases, sticking coefficients are near unity on bare surfaces, but much smaller sticking coefficients are often observed in chemisorption.

In adsorption from solution, sticking coefficients are unity at liquid surfaces; but at solid surfaces, the sticking coefficients decrease rapidly with increasing surface coverage, often to 10^{-7} or 10^{-8} for the last 20% of a monolayer. Thus, several hours are usually required for adsorption of a complete monolayer of surfactant on solid surfaces, depending on the nature of the system and especially on the temperature.

Kinetics of adsorption have received little attention so far, and they are not very well understood; however, they are very important in many industrial processes. SEE CHEMICAL DYNAMICS; INTERMOLECULAR FORCES.

Bibliography. A. W. Adamson, *The Physical Chemistry of Surfaces*, 4th ed., 1982; D. A. King and D. P. Woodruff (eds.), *Adsorption at Solid Surface*, 1983; J. Oscik, *Adsorption*, 1981; M. J. Rosen, *Surfactants and Interfacial Phenomena*, 1978

DESORPTION
M. L. KNOTEK

A process in which atomic and molecular species residing on the surface of a solid leave the surface and enter the surrounding gas or vacuum. In stimulated desorption studies, species residing on a surface are made to desorb by incident electrons or photons. Measurements of these species provide insight into the ways that radiation affects matter, and are useful analytical probes of surface physics and chemistry. In thermal desorption studies, adsorbed surface species are caused to desorb as the sample is heated under controlled conditions. These measurements can provide information on surface-bond energies, the species present on the surface and their coverage, the order of the desorption process, and the number of bonding states or sites.

Stimulated desorption. Stimulated desorption from surfaces is initiated by electronic excitation of the surface bond by incident electrons or photons. The classical model of desorption is an adaptation of the theory of gas-phase dissociation, in which desorption results from excitation from a bonding state to an antibonding state. In this model the excitations for gas-phase dissociation and desorption from surfaces are identical, but the surface, which can absorb energy from the excited adsorbate before it can desorb, dramatically reduces the overall dissociation yield from surfaces. Thus, processes which are strong in the gas phase can be quenched on the surface, leading to a marked difference in the relative importance of competing mechanisms in the two environments. This model applies predominantly to neutral desorption of adsorbed molecular species.

Another model, proposed by M. L. Knotek and P. J. Feibelman, which is more applicable to the phenomenon of ion desorption, was first observed in studies of the desorption of positively ionized oxygen (O^+) from the surface of titanium(IV) oxide (TiO_2). Here it is found that O^+ is desorbed not by valence level excitation, but by ionization of the titanium and oxygen core levels. These levels, of course, have little to do with bonding. Furthermore, the fact that the oxygen is desorbed as an O^+ ion (whereas it is nominally at O^{2-} on the surface) implies a large (three-electron) charge-transfer preceding desorption.

Titanium(IV) oxide is an ionic solid, and more importantly it is a maximal valency ionic compound. Maximal valency means that in the solid the titanium (Ti) is ionized to the noble-gas configuration Ti^{4+}, so that there are effectively no valence electrons on the Ti ion in the solid; its highest occupied orbital is $3p$ at about 30 eV below the Fermi level. The important thing about maximal valency ionic materials is that if a core level is ionized in either the cation or the anion, higher-lying valence electrons will decay into the deep hole left in the core level and give up the energy gained by emitting multiple-valence electrons from the solid (multiple Auger emission). Since valence electrons are only on the anion, the anion is selectively stripped in the process. In TiO_2 one electron is lost from O^{2-} by decay into the core hole; in about 10% of the events two Auger electrons are emitted, transforming the O^{2-} to an O^+ ion. An O^+ ion in a highly repulsive potential for positive ions results, which leads to O^+ desorption. In the general case of an ionic bond M^+A^- (**Fig. 1**), the incident radiation ionizes the core level of the M^+A^- pair. Auger decay of the core hole selectively removes charge from the A^- ion and results in its being transformed into an A^+, and the resulting repulsion causes the A^+ to desorb. This model applies to a wide range of ionic materials, and explains why maximal valency materials like TiO_2, vanadium (V) oxide (V_2O_5), tungsten(VI) oxide (WO_3), molybdenum(VI) oxide (MoO_3), sodium fluoride (NaF), lithium fluoride (LiF), sodium chloride (NaCl), and so forth, are decomposed by irradiation.

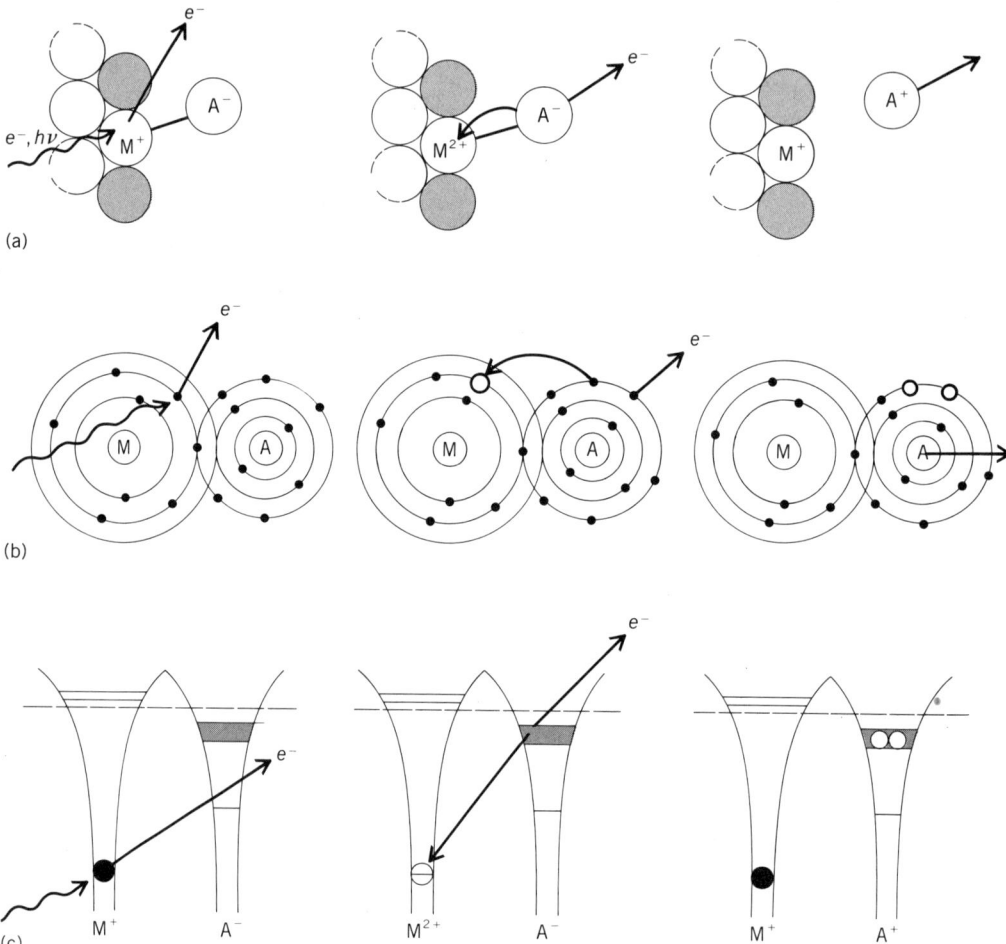

Fig. 1. Sequence of events leading to stimulated desorption by an electron (e^-) or photon ($h\nu$): (a) in the structural model, (b) in terms of orbitals, and (c) in terms of energy levels.

This mechanism for desorption can also be effective for covalently bonded surface species. Covalently bonded structures differ from ionically bonded structures in that the electronic charge which bonds the atoms together is distributed more evenly through the structure. The electronic interaction between atoms is strong so that an electronic excitation produced on one atom can quickly move to a neighbor. Thus, a simple electronic excitation can quickly move off the site where it was created before dissociation can occur. It typically takes 10^{-12} s for bond dissociation to occur, whereas the motion of simple electronic excitations can occur in 10^{-16} s. However, an important attribute of states where two or more electrons are removed from the atom is that such multiply excited states are much more difficult to move to a neighbor and hence are localized for times long enough for desorption to occur. Several features of a covalent bond can enhance this tendency to localize on the atom—for example, reducing the number of neighbors (and hence bonds), or reducing the bond's ability to transport charge to the desorbing species to neutralize the excited state.

Applications. Stimulated desorption studies are finding wide use. First, they can show the ways in which radiation affects the structure of solids. This will have important applications in the areas of radiation-induced damage and chemistry. Second, as an analytical tool, they offer a unique new way to study the physics and chemistry of atoms on surfaces which, when combined with the many other surface techniques based largely on electron spectroscopy, can provide new insight. Finally, models of the surface bond are put to a much sterner test in attempting to explain desorption phenomena.

Analysis of surface structure. The techniques of electron- and photon-stimulated desorption can be very useful in the study of how atoms and molecules bond to surfaces. Since ions are desorbed only from the topmost layers of the solid, these techniques study only that outermost layer. By mass-analyzing the desorbed atomic and molecular species, the nature of the chemical species on the surface can be deduced. More importantly, an adsorbed species will be desorbed when the core level of its bonding-site atom is ionized. Since the energy of the electron or photon necessary to ionize a given atom's core level is well known and characteristic of that atom, the bonding sites of each of the desorbing species can be determined. Thus surface-, site-, and adsorbate-specific information can be obtained. Elaborate methods for deducing how the bonding site and adsorbate atoms are electronically and structurally configured can be utilized with this technique. The method is almost the only spectroscopy capable of detecting hydrogen, a classic problem in this field.

Ion angular distributions. An additional important discovery is that ion angular distributions from stimulated desorption are not isotropic, but show that ions are emitted in relatively narrow cones which project along the nominal ground-state bond directions. An angular display shows an azimuthal pattern which reflects the symmetry of the bonding site, and a polar angular distribution which reflects the bonding angle relative to the surface normal. Thus this technique provides a direct display of the surface-bonding geometry.

There are a number of ways to measure the angular distributions of desorbed ions. One of the simplest employs an image intensifier. Individual ion events result in light pulses at a phosphor screen, and time-averaged photographs of the screen provide visual images of the distributions and thus allow a direct view of the arrangement of atoms and molecules on the surface (**Fig. 2**). The directness and simplicity of the technique make it particularly attractive to measure compli-

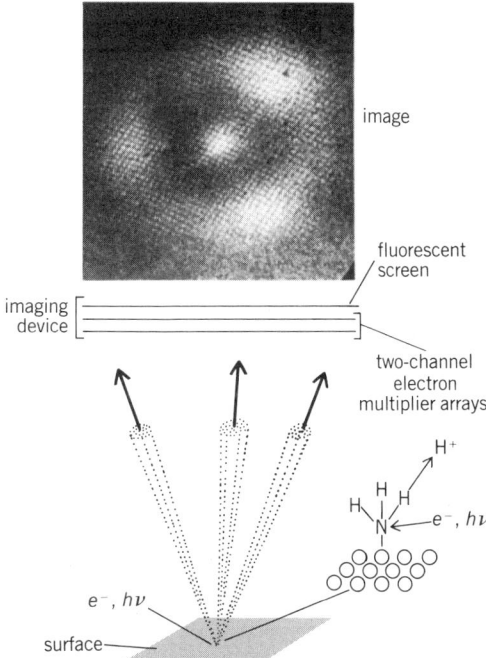

Fig. 2. Measurement of the angular distribution of ions leaving a surface in stimulated desorption.

cated surface adsorbate structures or chemical configurations which do not contain the long-range order necessary in many other techniques for measuring structure.

Thermal desorption. Thermal desorption mass spectroscopy is possibly the oldest technique for the study of adsorbates on surfaces. Three primary forms of the thermal desorption experiment involve measurement of (1) the rate of desorption from a surface during controlled heating (temperature-programmed thermal desorption), (2) the rate of desorption at constant temperature (isothermal desorption), and (3) surface lifetimes and diffusion under exposure to a pulsed beam of adsorbates (molecular-beam experiments). Of the three, temperature-programmed thermal desorption is by far the most widely applied, and the discussion below is confined to it.

In thermal desorption measurements, a surface is first dosed with an adsorbate gas, and then the sample is heated at a controlled rate β while the rate of desorption of the adsorbed gas is measured with a mass-filter gas analyzer. The most straightforward information provided is the nature of the desorbed species from mass analysis, and a determination of the absolute coverage by the adsorbate, which is very difficult to obtain with other techniques. Coverage is determined by measuring the total amount of gas desorbed. The technique can also provide several important kinetic parameters of the desorption process as follows.

1. Since adsorption reactions are usually nonactivated, the desorption activation energy E, which thermal desorption can provide, is a good approximation to the differential heat of adsorption. Thus these studies provide information on surface-bond energies. *See Adsorption.*

2. Desorption kinetics can be quite complex, especially when there are strong interactions between adsorbates. The general form of the mth-order desorption rate equation is given by Eq. (1), where $R(t)$ is the rate of desorption as a function of time, T_s is the substrate temperature, $\nu_0^{(m)}$

$$R(t) = \nu_0^{(m)} n^m \exp(-E/kT_s) \qquad (1)$$

is the preexponential factor of the desorption-rate coefficient for order m, and n is the two-dimensional adsorbate concentration. Desorption has been observed to obey zero-, first-, and second-order kinetics. Zero-order kinetics is observed for multilayers where removal of an adsorbed species does not lower the effective coverage feeding the desorption process. First-order kinetics is observed for simple desorption of an atomic or molecular species from a submonolayer of adsorbate. Second-order kinetics is often observed in systems where dissociative adsorption has occurred and recombination reactions occur before the desorption.

3. Desorption spectra can determine the number of binding states of an adsorbate on a surface, from the number of desorption peaks observed, and, coupled with structural models of the surface, can help determine both the nature of the bonding sites and the bonding geometries.

4. The preexponential factor of the desorption-rate coefficient, equivalent to an attempt frequency, the rate at which the desorbing particle attempts to leave the surface, not only is necessary to understand desorption kinetics, but helps in understanding the general problem of surface chemical dynamics.

5. The existence of indirect and direct interactions between coadsorbed species can be determined from thermal desorption. Knowledge of such interactions is directly applicable to such important problems as catalysis and corrosion. *See Catalysis.*

There are several methods to derive the kinetic parameters of Eq. (1) experimentally. If a value is assumed for the preexponential factor ν_0 (usually of the order of 10^{13}), the desorption activation energy is given by Eq. (2) for first-order kinetics, and by Eq. (3) for second-order kinet-

$$E/kT_p = (\nu_0^{(1)}/\beta) \exp(-E/kT_p) \qquad (2) \qquad E_k T_p^2 = (\nu_0^{(2)} n_0/\beta) \exp(-E/kT_p) \qquad (3)$$

ics. Here β is the heating rate, n_0 is the initial adsorbate concentration, and T_p is the temperature of peak desorption rate. Alternatively, β can be varied and E can be derived from Eq. (4) for first

$$E/kT_p = d(\ln \beta)/d(\ln T_p) \qquad (4)$$

order; for second order, E/k is determined from the slope of a plot of $\ln(n_0 T_p^2)$ versus $1/T_p$. It is also possible to derive the desorption energy from the width of the desorption peak.

One of the simplest of surface systems, hydrogen adsorbed on the (100) surface of tungsten (W), provides graphic examples how the above types of information manifest themselves and are analyzed. **Figure 3** shows molecular desorption of hydrogen as a function of sample temperature for heating rates β as shown in the figure. At low hydrogen coverages a single β_2 state is desorbed

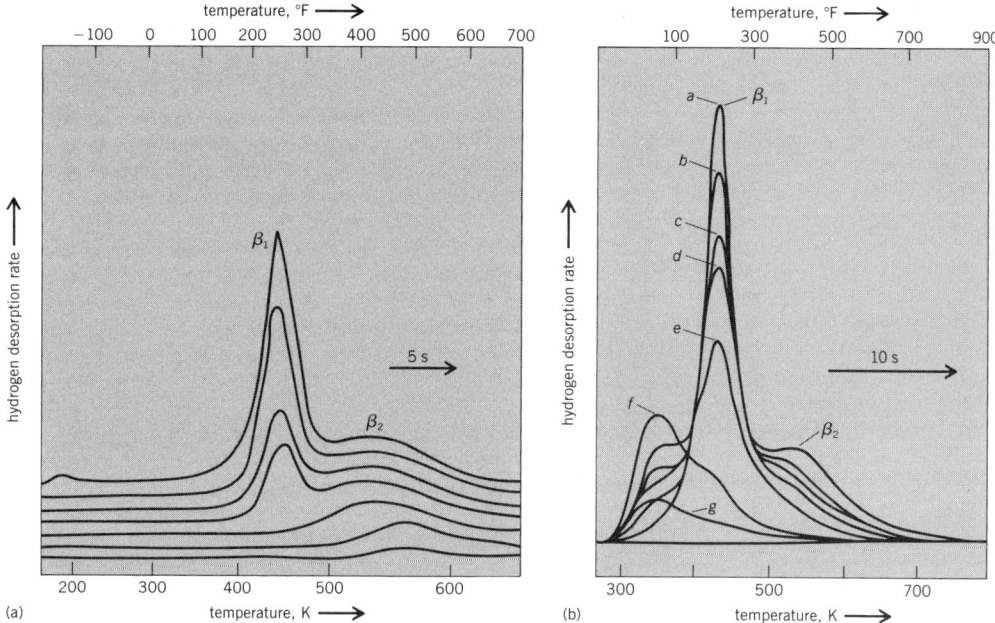

Fig. 3. Hydrogen desorption rate as a function of sample temperature for a heating rate of approximately 10°C/s. (*a*) Successive curves from bottom to top represent increasing hydrogen coverage. (*b*) Desorption when carbon monoxide (CO) is added to the hydrogen-preadsorbed surface. Successive curves from *a* to *g* represent increasing carbon monoxide coverage. (*After T. E. Madey and J. T. Yates, Surf. Sci., 63:203–231, 1977*)

with second-order kinetics. At coverages above 0.2 monolayer of hydrogen, the β_1 state desorbs with first-order kinetics. When carbon monoxide (CO) is added to the preadsorbed hydrogen layer, it causes the depopulation of both the β_1 and β_2 states and introduces a weak state of lower binding energy. The β_2 state is now known to be hydrogen bound at reconstructed sites on the W(100) surface, while the β_1 state is hydrogen at unreconstructed sites.

While the thermal desorption techniques are among the simplest of surface probes, they remain indispensable because of their directness and the variety of information they convey. Thus while surface science moves to detailed methods involving extremely sophisticated apparatus, the simple thermal desorption methods remain an important part of the overall picture.

MONOMOLECULAR FILM
George L. Gaines, Jr.

A film one molecule thick; often referred to as a monolayer. Films that form at surfaces or interfaces are of special importance. Such films may reduce friction, wear, and rust, or may stabilize emulsions, foams, and solid dispersions. Thin films on water surfaces reduce evaporation losses, which are important in arid regions throughout the world. Nevertheless, the removal of thin films of contaminants is one of many problems in the control of pollution. The broad field of catalysis, which is basic to petroleum refining and many chemical industries, involves chemical reactions that are accelerated in the thin films of reactants at interfaces. Moreover, thin films containing proteins, cholesterol, and related compounds constitute biological membranes, the internal interfaces that control the complex processes of life. *See* CATALYSIS.

In all of these areas, a single monomolecular layer at the interface is the most important. It is held to the adsorbing surface by forces stronger than those that hold any succeeding layer. On solid surfaces, it is the only layer that can be chemisorbed. It may be the site of enhanced chemical reactivity, or the last line of defense.

Monolayers on solids, or at liquid interfaces, may be formed by adsorption from the adjacent bulk phases; the process may show high specificity for particular chemical species. Measurements of the extent of adsorption have historically provided information on the composition and structure of monolayers formed in this way. A variety of surface-sensitive intrumental techniques, such as diffraction and scattering of low-energy electrons, neutrons, and ions, and spectroscopy of adsorbed species, have been brought to bear to obtain information about the structure of the surface layer and chemical perturbations in it. SEE ADSORPTION.

In addition, monolayers of a wide variety of substantially insoluble substances can be formed at a liquid-gas interface by allowing them to spread over the surface. The pioneering studies of I. Langmuir and W. D. Harkins in the United States, and N. K. Adam and E. K. Rideal in England, showed how to manipulate, control, and measure the properties of such films at the water-air interface in simple and elegant ways. In their research, and that of many subsequent workers, a variety of specialized experimental techniques have been developed to study these insoluble monolayers.

In order to form spread monolayers which are sufficiently stable to study, a substance must combine low solubility and volatility with some moiety which attracts it to the liquid surface; for films on water, this generally means one or more polar functional groups. Totally nonpolar substances, such as the higher-molecular-weight paraffin hydrocarbons, will not spread on water (although they can spread on liquids of very high surface tension, such as mercury). Typical among the large group of substances which do form insoluble monolayers on water are the long-chain fatty acids and their derivatives such as glycerides, sterols, and many lipid substances of biological origin, including the fat-soluble vitamins and natural pigments such as chlorophyll. Many polar synthetic polymers, including polyvinyl acetate and polymethyl methacrylate, can be made to spread as monolayers on water; so can many proteins, because their tertiary structure unfolds at the air-water interface.

Experimental techniques. The film balance provides basic information on molecular geometry and orientation, location and strength of polar groups, and forces of cohesion and adhesion. With this instrument, the surface pressure (or lowering of surface tension) is measured as a function of the area available to the film-forming molecules, or in other words, their concentration at the surface or proximity to one another.

The apparatus consists essentially of a long, shallow trough filled with high-purity water on which the monolayer is spread, and a float system for measuring the surface pressure (**Fig. 1**). The float is a strip of mica attached to the sides of the trough by thin, flexible platinum foils and to an aluminum stirrup by an unspun silk thread. The stirrup is fixed to a calibrated torsion wire that indicates the surface pressure. Small brass bars or barriers are used for sweeping the water surface free of contamination. A small amount (approximately 3.5×10^{-7} oz or 0.01 mg) of the film-forming compound in the volatile solvent is spread between the float and the large or main barrier. The barrier is moved gradually toward the float to compress the film. Compression is continued until the pressure remains constant or falls. This procedure at constant temperature provides data for plotting pressure-area isotherms which characterize the films.

Before each experiment the entire apparatus—trough, barriers, float, and platinum foils—

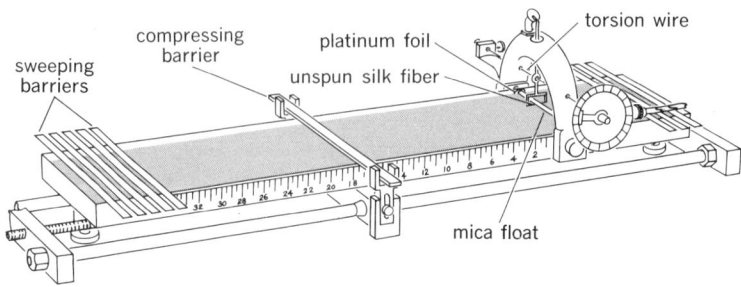

Fig. 1. Schematic drawing of film-balance apparatus.

are thoroughly cleaned and coated with high-melting paraffin wax (or with Teflon). The trough is filled with water to a height well above the rim. This height is necessary if the sweeping procedure is to be effective and if the monolayer is to be contained and controlled.

Three representative polar organic molecules oriented at the water-air interface are shown in **Fig. 2**. Stearic acid is the classical compound in monolayer studies; it is the simplest structure representative of thousands of important film-forming compounds. The stearic acid molecule consists of a long, straight hydrocarbon chain and a polar group at one extremity. Also included in Fig. 2 is the structural formula of a very similar molecule, isostearic acid, and a very dissimilar molecule, tri-p-cresyl phosphate. The difference between isostearic and stearic acid is very slight—the displacement of the small methyl group (CH_3) at the end of the molecule opposite the polar group. Tri-p-cresyl phosphate is greatly different; it has a bulky three-ring hydrocarbon portion attached to a strongly polar phosphate group.

Figure 3 shows the pressure-area isotherms for the three compounds. The surface pressure in millinewtons per meter (or dynes/cm) is plotted against the average area per molecule in square nanometers. Extrapolation or extension of the steepest part of an isotherm to zero pressure is often used as a measure of molecular area. The point at which the pressure falls or remains

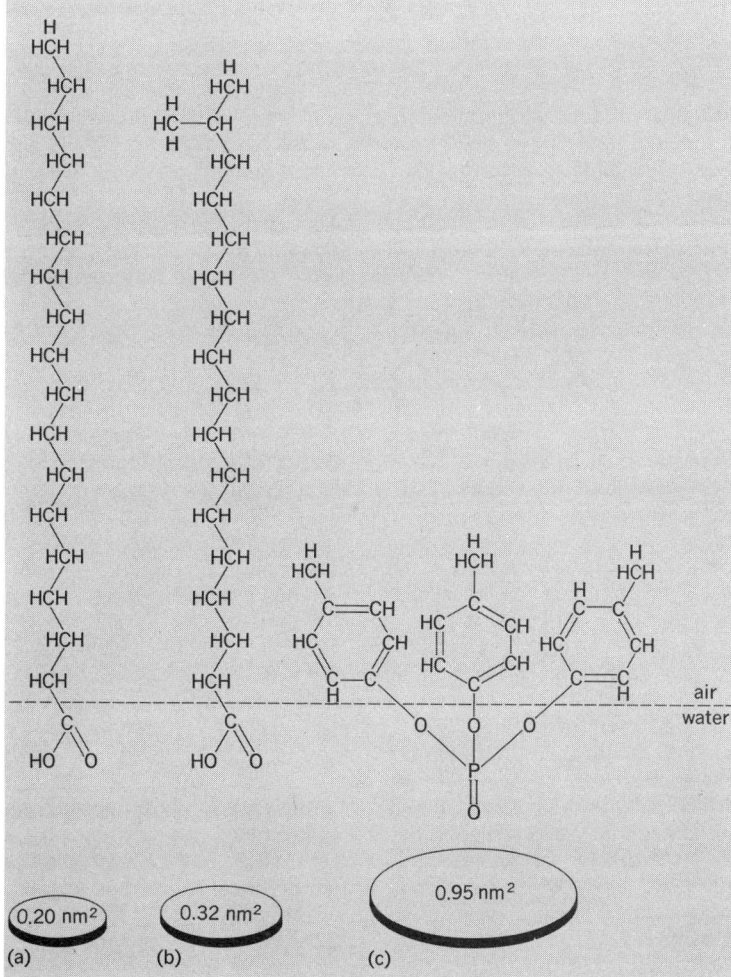

Fig. 2. Molecular orientation and cross-sectional areas of three representative polar organic molecules that are oriented at the water-air interface. (*a*) Stearic acid. (*b*) Isostearic acid. (*c*) Tri-*p*-cresyl phosphate.

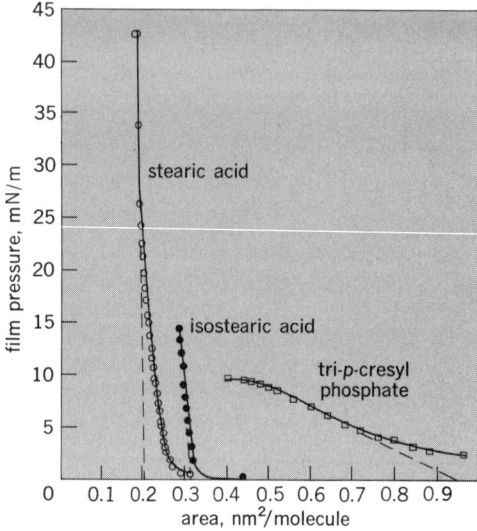

Fig. 3. Pressure-area isotherms.

constant is called the collapse pressure. Compressibility of the monolayer may be calculated from the slope of the isotherm. Thickness of the monolayer, or the length of the vertically oriented molecules, may be estimated by assuming a density for the monolayer; the volume and area then yield the thickness.

Comparison of the isotherms for stearic and isostearic acids in Fig. 3 demonstrates that the single, small side chain of isostearic acid has increased the cross section from 0.20 square nanometer for the stearic acid molecule to 0.32 square nanometer for isostearic acid, an increase of more than 50%. Collapse pressure falls from 42 mN/m to one-third of this value, 14 mN/m. These are indeed striking differences between molecules that are extremely difficult to distinguish by most chemical methods.

The curve for tri-p-cresyl phosphate reflects a very different molecular structure. The extrapolated area, 0.95 nm^2/molecule, shows the bulkiness of the three-ring group held close to the surface. The low collapse pressure, 9 mN/m, reflects the weakness of such a film. The gradual slope of the curve, or the high compressibility of the film, indicates poor packing of the molecules.

In addition to changing the surface tension, the presence of a spread monolayer alters other properties of the liquid surface. The electric field, flow properties, and optical reflectivity are all more or less altered. Special techniques have been devised to measure the surface potential difference, surface viscosity, and reflectance change due to spread monolayers. Among other properties which have been measured are the ability to damp ripples or to suppress evaporation. While various film-forming substances differ in their effect on all these properties, they are still not understood well enough for the differences to be related to molecular structure in most cases.

Absorption and emission spectra of monolayers of colored and fluorescent substances, such as chlorophyll, have also been measured. Because the molecules in a spread monolayer are held in a preferred orientation at the air-water interface, and are packed close together when the film is compressed, the spectra are sometimes very different from those obtained from solutions. **Figure 4** shows an extreme example of this effect. The two isomers of the porphyrin differ only in the position of the substituents at the periphery of the ring system, and their spectra in solution are indistinguishable. When they are spread as monolayers, the absorption spectra differ markedly.

Transfer of spread monolayers. Once a monolayer has been formed at the water-air interface, it is possible to transfer it quantitatively to another surface, such as that of a smooth solid. Dipping a clean glass slide, for example, into and out of the film balance trough while the monolayer on it is held at constant surface pressure (by advancing the confining barrier) leads to such a quantitative transfer. The area of monolayer taken up is the same as the surface area of

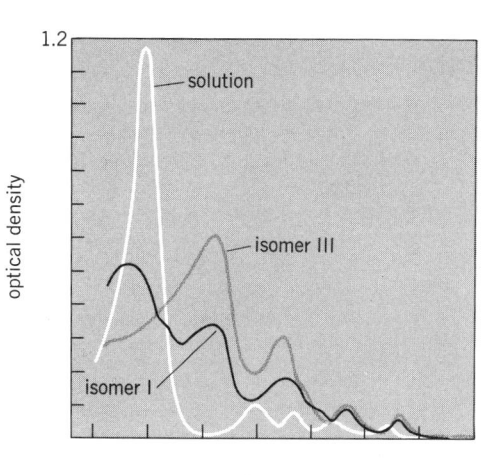

Fig. 4. Absorption spectra of monomolecular films of two isomers of coproporphyrin tetramethyl ester. While the solution spectra are identical, the monolayer spectra are very different, probably because the neighboring polar (—COOCH₃) groups on isomer III (shaded) provide a preferred orientation at the water surface.

the dipped slide. Properties of the transferred layer such as its wettability and electron diffraction pattern indicate that the molecules on the solid surface retain the preferred orientation which they had on the water surface. Monolayers transferred to solids in this way bear close resemblance in structural details both to the precursor liquid-supported monolayer and to films formed on solids by other processes such as adsorption from solutions, but there are usually some differences. The differences, in structure, tightness of bonding, and so on, depend on the nature of both the monolayer and the solid support.

It is also possible to deposit certain types of monolayers (especially heavy-metal soaps of long-chain fatty acids) sequentially on solid surfaces to form built-up films or multilayers (**Fig. 5**). Since each monolayer is extremely thin but uniform (for barium stearate, for example, almost exactly 2.5 nm), such layer structures are very useful as spacers or thickness gages. Films more than 1000 layers thick have been made. In recent years, much interest has developed in the optical and electrical properties of such structures. It is possible to assemble multicomponent structures in many ways, both by using monolayers containing more than one chemical species and by changing from one kind of monolayer to another at different cycles in the dipping process. Locating different molecules at known small distances and in controlled relative orientation has permitted the study of such processes as energy transfer and electron transfer between them.

Chemical reactions. The orientation and close packing of molecules in monolayers may alter patterns of chemical reactivity. Monolayers of fatty acids show highly preferential incorporation of cations (soap formation) with certain salts in the underlying water. For example, a stearic acid film spread on a solution with equimolar concentrations of calcium and magnesium salts may contain 10 times as much calcium stearate as magnesium stearate.

At the air–aqueous solution interface, the electric field, and hence distribution of ions near the liquid surface, is unsymmetrical. This can have a controlling effect on the rates of chemical reactions involving these ions. An interesting example is the rate of hydrolysis of long-chain esters in monolayers, which involves the negatively charged hydroxyl ion from the underlying solution. To begin with, the rate of the reaction can be reduced by increasing the surface pressure, which forces the ester molecules closer together and also changes the surface electrical potential. If charged mono-layer-forming molecules, such as long-chain quanternary ammonium compounds

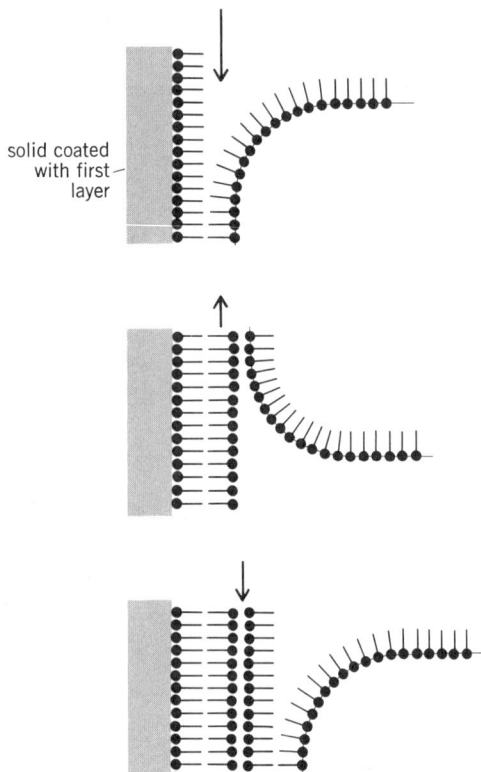

Fig. 5. Successive folding back and forth of a monolayer onto a solid plate, as it is dipped into and out of the liquid in the Langmuir-Blodgett multilayer deposition method. (*After G. L. Gaines, Jr.,* Insoluble Monolayers at Liquid-Gas Interfaces, *John Wiley and Sons, 1966*)

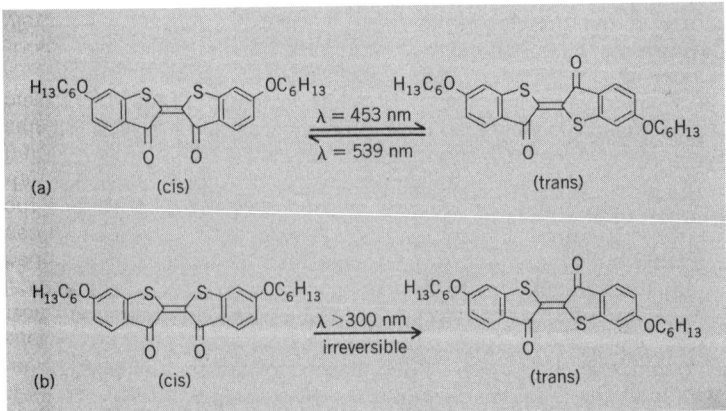

Fig. 6. Thioindigo dye undergoing a photoinduced cis-trans isomerization. (*a*) In solution, blue light (λ = 453 nm) converts it to the trans form, while green light (λ = 539 nm) reverses the reaction. (*b*) In a monolayer, only the cis→trans conversion can occur.

(with positive charges), are incorporated in the ester film, the reaction rate can be increased markedly. This results from the increased concentration of OH⁻ ions induced by the positive charge at the surface. Reactions of this kind are of special interest because they resemble many of the catalytic reactions which occur at membrane surfaces in biological systems.

In extreme cases, even the reaction products which can be formed are altered when molecules are constrained in the special environment fo a monolayer or built-up multilayer. For example, certain thioindigo dyes (**Fig. 6**) undergo reversible light-induced cis-trans isomerization in solution. In monolayer assemblies, however, only the cis to trans transformation seems to be possible; once the trans isomer is formed, illumination cannot reverse the reaction. This effect apparently results from the fact that the trans form occupies less space than the cis form. The isomerization which does occur, therefore, involves a contraction in volume (on a molecular scale), and is irreversible in the built-up film. The surface pressure–area isotherms of the two forms indicate the same thing, since the trans isomer requires less area at the air-water interface, and if a monolayer of the cis isomer is illuminated (at constant surface pressure), it shrinks.

Reactions of the kind just described are fairly easy to study because the reactants and products have intense and very different absorption spectra. In the past, analytical techniques capable of measuring the very small amounts of material present in any convenient area of a single monomolecular layer (typically less than a microgram per square centimeter) were very limited. Modern analytical methods, however, have greatly improved sensitivity, and are being increasingly applied to study monomolecular films. Coupled with improved understanding of physical properties of the films and of liquid interfaces, these newer techniques promise new insights into the behavior of molecules at surfaces and the important biological and technological processes which they control. SEE COLLOID; INTERFACE OF PHASES; SURFACE TENSION.

Bibliography. A. W. Adamson, *Physical Chemistry of Surfaces*, 4th ed., 1982; G. L. Gaines, Jr., *Insoluble Monolayers at Liquid-Gas Interfaces*, 1966; E. D. Goddard (ed.), *Monolayers*, Advances in Chemistry Series, no. 144, 1975; M. J. Jaycock and G. D. Parfitt, *Chemistry of Interfaces*, 1981; A. Weissberger and B. W. Rossiter (eds.), *Physical Methods of Chemistry*, pt. IIIB, 1972.

TRANSPORT PROCESSES

Transport processes	**120**
Diffusion in gases and liquids	**122**
Osmosis	**128**
Viscosity	**129**
Dialysis	**132**

TRANSPORT PROCESSES
W. A. WAKEHAM

The processes whereby mass, energy, or momentum are transported from one region of a material to another under the influence of composition, temperature, or velocity gradients. If a sample of a material in which the chemical composition, the temperature, or the velocity vary from point to point is isolated from its surroundings, the transport processes act so as to eventually render these quantities uniform throughout the material. The nonuniform state required to generate these transport processes causes them to be known also as nonequilibrium processes. Associated with gradients of composition, temperature, and velocity in a material are the transport processes of diffusion, thermal conduction, and viscosity, respectively. For a large class of materials, the laws which govern the transport processes are quite simple.

Diffusion. Figure 1 shows a sample of a material which is composed of two chemical species. The sample is stationary and has a uniform temperature throughout, but a composition difference is maintained across its two ends, and in this steady state the two species continuously migrate down their concentration gradients. Expressing the composition of the material by means of the molar concentration of one species, c_1 (moles/m^3), it is found that the number of moles of this species which cross unit area of the sample perpendicular to the z direction in unit time known as the flux of mass (J_1), is given by Eq. (1) which is Fick's law of diffusion. The constant

$$J_1 = -D \frac{dc_1}{dz} \qquad (1)$$

of proportionality, D, between the mass flux and the concentration gradient, which depends upon the nature of the material, its temperature, pressure, and composition, is called the diffusion coefficient.

The phenomenon of diffusion occurs widely in nature, and it is frequently important in technological applications. For example, the transpiration of the leaves of plants, in which they absorb carbon dioxide from the atmosphere and give off water vapor, is controlled by a diffusion process. The rates of many chemical reactions in fluids which are promoted by catalysts may similarly be controlled by the diffusion of reactants to the active catalyst sites. SEE DIFFUSION IN GASES AND LIQUIDS.

Thermal conduction. In **Fig. 2**, a sample of a material is subjected to a steady temperature difference between two faces perpendicular to the z direction. Under these conditions, energy is continually transported from the hotter face to the colder, and the energy flux, J_q, in the z direction (the energy crossing unit area in unit time) is given by Fourier's law as Eq. (2). The

$$J_q = -\lambda \frac{dT}{dz} \qquad (2)$$

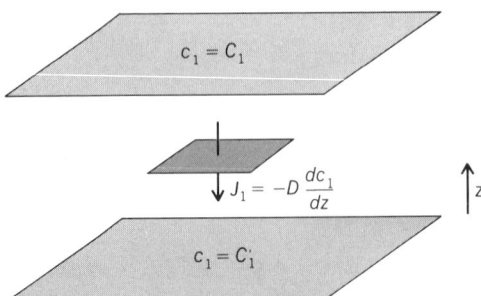

Fig. 1. Diffusion in a sample of material composed of two chemical species, C_1 and C_1' represent the molar concentration of one of the species in the two planes bounding the material.

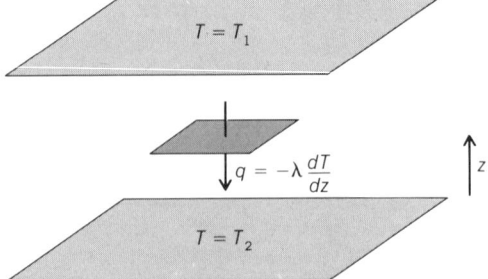

Fig. 2. Thermal conductivity in a sample of material subjected to a steady temperature difference.

constant of proportionality λ between the flux and the temperature gradient, dT/dz, is the thermal conductivity coefficient, which again depends upon the material as well as its temperature, pressure, and composition.

Viscosity. The phenomenon of viscosity is associated with the gradient of velocity in a material. Since it is difficult to maintain velocity gradients in solids, the phenomenon is only readily observed in fluids. Because velocity is a vector quantity, **Fig. 3** shows a fluid whose upper surface is in contact with a solid boundary which moves with a steady velocity U in the x direction only; the lower surface of the fluid is held stationary. As a result, various layers of the fluid in the z direction move with different x-direction velocities, u. Associated with the motion in the x direction, the fluid possesses a momentum, and the x-direction, momentum is transported down the velocity or momentum gradient. The flux of x momentum in the z direction, J_m, is equivalent to a tangential shear stress, τ_{xz}, acting in the negative x direction on each layer of the fluid. This means that a tangential force must be applied to the upper plate to keep it in steady motion. Again the flux is proportional to the imposed gradient and is given by Eq. (3), which is Newton's

$$J_m = \tau_{xz} = -\eta \frac{du}{dz} \qquad (3)$$

law of viscosity. The proportionality constant η is the viscosity coefficient for the material, and it too depends on the thermodynamic state of the material. The phenomenon of viscosity is revealed whenever a fluid flows near a solid boundary, and it is therefore of significance in almost every aspect of enginering. SEE VISCOSITY.

Thermal processes. Other, more subtle transport processes can occur. For example, in a mixture of two chemical species, the imposition of a temperature gradient leads not only to energy transport, but also to a mass transport which causes a partial separation of the mixture. This phenomenon is called thermal diffusion. Conversely, when diffusion takes place in an initially isothermal mixture as a result of a composition gradient, small temperature gradients can be observed in the material arising from an energy transport. This is the diffusion thermoeffect.

Transport coefficients. The coefficients D, λ, and η are known collectively as transport coefficients. The measurement of these coefficients for materials in solid, liquid, and gaseous phases has been the object of a considerable research effort for many years. The measurements can only rarely be carried out by directly implementing the situations envisaged in Figs. 1 to 3. This is because it is difficult to achieve the one-dimensional gradients of the quantities required when the sample is of a finite size. The exceptions to this are the measurement of thermal conductivity and diffusion in solids, where simple methods have proved effective. In fluids, the diffusion coefficient has most often been determined in a time-dependent experiment in which an initial concentration gradient in a mixture is allowed to decay in a closed vessel of known geometry. The approach to equilibrium, which is governed by the diffusion coefficient, is observed with a suitable concentration monitor.

The coefficient of thermal conductivity in fluids is also most accurately determined in a transient experiment. The fluid surrounds a thin, vertical wire which is suddenly heated by an electric current; the rate of the temperature rise of the wire is observed, and the thermal conductivity of the fluid is deduced from it.

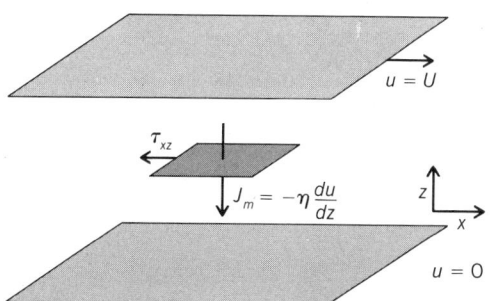

Fig. 3. Viscosity in a fluid whose upper surface is in contact with a solid boundary.

The viscosity coefficient of gases and liquids is generally determined by one of two techniques. In the first, the fluid flows through a capillary tube of known geometry, and the pressure difference across the ends of the tube necessary to maintain a given flow of rate is determined. This pressure difference is then proportional to the viscosity coefficient of the fluid. In the second method, the damping of the torsional oscillations of a thin, solid, horizontal, circular disk is observed when the disk is suspended in the fluid. The measurement of the logarithmic decrement of the oscillations serves to determine the viscosity coefficient. SEE QUASIELASTIC LIGHT SCATTERING.

The results of measurements of the transport coefficients of material are of importance since there are few technological activities which do not involve one or more of the transport processes. However, because the transport coefficients derive their values from the properties and behavior of the atoms and molecules which make up the material, they are also of more fundamental significance. In the particular case of gases at low density (near atmospheric pressure), the kinetic theory of gases has provided an almost complete description of their transport coefficients. In such gases the sole mechanism for the transport of mass, energy, or momentum is by means of free molecular motion. The transport coefficients are therefore a direct measure of the ease of this free molecular motion. Since the molecular motion is hindered only by collisions between pairs of molecules, the transport coefficients are sensitive to the details of such collisions and thereby to the potentials describing the interactions of molecules. Indeed, the transport coefficients of dilute gases have proved a valuable source of information about these interactions. The viscosity and thermal conductivity of coefficients of moderately dense gases are essentially independent of pressure, but the diffusion coefficient is inversely proportional to it. The thermal conductivity and viscosity increase with temperature roughly proportionally, whereas the temperature dependence of the diffusion coefficient varies more nearly as the square of the temperature at constant pressure. These temperature dependencies arise principally from the temperature dependence of the velocity of the free molecular motion.

As the density of the material is increased toward that of a liquid and finally to that of a solid, significant changes in the molecular mechanism of the transport processes occur. The transport by free molecular motion becomes a smaller contribution as the volume available for such motion decreases. In addition, the attractive forces between molecules, which become increasingly significant, tend to inhibit molecular motion. Thus, on the one hand, the diffusion coefficient in condensed phases, which is still determined by molecular motion, is very much smaller (about 10^4 times) than that in low-density gases. On the other hand, the viscosity of liquids is very much greater than that in gases because the attractive forces between molecules make the relative motion of various layers in the fluid much more difficult to achieve. Because increasing the temperature of a liquid increases the average separation of the molecules as well as their energy, the diffusion coefficients for liquids increase rapidly with temperature and, for the same reasons, the viscosity decreases. In the solid the molecules acquire almost fixed positions, and the diffusion coefficient consequently becomes even smaller, whereas the viscosity is practically infinite.

Bibliography. R. B. Bird, W. E. Stewart, and E. N. Lightfoot, *Transport Phenomena*, 1960; S. R. de Groot and P. Mazur, *Non-Equilibrium Thermodynamics*, 1984; R. DiPippo, J. Kestin, and J. H. Whitelaw, A high-temperature oscillating disc viscometer, *Physica*, 32:2064–2080, 1966; R. Fahien, *Fundamentals of Transport Phenomena*, 1983; G. C. Maitland et al., *Intermolecular Forces: Their Origin and Determination*, 1982.

DIFFUSION IN GASES AND LIQUIDS
CHARLES R. WILKE

The spreading or scattering of matter under the influence of a concentration gradient.

Molecular diffusion. Consider a fluid confined in a space of dimensions which are large compared to the mean free path of the fluid molecules. For gases at atmospheric pressure and room temperature, the mean free path is on the order of 4×10^{-6} in. (10^{-5} cm) and varies inversely with pressure. Liquids in general have much smaller free paths. Assume that one of the components of the fluid exists initially at different concentrations, or more rigorously at different activities, in two or more different locations in the confining space. At constant temperature, and in the absence of external forces, there will be a spontaneous movement, that is, diffusion, of the component in the direction of establishing a uniform concentration of that component in all parts of the enclosure.

In a very qualitative way the cause of this spontaneous mixing may be interpreted as follows. As a consequence of thermal agitation, molecules of a fluid are in constant motion in all directions. The number of molecules of a given kind moving in any given direction at a particular point in the fluid is proportional to the number of these molecules present per unit volume. In the absence of a concentration gradient, on the average as many molecules per unit time leave any hypothetical plane in a given direction as return to the plane from the opposite direction. However, if the number of molecules per unit volume decreases in a given direction, more molecules on the average move into the region of lower concentration than return from it. This results in a net transfer of molecules of that kind toward the region of lower concentration.

In general, the diffusion of one component of a mixture will be accompanied by diffusion of one or more other components in the opposite direction to maintain the volume of fluid constant in the region under consideration. This may not be the case when special restrictions are placed on the system, for example, in the selective absorption of one component from a gas by a liquid acting as a boundary of the confining space, or for liquids when there is a volume change on mixing.

The rate law for equivolume diffusion was proposed by Adolf Fick in 1855. For diffusion of component A in a mixture of A in component B in the z direction across a plane of unit cross section perpendicular to the direction of diffusion, the rate is given by Eq. (1). Here N_{A_d} is the

$$N_{A_d} = -D_{AB} \frac{dC_A}{dz} \tag{1}$$

rate of diffusion in moles/(cm^2) (s); D_{AB} is the diffusion coefficient of A in B, cm^2/s; C_A is the concentration of component A, moles/cm^3; and z is the distance in direction of diffusion in centimeters. Similar equations may be written for concentration gradients in the other coordinate directions.

At room temperature and atmospheric pressure, diffusion coefficients for most gases and vapors lie in the range of 0.1 to 1 cm^2/s. In liquids of about 1 centipoise viscoity, diffusivities are of the order 10^{-4} to 10^{-5} cm^2/s.

Eddy diffusion. Fluid motion may be viscous or turbulent. Viscous flow is characterized by movement of the fluid in essentially smooth streamlines parallel to the walls of the pipe through which the fluid is flowing, or more generally, parallel to any surface over which flow occurs. Turbulent flow, on the other hand, is characterized by movement of fluid in the form of statistically defined lumps or "eddies" which fluctuate randomly in velocity perpendicular to the surface as well as parallel to the direction of flow. Turbulence develops when fluid flows past a surface at a sufficiently high velocity so that the Reynolds number exceeds some critical value for the system. In viscous flow, mass transfer resulting from a concentration gradient occurs by molecular diffusion. In turbulent flow, transfer of material occurs by the more rapid process of mixing of the swirling eddies of fluid, a process termed turbulent diffusion or eddy diffusion. It is customary to define the mass-transfer flux in eddy diffusion by means of an eddy-diffusion coefficient E as in Eq. (2), where E is roughly proportional to the size of the eddies and to the magnitude of the

$$N_A = -E_A \frac{dC_A}{dz} \tag{2}$$

velocity fluctuations. Eddy-diffusion coefficients are usually much larger than molecular coefficients. In flow of air in a duct, for example, E has been observed to vary from 3 to 40 cm^2/s over a range of Reynolds numbers from 10,000 to 175,000. Liquid-phase values would be lower by a factor of about 10^{-3}. However, in the vicinity of a rigid boundary, turbulence is suppressed and much lower values of E may exist, so that the molecular and turbulent coefficients may become similar in magnitude, and immediately at the interface, E may become negligible. Finally, description of turbulent mixing with a simple diffusion law is a convenient oversimplification of an extremely complex fluid-mechanical problem which is not fully understood.

Thermal diffusion. S. Chapman demonstrated in 1916 as a consequence of the kinetic theory that a temperature gradient in a mixed gas might give rise to a flow of one constituent relative to the mixture as a whole. His finding was verified experimentally by L. Chapman and

F. W. Dootsen, who placed various mixtures in a bulb which was heated at one end and cooled at the other. A similar effect had been observed for liquids by Ludwig in 1856 and by J. L. Soret in 1879, and is usually called the Soret effect. The phenomenon as it pertains to fluids generally is called thermal diffusion. The thermal diffusion flux of a given component of a binary mixture is expressed by Eq. (3), where N_{A_T} is the number of moles/(cm^2)(s), D_T is the coefficient of thermal

$$N_{A_T} = D_T \rho' \frac{d \ln T}{dz} \tag{3}$$

diffusion in cm^2/s, T is temperature in kelvins, z is the distance in direction of diffusion in centimeters, and ρ' is the fluid density in g-moles/cm^3.

Magnitude of the thermal-diffusion coefficient varies widely, depending upon the sizes and chemical nature of the molecules. However, it seldom has a value greater than 30% of the molecular diffusion coefficient and is usually much smaller. Hence, unless temperature gradients are quite large and the fluid is nonturbulent as well, thermal diffusion is not an important factor in most mass-transfer operations. Specific equipment and processes have been developed to utilize thermal diffusion for separation of isotopes and other substances that may not yield to less costly separation techniques.

Forced diffusion. An external force acting upon each of a group of molecules may induce molecular movement analogous to diffusion. Such movements, which may result from gradients of pressure within a fluid, or from electric or magnetic fields, may be termed forced diffusion. An important example is the process of electrolytic migration of ions under an applied potential. Movement of materials between a working electrode and the solution with which it is in contact may result predominantly from migration in the absence of an inert electrolyte to carry the current. Rate of migration at a given point within a fluid is given by Eq. (4), where U_A is the mobility of A

$$N_{A_m} = C_A U_A \frac{di}{dz} \tag{4}$$

in cm^2/(s)(V) and i is the potential in volts. Ionic mobilities depend upon the particular ion, the nature of the solvent, and the temperature. For dilute aqueous solutions at 77°F (25°C) most ions with the exception of hydrogen and hydroxyl ions have mobilities in the range of 3 × to 8 × 10^{-4} cm^2/(s)(V).

Convection or bulk flow. If the fluid as a whole is moving in the direction of diffusion with velocity V_z, component A is carried with it at a rate shown by Eq. (5).

$$N_{A_C} = V_z C_A \tag{5}$$

General equation for mass-transfer flux. It is generally assumed that all mechanisms act simultaneously in an additive manner; therefore, the equation for the net flux of a given component may be written as the sum of the fluxes due to each mechanism. For the flux of component A passing in the z direction through a unit cross section at any point within the fluid and at any particular instant of time, Eq. (6) holds. Similar equations can be written for the fluxes

$$(N_A)_z = -D_A \frac{\partial C_A}{\partial z} - E \frac{\partial C_A}{\partial z} + D_T \rho' \frac{\partial \ln T}{\partial z} + C_A U_A \frac{\partial i}{\partial z} + V_z C_A \tag{6}$$

in other coordinate directions to give a set of general differential equations describing mass transfer in a fluid. These equations coupled with the equations for conservation of mass and energy offer the basis for general solution of mass-transfer problems. Unfortunately, these equations seldom can be solved for situations of practical importance because knowledge of all coefficients, fluid velocities, and other variables must be known as a function of concentration and position.

In the absence of thermal diffusion and migration, Eq. (6) reduces to the diffusion-convection equation shown as Eq. (7). The diffusion-convection equations have been solved for a few

$$(N_A)_z = -D_A \frac{\partial C_A}{\partial z} + V_z C_A \tag{7}$$

situations such as the case of mass transfer from the wall of a tube to a fluid passing through it in viscous flow or for laminar flow over a flat plate. In these cases fluid velocities are known as a function of position in the fluid, and the diffusion coefficient is constant so that the differential equations may be integrated.

Mass-transfer coefficients. Mass transfer through a turbulent fluid to an interface formed by some other fluid with which it is immiscible or by a solid surface is commonly encountered in processes of industrial importance. For such situations it is necessary to resort to empirical equations to describe mass-transfer rates. An an example, consider the dissolution of a slightly soluble solute, such as benzoic acid, from the wall of a pipe into water in turbulent flow. Solute concentrations in the fluid in the vicinity of the wall are illustrated qualitatively in **Fig. 1**. At the interface the water is saturated with solute at concentration C_i corresponding to the solubility of the solute. The concentration falls progressively from the wall until a value corresponding essentially to the value in the bulk liquid, C_0, is reached. Actually the concentration will ultimately reach some minimum value at the center of the pipe which will be very close to C_0 because of the flat concentration profile which exists in the turbulent core. The exact shape of the concentration profile will depend upon the magnitude of the molecular- and eddy-diffusion coefficients throughout the fluid. Very near the wall, where molecular diffusion is more nearly controlling, the profile will be steep, and in the turbulent core, in which the eddy diffusivity is very large compared to the molecular value, the profile will be relatively flat. In the steady state the mass-transfer flux at a given point is commonly expressed by a mass-transfer coefficient, k_C, as in Eq. (8). Similar

$$N_A = k_C(C_i - C_0) \tag{8}$$

concepts can be applied to both gases and liquids and for surfaces other than pipe walls. Various mass-transfer coefficients may be defined corresponding to the units used for mass-transfer rates, concentrations, and distance.

The most common concentration differences on which the various coefficients are based are partial pressure (for gases), molar concentration, and mole fraction. Thus for transfer in the gas phase of component A through a fluid to an interface, Eq. (9) holds. Here r_A is the rate of

$$r_A = k_{g_A}S(p_{A_0} - p_{A_i}) = k_{Y_A}S(Y_{A_0} - Y_{A_i}) = k_{C_A}S(C_{A_0} - C_{A_i}) \tag{9}$$

transfer of component A in lb-moles/h; k_{g_A} is the gas-phase mass-transfer coefficient for component A in lb-moles/(h)(atm)(ft^2); S is the interfacial area available for mass transfer in ft^2; p_{A_0}, p_{A_i} are the partial pressure of component A in the fluid bulk and interface, respectively, in atm; k_{Y_A} is the gas-phase mass-transfer coefficient, mole fraction basis, in lb-moles/(h)(ft^2); Y_{A_0}, Y_{A_i} are the mole fractions of component A in the fluid bulk and interface, respectively; k_{C_A} is the gas-phase mass-transfer coefficient, concentration basis, in ft/h; and C_{A_0}, C_{Ai} are the concentration of A in the bulk and interface, respectively.

For liquid-phase transfer, similar equations may be written, as in Eq. (10). Here k_{X_A} is the

$$r_A = k_{X_A}S(X_{A_0} - X_{A_i}) = k_{L_A}S(C_{A_0} - C_{A_i}) \tag{10}$$

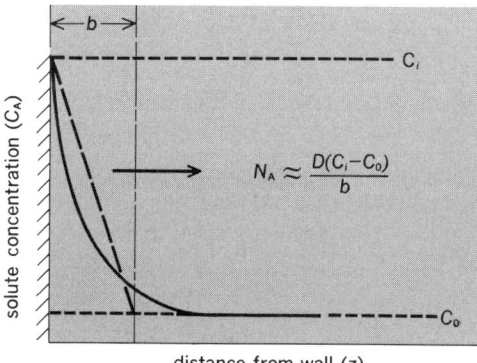

Fig. 1. Concentration profile for mass transfer from pipe wall into turbulent fluid.

Mass-transfer coefficients in fluid systems

System, at 77°F (25°C), 1 atm (100 kPa)	Gas-phase water vapor to air stream k_y, lb-moles/ (h)(ft²)*	Liquid-phase benzoic acid to water stream k_x, lb-moles/ (h)(ft²)
Pipe wall to fluid in 1-in. or 2.54-cm (interior diameter) tube at Reynolds number = 30,000	3.0	1.3
Pipe wall to fluid in 1-in. or 2.54-cm (interior diameter) tube at fluid velocity = 15 ft/s (4.58 m/s)	0.92	4.2
Packed spheres to fluid ¼-in. or 0.64-cm spheres at modified Reynolds number = 1000	6.2	3.5

*1 lb-mole/(h)(ft²) = 4903.2 g-moles/(h)(m²).

liquid-phase mass-transfer coefficient, mole-fraction basis, lb-mole/(h)(ft²); X_{A_0}, X_{A_i} are the mole fraction of A in the fluid bulk and interface, respectively; and k_{L_A} is the liquid-phase mass-transfer coefficient, concentration basis, in ft/h. Other symbols are the same as in the above equations for gases. In many types of equipment, fluid concentrations change over the mass-transfer area and an appropriate average concentration difference between bulk fluid and interface must be employed.

To illustrate the order of magnitude of mass-transfer coefficients obtained under various conditions, a range of typical values is listed in the **table**.

Equivalent film thickness. An alternate method of describing the rate of mass transfer is to define an equivalent film b of stagnant fluid which offers the same resistance to transfer (that is, requires the same total drop in concentration between the interface and bulk liquid) as the actual fluid region which is not completely stagnant (Fig. 1). Integration of Eq. (1) for steady-state diffusion gives the relation between the mass-transfer flux and the film thickness for equivolume diffusion, Eq. (11).

$$N_A = \frac{D_A(C_i - C_0)}{b} \quad (11)$$

Correlation of mass-transfer coefficients. Theories for mass transfer in turbulent flow in pipes have yielded general correlations for the prediction of mass-transfer coefficients. All of these depend upon a knowledge of fluid-velocity distribution and an assumption with respect to the relationship between transfer of momentum and of mass. Momentum transfer in turbulent flow can be described in terms of the interdiffusion of fluid eddies by means of an eddy-diffusion coefficient for momentum analogous to the similar quantity for diffusion of mass. One of the most successful of these semiempirical methods is that of C. S. Lin and associates, in which the eddy-diffusion coefficients for mass and momentum are assumed to be equal and to vary as the cube of the distance from the wall until the fully turbulent core is reached. The resulting correlation has the form of Eq. (12). In Eq. (12) relation (13) holds, and x_f is a term for the effect of net fluid motion

$$\frac{fU}{2k_c x_f} = 1 + \sqrt{\frac{f}{2}} \left[\frac{14.5}{3} \left(\frac{\mu}{\rho D}\right)^{2/3} F\left(\frac{\mu}{\rho D}\right) + 5 \ln \frac{1 + 5.64(\mu/\rho D)}{6.64[1 + 0.41\,(\mu/\rho D)]} - 4.77 \right] \quad (12)$$

$$F\left(\frac{\mu}{\rho D}\right) = \frac{1}{2} \ln \frac{[1 + (5/14.5)(\mu/\rho D)^{1/3}]^2}{1 - (5/14.5)(\mu/\rho D)^{1/3} + (5/14.5)(\mu/\rho D)^{2/3}}$$

$$+ \sqrt{3} \tan^{-1} \frac{(10/14.5)(\mu/\rho D)^{1/3} - 1}{\sqrt{3}} + \frac{\pi\sqrt{3}}{6} \quad (13)$$

in the direction of mass transfer, approximately unity for dilute solutions or for equivolume diffusion, U is the average fluid velocity in the pipe, f is the Fanning friction factor (a function of Reynolds number), μ is the fluid viscosity, ρ is the fluid density, and D is the diffusion coefficient of the component transferred.

It is apparent that even for apparatus as simple as a pipe the relationship of variables is quite complex. For turbulent flow in most systems, the velocity distribution and other factors necessary for a similar analysis are unknown, but it is generally assumed that the correlating variables for mass transfer appear as functions of the same dimensionless groups as for pipes. A representative example is the correlation of the mass-transfer factor j_d, as a function of the modified Reynolds number for transfer between solid particles in packed beds and fluids flowing through the beds as shown in **Fig. 2**. The mass-transfer factor and Reynolds number N_{Re} are defined in Eqs. (14) and (15), where U is the superficial fluid velocity through the bed, $G = U\rho$ is

$$j_d = \frac{k_c x_f}{U}\left(\frac{\mu}{\rho D}\right)^{2/3} \quad (14) \qquad N_{Re} = \frac{D_p G}{\mu} \quad (15)$$

the fluid-mass velocity through the bed, and D_p is the diameter of spherical particle having the same external surface area as the particles in the bed.

Unsteady-state diffusion and convection. For unsteady-state processes, variation with time must be considered. For diffusion and convection, equations such as Eq. (7) may be written for each coordinate direction and combined with material balances over a fluid element during an interval of time $d\theta$ to give a partial differential equation which expresses the variation of concentration of a solute with time at a point within the fluid. For an incompressible fluid with constant D_A, Eq. (16) holds. Here V_x, V_y, and V_z are the components of fluid velocity in the coordinate directions x, y, and z.

$$\frac{\partial C_A}{\partial \theta} = D_A \left[\frac{\partial^2 C_A}{\partial x^2} + \frac{\partial^2 C_A}{\partial y^2} + \frac{\partial^2 C_A}{\partial z^2}\right] - V_x \frac{\partial C_A}{\partial x} - V_y \frac{\partial C_A}{\partial y} - V_z \frac{\partial C_A}{\partial z} \quad (16)$$

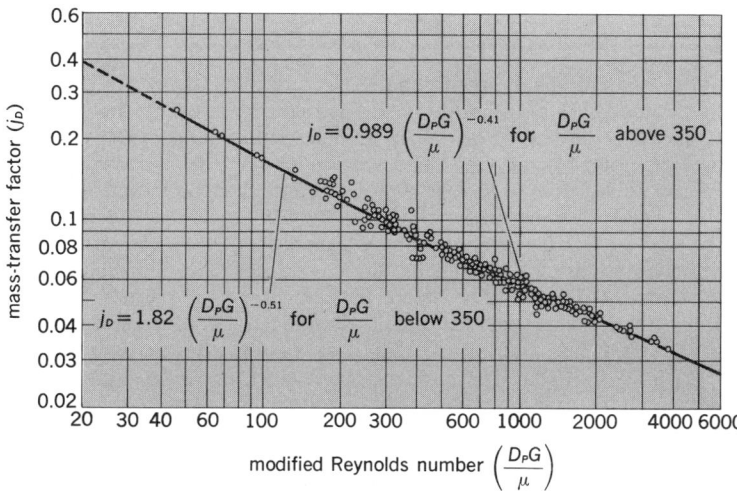

Fig. 2. Graph showing mass transfer in the flow of fluids through granular beds. (*After C. R. Wilke and O. A. Hougen, Mass transfer in flow of gases through granular solids, Trans. A. I. Ch. E., 41:445–451, 1945*)

Equation (15) may be combined with suitable relations describing the state of fluid motion to provide an analytical solution to mass-transfer problems in a few cases involving fluids flowing in laminar motion. If the fluid is at rest the last three terms on the right become zero and the problem becomes one of diffusion only, which has been solved for many of the commonly encountered boundary conditions.

Interdiffusion of two fluids. As an example of unsteady-state motion, consider two pure fluids, A and B, initially separated by a diaphragm at the midpoint of a cylindrical container of total length L, which are allowed to mix by diffusion. For this case, Eq. (15) reduces to Eq. (17).

$$\frac{\partial C_A}{\partial \theta} = D \frac{\partial^2 C_A}{\partial z^2} \qquad (17)$$

This equation may be solved to give the mole fraction of fluid A in each half of the cylinder at time θ after removal of the diaphragm, as shown by Eq. (18), where Y_{A1} is the mole fraction of A

$$Y_{A1} - Y_{A2} = \frac{8}{\pi^2}\left\{\exp\left[-\left(\frac{\pi}{L}\right)^2 D\theta\right] + \frac{1}{9}\exp\left[-9\left(\frac{\pi}{L}\right)^2 D\theta\right] + \frac{1}{25}\exp\left[-25\left(\frac{\pi}{L}\right)^2 D\theta\right] + \ldots\right\} \qquad (18)$$

in the half of the cylinder originally containing A, Y_{A2} is the mole fraction of A in the half of the cylinder originally containing B, and D is the diffusion coefficient for A in B. This equation has been applied by A. S. Smith to the estimation of time required for mixing of gases in commercial cylinders of approximately 1.5 ft³ (0.0425 m³) capacity, and 4.1 ft (1.25 m) length at 77°F (25°C). For example, with butane-air at 5 atm (500 kilopascals) pressure, $Y_{A1} - Y_{A2}$ is still 0.25 after 27 h, and with methane-helium at 100 atm (10 megapascals), the difference in mole fractions is 0.45 after nearly 280 h. These results indicate that considerable caution must be exercised to assure complete mixing in the blending of gases. Similar considerations apply for liquids in which diffusion times for equal degrees of mixing are much longer than for gases.

Bibliography. M. Benedict et al., *Nuclear Chemical Engineering*, 2d ed., 1981; R. E. Cunningham and R. J. Williams, *Diffusion in Gases and Porous Media*, 1980; D. K. Edwards et al., *Transfer Processes*, 2d ed., 1979; J. R. Ockendon and W. R. Hodgkins (eds.), *Moving Boundary Problems in Heat Flow and Diffusion: University of Oxford Conference, March 25–27, 1974*, 1975; C. W. Satterfield and T. K. Sherwood, *Role of Diffusion in Catalysis*, 1963; T. K. Sherwood et al., *Mass Transfer*, 1975; R. E. Treybal, *Mass Transfer Operations: Chemical Engineering*, 1980.

OSMOSIS
Francis J. Johnston

The transport of solvent through a semipermeable membrane separating two solutions of different solute concentration. The solvent diffuses from the solution that is dilute in solute to the solution that is concentrated. The phenomenon may be observed by immersing in water a tube partially filled with an aqueous sugar solution and closed at the end with parchment. An increase in the level of the liquid in the solution results from a flow of water through the parchment into the solution. The process occurs as a result of a thermodynamic tendency to equalize the sugar concentrations on both sides of the barrier. The parchment permits the passage of water, but hinders that of the sugar, and is said to be semipermeable. Specially treated collodion and cellophane membranes also exhibit this behavior. These membranes are not perfect, and a gradual diffusion of solute molecules into the more dilute solution will occur. Of all artificial membranes, a deposit of cupric ferrocyanide in the pores of a fine-grained porcelain most nearly approaches complete semipermeability.

The flow of liquid through such a barrier may be stopped by applying pressure to the liquid on the side of higher solute concentration. The applied pressure required to prevent the flow of solvent across a perfectly semipermeable membrane is called the osmotic pressure and is a characteristic of the solution. The walls of cells in living organisms permit the passage of water and certain solutes, while preventing the passage of other solutes, usually of relatively high molecular weight. These walls act as selectively permeable membranes, and allow osmosis to occur between the interior of the cell and the surrounding media. *See Solution.*

VISCOSITY
Norman H. Nachtrieb

The resistance that a gaseous or liquid system offers to flow when it is subjected to a shear stress. Viscosity is a measure of the internal friction that arises when there are velocity gradients within the system. For fluids (gases and liquids) its meaning is conceptually and operationally well defined. In the regime of laminar or streamline flow the force required to maintain a velocity gradient, (dv/dx), between planes of fluid of area A is described by Newton's equation (1) and **Fig. 1**. The

$$f = \eta A \left(\frac{dv}{dx} \right) \tag{1}$$

proportionality constant η is called the viscosity coefficient. Its dimensions are $(\text{mass})(\text{length})^{-1}(\text{time})^{-1}$, and in the cgs system the unit of viscosity is the poise (1 g · cm^{-1} s^{-1}). It is the force per unit area (dynes cm^{-2}) required to sustain a unit velocity gradient (cm s^{-1} cm^{-1}) normal to the flow direction. In the International System (SI) the unit of viscosity is kg · m^{-1} s^{-1}, and is hence larger than the poise by a factor of 10; conversely 1 kg · m^{-1} s^{-1} = 10 poise. In the British absolute system of units, the unit of viscosity is 1 lbf · s · ft^{-2} = 1 slug · ft^{-1} · s^{-1} = 47.88 kg · m^{-1} s^{-1} = 478.8 poise. Conversely, 1 kg · m^{-1} s^{-1} = 2.088 × 10^{-2} lbf · s · ft^{-2} and 1 poise = 2.088 × 10^{-3} lbf · s · ft^{-2}.

Simple gases typically have viscosities in the range of 100 to 200 micropoise at standard temperature and pressure (273 K, 1 atm or 101,325 pascals), whereas simple liquids under the same conditions have coefficients of viscosity about two orders of magnitude larger. The **table** lists values for the coefficients of viscosity of selected gases and liquids. The flow characteristics of gases and simple liquids such as water, carbon tetrachloride, and ethyl alcohol are accurately described by Eq. (1), and such fluids are called newtonian fluids. Aqueous suspensions, such as clays, gelatin, and agar, are termed non-newtonian fluids because their viscosities may depend upon the rate of shear and prior treatment. Hydrophilic sols often form extended networks involving water, and their nonnewtonian behavior is believed to be due to the breakdown of their structure under shear.

Molecular basis of viscosity in gases. The origin of internal friction (viscosity) at the molecular level is the net transfer of momentum between layers of fluid moving with different velocities in parallel flow by the mechanism of molecular collisions. In this process the directed energy of fluid flow is degraded to random thermal energy (heat).

It was one of the early triumphs of the kinetic theory of gases that established the relationship between the viscosity of a hard-sphere gas and the mean speed of its molecules. If x in Fig. 1 represents the mean free path λ of molecules in hypothetical planes that move with velocities equal to v and v', respectively, an exchange of molecules between the planes will result in the net transfer of momentum per unit time equal to $\frac{1}{3} A n \bar{c} (mv - mv')$, where n is the number of molecules per unit volume, m is the mass of a molecule, mv and mv' are the additional momenta of molecules in the planes in consequence of their shear velocities, and $\bar{c}$ is the mean speed of

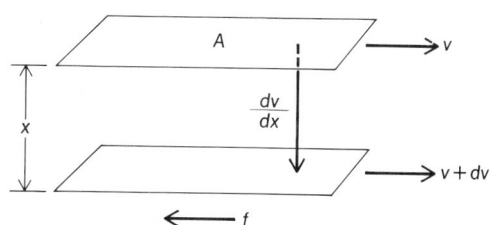

Fig. 1. Viscous shear in fluids.

Coefficients of viscosity of selected gases and liquids

Substance	Temperature, °F (°C)	η, poise*
Hydrogen	32 (0)	84.2×10^{-6}
Helium	32 (0)	186×10^{-6}
Nitrogen	32 (0)	167×10^{-6}
Oxygen	32 (0)	181×10^{-6}
Water (liquid)	68 (20)	10.1×10^{-3}
Ethyl alcohol	68 (20)	12.0×10^{-3}
Diethyl ether	68 (20)	2.5×10^{-3}
Carbon tetrachloride	68 (20)	9.8×10^{-3}
Mercury	68 (20)	15.5×10^{-3}
Glycerin	68 (20)	10.69
Glass	752 (400)	10^{13}
Glass	1472 (800)	10^{7}

*1 poise = $0.1 \text{ kg} \cdot \text{m}^{-1} \text{s}^{-1}$.

molecules $(8kT/\pi m)^{1/2}$, where k is Boltzmann's constant and T is the absolute temperature. Since the gas density if $\rho = nm$, the retarding force that resists the shear is given by Eq. (2). Combining Eqs. (1) and (2) gives Eq. (3).

$$f = \tfrac{1}{3} A \rho \bar{c} \left(\frac{dv}{dx}\right) \lambda \quad (2) \qquad \eta = \tfrac{1}{3} \bar{c} \rho \lambda \quad (3)$$

Substitution for the mean free path $[\lambda = (n\sqrt{2}\pi d^2)^{-1}$, where d is the average diameter of a molecule] permits restatement of Eq. (3) as Eq. (4).

$$\eta = \frac{m\bar{c}}{3\sqrt{2}\pi d^2} \quad (4)$$

More refined calculations for hard-sphere gases replace the factor ⅓ in Eq. (4) by 0.499, but the functional form is correct. It predicts that the viscosity of gases should be independent of pressure because the mean free path and gas density are affected in opposite ways by pressure, and this prediction is in accord with experiment up to moderately high pressures. Equation (4) also predicts a $T^{1/2}$ dependence of the gas viscosity, which is in fair agreement with experiment. Real gases show a somewhat stronger temperature dependence because their molecules are not ideal hard spheres. Equation (4) has had widespread application to the determination of the diameters of molecules, and the agreement is good but not perfect. Such minor discrepancies as do exist with molecular diameters determined by other methods (such as molar refraction, equation of state, and electron diffraction) are due to the fact that each method probes a somewhat different region of the potential surface of real molecules.

Viscosity of liquids. Momentum transfer between shearing layers also underlies the viscous behavior of simple liquids, but since the mean free path has little meaning for liquids, no simple relation such as Eq. (4) exists for them. In contrast to the behavior of gases, temperature decreases the viscosities of simple liquids and its effect is much larger. The temperature dependence of the viscosity of simple liquids bears no simple relationship to gas kinetic theory, but instead generally follows an exponential law of the form of Eq. (5), where A and B are parameters characteristic of the liquid and are reasonably constant over finite ranges of temperature; R is the gas constant, $8.314 \text{ J} \cdot \text{mol}^{-1} \cdot \text{K}^{-1}$. The form of Eq. (5) is the same as that typically found for

$$\eta = A \exp(B/RT) \quad (5)$$

transport properties in the defect crystalline state, where the concept of a simple thermally activated process is generally accepted. This similarity in temperature dependence has in the past led to various "hole" theories of the liquid state. These have been based upon analogy with the vacancy model of defect crystals, and the parameter B has been identified with the energy required to create a void of molecular dimensions in the liquid and to move a nearby molecule into it. Such hole theories are now generally thought to be oversimplified, and viscous flow, like diffusion in liquids, is thought to be a highly complex process in which many molecules participate.

Measurements of viscosity as a function of pressure likewise show completely different behavior for gases and liquids. Whereas the former show little dependence of viscosity on pressure in the low-density region, very high hydrostatic pressure generally increases the viscosity of liquids, sometimes quite markedly. In the region of laminar flow, nevertheless, Newton's equation accurately describes the viscous behavior of simple liquids, and the presumption is that the transfer of momentum between shearing layers involves a high degree of correlated molecular motions.

When the shear velocity exceeds a critical value in vessels of a given radius, streamline flow is replaced by turbulent flow. The criterion for the onset of turbulence in a tube of radius r is that the Reynolds number $2r\rho v/\eta$ exceed a certain value (approximately 2000 for normal liquids).

Measurement of viscosity. The laminar flow of both gases and liquids in long narrow tubes is described by Poiseuille's equation (6), where η is the fluid viscosity (lbf · s · ft^{-2}, or poise

$$\eta = \frac{\pi(p_1 - p_2)r^4 t}{8Vl} \tag{6}$$

or kg · m^{-1} s^{-1}), where r is the radius of the tube (ft, or cm or m), l is its length (ft, or cm or m), $(p_1 - p_2)$ is the pressure drop (lbf · ft^{-2}, or dynes cm^{-2} or Pa) across the tube, and V is the volume of fluid (ft^3, or cm^3 or m^3) that flows through the tube in time t (s). Poiseuille's equation is based on the assumption that the layer of fluid in contact with the tube wall is stationary. It provides a basis for the absolute measurement of the viscosity of both gases and liquids.

More commonly for liquids, relative measurements of viscosity are made by use of either the Ostwald viscometer or the falling-sphere viscometer. The former (**Fig. 2**) consists of two glass bulbs separated by a length of capillary tubing. Liquid is drawn up into the upper bulb, and the time required for its meniscus to fall between calibration marks above and below the upper bulb is accurately measured. A similar measurement is made with a liquid of known viscosity. From Eq. (6), Eq. (7) follows. Here η_1 and η_2 are the viscosities of the two liquids, ρ_1 and ρ_2 are their

$$\frac{\eta_1}{\eta_2} = \frac{\rho_1 t_1}{\rho_2 t_2} \tag{7}$$

densities, and t_1 and t_2 are the corresponding flow times. Equation (7) takes an even simpler form if the viscosity is divided by the density of the liquid. This quantity, called the kinematic viscosity, is measured in units termed stokes (cm^2 s^{-1}) in cgs units, in units of m^2 s^{-1} in SI, and in units of ft^2 s^{-1} in the British absolute system.

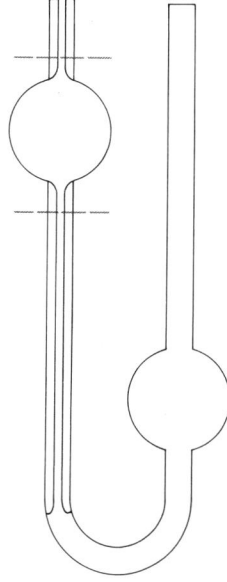

Fig. 2. Ostwald viscometer.

The falling-sphere viscometer is based upon Stokes' law for the frictional force on a spherical body of radius r falling with constant velocity in a fluid of viscosity η in an unbounded space, Eq. (8). This force is equal and opposite to the net force of gravity acting on the sphere, as in Eq. (9), where ρ and ρ' are the densities of a metal sphere and the fluid, and g is the acceleration of gravity. Equations (8) and (9) lead to the absolute viscosity of the fluid, Eq. (10).

$$rf = 6\pi\eta vr \quad (8) \qquad f = \tfrac{4}{3}\pi r^3(\rho - \rho')g \quad (9) \qquad \eta = \frac{2gr^2(\rho - \rho')}{9v} \quad (10)$$

As with the Ostwald viscometer, it is simpler to compare the times of fall of the sphere in fluids of known and unknown viscosity, and to use Eq. (11).

$$\frac{\eta_1}{\eta_2} = \frac{t_1(\rho - \rho'_1)}{t_2(\rho - \rho'_2)} \quad (11)$$

Other methods for the absolute or relative measurement of fluid viscosities are based upon the determination of the torque exerted upon a cylinder immersed in a fluid when a coaxial cylinder is rotated with constant velocity, or the damping of the amplitude of an oscillating disk suspended in the fluid by a torsion fiber.

Flow behavior of complex fluids. Many fluids display flow behavior that deviates profoundly from that of simple gases and liquids. This subject is normally treated by the field of rheology. A few examples will serve to indicate the complexity of flow behavior in some fluids. Monoclinic sulfur, whose molecules consist of puckered rings of eight sulfur atoms, melts at 203.9°F (95.5°C) to form a simple liquid (S_λ) of the same molecularity. Its viscosity is low enough to classify it as a normal liquid, and its viscosity decreases with temperature in the normal manner. Between 320 and 356°F (160 and 180°C), however, the viscosity increases dramatically by many orders of magnitude, and it appears that ring opening occurs followed by the formation of long-chain polymers. Above this temperature interval, the viscosity again decreases as thermal energy breaks up the long chains into smaller units. The process is highly irreversible, and crystalline sulfur may be recovered only by condensation from sulfur vapor.

Various colloidal dispersions of solids in oil or aqueous media decrease their viscosity when stirred at constant temperature, and revert to their former state of higher viscosity when the shear stresses are reduced. This phenomenon of thixotropy is an essential property of paints that contain solid pigments.

The flow of blood in mammalian vascular systems is non-newtonian, and Poiseuille's law is not obeyed. In part this behavior is attributable to the presence of red corpuscles and other suspended bodies, but the phenomenon is very complex.

Glasses are amorphous solids, structurally much closer to liquids than to crystals. Even at ordinary temperatures they deform under stress over long periods of time, and their viscosity varies over tens of orders of magnitude as the temperature is raised to the softening point. Profound structural changes in the random three-dimensional network and of the dynamical modes of local structural elements take place as the temperature of a glass is increased. SEE AMORPHOUS SOLID.

Some adhesives exhibit flow along directions that are not parallel to the direction of stress. Such fluids are anisotropic, and their flow properties are tensors. SEE LIQUID.

Bibliography. P. W. Atkins, *Physical Chemistry*, 3d ed., 1986; J. O. Hirschfelder, C. Curtiss, and R. B. Bird, *The Molecular Theory of Gases and Liquids*, 1964; W. J. Moore, *Basic Physical Chemistry*, 1983.

DIALYSIS
QUENTIN VAN WINKLE

A process of selective diffusion through a membrane by dissolved solutes in liquid solution. As dialysis is usually carried out, the membrane permits the diffusion of low-molecular-weight solutes (crystalloids) but prevents the passage of colloidal and high-molecular-weight solutes (macromolecules). Membranes suitable for this purpose include vegetable parchment, animal parchment, goldbeater's skin (peritoneal membranes of cattle), fish bladders, dialyzing cellophane (Visking sausage casing), and collodion (nitrocellulose deposited from alcohol-ether solution).

The solution is contained within such a membrane. The low-molecular-weight solutes are removed by placing pure solvent outside the membrane. This solvent is changed periodically or continuously until the concentration of diffusible solutes in the solution is reduced to near zero. The technique is used extensively in separating and purifying macromolecules of biological origin (see **illus.**).

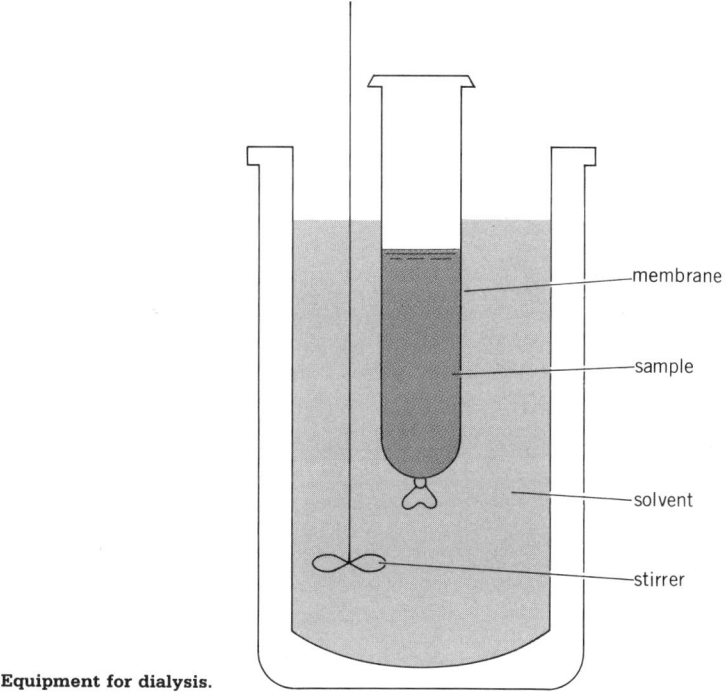

Equipment for dialysis.

Dialysis rates of ionic low-molecular-weight solutes may be increased greatly by applying an electric field to achieve an electrophoretic movement of ions through the dialytic membrane. This combined process of dialysis and electrophoretic transport of solutes through a membrane is known as electrodialysis. SEE ELECTROPHORESIS.

Ions do not migrate readily through membranes that carry in their pores charges of the same sign as the ion; ions of charge opposite to that of the membrane are not prevented from passing through. Such a selective permeability to ions by charged membranes has been used to improve the efficiency of electrodialysis. Membranes of collodion, vegetable parchment, and cellophane carry negative charges in contact with aqueous solutions. Animal membranes show positive charges at low pH and negative charges at high pH.

Membranes of ion-exchange resins are commercially available in cationic and anionic forms. The ion-exchange membranes show a high selectivity for ions of one charge type. SEE ION-SELECTIVE MEMBRANES AND ELECTRODES.

For desalting small quantities of solutions for chromatography and radioisotope tracer studies, micro- and semimicroelectrodialyzers have proved valuable. On a larger scale, electrodialysis combined with ion-exchange membranes has been used to remove salt from sea water. SEE COLLOID.

5 MATTER: STRUCTURE AND PROPERTIES

Combining volumes, law of	137
Chemical bond theory	137
Chemical bonding	140
Bond angle and distance	142
Crystal field theory	144
Conjugation and hyperconjugation	148
Intercalation compounds	150
Hydrogen bond	152
Molecular orbital theory	154
Molecular isomerism	162
Molecular structure and spectra	168
Stereochemistry	179
Electron affinity	184
Electronegativity	185
Valence	187
Intermolecular forces	191
Free radical	194
Molecule	198
Ion	198
Gas	198
Liquid	202
Amorphous solid	204
Liquid crystals	205
Colloid	208
Colloidal crystals	211

Micelle	213
Gel	214
Emulsion	215
Solution	216
Solid solution	222
Crystallization	223
Avogadro's number	225
Gas constant	227
Mass	227
Density	230
Relative atomic mass	230
Atomic weight	230
Relative molecular mass	232
Molecular weight	232
Gram-equivalent weight	236
Gram-molecular weight	236
Boiling point	237
Melting point	238
Transition point	238
Triple point	239
Isoelectric point	240
Specific heat	240

COMBINING VOLUMES, LAW OF
Thomas C. Waddington

The principle that when gases take part in chemical reactions the volumes of the reacting gases and those of the products, if gaseous, are in the ratio of small whole numbers, provided that all measurements are made at the same temperature and pressure. The law is illustrated by the following reactions:
    1. One volume of chlorine and one volume of hydrogen combines to give two volumes of hydrogen chloride.
    2. Two volumes of hydrogen and one volume of oxygen combine to give two volumes of steam.
    3. One volume of ammonia and one volume of hydrogen chloride combine to give solid ammonium chloride.
    4. One volume of oxygen when heated with solid carbon gives one volume of carbon dioxide.
    It should be noted that the law applies to all reactions in which gases take part, even though solids or liquids are also reactants or products.
    The law of combining volumes was put forward on the basis of experimental evidence, and was first explained by Avogadro's hypothesis that equal volumes of all gases and vapors under the same conditions of temperature and pressure contain identical numbers of molecules. SEE AVOGADRO'S NUMBER.
    The law of combining volumes is similar to the other gas laws in that it is strictly true only for an ideal gas, though most gases obey it closely at room temperatures and atmospheric pressure. Under high pressures used in many large-scale industrial operations, such as the manufacture of ammonia from hydrogen and nitrogen, the law ceases to be even approximately true. SEE GAS.

CHEMICAL BOND THEORY
Robert G. Parr

Chemical bonds are the forces that hold atoms together in molecules and solids. Chemical bond theory is the explanation of the physical basis of chemical bonds and of the relationship between chemical bonds and the properties of substances.
    The simplest chemical bonds to describe are those resulting from direct coulombic attractions between ions of opposite charge, as in most crystalline salts. These are termed ionic bonds.
    Other chemical bonds are of a wide variety of types, ranging from the very weak van der Waals attractions, which bind Ne atoms together in solid neon, to metallic bonds or metallike bonds, in which very many electrons are spread over a lattice of positively charged atom cores and give rise to a stable configuration for those cores. The theory of many of these bond types has been well developed by theoretical chemists. SEE CHEMICAL BONDING; MOLECULAR ORBITAL THEORY.
    Covalent bond. Since the normal convalent bond, in which two electrons bind two atoms together, as in

$$\text{H—H, H—Cl, F—F,} \quad \overset{\text{H}}{\underset{\text{H}}{\diagdown\diagup}}\text{O,} \quad \text{or} \quad \overset{\text{H} \quad \text{H}}{\underset{\text{H} \quad \text{H}}{\diagdown\diagup \atop \diagup\diagdown}}\text{C}$$

is the most characteristic link in chemistry, an adequate theory to account for it is the central problem in chemical bond theory. The characteristic physical and chemical properties of any molecule are direct consequences of its particular detailed electronic structure. Yet the theory of any one covalent chemical bond, for example, the O—H bond in the water molecule, has much in common with the theory of any other covalent bond, for example, the C—H bond in the methane molecule. The accurate theory of covalent bonds that now exists is capable both of treating their qualitative features and of quantitatively accounting for the molecular properties which are a consequence of those features. The theory is a branch of quantum theory. SEE QUANTUM CHEMISTRY.

Hydrogen molecule. A brief outline of the application of quantum theory to the bond in the hydrogen molecule H—H follows. Here two electrons, each of charge $-e$, bind together two protons, each of charge $+e$, with the electrons much lighter than the protons. What must be explained, above all else, is that these particles form an entity with the protons 0.074 nanometer apart, more stable by $D = 109$ kcal (456 kilojoules) per mole than two separate hydrogen atoms, where D is the binding energy. In more detail, a molecular energy is involved (ignoring nuclear kinetic energy) that depends on internuclear distance, as shown in the **illustration**. This curve

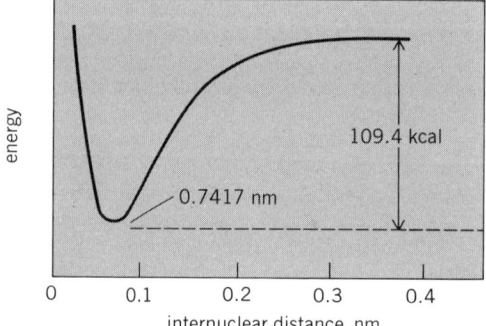

Potential energy of the hydrogen molecule.

can be determined experimentally, and it can be used to interpret the characteristic spectroscopic properties of hydrogen gas.

The quantum theory accounts for the properties of isolated atoms by assigning atomic orbitals for individual electrons to move in, not more than two electrons at a time. For the hydrogen atom, the orbitals are labeled $1s$, $2s$, $2p_x$, $2p_y$, $2p_z$, and so on, with $1s$ the one having the lowest energy. For the molecule H_2 one electron, say electron 1, might be assigned to a $1s$ orbital on proton A, with $1s_A(1)$ written to signify this; similarly electron 2 might be assigned to the same kind of orbital on proton B, written as $1s_B(2)$. Since independent probabilities multiply and orbitals represent probability amplitudes, the description for the combined system shown in Eq. (1) is arrived at.

$$\phi(1,2) = 1s_A(1)1s_B(2) \tag{1}$$

Unfortunately, this fails to account for the bond properties; it gives a binding energy of only 10 kcal (41.9 kJ) per mole.

An essential defect of Eq. (1) is the numbering of the electrons; it puts electron 1 on proton A, electron 2 on proton B. Electrons cannot be distinguished experimentally, so they should not be given unique numbers; the function $1s_A(2)1s_B(1)$ would be just as good as the foregoing. It is necessary to use a description that is not affected by interchange of electron labels, as in the additive combination of Eq. (2). (The difference combination also is an acceptable description, but

$$\phi(1,2) = 1s_A(1)1s_B(2) + 1s_A(2)1s_B(1) \tag{2}$$

it represents an excited state of the molecule.)

Any complete molecular electronic wave function should include electron spin. Symmetric space wave functions like Eq. (2) must be multiplied by antisymmetric spin wave functions to give total wave functions that are antisymmetrical with respect to interchange of electrons. For the ground state of hydrogen, and for the normal covalent bond elsewhere, this requirement means that the electron spins must be paired to give a net electron spin of zero.

The simple relationship described by Eq. (2) qualitatively accounts for the existence of the covalent bond; the predicted binding energy is $D = 74$ kcal (310 kJ) per mole; and the shape of the curve, with the minimum appearing at 0.080 nm, is right.

The description of Eq. (2) can be systematically improved. The charge acting on the electron may be changed from ^+1e to the larger value, ^+Ze, which is more realistic for the actual molecule. With $Z = 1.17$ this gives $D = 87$ kcal (364 kJ) per mole. Polarization effects may be introduced by taking Eq. (3), where $1\sigma_A = 1s_A + \lambda 2pz_A$ and $1\sigma_B = 1s_B + \lambda 2pz_B$. This gives D

$$\phi(1,2) = 1\sigma_A(1)1\sigma_B(2) + 1\sigma_A(2)1\sigma_B(1) \tag{3}$$

$= 93$ kcal (389 kJ). Ionic terms may be introduced, acknowledging the possibility that both electrons may be on one atom, by taking Eq. (4).

$$\phi(1,2) = \sigma_1[1s_A(1)1s_B(2) + 1s_A(2)1s_B(1)] + \sigma_2[1s_A(1)1s_A(2) + 1s_B(1)1s_B(2)] \tag{4}$$

This also gives (with $Z = 1.19$) $D = 93$ kcal (389 kJ). Another possible approach is to include both ionic terms and polarization effects, and other terms involving $2s$, $3d$, $4f$, and other orbitals. If this is done, eventually one obtains the observed D value and a potential curve that is in excellent agreement with experiment.

The linear mixing of terms such as $1s_A(1)1s_B(2)$ is called resonance; the method of mixing covalent and ionic structures is called the valence bond (VB) method. The particular mixing coefficients can be found by the variational principle: The best values for such parameters are those that make the total energy of the molecule, properly computed from quantum mechanics, a minimum. The energy expression only contains terms that have a direct classical interpretation: the kinetic energy of the electrons, their energy of repulsion for one another, their energy of attraction for the nuclei, and the nuclear-nuclear repulsion energy. The breakdown of the binding energy is in accord with the virial theorem: As the molecule is formed from the atoms, the kinetic energy increases by amount D and the potential energy decreases by $2D$.

Alternative descriptions of H_2 are possible, of which the most important is provided by the molecular orbital (MO) method. Here one puts electrons one at a time into orbitals which are spread over the whole molecule, usually approximating these orbitals by linear combinations of atomic orbitals (LCAO). For H_2 the lowest molecular orbital is $\phi_1 \approx 1s_A + 1s_B$, the next $\phi_2 \approx 1s_A - 1s_B$. The simplest molecular orbital description is displayed in Eq. (5), which represents an

$$\phi(1,2) = \phi_1(1)\phi_2(2) \tag{5}$$

equal weighting of covalent and ionic structures; it gives $D = 61$ kcal (255 kJ) for $Z = 1.00$ and $D = 80$ kcal (335 kJ) for $Z = 1.20$. More suitable is a mixture of this function with the function obtained by promoting both electrons from ϕ_1 to ϕ_2. The result of this configuration interaction process has the form of Eq. (6), and it is identical with the valence bond function of Eq. (4). In

$$\phi(1,2) = D_1\phi_1(2) + D_2\phi_2(1)\phi_2(2) \tag{6}$$

this manner more terms can be added, using more orbitals, until, again, the accurate potential energy curve is obtained.

The most accurate description known for the chemical bond in H_2 is a very complicated electronic wave function. It accounts for all known properties of hydrogen to high accuracy and confirms that the nonrelativistic quantum mechanics of E. Schrödinger will suffice for most chemical purposes. The calculated and observed values of D agree absolutely. SEE HYDROGEN BOND.

Complex molecules. The development of a quantitative treatment of chemical bonds in molecules that are more complicated than H_2 has many inherent difficulties. It constitutes, however, an active and useful field of research which was stimulated by the development of large and fast electronic computers and advanced numerical and analytical techniques for handling atomic orbitals on many different nuclei. The qualitative and semiquantitative theory preserves the use of many chemical concepts that predate quantum chemistry itself; among these are electrostatic and steric factors, tautomerism, and electronegativity. The quantitative theory provides both a physical basis for chemical concepts if they are valid and a tool for weeding them out if they are invalid.

In molecules containing many chemical bonds, it should be possible in many instances to construct accurate descriptions of the whole electronic system from descriptions of the separate electron pairs. If $\phi_A(1,2)\phi_B(3,4), \ldots,$ are wave functions for pairs A, B, ..., including spin, then it may be that the function shown in Eq. (7), where $\mathcal{A}$ represents anti-symmetrization with

$$\phi = \mathcal{A}[\phi_A(1,2)\phi_B(3,4)\ldots] \tag{7}$$

Spectroscopic properties of CO

Property*	Observed value	Calculated value
r_e (nm)	0.1128	0.1119
ω_e (cm^{-1})	2170.	2357.
$\omega_e x_e$ (cm^{-1})	13.5	11.1
B_e (cm^{-1})	1.93	1.97

*r_e = internuclear distance at which potential energy is minimum.
ω_e = a quantity whose square is proportional to the curvature of the potential energy at its minimum.
x_e = anharmonic constant.
B_e = rotational constant.

respect to electron interchange, is an accurate wave function for the molecule. If this is the case and if certain auxiliary conditions are satisfied, the description is one in which the traditional concept of separate chemical bonds has been preserved. The functions $\phi_A, \phi_B, \ldots$, are called geminals; they are important entites in the theory of chemical bonding in large molecules.

To illustrate the level of accuracy of contemporary quantum-chemical calculations, the **table** gives observed and calculated values for certain spectroscopic properties of the carbon monoxide molecule. *See* MOLECULAR ORBITAL THEORY.

Bibliography. A. L. Companion, *Chemical Bonding*, 2d ed., 1979; H. F. Hameka, *Quantum Theory of the Chemical Bond*, 1975; J. Murrell et al., *The Chemical Bond*, 1978; L. Pauling and E. B. Wilson, Jr., *Introduction to Quantum Mechanics*, 1935.

CHEMICAL BONDING
E. BRIGHT WILSON, JR.

Atoms contain electric charges—a small, positively charged nucleus surrounded by a cloud of moving, negatively charged electrons. All chemical bonding is caused by the mutual attractions and repulsions of these electric charges. Other types of forces—gravitational, magnetic, nuclear—have negligible direct effect. The electric forces are governed by Coulomb's law. On the other hand, the motions and distributions of the electrons in the atoms are controlled by the laws of quantum mechanics.

Chemical bonds are very strong. To break one bond in each molecule in a gram mole of material will typically require an energy of many tens of kilocalories per mole. Most of the energy used by humans is chemical energy, derived from changing chemical bonds in food or fuel.

It is convenient to classify chemical bonding into several types, although all real cases are mixtures of these idealized, purely electrical cases.

Ionic bonding. This is the simplest type of bonding, in which one or more electrons are transferred completely from one atom to another, thus converting the neutral atoms into electrically charged ions. These ions are approximately spherical in shape and attract one another because of their opposite charges. The ions are drawn together until their spherical electron clouds sufficiently interpenetrate and repel one another to balance the force of attraction. Molecules can consist of two or more such ions. Many inorganic crystals can be considered as giant molecules made up of ions. Thus common salt, sodium chloride, consists of a lattice of Na$^+$ ions (positive sodium) each surrounded by six Cl$^-$ ions (negative chlorine) and vice versa.

In a purely ionic compound or crystal, the ions pack together in a geometrical arrangement which minimizes the total energy. Since unlike charges attract, and like charges repel each other, the positive ions will normally be next to negative ions and vice versa. The radii of ions are characteristic of the element and of the charge. Negative ions (anions) are usually larger than positive ions. The number of negative ions surrounding a positive ion (cation) is mainly determined by the ratio of the radii of the two ions. Thus, a small cation will not be surrounded by as many negative ions as will a larger cation. The number of positive ions surrounding a negative ion is largely governed by the importance of maintaining a local balance of positive and negative charge. Ions act as if the attractive and repulsive forces have no specific directional properties. The number of charges on an ion is called the electrovalence of the element. Sodium, for example, normally carries a single positive charge in its ionic compounds and thus has an electrovalence of $+1$. Chlorine in most inorganic compounds is an ion with one negative charge, hence it has an electrovalence of -1. Many elements can form several different ions. Thus iron commonly occurs as ferrous iron, Fe^{2+}, or as ferric iron, Fe^{3+}, and consequently displays electrovalencies of either $2+$ or $3+$. Since completely filled electron shells are especially stable, the ionic state which yields a completed shell is for many elements the most stable one.

Covalent bonding. This is another limiting type of chemical bonding. Here each atom of a bonded pair contributes one electron to form a pair of electrons which move in such a manner as to increase the density of electric charge in the space between the two atoms. The negative charge in the region between the two atoms attracts the two positive nuclei. This type of bond is also called an electron pair bond because its essential feature is the formation of a pair of electrons which spend much time in the region between the two atoms. These two electrons have their spins pointing in opposite directions; in other words, they are paired. If this were not the case, the two electrons could not both have a low kinetic energy (which is important for the stability of the molecule) and also spend much time in the region between the two atoms. This requirement of pairing is due to the basic principles of quantum mechanics, particularly the Pauli exclusion principle.

The number of covalent bonds which an atom can form is called the covalence and is governed by the detailed electron configuration of the atom. The meaning of the historical word valence has been gradually modified so that it now usually refers to the electrovalence or the covalence of an atom. An extremely important case is that of carbon. In most of its compounds, carbon forms four covalent bonds. When these connect it to four other atoms, the directions of the bonds to these other atoms will normally make angles of about 109° to one another, unless the attached atoms are crowded or constrained by other bonds. In other words, covalent bonds have preferred directions. However, to preserve the idea that carbon forms four bonds, it is necessary to introduce the notion of double and triple bonds. Thus in the structural formula of ethylene, C_2H_4 (I), all lines denote covalent bonds, the double line connecting the carbon atoms being a double bond. This idea has physical reality because such double bonds are distinctly shorter, almost twice as stiff, and require considerably more energy to break completely than do single bonds. However, they do not require twice as much energy to break as a single bond, so it is energetically advantageous for a molecule to open one component of a double bond and add two atoms; for example, H_2 adds to C_2H_4 to form ethane (II). Similarly, acetylene (III) is written with a triple

bond, which is still shorter than a double bond. A carbon-carbon single bond has a length close to 1.54×10^{-8} cm, whereas the triple bond is about 1.21×10^{-8} cm long.

In many compounds the rules for writing bond formulas are not unique. For example, benzene, C_6H_6 (IV), can be written in two forms. Evidence proves that all six C-C bonds are equivalent, so neither formula can be correct. Quantum mechanical arguments show that the correct picture is a blend of the two, in which the bonds have many properties intermediate between those of double and single bonds but in which the whole molecule displays an additional stability. This phenomenon, called resonance, occurs whenever the structure is such that two or more different bond formulas can legitimately be drawn for the same geometry.

Many substances have some bonds which are covalent and others which are ionic. Thus in crystalline ammonium chloride, NH_4Cl, the hydrogens are bound to nitrogen by electron pairs, but the NH_4 group is a positive ion and the chlorine is a negative ion. On solution in water, ionic crystals undergo dissolution into their separate ions.

Both electrons of a covalent bond may come from one of the atoms. Such a bond is called a coordinate or dative covalent bond or semipolar double bond and is one example of the combination of ionic and covalent bonding. Actually, the electron pair does not have to be symmetrically located in any case; so all degrees of mixing of ionic and covalent character may occur.

The hydrogen bond is a special type of a bond in which a hydrogen atom links a pair of other atoms. The linked atoms are normally oxygen, fluorine, chlorine, or nitrogen. These four elements are all quite electronegative, a fact which favors a partially ionic interpretation of this kind of bonding. SEE HYDROGEN BOND.

Metallic bonding. This is a third type of chemical bonding, which is exemplified in the common metals. There are several ways of looking at this bonding, but perhaps the simplest is to consider the crystal as consisting of positive ions of the metallic element immersed in a sea of electrons. The attraction of the positive ions for the electrons holds the crystal together. Some of the electrons are free to move about the whole crystal of the metal, and this is what makes the metal an electrical conductor.

Van der Waals forces. Although many crystals are giant molecules whose atoms are completely linked together by strong ionic or covalent bonds, others consist of discrete molecules, strongly bonded internally, but held to each other by much weaker forces. Most organic crystals are soft and have low melting points. Their discrete molecules are held together by what are called van der Waals or dispersion forces. Although these forces are electrical in nature, they are too weak to be considered as true chemical bonding. The picture is as follows: The electrons in the cloud of negative electricity in a molecule are in rapid motion, and the charge distribution may therefore be thought of as fluctuating in time. At any instant, the distribution in a given molecule may be quite unsymmetrical so that an electric dipole moment results, even though there need not be any average dipole moment. This dipole moment, however, may reverse rapidly or change in direction or magnitude as the electrons move. At any instant, the dipole in one molecule will cause an electric field to act on a neighboring molecule. This electric field will disturb the motion of the electrons in the second molecule in such a way as to produce an instantaneous induced dipole moment in the second molecule. The interaction of the two dipole moments, namely, the original instantaneous one in the first molecule and the induced one in the second, will result always in a net but weak attractive force between the two molecules. This force will then bring the molecules together until the repulsion of their electron clouds for one another keeps them at an equilibrium distance of separation. SEE INTERMOLECULAR FORCES; MOLECULAR STRUCTURE AND SPECTRA; QUANTUM CHEMISTRY; VALENCE.

Bibliography. A. L. Companion, *Chemical Bonding*, 2d ed., 1979; R. L. DeKock and H. B. Gray, *Chemical Structure and Bonding*, 1980; H. Gray, *Chemical Bonds: An Introduction to Atomic and Molecular Structure*, 1975; J. N. Murrell et al., *The Chemical Bond*, 1978; L. Pauling, *Nature of the Chemical Bond and the Structure of Molecules and Crystals*, 3d ed., 1960.

BOND ANGLE AND DISTANCE
VICTOR W. LAURIE

In molecules two atoms are said to be bonded to each other when there are strong attractive forces between them. These two atoms constitute a chemical bond. One important characteristic of a chemical bond is the separation (called the bond length or distance) of the two nuclei of the

Table 1. Variation of bond length with bond multiplicity

Bond	Molecule	Length, nm	Multiplicity
C—C	CH_3CH_3	0.153	Single
C—C	CH_2CH_2	0.133	Double
C—C	CHCH	0.121	Triple
C—O	CH_3OH	0.143	Single
C—O	CH_3COCH_3	0.121	Double
C—N	CH_3NH_2	0.147	Single
C—N	CH_3CN	0.116	Triple

Table 2. Variation of C-C single-bond length with hybridization

Molecule	Length, nm	Nominal hybridization
CH_3CH_3	0.153	sp^3–sp^3
$CH_3CH{=}CH_2$	0.151	sp^3–sp^2
$CH_3C{\equiv}CH$	0.148	sp^3–sp

atoms involved. A given atom may be involved in more than one chemical bond, and another important characteristic is the angle between bonds sharing one common atom. It is also often important to know the angle between two bonds which are connected by a third bond. This angle is called the dihedral or torsional angle. *See* CHEMICAL BONDING.

Physical properties. An obvious reason for the importance of bond angles and lengths is that they determine the shape and size of a molecule, and thus determine many physical properties. For example, molecules which are spherical in shape will pack in a crystal in a different manner from long, rodlike molecules, and will likely have a different density and crystal structure. Similarly, the molar volumes in the liquid are likely to be different as well as the diffusion coefficients. Another example is the observation that the melting points for normal alkanes with an even number of carbons behave differently from those with an odd number. This arises because of their different shapes and the way in which they interact in the solid.

Bonding information. Bond angles and distances are functions of the type of chemical binding present and give information about the electronic structure of the molecule. For example, bond lengths shorten considerably in going from a single bond to a triple bond (**Table 1**), and are thus dependent on the hybridization present. This dependence on hybridization also manifests itself in the variation of the length of C-C single bonds as the hybridization of the C atoms is changed (**Table 2**).

Since the strength of a bond increases with its multiplicity, there is an inverse relation between the strength of a bond and its length. Related to the strength of a bond is its stretching-force constant, and in fact general relations exist between the stretching-force constant for a bond and its length. One of the simplest is Badger's rule, shown in the equation below, where k is the

$$k = 1.86 \, (r - d_{ij})^{-3}$$

force constant, r is the bond length, and d_{ij} is a constant for atoms from any two rows i and j of the periodic table. This relationship was first enunciated by Richard Badger in 1934 and was applied only to diatomics, but it has since been found to hold for polyatomics as well. *See* MOLECULAR STRUCTURE AND SPECTRA.

Bond angles are also dependent on electronic structure. **Table 3** shows the nominal values for different hybridization. These values for the hypothetical ideal cases can be compared with those given in **Table 4**. If s and p atomic orbitals are assumed to be the only ones involved in the

Table 3. Variation of bond angle with hybridization

Hybridization	Angle
sp	180°
sp^2	120°
sp^3	109.5°
p	90°

Table 4. Comparison of observed bond angles with nominal hybridization

Angle	Molecule	Value	Nominal hybridization
CCH	HC≡CH	180°	sp
HCH	$H_2C{=}CH_2$	117°	sp^2
CCH	H_3CCH_3	110°	sp^3
ClPCl	PCl_3	100°	p

bonding, the bond angles around a central atom can be used to determine the hybridization of the central atom.

Biological effects. Dimensions and structure are exceedingly important in biological molecules. For example, the functioning of a molecule such as deoxyribonucleic acid depends very precisely on the structure of the individual, small components which make up this very large molecule. Another example is the great specificity of enzymes, which is believed to be associated with the exact shape and size of the enzyme. SEE HYDROGEN BOND.

Determination. Since bond angles and lengths are important molecular parameters, considerable experimental effort is devoted to their determination, and extensive compilations of values for individual molecules have been made. The experimental methods which are used to determine molecular structure can be divided into two groups. One group is made up of diffraction methods, and the other is composed of spectroscopic methods.

Diffraction methods make use of the fact that not only photons but also particles can behave as waves and can be diffracted. If the de Broglie wavelength of a particle or the wavelength of a photon is of the same order of magnitude as interatomic distances, then they will be diffracted by molecules. Either electrons or neutrons may be used as particles, and x-ray photons are of the appropriate wavelength. X-ray and neutron diffraction are applied to solids, whereas electron diffraction is used on gases.

Spectroscopic methods, with a few exceptions, make use of the quantization of the rotational energies of molecules in the gas phase. The rotational energy of a molecule is determined by its moments of inertia, and these are in turn functions of bond angles and lengths. Rotational transitions are observed either as fine structure in infrared, visible, and ultraviolet spectroscopy, or directly in microwave and far-infrared spectroscopy.

Bibliography. M. Balaban (ed.), *Molecular Structure and Dynamics*, 1981; J. N. Murrell et al., *Valence Theory*, 2d ed., 1970; L. C. Pauling, *The Nature of the Chemical Bond*, 3d ed., 1960.

CRYSTAL FIELD THEORY
R. A. D. WENTWORTH

An essentially ionic approach to chemical bonding which is often used with coordination compounds. These compounds consist of a central transition metal ion that is surrounded by a regular array of coordinated atoms or ligands. Accordingly, the ligands are assumed to be sources of negative charge which perturb the energy levels of the central metal ion. In this respect the ligands subject the metal ion to an electric field which is analogous to the electric or crystal field produced by the regular distribution of nearest neighbors within an ionic crystalline lattice. For example, the crystal field produced by the Cl ion ligand in octahedral $TiCl_6^{3-}$ is considered to be similar to that produced by the octahedral array of the six Cl ions about each Na ion in NaCl. The Na ion with its rare-gas configuration has an electronic charge distribution which is spherically symmetric both within and without the crystal field. The paramagnetic Ti(III) ion, which possesses one 3d electron (d^1), has a spherically symmetric charge distribution only in the absence of the crystal field produced by the ligands. The presence of the ligands destroys the spherical symmetry and produces a more complex set of energy levels within the central metal ion. The crystal field theory allows the energy levels to be calculated and related to experimental observation.

To illustrate the results of a typical crystal field calculation, assume that the single d electron in the Ti(III) ion will experience a coulombic repulsion with each of the six nearest neighbor Cl ions which are taken as point negative charges. This model for coulombic repulsion may be described mathematically as the summation $eq\Sigma r^{-1}$, where e and q are the electronic and ligand charges, respectively, and r is the distance between the electron and the ligand. The summation extends over all the ligands. A detailed quantum mechanical calculation can then be made. Fortunately, it is possible to arrive at identical results by a very qualitative procedure. This method considers the spatial orientation of d orbitals with relation to the ligands when both are viewed within the same coordinate system, as shown in **Fig. 1**.

The d_{z^2} orbital is oriented along the z axis with a ring in the xy plane, while the $d_{x^2-y^2}$ orbital is oriented only along the x and y axes. Although it is not visually obvious, both are

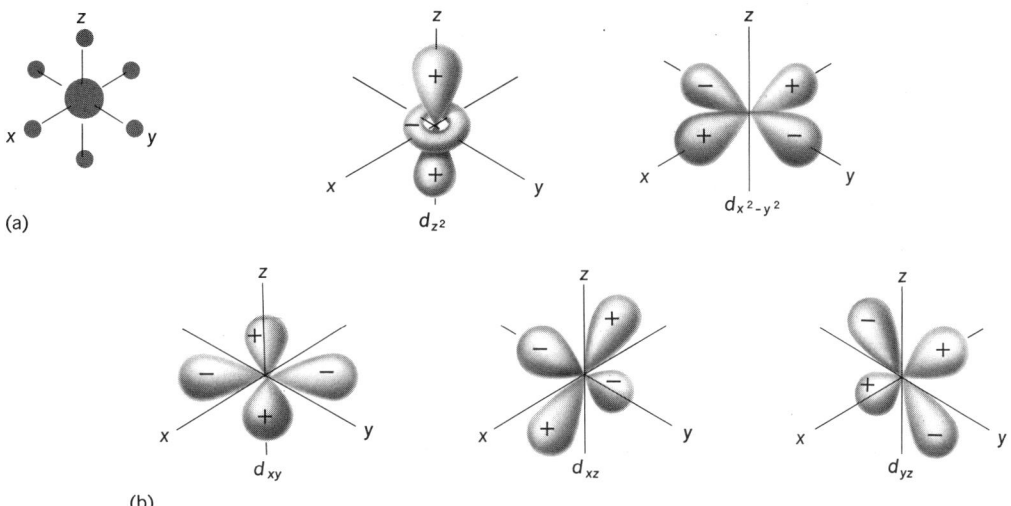

Fig. 1. Spatial orientation of d orbitals with relation to ligands in $TiCl_6^{3-}$. (a) The coordinate system for the octahedral $TiCl_6^{3-}$ ion. (b) The five d orbitals. The d_{z^2} orbital is symmetric with respect to the z axis.

equivalent. The d_{xy}, d_{xz}, and d_{yz} orbitals are oriented between the x, y, and z axes and are geometrically equivalent. The ligands are placed along the coordinate x, y, and z axes. An electron in any of the five orbitals will experience a coulombic repulsion due to the crystal field of the ligands. However, since the d_{z^2} and $d_{x^2-y^2}$ orbitals are directed toward the ligands, an electron in these orbitals will undergo far more repulsion than one in either of the three other orbitals. The imposition of six point negative charges, then, will not allow the five d orbitals to be equally energetic (degenerate) as they are in the bare Ti(III) ion, but causes three of these orbitals to be more stable than the remaining two. The difference in energy is termed $10Dq$. Thus, in the ground state of $TiCl_6^{3-}$, the single d electron will be found in the lower set of orbitals, whose energy is $-4Dq$. The crystal field stabilization energy (CFSE) is then said to be $4Dq$ (**Fig. 2**).

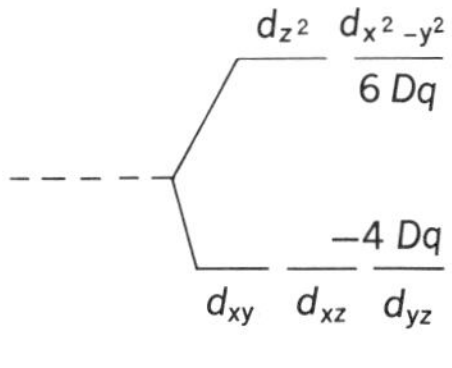

Fig. 2. The splitting of the five d orbitals because of an octahedral crystal field.

Coordination compounds are often colored. Crystal field theory suggests that in the case of $TiCl_6^{3-}$ its color is a result of an electronic excitation of the electron from the threefold set of orbitals into the twofold set. In the spectrum of $TiCl_6^{3-}$, an absorption band maximum is found at 13,000 cm^{-1}, which is then the energy associated with the transition, or $10Dq$. SEE MOLECULAR STRUCTURE AND SPECTRA.

Several electrons. When more than one d electron is present, the spectroscopic evaluation of Dq is not as simple, but can generally be accomplished. Nevertheless, the CFSE may be easily and formally obtained for these cases, as shown in the **table**, since this energy is simply the number of electrons occupying the orbital multiplied by the orbital energy. Thus, the CFSE amounts to $8Dq$ in an octahedral d^2 complex, and $(6 \times 4) - (2 \times 6) = 12Dq$ in a similar d^8

Crystal field stabilization energy		
	Octahedral CFSE (Dq)	
d^n	Weak	Strong
1	4	4
2	8	8
3	12	12
4	6	16
5	0	20
6	4	24
7	8	18
8	12	12
9	6	6

complex. With d^4, d^5, d^6, and d^7, however, two possibilities exist. If the crystal field is sufficiently strong to overcome the repulsion energy which will result from pairing the electrons in the lower set of orbitals, the maximum number of electrons will be found in the lower set. This situation is termed the strong-field case. In the weak-field case, the electron-pairing (repulsive) energy is greater than the crystal field (attractive) energy and the maximum number of unpaired electrons will result. Thus, in a strong crystal field a d^5 ion should have only one unpaired electron and a CFSE of $20Dq$. This is found in the $Mn(CN)_6^{4-}$ ion, which has been shown experimentally to possess only one unpaired electron. In most complexes of Mn(II), such as $Mn(H_2O)_6^{2+}$, the crystal field is weak so that five unpaired electrons result and the CFSE is zero.

Simple considerations such as these have enabled inorganic chemists to understand why certain coordination compounds containing a given metal ion may exhibit full paramagnetism, while others containing the metal in exactly the same formal oxidation state may show either a much weaker paramagnetism or none at all. A striking example of this is provided by paramagnetic CoF_6^{3-}, which possesses six unpaired d electrons and the diamagnetic $Co(NH_3)_6^{3+}$, in which none of the d electrons are unpaired. Both contain Co(III).

Tetrahedral array. Tetrahedral complexes may be treated in a similar fashion. The results indicate that the only difference, when compared to the octahedral complex, is in the orbital splitting pattern and the relative magnitude of Dq. The tetrahedral array of ligands causes an inversion of the pattern such that the d_{z^2} and $d_{x^2-y^2}$ orbitals lie lowest. The difference in energy between the two sets of orbitals is now found to be 4/9 of that in an octahedral complex, or Dq(tetrahedral) = $(4/9)Dq$(octahedral). This particularly simple result has led to the understanding of many stereochemical phenomena. An important early example was found in the cation distribution of normal and inverse spinels. The former are double oxides having the general formula $M(II)[M'(III)]_2O_4$ in which the oxygens lie in a close-packed system. The divalent metal ions, M(II), occupy one-eighth of the tetrahedral holes. In the inverse spinel $M'(III)[M(II)M'(III)]O_4$ the divalent metal ions have changed places with one-half of the trivalent ions.

Experimentally, it has been found that $MnCr_2O_4$ [containing Mn(II) with five unpaired electrons] and $NiFe_2O_4$ [containing Fe(III) with five unpaired electrons] have the normal and inverse structures, respectively. A simple application of crystal field theory results in exact agreement with experiment. In $MnCr_2O_4$ the ion at each octahedral site has a stability of $12Dq$, while the tetrahedral site has no CFSE. The total CFSE is then $24Dq$. If the structure were inverse, the total CFSE would be only $(4/9) \times 12 + 12 = 17.3Dq$. With $NiFe_2O_4$ the CFSE of the normal spinel is only $5.3Dq$, but in the inverse structure the CFSE increases to $12Dq$. In general, agreement between predicted cation distribution and that experimentally observed is good.

The application of these methods to conventional coordination compounds also meets with a fair amount of success. For example, with only one or two exceptions octahedral coordination of Cr(III) and diamagnetic Co(III) prevails in all compounds, as one would predict from crystal field theory. Important exceptions to these rules do exist and point out that other factors, such as ligand-ligand repulsions, sometimes outweigh the CFSE. Octahedral coordination of Ni(II) is favored over the tetrahedral arrangement insofar as CFSE is concerned, yet the tetrahedral $NiCl_4^{2-}$, $NiBr_4^{2-}$, and NiI_4^{2-} ions are well known.

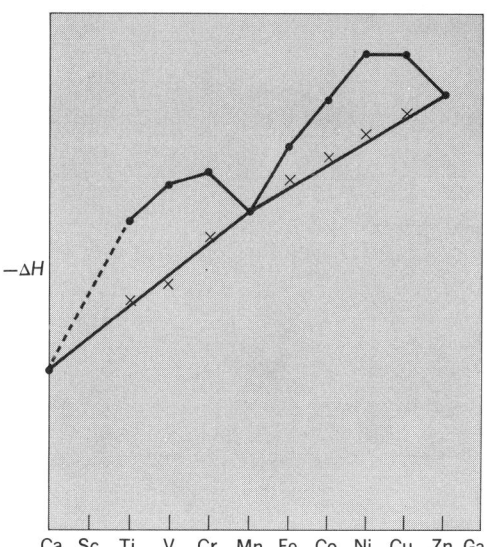

Fig. 3. Variation of heat of hydration $-\Delta H$ of divalent ions of first transition series. Experimental values are given by the dots, and the corresponding values, after correction for CFSE, are given by the x's.

Anomalous effects. Heats of hydration, lattice energies, crystal radii, and oxidation potentials of transition metal ions and their complexes contain apparent anomalies which are best explained in terms of an effect due to the crystal field. As an example, when the heats of hydration of the ions of the first transition series are plotted in **Fig. 3** with respect to atomic number, a peculiar double-humped curve is obtained. The results for those ions not possessing any CFSE, that is, Ca, Mn, and Zn, lie nearly on a straight line. In the absence of any other effect, it might be expected that with the successive addition of each nuclear charge, a monotonic increase in the heat of hydration would occur. Instead, it was found that the heat of hydration of those metal ions possessing CFSE is far more exothermic than would be predicted by arguments pertaining only to the successive increase in atomic number. In fact, the excess heat is best accounted for in terms of the CFSE. The total heat of hydration may be written as $\Delta H = \Delta H^0 + \text{CFSE}$, where ΔH^0 is the heat of hydration that would be expected for a hypothetical metal ion which ignored the crystal field. The change in ΔH^0 with atomic number would then be expected to be monotonic. When the observed heats of hydration are corrected for the CFSE obtained from spectroscopic data for the resulting $M(H_2O)_6^{2+}$ complexes, the expected monotonic increase is observed. Similar double-humped curves occur with the lattice energies and crystal radii and can be explained by including the effects due to the crystal field.

Stereochemical anomalies can also often be explained through the judicious use of simple arguments. A particularly important example is found in complexes of Cu(II). X-ray crystallography has established that most "octahedral" complexes containing that ion are in fact elongated along one axis. According to a theorem due to H. A. Jahn and E. Teller, this behavior is not unexpected. The theorem states that a system possessing a degenerate ground state will distort in some unspecified manner to remove the degeneracy.

The degeneracy in a regular octahedral complex of Cu(II) is easily illustrated by the possibility of writing the electronic configuration in two distinct, but equally energetic, ways:

$$(d_{xy})^2(d_{xz})^2(d_{yz})^2(d_{z^2})^2(d_{x^2-y^2})^1 \qquad (d_{xy})^2(d_{xz})^2(d_{yz})^2(d_{x^2-y^2})^2(d_{z^2})^1$$

For each, the CFSE is $6Dq$. In addition to the two-fold degeneracy, there are two separate means by which the degeneracy may be removed. If the ligands along the z axis of the octahedron move away from the metal ion while those in the xy plane move toward the center of the octahedron, then according to simple crystal field arguments, this movement will result in stabilizing the d_{z^2} orbital with respect to the $d_{x^2-y^2}$ orbital. Alternatively, the completely opposite movement will

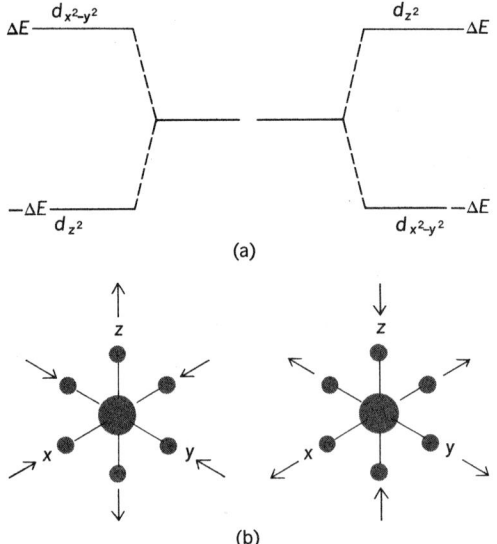

Fig. 4. The splitting of the $d_{x^2-y^2}$ and d_{z^2} orbitals by an axial distortion of the octahedron. (a) Alternate possiblities for writing the electronic configuration. (b) The two modes of distortion.

stabilize the $d_{x^2-y^2}$ orbital with respect to the d_{z^2} orbital (**Fig. 4**). In either case, an additional increment of energy is added to the CFSE of the Cu(II) complex: $6Dq + \Delta E$. Thus, the driving force the distortion is the additional stabilization energy ΔE. Crystal field theory is unable to predict which mode of distortion will occur, but it clearly predicts that a distortion should occur. From structure determinations through the use of x-rays, it is found that elongation along one axis is by far the most predominant mode.

Complete theory. The many successes of crystal field theory in the interpretation of the natural phenomena associated with transition metal compounds should not lead one to the conclusion that the bonding within these compounds can be truthfully represented by a strictly ionic model. Crystal field theory is essentially a very specialized form of the more complete molecular orbital theory. The need for the more complete theory becomes obvious when an attempt is made to rationalize the absolute value of Dq, the observation of ligand-to-metal electronic transitions, and certain observations in both nuclear magnetic resonance and electron spin resonance experiments, which can only be interpreted in terms of some covalent bonding. SEE MOLECULAR ORBITAL THEORY.

Bibliography. R. L. DeKock and H. B. Gray, *Chemical Structure and Bonding*, 1980; T. M. Dunn, D. S. McClure, and R. G. Pearson, *Some Aspects of Crystal Field Theory*, 1965; M. Gerloch and R. C. Slade, *Ligand-Field Parameters*, 1973; H. L. Schlafer and G. Glieman, *Basic Principles of Ligand Field Theory*, 1969.

CONJUGATION AND HYPERCONJUGATION
ROBERT S. MULLIKEN

An arrangement of bonds in a molecule such that a single bond lies directly between two multiple (that is, double or triple) bonds or between a multiple bond and a group containing a lone π electron, a π-electron pair or quartet, or a π-electron vacancy makes the molecule show unusual chemical behavior and physical properties. Such molecules and such multiple bonds are called conjugated. Two simple examples are 1,3-butadiene ($H_2C=CH-CH=CH_2$) and allylamine ($H_2C=CH-\ddot{N}H_2$).

The unusual properties of conjugated molecules can be understood theoretically, according to quantum mechanics, in terms of a bond structure in which, in the two examples mentioned, small proportions of $H_2\dot{C}-CH=CH-\dot{C}H_2$ or $H_2C^--CH=N^+H_2$, respectively, "resonate with"

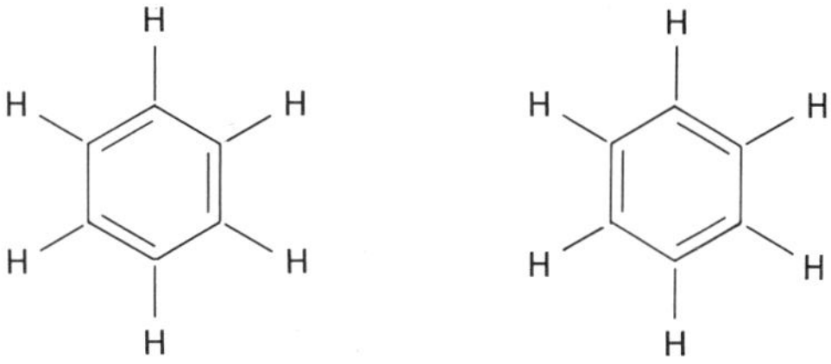

Fig. 1. Isovalent structures of benzene.

the main bond structures given above. In these examples one speaks of sacrificial conjugation, since in the minor resonance structures there is one less bond, or else much energy is required to transfer a charge. *See* CHEMICAL BONDING.

Further examples of sacrificial conjugation include $HC{\equiv}C{-}C{=}CH_2$ and diacetylene ($HC{\equiv}C{-}C{\equiv}CH$). In the latter, both π bonds of the triple bond are conjugated. In the former, only one π bond of the triple bond is conjugated (minor resonance structure $H\dot{C}{=}C{=}C{-}\dot{C}H_2$) with the one π bond of the double bond.

Stronger resonance effects occur in conjugated molecules in which alternative structures with equal numbers of bonds can be written (isovalent conjugation), for example, in benzene (**Fig. 1**) or the allyl ion ($H_2C{=}CH{-}C^+H_2$ and $H_2C^+{-}CH{=}CH_2$) or radical $H_2C{=}CH{-}\dot{C}H_2$ and $H_2\dot{C}{-}CH{=}CH_2$). Conjugation brings about energy stabilization, but much more so in isovalent than in sacrificial conjugation.

Hyperconjugation. This phenomenon is similar to conjugation in its formulation and manifestations, but the effects are weaker. It occurs when a CH_2 or CH_3 group (or in general an AR_2 or AR_3 group, where A may be any polyvalent atom and R any atom or radical) is adjacent to a multiple bond or to a group containing an atom with a lone π electron, a π-electron pair or quartet, or a π-electron vacancy. Hyperconjugation can be sacrificial (this is relatively weak) or isovalent (stronger).

Examples of molecules exhibiting first-order sacrificial hyperconjugation are:

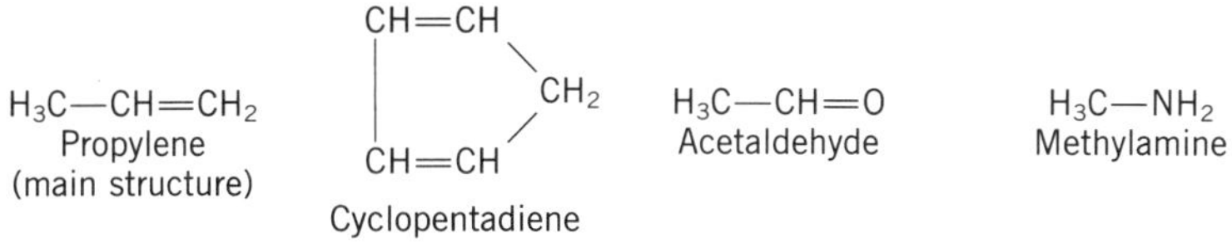

$H_3C{-}CH{=}CH_2$
Propylene
(main structure)

Cyclopentadiene (with CH=CH / CH=CH ring and CH₂)

$H_3C{-}CH{=}O$
Acetaldehyde

$H_3C{-}NH_2$
Methylamine

The analogy to conjugation is brought out by rewriting the main structures as

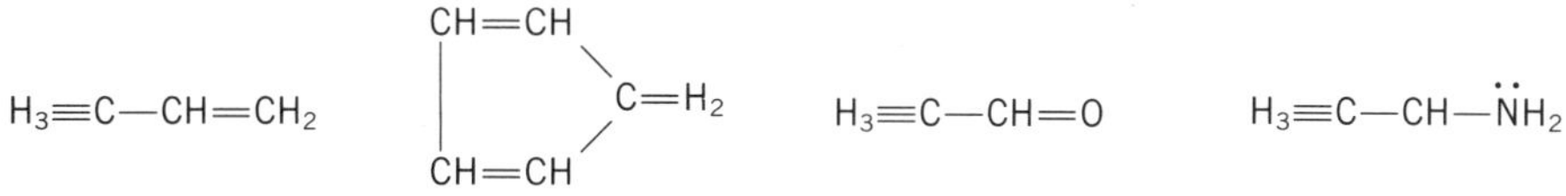

$H_3{\equiv}C{-}CH{=}CH_2$

(cyclopentadiene analog with $C{=}H_2$)

$H_3{\equiv}C{-}CH{=}O$

$H_3{\equiv}C{-}CH{-}\ddot{N}H_2$

with minor resonance structures

$\dot{H}_3{=}C{=}CH{-}\dot{C}H_2$

(cyclopentadiene analog with $C{-}\dot{H}_2$, two structures)

$\dot{H}_3{=}C{=}CH{-}\dot{O}$
(also $H_3{\equiv}C{=}CH{-}O$)

$H_3C{=}C{=}NH_2$

(two structures)

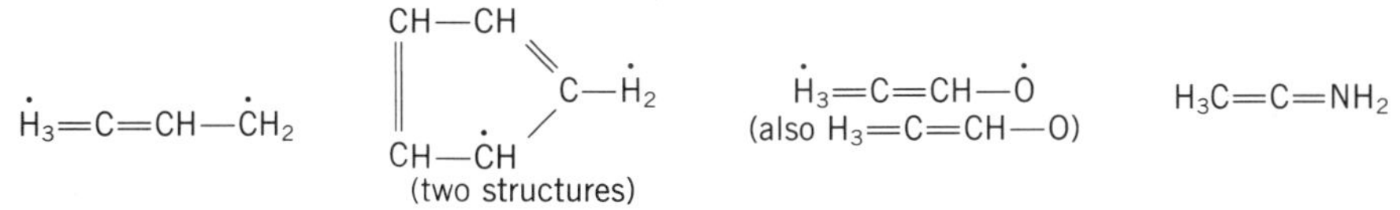

A still weaker effect is second-order sacrificial hyperconjugation, as in ethane ($H_3{\equiv}C{-}C{\equiv}H_3$; compare $HC{\equiv}C{-}C{\equiv}CH$) or in ethylene ($H_2{=}C{=}C{=}H_2$; compare $H_2C{=}C{=}C{=}CH_2$).

Examples of isovalent hyperconjugation are $H_3{\equiv}C{-}C{\equiv}H_2$ (with hyperisovalent resonance structure $H_3^+{\equiv}C{=}CH_2$) and $H_3{\equiv}C{-}\dot{C}H_2$ (resonance structure $\dot{H}_3{=}C{=}CH_2$). The number of bonds is the same in the two resonance structures, just as in isovalent conjugation, but the second structure is energetically less favorable than the first structure; for this reason the resulting stabilizing resonance energy is less than in isovalent conjugation.

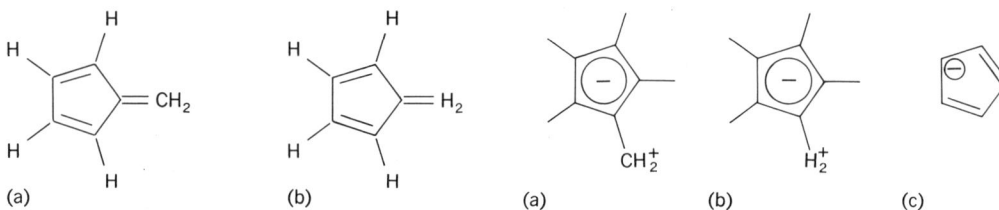

Fig. 2. Major resonance structures of the molecules of (a) fulvene and (b) cyclopentadiene.

Fig. 3. Minor resonance structures of the molecules of (a) fulvene and (b) cyclopentadiene. The pentagons in a and b represent the cyclopentadienyl anion, which is particularly stable because of resonance among five equivalent structures of the type shown by c.

Physical properties. The most notable differences in physical properties for sacrificially conjugated as compared with unconjugated molecules are as follows: (1) Single bonds lying between two multiple bonds are shortened. For example, the lengths of the C-C single bonds in diacetylene and 1,3-butadiene are 0.138 nanometer and 0.148 nm, as compared with 0.154 nm for ethane. The lengths of the double or triple bonds, however, are scarcely affected. According to quantum theory, the observed shortenings result partly from conjugation and partly from differences in the hybridization of the carbon atom in forming single, double, and triple bonds. (2) Electronic absorption spectra begin at considerably longer wavelengths for conjugated than for related unconjugated molecules. (3) Ionization potentials are lower than for isomeric unconjugated molecules. (4) Related to points 2 and 3 is the fact that refractivities and polarizabilities are larger for conjugated than for corresponding unconjugated dienes. (5) As compared with predictions from formulas for unconjugated molecules, conjugated molecules show lower energies, for example, about 6 kcal/mole (25 kilopascals/mole) lower for 1,3-butadiene than for unconjugated dienes; the excess stability is understandable theoretically as resonance energy. SEE BOND ANGLE AND DISTANCE.

More or less similar, but larger, effects occur for isovalent conjugation. Taking benzene as an example, interaction between two resonance structures removes the distinction between single and double bonds and leaves all bonds of length 0.139 nm and yields a resonance energy of about 40 kcal/mole (167 kilojoules/mole).

In sacrificial and in isovalent hyperconjugation, the effects are similar to those in conjugation, but in general they are smaller. However, in radicals and especially in radical ions, large-energy stabilizations can occur (for example, 16 and 84 kcal/mole (67 and 351 kJ/mole) in the t-butyl radical and cation, respectively).

Observed small dipole moments often can be explained in part by hyperconjugation. Thus for propylene, besides the minor resonance structure H_3=C=CH—$\dot{C}H_2$ already mentioned, one has some $H_3{}^+$=C^-—C^+H—C^-H_2 and H_3=C^+—C^-H—$^+H_2$, and if the first-named of these two predominates, a dipole moment in the direction $CH_3{}^+$—CH—$CH_2{}^-$ results, in agreement with experimental indications. However, other factors, including differences in hybridization, must also contribute.

Clear-cut examples of molecules where conjugation or hyperconjugation, respectively, are of primary importance in creating dipole moments are fulvene and cyclopentadiene, with major resonance structures show in **Fig. 2** and minor structures shown in **Fig. 3**. The minor structures are in addition to the nonpolar structure already mentioned. Fulvene has a dipole moment of 1.2 D, which is explained in terms of the minor structures, and one can safely predict an analogous but smaller dipole moment for cyclopentadiene.

INTERCALATION COMPOUNDS
FRANK J. DISALVO

Compounds that are formed by inserting extra atoms or molecules into a host structure, without disrupting the strong chemical bonds of the host material. For example, graphite is such a host

MATTER: STRUCTURE AND PROPERTIES

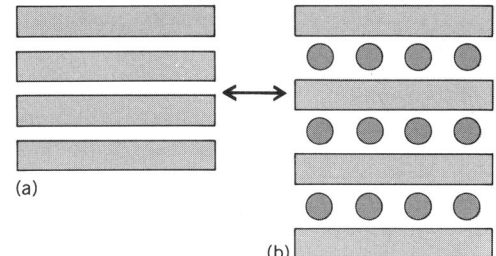

Formation of an intercalation compound. (*a*) Layered host compound, such as graphite. (*b*) Intercalation compound, with extra (intercalant) atoms or molecules inserted between the layers.

material, in which the carbon atoms are covalently bonded to form layers one atom thick. The layers are held together only weakly by van der Waals forces. Lithium atoms can be inserted between the layers by immersing the graphite in liquid lithium at 400–750°F (200–400°C) to form a new structure with alternating, one-atom-thick layers of carbon and lithium (see **illus**.). Intercalation compounds that consist of a periodic-structure of n layers of host material and one layer of intercalant can be formed from many layered hosts. These compounds are called nth-stage intercalation compounds. The example in the illustration is a first-stage compound. SEE CHEMICAL BONDING; INTERMOLECULAR FORCES.

Intercalation reaction. Since no strong chemical bonds are broken in the intercalation reaction, the kinetic barrier to reaction is usually small; that is, intercalation reactions proceed rapidly at low temperature, even at room temperature. The reaction takes place at the edges of the layers of the solid host compound, and the intercalant (the extra atoms or molecules) rapidly diffuses into the interior of the host. There must be a lowering of the free energy due to intercalation for the reaction to spontaneously proceed. This energy decrease is chemical or electrostatic in nature, and it must be large enough to overcome the small energy necessary to pry the layers apart. For example, for intercalants such as lithium, the electron affinity of the layers must be large enough to take the valence electron away from the lithium. The electrostatic attraction of the now negatively charged layers and the positive lithium ions provides enough energy to make the reaction proceed spontaneously. In the case of graphite, the layers are electron acceptors when the intercalant is a strong electron donor, but the layers become electron donors when the intercalant is a strong electron acceptor; that is, graphite is amphoteric. Intercalation into graphite does not occur when the intercalant is only a moderate or weak donor or acceptor. Other layered compounds, such as tantalum disulfide, are strong electron acceptors and can be intercalated only by electron donors. SEE CHEMICAL THERMODYNAMICS; ELECTRON AFFINITY.

Usually, the intercalation process is reversible, even at room temperature, if a chemical or electrochemical driving force is applied. For example, if an intercalation compound formed by inserting an electron donor such as lithium is immersed in a solution containing an oxidizing agent whose electron affinity is higher than that of the host, the intercalant will diffuse out of the host to react with the agent in solution. This reversibility, coupled with the high rate of reaction that is possible at room temperature due to the large intercalant diffusivity between the layers, makes possible rechargeable high-energy-density batteries based upon the intercalation reaction.

Types of hosts. Hosts for insertion reactions include not only layered compounds (such as graphite, micas, clays, and transition-metal dichalcogenides) but also linear compounds, in which chains having strong bonding only along one direction are bonded weakly to each other, and also framework compounds, which have a structure with open channels into which intercalants smaller than the channel can be inserted. In the last case, the compounds are more frequently called insertion compounds.

Physical properties. The physical properties of the host or the intercalant are often greatly changed by forming an intercalation compound. For example, tantalum disulfide is metallic; that is, it conducts electricity very well parallel to the layers and almost as well perpendicular to them. When intercalated with certain organic molecules, the individual 0.6-nanometer-thick layers can be separated by nonconducting organic layers of 5 nanometers or more in thickness. This produces a material with enormous electrical anisotropy—conducting parallel to the layers but insulating perpendicular to them. This process also produces the thinnest known two-dimensional metallic sheets of significant area.

Since there are such a large number of hosts and possible intercalants that form intercalation compounds, their physical and chemical properties are quite varied and will continue to be examined to find both novel phenomena and new technological applications.

Bibliography. E. Mooser (ed.), *The Physics and Chemistry of Materials with Layered Structures*, vol. 1–6, 1977–1984; D. W. Murphy and P. A. Christian, Solid state electrodes for high energy batteries, *Science,* 205:651–656, 1979; Y. Nishina, S. Tanuma, and H. W. Myron (eds.), Proceedings of the 4th Yamada Conference on Physics and Chemistry of Layered Materials, September 1980, *Physica,* vols. 105B and C, 1981; M. S. Whittingham, Chemistry of intercalation compounds: Metal guests in chalcogenide hosts, *Prog. Solid State Chem.*, 12:1–40, 1978; M. S. Whittingham and A. J. Jacobsen, *Intercalation Chemistry*, 1982.

HYDROGEN BOND
JACK M. WILLIAMS

The interaction which occurs when a hydrogen atom, covalently bonded to an electronegative atom (as in A—H), interacts with another atom to form the aggregate A—H $\cdots$ Y. The shortest and strongest bond is indicated as A—H, while the secondary and weaker interaction is written as H $\cdots$ Y. Thus A—H is a proton donor, while (Y) is a proton acceptor which often contains lone pair electrons and can act as a base. The strongest hydrogen bonds are formed between the most electronegative (A) atoms such as fluorine, nitrogen, and oxygen which interact with (Y) atoms having electronegativity greater than that of hydrogen (C, N, O, S, Se, F, Cl, Br, I). The weakest of hydrogen bonds are formed by acidic protons of C—H groups, as in chloroform and acetylene, and by olefinic and aromatic π-electrons acting as (Y).

Bond energies. The majority of hydrogen bonds have energies in the range 4–6 kcal/mole (17–25 kilojoules/mole) and involve those between O—H functional groups (as in water, alcohols, or acids) or N—H groups (as in amides or amines) and oxygen atoms (as in water, alcohols, carbonyls, or esters). The strongest hydrogen bond known is that found in the hydrogen difluoride ion, (F—H—F)$^-$, which has been variously estimated at 37–55 kcal/mole (155–230 kJ/mole). Therefore, the average hydrogen bond is of much lower energy than a normal chemical bond (>100 kcal/mole or 418 kJ/mole). Although hydrogen bonding gives rise to a specific interaction between atoms, resulting in a complex with characteristic A—H $\cdots$ Y distances and angles, especially in the solid state, it is difficult to establish a lower limit for the H—bond enthalpy because experimental methods of detection are becoming increasingly more sensitive and accurate.

The weaker the hydrogen bond, the shorter the lifetime of the complex it forms. The detection of weak hydrogen bonds often amounts to measuring shorter and shorter lifetimes of rapidly associating and dissociating species in equilibrium. This is a difficult problem because proton transfer along hydrogen bonds belongs to the fastest known chemical reactions, and in most experimental studies only mean values and "average" structures are determined.

An important aspect of weak hydrogen bond formation is that the different molecular aggregates which do form can be easily and reversibly transformed. Thus the small energy changes resulting in the rapid making and breaking of hydrogen bonds in biological systems are of great importance; for example, hydrogen bonding determines the configuration of the famous α-helix, and the structures of most proteins, thereby serving an important function in determining the nature of all living things.

Spectroscopy. Even though slight energy changes are usually involved in hydrogen bond formation, the aggregate once formed changes almost every measureable physical property of the original species. When investigating hydrogen bonding, the most frequently used techniques are infrared and nuclear magnetic resonance spectroscopy. In the infrared method the A—H stretching frequency is shifted to lower values and is accompanied by band broadening and increased intensity. Such changes are usually easily discernible, but in the case of very strong hydrogen bonds, such as (F—H—F)$^-$, the shift and broadening is so drastic that it is difficult to assign the new frequencies correctly. In the case of nuclear magnetic resonance, hydrogen bonding usually shifts the proton resonance to lower fields.

Neutron diffraction. While infrared and nuclear magnetic resonance techniques can yield considerable information about hydrogen bond formation, far greater information content is

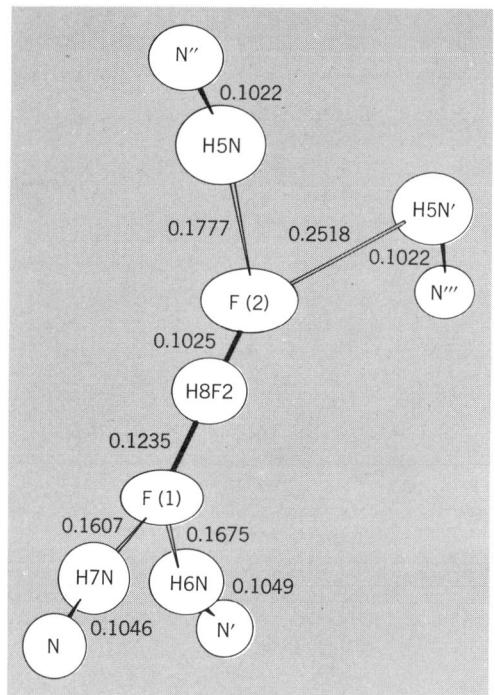

Hydrogen difluoride ion, (F—H—F)⁻, geometry. Distances are in nanometers, and atoms are represented as ellipsoids of 50% probability. Numbers following symbols for the elements signify that like atoms are not equivalent structurally. The primes on N atoms indicate that they are structurally equivalent and related by symmetry operations.

derived from diffraction studies of crystalline solids. The method of choice is neutron diffraction (versus x-ray diffraction), because in the former case hydrogen atoms scatter almost as well as any other atom while their scattering is often swamped out in the x-ray case. Neutron diffraction crystal structure analysis has become the best probe available for the study of the geometry of A—H · · · Y bonds. Using this technique, it is generally observed that the A—H separation is ~0.10 ± 0.01 nanometer and is much less than the H · · · Y separation; that is, hydrogen bonds are usually asymmetric. In the extreme case the hydrogen atom may be equally bonded to both atoms, as in certain O—H—O and (F—H—F)⁻ containing systems where the atomic environment around A and Y is identical and symmetric. In such systems strong hydrogen bond formation is indicated when the A · · · Y separation is ~0.02–0.03 nm less than the sum of the van der Waals radii. However, contrary to prior belief, even the shortest and strongest hydrogen bond known, (F—H—F)⁻, may be asymmetric. This was demonstrated in a neutron diffraction study of *p*-toluidinium hydrogen difluoride in which the two terminal F atoms exist in vastly different F · · · H—N hydrogen bonding environments (see **illus**.). The F—H distances are unequal, and the (F—H—F)⁻ ion is asymmetric, because of the very different N—H · · · F hydrogen bonding environments around the F atoms. Thus it seems that the H atom of a strong (X—H—X)⁻ bond is a probe of the X atom environment. Indeed, the hydrogen dichloride ion, $(CHl_2)^-$, with a bond energy of about 12 kcal/mole (49 kJ/mole) appears to be symmetric (centered H atom) in some salts and asymmetric in others. Other types of hydrogen atom interactions such as those of M—H · · · M and C—H · · · M, where M is a metal, are becoming increasingly important in catalysis.

Theory. Developments in theory have made it possible to better define certain contributions to hydrogen bond energies. The relative importance of forces of different origin (dispersion, polarization, exchange, coulomb, and so on) have become possible to estimate by using both molecular orbital methods and perturbation theory. In general, it appears that quantum theory gives reliable descriptions of isolated imers and trimers, but fails when dealing with large clusters of the type found in condensed phases.

MOLECULAR ORBITAL THEORY
Jeremy K. Burdett

A quantum-mechanical model concerned with the description of the discrete energy levels associated with electrons in molecules. One useful way to generate such levels is to assume that the molecular orbital wave function (ψ_j) may be written as a simple weighted sum of the constituent atomic orbitals (χ_i). This [Eq. (1)] is called the linear combination of atomic orbitals approximation. The c_{ij} coefficients may be determined numerically by substitution of Eq. (1) into the Schrödinger

$$\psi_j = \Sigma c_{ij} \chi_i \qquad (1)$$

equation and application of the variational theorem. The theorem states that an approximate wave function will always be an upper bound to the true energy; thus minimization of the energy of the system given by the wave function of Eq. (1) will provide the best values of c_{ij}. Once the wave function is known, its associated energy may be calculated. The energies of the occupied orbitals in molecules may be probed by using photoelectron spectroscopy, which gives a good check on the accuracy of the theory. There are some simple concepts which contribute to a qualitative understanding of these molecular orbital energy levels and hence an insight into chemical bonding in molecules. They may be illustrated with reference to the hydrogen molecule. *See* Chemical Bonding.

Hydrogen molecule. First, the basis orbitals (χ_i) used in the expansion of Eq. (1) can usefully be restricted to include the valence orbitals only. For molecular hydrogen (H_2) the 1s orbitals on the two hydrogen atoms are then the only two orbitals to be included. Second, since hydrogen atoms are chemically identical, any observable characteristic whose value might be computed with Eq. (1) must be the same for both atoms. This leads to the requirement that $c_{1j}^2 = c_{2j}^2$, where the labels 1, 2 refer to hydrogen atoms 1 and 2. As a consequence $c_{1j} = \pm c_{2j}$.

When the signs of the two coefficients are the same, the two hydrogen orbitals are mixed in phase; when they are different, the two hydrogen orbitals are mixed out of phase. **Figure 1** shows how the wave function ψ_j and its square (giving the electron probability distribution) may

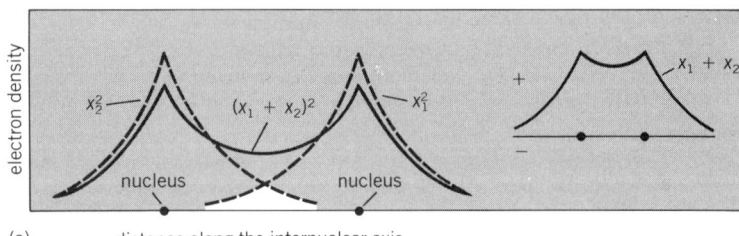

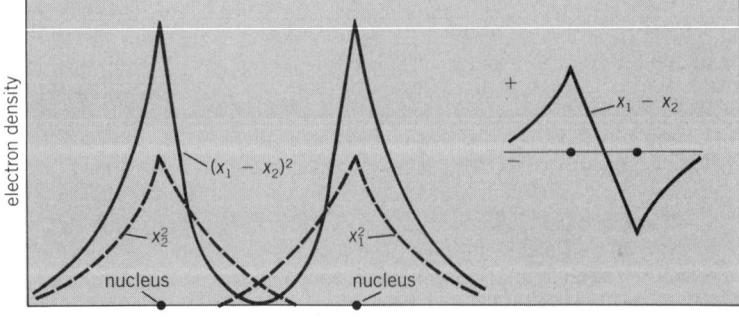

Fig. 1. In-phase and out-of-phase addition of atomic hydrogen wave functions (x_1, x_2) to give (a) bonding and (b) antibonding molecular orbitals.

be constructed for those two cases. When the atomic orbitals are mixed in-phase, then electron density is built up between the two hydrogen nuclei and the potential energy of the nuclei and electrons is lowered. In fact, a reduction of kinetic energy also occurs. An electron lying in the molecular orbital corresponding to Fig. 1a is then of lower energy than an electron associated with an isolated hydrogen 1s orbital. It is called a bonding orbital. The increase in electron density between the two nuclei is the electronic "glue" holding the nuclei together. When the atomic orbitals are mixed out of phase, the opposite behavior occurs (Fig. 1b). Electron density is removed from the region between the two nuclei, resulting in an increase of both potential and kinetic energy of the electrons. An electron lying in such a molecular orbital would experience an energetic destabilization relative to an electron associated with an isolated hydrogen 1s orbital.

Such a molecular orbital is called an antibonding orbital. **Figure 2** shows how this information may be collected together on a molecular orbital diagram. The shading convention of the orbitals has been adopted to indicate the in- and out-of-phase mixing of the basis orbitals. Just as the energy levels of atoms are filled in an Aufbau process, so the orbitals of the molecule may be analogously filled up with electrons, each level accommodating two electrons of opposite spin. In H_2 there are two electrons to be accounted for. They lie in the bonding orbital, and the stabilization energy relative to two isolated hydrogen atoms (the bond energy) is $2x$. Antibonding orbitals are invariably destabilized more than their bonding counterparts are stabilized. This is shown in Fig. 2 by making $y > x$. With four electrons to be accommodated in this collection of orbitals (this would correspond to the hypothetical case of the He_2 molecule), one electron pair resides in the bonding orbital and one pair in the antibonding orbital. Since $y > x$, this molecule is less stable relative to two isolated helium atoms, and as a result the molecule does not exist as a stable entity. He_2^+, however, with only three electrons is known.

The size of the interaction energy associated with two atomic orbitals (x in Fig. 2) is controlled by the extent of their spatial overlap. This overlap integral is proportional to the shaded region of Fig. 1a, and its magnitude clearly depends upon the internuclear separation. The equilibrium bond length in the hydrogen molecule (and indeed in all molecules) is then a balance between the attractive forces associated with bonding orbital formation and the electrostatic repulsion between the nuclei. Such a molecular parameter is amenable to numerical calculation.

The description of the bonding in the H_2 molecule using this model is one where two electrons occupy a bonding orbital and give rise to a simple two-center–two-electron bond traditionally written as H—H. Since the electron density associated with the bonding orbital is cylindrically and symmetrically located about the H—H axis, this bond is called a σ bond. In Fig. 2 the bonding orbital is labeled with a σ and the corresponding antibonding orbital with a σ*.

First-row diatomics. Ideas similar to those above are readily extended to diatomic molecules from the first row of the periodic table, such as N_2 and O_2, where the valence orbitals to be considered are the one 2s and the three 2p orbitals of the atoms. The 2s orbitals lie deeper in

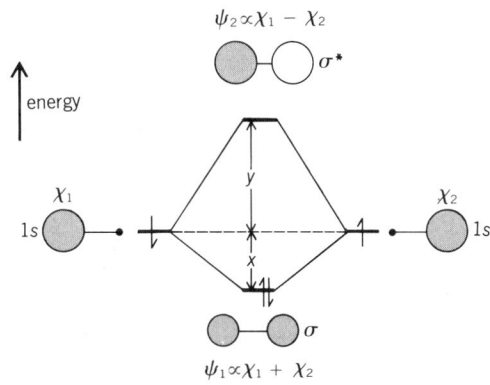

Fig. 2. Molecular orbital diagram of H_2.

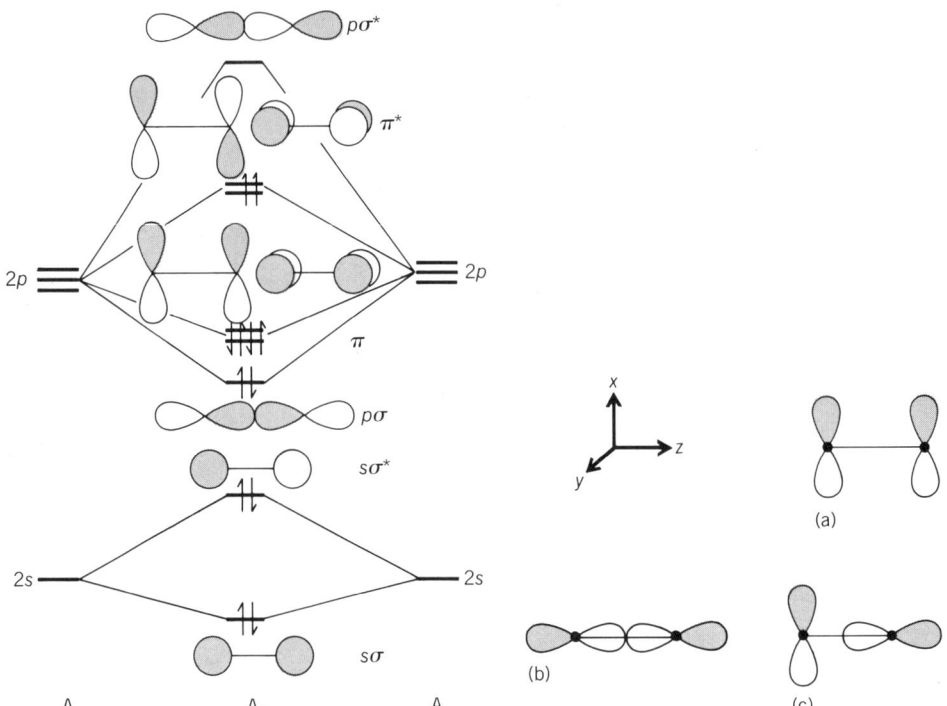

Fig. 3. Molecular orbital diagram for a first-row diatomic molecule (A_2). The electronic configuration corresponds to that of O_2.

Fig. 4. Possible orientations of the p orbitals on adjacent atomic centers. (**a**) End-on overlap. (**b**) Sideways overlap. (**c**) Zero overlap.

energy than the triply degenerate 2p orbitals. As shown in **Fig. 3**, the atomic 2s orbitals form bonding and antibonding orbitals (labeled $s\sigma$ and $s\sigma^*$) just as in the case of elemental hydrogen described above, but the behavior of the 2p orbitals is a little different. Here there are three possible types of interaction between the p orbitals on one center and those on the other. The end-on overlap of two p orbitals gives rise to a σ interaction (**Fig. 4a**), and the sideways overlap of two p orbitals gives rise to a π interaction (Fig. 4b). The interaction in Fig. 4c can be ignored since the overlap between the two orbitals in this orientation can be seen to be identically zero. The result shown in Fig. 3 is a σ-bonding orbital and a σ-antibonding orbital (labeled $p\sigma$ and $p\sigma^*$), and a pair of π-bonding and a pair of π-antibonding orbitals (labeled π and π^*). The larger interaction energy associated with $p\sigma$ compared to $p\pi$ is due to the larger σ overlap compared to π overlap in Fig. 4.

Filling these orbitals with electrons allows comment on the stability of the resulting diatomics. The molecule Li_2 $(s\sigma)^2$ is known and, like H_2, may be written as Li—Li to emphasize the single, two-center, two-electron bond between the nuclei. The molecule Be_2 which would have the configuration $(s\sigma)^2(s\sigma^*)^2$ is unknown since, just as in He_2, $s\sigma^*$ is destabilized more than $s\sigma$ is stabilized relative to an atomic 2s level. If the molecular orbital bond order is written as expression (2), then the bond order in Li_2 is one but the bond order in Be_2 is zero.

$$\frac{\text{Molecular}}{\text{bond order}} = \frac{(\text{number of bonding}}{\text{electron pairs})} - \frac{(\text{number of antibonding}}{\text{electron pairs})} \qquad (2)$$

By filling up the molecular orbital levels derived from the 2p orbitals, the bond order associated with the other diatomics may be generated: $B_2(1)$, $C_2(2)$, $N_2(3)$, $O_2(2)$, $F_2(1)$, and $Ne_2(0)$. All of these species are known except Ne_2, which is predicted, like He_2 and Be_2, to have a zero bond order and therefore not to exist as a stable molecule. The molecular orbital bond orders for the three best-known diatomics are consistent with their traditional formulation as N≡N, O=O, and F—F. N_2, for example, would be described as having one σ and two π bonds. Figure 3 shows the electron occupancy for O_2. With the configuration $(s\sigma)^2(s\sigma^*)^2(p\sigma)^2(\pi)^4(\pi^*)^2$, there are four bonding pairs of electrons and two antibonding pairs giving rise to a net bond order of two. The pair of π^* orbitals is only doubly occupied, whereas there is space for four electrons. Hund's rules (which for the electronic ground state maximize the number of electrons with parallel spins) identify the lowest-energy arrangement as the one where each of the degenerate π^* components is singly occupied, the spins of the two electrons being parallel. Unpaired electrons give rise to paramagnetic behavior, and gaseous oxygen is indeed paramagnetic.

Multicenter bonding. Not all molecules can be described as being held together by simple two-center, two-electron bonds. Multicenter delocalized bonding, for example, is the best way of describing the π bonding which occurs in organic acyclic and cyclic polyenes. **Figure 5** shows the molecular orbitals which result from interaction of the $p\pi$ orbitals which lie perpendic-

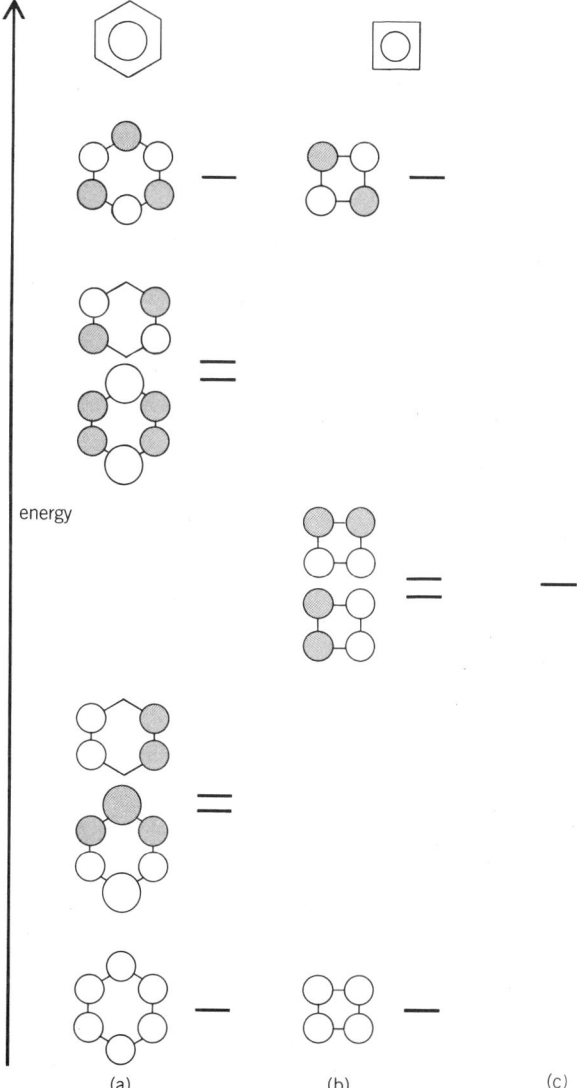

Fig. 5. Molecular π orbitals and energy levels (indicated by short horizontals) for (a) benzene and (b) cyclobutadiene viewed from above so that the relative phases only of the upper half of the $p\pi$ orbitals are shown. (c) The energy level of an isolated $p\pi$ orbital.

ular to the molecular plane in benzene and cyclobutadiene. For benzene there are three bonding orbitals, one lying deeper than the other degenerate pair and three analogous antibonding orbitals. In cyclobutadiene there is a bonding and antibonding orbital in addition to a degenerate pair of nonbonding orbitals. A nonbonding molecular orbital has the same energy as the isolated atomic orbitals from which it is constructed. In these examples of neutral hydrocarbons the number of molecular orbitals is equal to the number of atomic orbitals used in synthesizing the diagram (Fig. 5). Each carbon atom contributes one electron to this π-orbital manifold. Thus in benzene the lower three orbitals are doubly occupied. As far as the π manifold of benzene is concerned, there are then three bonding pairs but six close contacts between carbon atoms. As a result, the π-bond order is one-half, and the π bonding in this molecule is referred to as being delocalized over the carbon framework. Similar ideas employing delocalized bonding can be applied to three-dimensional molecules such as the boranes, carboranes, and transition-metal cluster compounds.

There is a simple rule that has its basis in the π orbital structure of these hydrocarbons. If a molecule possesses $4n + 2$ π electrons ($n = 0, 1, 2 \ldots$), then it will be stable as a planar aromatic molecule. If the π-electron count is otherwise, then the planar symmetric molecule will be antiaromatic and unstable. This is called Hückel's rule. Benzene, with six electrons, satisfies the rule ($n = 1$), but cyclobutadiene does not. The latter molecule distorts away from the square geometry, probably to a rhombus, and is kinetically unstable. C_8H_8 is another species which does not satisfy the rule, and this molecule is nonplanar. $C_8H_8^{2-}$ and $C_4H_4^{2-}$ do satisfy Hückel's rule and are found as planar, symmetric units bound to transition metals. The species S_4^{2+}, isoelec-

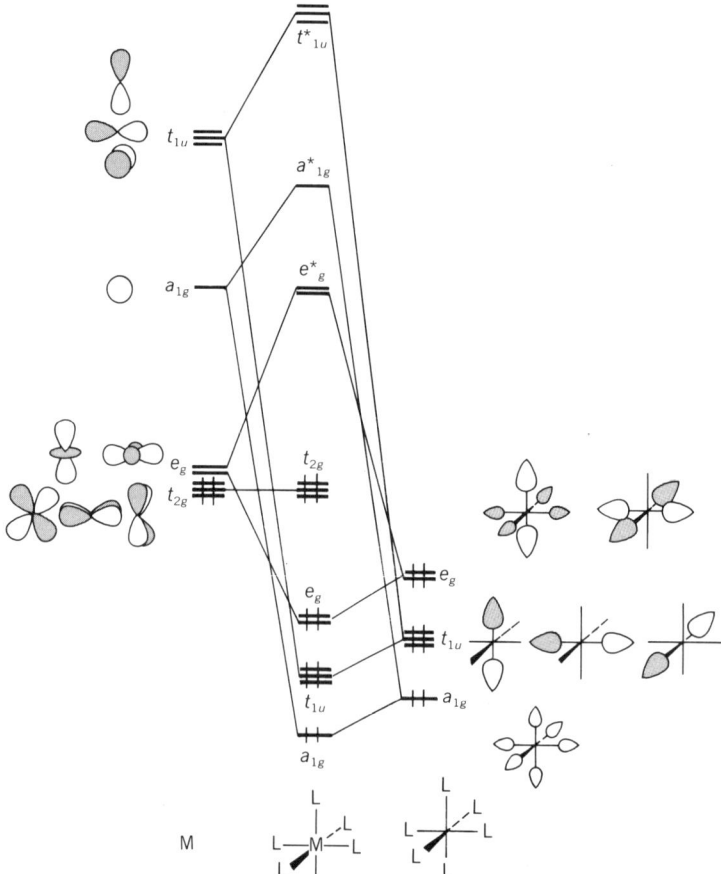

Fig. 6. Molecular orbital diagram for an ML_6 transition-metal complex. The labels a_{1g}, e_g, t_{1u}, and t_{2g} describe the symmetry properties of the orbital levels.

tronic with $C_4H_4^{2-}$, is also a stable, square molecule. Hückel's rule has its foundation in the requirement that all bonding and nonbonding orbitals of the π network be full of electrons for structural stability.

Transition-metal complexes. Similar orbital ideas applied to transition-metal complexes have led to an understanding concerning their structures and reactivity. Molecular orbital models have virtually replaced the older concepts of crystal field and valence bond theory in the area. **Figure 6** shows a molecular orbital diagram for an octahedral ML_6 complex, where L is a ligand which possesses an accessible orbital with which to bond in a σ fashion to the metal. (Examples are hydride, water, and halide.) This orbital diagram is different in several ways from those previously described. The ionization potential for an electron located in a ligand orbital is larger than for a metal-located electron. This is shown by drawing the metal atomic levels higher in energy than the ligand levels. The six ligand orbitals split into three sets when the ligands are coordinated to the metal. (The reasons for this are described in the next section.) The five atomic metal d orbitals split into two sets in ML_6, one doubly degenerate and one triply degenerate. The set labeled t_{2g} is nonbonding between metal and ligands, whereas the set labeled e_g is metal-ligand antibonding. The energy separation between the two is conventionally labeled Δ. Electronic transitions between these orbital sets occur in the visible part of the spectrum for many first-row transition-metal complexes (scandium through copper) and are thus the source of their color. The number of electrons which occupy these orbitals is simply given by counting the number of d + s electrons of the corresponding gaseous ion. Thus Cr(II) is a d^4 system while Zn(II) is a d^{10} system. In the latter, the e_g and t_{2g} orbitals are completely filled. In organometallic compounds where Δ is large, a rule similar to Hückel's rule is found to apply. If all bonding and nonbonding orbitals are occupied, then kinetic and structural stability is conferred upon the system. This occurs when a total of 18 electrons (or nine electron pairs) are associated with the metal. In Fig. 6 this would occur for six pairs of electrons in ligand-located orbitals plus three pairs of electrons in the nonbonding t_{2g} set. This 18-electron rule has proven to be very useful in understanding organometallic chemistry.

Symmetry considerations. Symmetry is very useful in understanding many aspects of molecular orbital theory. In qualitative terms, the more symmetric a molecule, the larger the number of elements of symmetry it possesses. **Figure 7** shows the symmetry elements—the mirror planes (σ_v), rotation axes (C_x, for a rotation of $360°/x$), and center of symmetry (i)—for the A_2 diatomic molecule shown in Fig. 3. Although only one mirror plane containing the internuclear axis is shown, there are an infinite number of them, just as there are an infinite number of C_2 axes perpendicular to the internuclear axis.

The mathematical theory which allows manipulation of such observations is called group theory. It can be seen that the directions x and y in Fig. 7 are symmetry-equivalent in such a

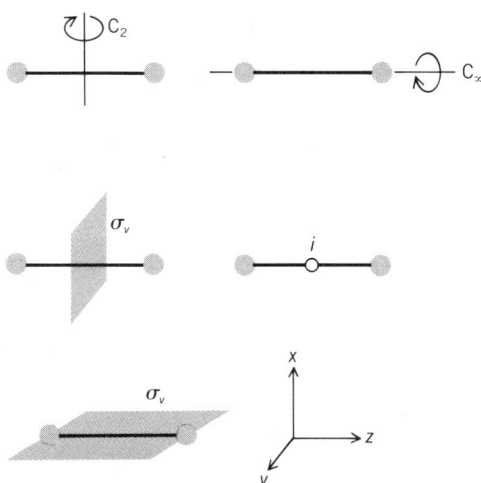

Fig. 7. Symmetry elements for the H_2 or A_2 diatomic molecule.

molecule, but that the z direction is unrelated to either x or y. A more formal way of putting this is to note that, although there are symmetry operations possessed by the molecule which will interconvert x and y, there is no such operation which will interconvert z with either x or y. A direct result of this observation is that the p_x and p_y orbitals will always be found in degenerate orbital situations, but that p_z will be energetically separate. In the linear molecule the symmetry label π is used to refer to such a double degeneracy. The nondegenerate p_z orbital is described by the symmetry label σ. The symmetry description of the three valence p orbitals on an A atom in the A_2 diatomic is then $\sigma + \pi$. As the number of symmetry elements associated with a molecule increases, the size of possible degeneracies also increases. In octahedral and tetrahedral molecules, for example, double and triple degeneracies are possible, and in icosahedral molecules fivefold degeneracies may occur. In the octahedral molecule shown in Fig. 6, the x, y, and z directions are symmetry-equivalent and are described by the symmetry label t_{1u} representing a triple degeneracy. (Greek letters are used for the symmetry labels of linear molecules, for example, σ or π, and Roman letters, for example, a_{2g} or t_{2u} for nonlinear molecules.)

Here a useful subscript is used to describe a function's behavior with respect to inversion. The label g applies to functions which are symmetric with respect to such an operation (s and d orbitals on the central atom, for example) and the label u for antisymmetric functions (p and f orbitals on the central atom, for example). The five d orbitals become of $e_g + t_{2g}$ symmetry in the octahedral molecule, just as the three p orbitals become of $\sigma + \pi$ symmetry in the diatomic molecule shown in Fig. 3.

There is one very powerful result which is extremely useful in constructing orbital diagrams. Orbitals of different symmetry have zero overlap and therefore do not interact with each other. For example, in Fig. 4b the two p_x orbitals have a nonzero interaction since they are both of π symmetry, but the overlap integral for the combination of Fig. 4c is zero since one orbital is of π and the other of σ symmetry. The molecular orbital diagrams shown in Figs. 2, 3, and 6 may then be generated simply by matching orbitals of the same symmetry at the left- and right-hand sides of the picture and constructing bonding and antibonding orbital combinations from each pair. In Fig. 6 the central metal atom orbitals of t_{2g} symmetry find no symmetry match with the orbitals of the ligands and so remain nonbonding. Symmetry arguments such as these are behind many of the basic orbital concepts in molecules.

Orbital symmetry. The Woodward-Hoffmann rules, which collect together within a theoretical framework many important organic reactions, are based on the symmetry control of orbital interaction. **Figure 8** shows orbital correlation diagrams linking the orbitals of a 1,4 substituted butadiene and the product of ring closure, the cyclobutene. Two different modes of closure may be envisaged, the conrotatory and disrotatory motion of the groups attached at the 1 and 4 positions. In the former a twofold rotation axis is preserved, and in the latter a mirror plane is preserved during the reaction. In Fig. 8a all the levels may then be classified as being either symmetric (S) or antisymmetric (A) with respect to the twofold rotation axis (C_2). Correspondingly in Fig. 8b the levels may be classified analogously according to their behavior with respect to the mirror plane operation. Orbitals of the same symmetry will interact with each other and will not cross along the reaction coordinate, but orbitals of different symmetry may cross.

The orbital correlation diagrams of Fig. 8 may thus be generated by connecting pairwise the orbitals at the left- and right-hand sides. Notice how in Fig. 8b the highest occupied molecular orbital (HOMO) of the reactant correlates with the lowest unoccupied molecular orbital (LUMO) of the product. For this reaction a high activation energy is expected as a result. Contrast the process in Fig. 8a which will experience no such energetic penalty. The thermal reaction of Fig. 8b is then orbitally forbidden, whereas that of Fig. 8a is orbitally allowed. On photochemical excitation of an electron from HOMO to LUMO of the reactant, the opposite is now true. The disrotatory motion is photochemically allowed, but the conrotatory motion is a photochemically forbidden process. This generalization has had a profound effect on organic chemistry. SEE STEREOCHEMISTRY.

Frontier orbitals. The use of valence orbitals alone is sufficient to build up the molecular orbitals and a qualitative picture of chemical bonding. This occurs because the core orbitals (for example, the 1s orbital on the oxygen atom) are very tightly bound and have a negligible overlap with the orbitals of a neighboring atom. When two molecules, or fragments are brought together, it is often found that their mutual interaction is qualitatively well described by considering, in turn, a rather small number of these valence-orbital–based molecular orbitals. These so-called

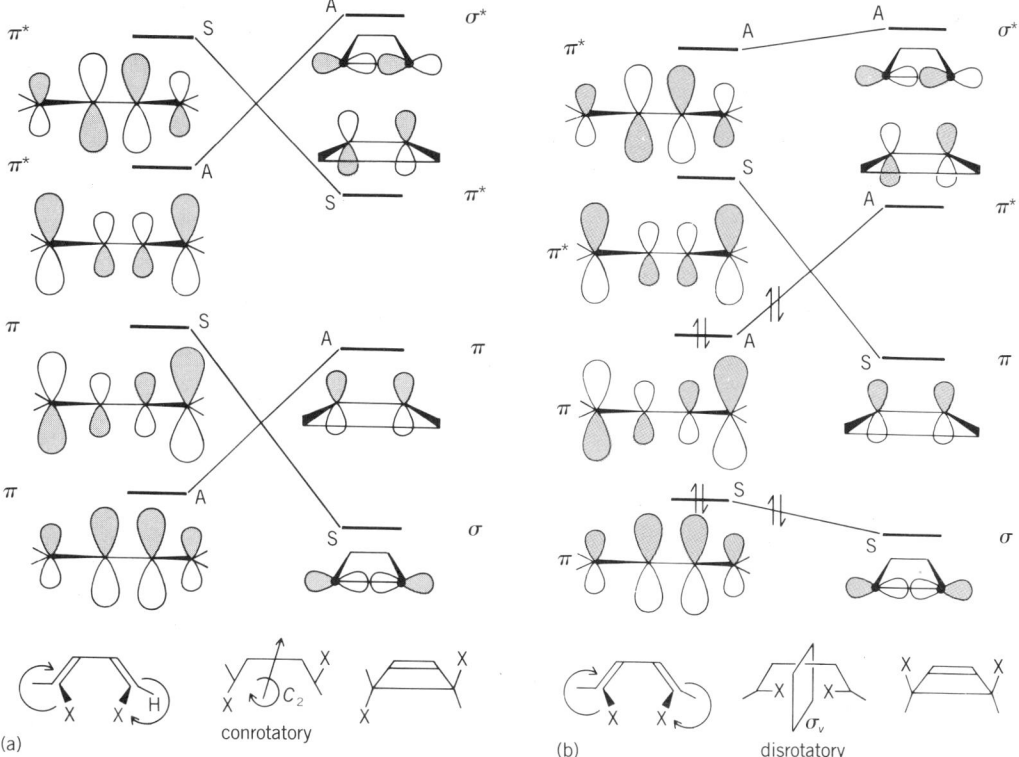

Fig. 8. Orbital correlation diagram for (a) the conrotatory and (b) the disrotatory ring closing of 1,4-butadienes.

frontier orbitals are the orbitals of the two molecules which are spatially arranged so that the overlap integral between them is significant. Other interactions are either zero by symmetry, or small because their geometrical arrangement precludes good overlap.

Many aspects of chemical reactivity and structure may be understood by considering the details of their interaction. It often turns out that the energetically most important interactions occur between the HOMO of one molecule and the LUMO of another, giving rise to a stabilizing interaction. **Figure 9** shows how consideration of the frontier orbitals of these molecules allows understanding of the observation that maleic anhydride reacts easily with butadiene but with difficulty with ethylene. The HOMO-LUMO overlap in the former case (Fig. 9a) is nonzero and gives rise to a stabilizing interaction. In the latter such overlap is exactly zero since the positive overlap at the left-hand side is canceled by the negative overlap at the right-hand side and the interaction of the reactants is zero (Fig. 9a). In the area of organic chemistry, frontier orbital theory allows explanation of many aspects of the Woodward-Hoffmann rules. In the areas of molecular geometry such ideas are very useful in understanding the relative orientation of different parts of a molecule by maximizing such interactions.

Present status. In general, numerical calculations are needed to generate molecular orbital energy levels, although a qualitative understanding of the level structure may also be achieved and is to be encouraged. Good calculations are at present able to accurately reproduce geometries (bond lengths and angles), vibrational force constants, activation energies, and reaction coordinates for small molecules containing a few first-row atoms. As the size of the molecule increases, the feasibility of such accurate calculations is lost and ab initio calculations on larger molecules are only approximate. Considerable effort has been made to develop methodology to accurately reproduce the photoelectron spectra and reaction pathways of larger molecules. In the

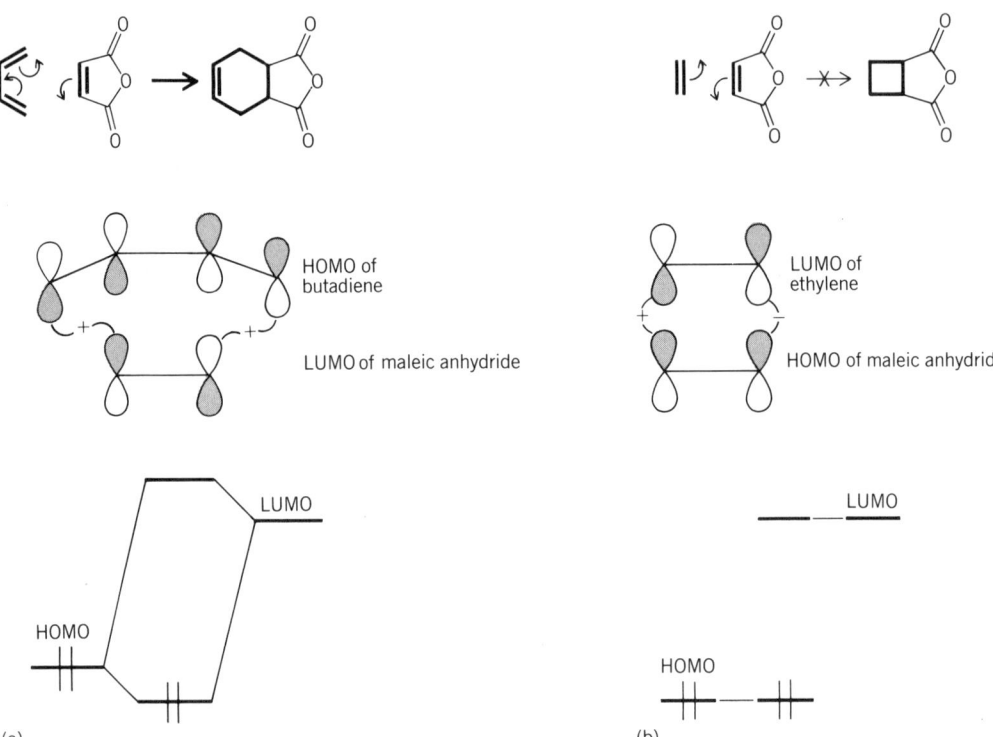

Fig. 9. HOMO-LUMO interactions between maleic anhydride and (a) butadiene and (b) ethylene.

realm of organic molecules there are several semiempirical methods, for example, complete neglect of differential overlap (CNDO), which, when used judiciously, are able to give good indications of the energetics of chemical processes. Considerable success has been achieved, however, in the use of qualitative molecular orbital ideas. Arguments based on symmetry, overlap, and electronegativity differences have provided a way by which to link together many aspects of chemistry. SEE CHEMICAL BOND THEORY; MOLECULAR STRUCTURE AND SPECTRA; QUANTUM CHEMISTRY.

Bibliography. T. A. Albright, J. K. Burdett, and M. H. Whangbo, *Orbital Interactions in Chemistry*, 1984; F. A. Cotton, *Chemical Applications of Group Theory*, 2d ed., 1971; I. Fleming, *Frontier Orbitals and Organic Chemical Reactions*, 1976; W. J. Hehre et al., *Molecular Orbital Theory*, 1986; H. F. Schaefer (ed.), *Methods of Electronic Structure Theory*, 1977; A. Streitweiser, *Molecular Orbital Theory for Organic Chemists*, 1961; R. B. Woodward and R. Hoffmann, *The Conservation of Orbital Symmetry*, 1970.

MOLECULAR ISOMERISM
ERNEST L. ELIEL

The property of compounds (isomers) which have the same molecular formula but different physical and chemical properties. The difference in properties is caused by a difference in molecular structure (that is, molecular architecture). A typical example is dimethyl ether, CH_3OCH_3, a chemically quite inert gas which condenses at $-24°C$ ($-11°F$), and ethyl alcohol, CH_3CH_2OH, a liquid of substantial chemical reactivity which boils at $78°C$ ($172°F$); both compounds have the molecular formula C_2H_6O.

Classification. Isomers may be classified as constitutional isomers or stereoisomers. Constitutional isomers differ in constitution or connectedness, relating to the question as to which atoms are linked to which others and how. Dimethyl ether and ethanol (**Fig. 1**) are constitutional isomers. In dimethyl ether each carbon is connected to three hydrogen atoms and the one oxygen atom; the two carbon atoms are thus equivalent. In ethyl alcohol (ethanol) one carbon is linked to three hydrogen atoms and the other carbon; the second carbon is linked to the first carbon, two hydrogens, and the oxygen atom which, in turn, in linked to the sixth hydrogen atom; the two carbon atoms are not equivalent. (Constitutional isomers can often be distinguished, and their constitution recognized, by carbon-13 nuclear magnetic resonance.) Stereoisomers, in contrast, have the same constitution but differ in the three-dimensional array of the atoms in space, called configuration. (In some cases, the difference in three-dimensional arrangement may, however, be due to rotation about single bonds, in which case it is spoken of as a difference in conformation.)

Constitutional isomers. Constitutional isomers have been subdivided into functional isomers, positional isomers, and chain isomers.

Functional isomers. Functional isomers (Fig. 1) differ in functional group, that is, the group (or groups) most material in determining chemical behavior. The ammonium cyanate–urea pair (CH_4N_2O) shown in Fig. 1 plays an important role in the history of chemistry inasmuch as the conversion of ammonium cyanate to urea by heating, effected by F. Wöhler in 1828, is considered the first example of an organic compound (urea) having been produced in the laboratory from a mineral one (ammonium cyanate). The third example shown in Fig. 1, that of propionaldehyde (propanal), allyl alcohol (2-propen-1-ol), and propylene oxide (methyloxirane), illustrates the fact that functional isomers do not necessarily come in pairs. The three compounds all correspond to the molecular formula C_3H_6O, but the first one has an aldehyde function, the second combines a double bond with an alcohol function, and the third one has an epoxide function. Indeed, the number of possible isomers corresponding to a given molecular formula is generally remarkably large. Thus, even for the relatively simple composition $C_{10}H_{22}$, a saturated hydrocarbon with 10 carbon atoms, there are 75 isomers, and for a hydrocarbon with twice as many carbon atoms, $C_{20}H_{42}$, the number is well over 300,000.

Positional and chain isomers. Positional isomers (**Fig. 2**) have the same functional group but differ in its position along a chain or in a ring. Closely related are chain isomers which

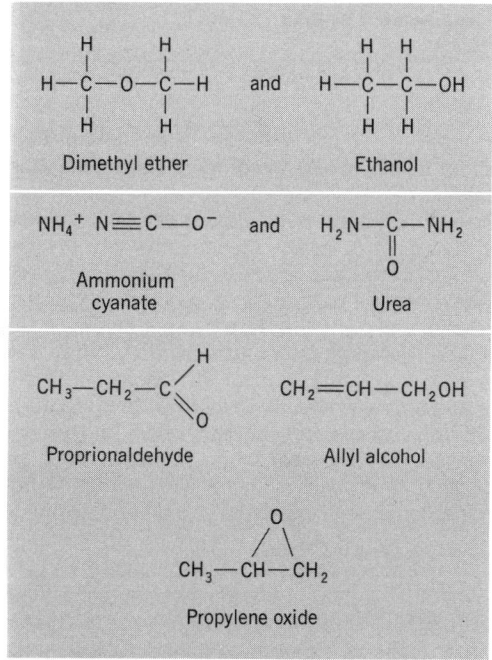

Fig. 1. Functional isomers.

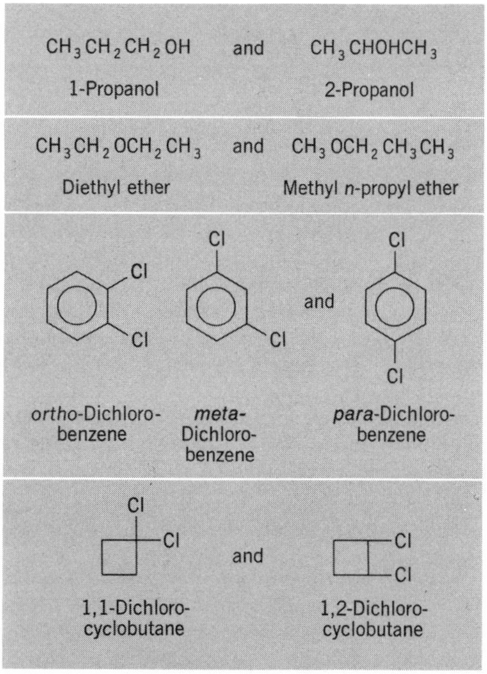

Fig. 2. Positional isomers.

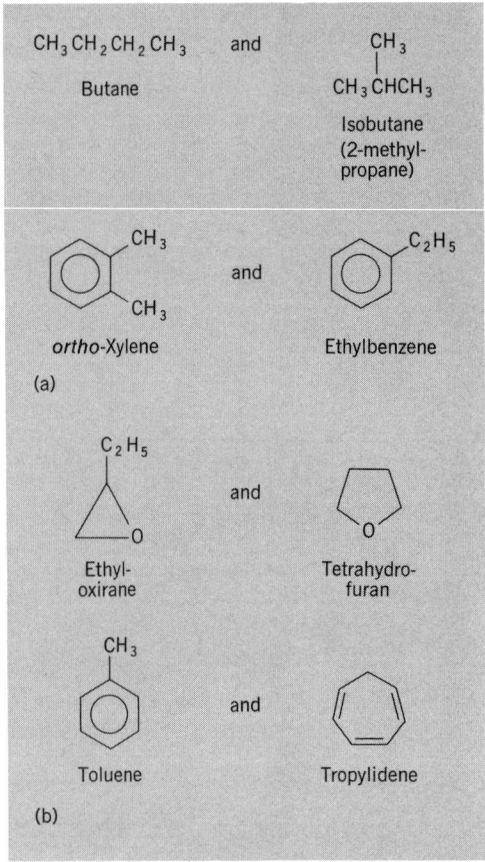

Fig. 3. Skeletal isomers. (a) Chain isomers. (b) Ring isomers.

also have the same functional group or groups but differ in the shape of the carbon chain (**Fig. 3**a); quite similar are ring isomers (Fig. 3b) which differ in the size of one or more rings. Ring and chain isomers together are sometimes called skeletal isomers.

Properties. It should be emphasized that these subclassifications of constitutional isomers are made for the convenience of the chemist rather than because of any fundamental importance. All constitutional isomers differ in physical and chemical properties, such as melting and boiling points, density, refractive index, and free energy, as well as in all kinds of spectral properties, such as ultraviolet, infrared and nuclear magnetic resonance spectra, and, to a lesser extent, mass spectra. If crystalline, isomers can generally be assigned their proper structure by x-ray diffraction analysis. The above differences tend to be greatest for functional isomers and more subtle for positional and skeletal isomers. However, the last statement is not universally true; thus, there is a considerable difference between the fairly reactive ethyloxirane and the fairly inert ring isomer tetrahydrofuran shown in Fig. 3b. Indeed, it is often not entirely clear when isomers should be called functional and when they should be called positional or chain isomers. Thus acetone (propanone; I) and propionaldehyde (propanal; II) may be considered positionally isomeric carbonyl

$$CH_3CCH_3 \quad\quad\quad CH_3CH_2CH=O$$
$$\underset{(I)}{\overset{\|}{O}} \quad\quad\quad (II)$$

compounds if one chooses not to distinguish between aldehyde and ketone functions. Yet these functions are substantially different (for example, aldehydes are readily oxidized to acids, while ketones are not), so an alternative might be to consider the ketone and the aldehyde as functional isomers. A similar situation occurs with amines: t-butylamine, $(CH_3)_3CNH_2$, and n-butylamine, $CH_3CH_2CH_2CH_2NH_2$, are clearly chain isomers, whereas diethylamine (III) and methylpropylamine (IV) should be classified as positional isomers. When one compares n-butylamine with diethylam-

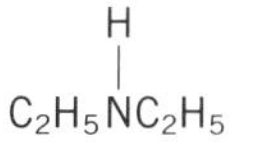

$$\underset{(III)}{C_2H_5\overset{\overset{H}{|}}{N}C_2H_5} \qquad \underset{(IV)}{CH_3\overset{\overset{H}{|}}{N}CH_2CH_2CH_3}$$

ine, however, the situation is less clear-cut. One could argue that these are positional isomers also, but in view of the functional difference (though slight) between primary and secondary amines, they are probably better classified as functional isomers. Such differences in classification are obviously quite tenuous.

Stereoisomers. Compounds which have not only the same molecular formula but also the same constitution (connectivity of the atoms) but which differ in the disposition of the atoms in space are called stereoisomers. Stereoisomers, in turn, are subdivided into two types; those that are mirror images of each other, called enantiomers, and those which are not mirror images, called diastereomers or diastereoisomers.

Enantiomers. These isomers are unique in that they always come in pairs (**Fig. 4**). Either a molecule is superposable with its mirror image, in which case it does not have an enantiomer, or it is not superposable with its mirror image, in which case it has one and only one enantiomer (since an object can have only one mirror image). Molecules which are not superposable with their mirror images are called chiral; those which are so superposable are called achiral. Enantiomers are much more alike than are other sets of isomers (constitutional isomers or diastereomers); thus they have the same melting point, boiling point, free energy, spectral properties, x-ray diffraction pattern, and so on. This is because their internal relationships are the same; for example, the distances between corresponding atoms are the same, much as the distances between corresponding fingers are the same in right or left hand. However, enantiomers differ in their behavior

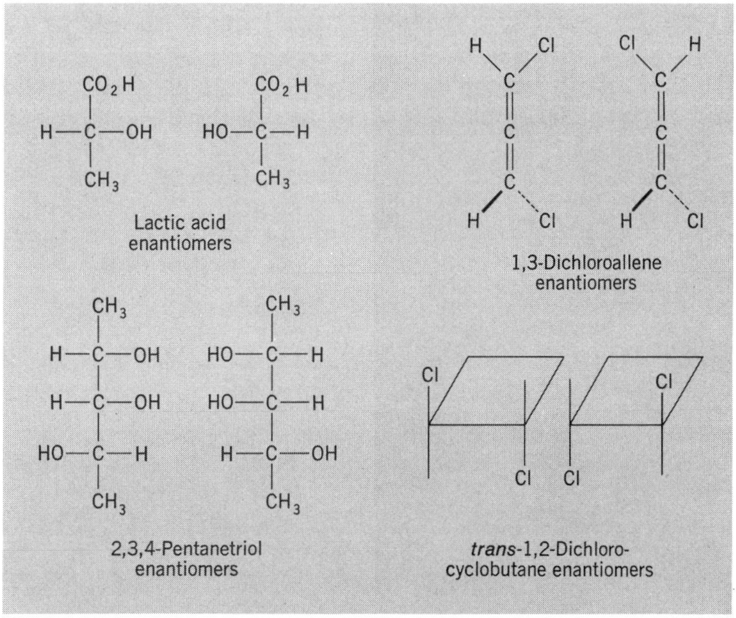

Fig. 4. Enantiomers, mirror-image isomers.

toward chiral reagents, much as a right hand and a left hand differ in their relation to a right glove, and they are also different in their behavior toward chiral physical agents, such as circularly polarized light. Thus they differ in circular dichroism and in the direction of rotation which they impart to plane polarized light (optical rotation, optical rotatory dispersion). Such differential physical properties have been called chiroptic properties.

Diastereomers. These isomers have the same constitution but different spatial arrangement and are not mirror images (**Fig. 5**). They resemble constitutional isomers in that there may be more than two isomers in a set and that their physical, energetic, and spectral properties are generally quite distinct. The example of cis- and trans-1,2-dibromoethene illustrates cis-trans isomerism in olefins. (It is recommended that the term geometrical isomers which was formerly used for this type of isomers be abandoned, just as optical isomers should no longer be used as a synonym for enantiomers.) 1,3-Dichlorocyclobutane and 1,2-dimethylcyclopropane illustrate diastereoisomerism (also of the cis-trans type) in cyclanes. The 1,3-dichlorocyclobutane exists in two achiral diastereomeric forms, whereas the 1,2-dimethylcyclopropane has a pair of (chiral) enantiomers (trans) which are diastereomeric with the (achiral) cis isomer (called meso because it is an achiral diastereomer in a set also containing chiral species). Pentane-2,3,4-triol illustrates a case with one chiral and two achiral (meso) diastereomers. The enantiomeric pair of the chiral 2,3,4-pentanetriols is shown in Fig. 4.

A set of stereoisomers containing n chiral centers will normally contain 2^n members. Since each member of the set has an enantiomer, there will be 2^{n-1} enantiomer pairs which, in relation to each other, are diastereomeric. However, when there is degeneracy, that is, when two or more of the chiral centers are equivalent (as in the 2,3,4-pentanetriols where the chiral centers at C-2 and C-4 are alike), there will be fewer isomers than the formula predicts. Thus there are only four stereoisomeric 2,3,4-pentanetriols instead of the eight (2^3) predicted by the formula (see Figs. 4 and 5). A general method for counting stereoisomers has been developed. SEE STEREOCHEMISTRY.

Limitations. Many isomers are quite stable and if they can be interconverted at all, the barrier between them is quite high (**Fig. 6**). For example, the barrier between cis- and trans-2-

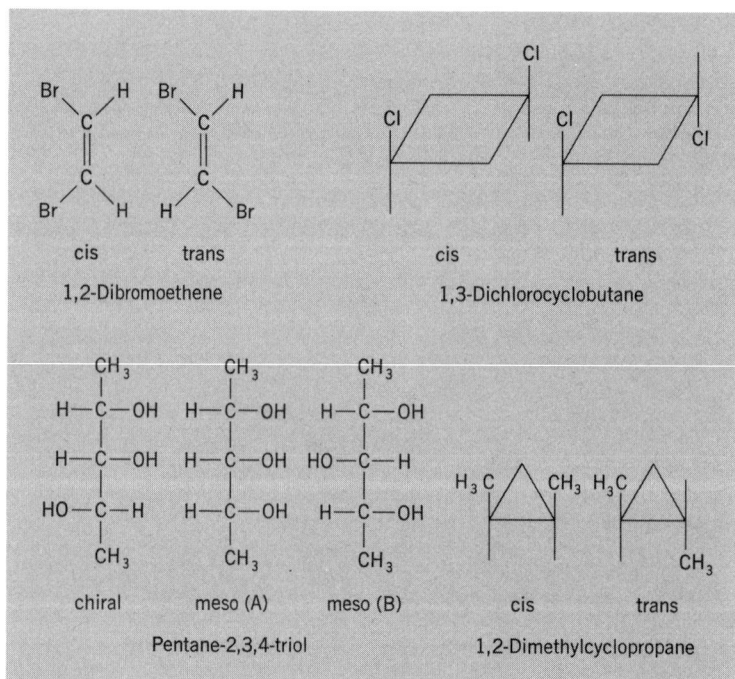

Fig. 5. Diastereomers, isomers having different spatial arrangements of the same groups of atoms.

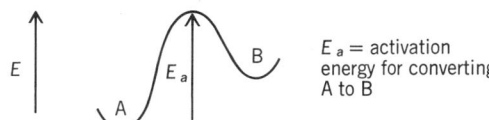

Fig. 6. Energy barriers between isomers A and B.

butene, $CH_3CH=CHCH_3$, is of the order of 65 kcal/mol (272 kilojoules/mol), and such isomers (the 1,2-dibromoethenes in Fig. 5 are another example) are spontaneously interconverted only at very high temperatures. In other cases, however, the barriers are intrinsically quite low or are readily lowered by deliberate or adventitious catalysis, frequently by acids or bases. An example relates to the keto and enol forms of ethyl acetoacetate (**Fig. 7a**). While these isomers can be separated by fractional distillation in clean quartz vessels, they are quickly interconverted in the presence of traces of acids or bases. Such easily interconvertible isomers which differ only in the position of an atom or group (in the case of ethyl acetoacetate, a hydrogen atom which is attached to carbon in the keto form and to oxygen in the enol form) are called tautomers, and the phenomenon is called tautomerism. A closely related type of isomerism in which only bonds shift and atoms remain in place (except for changes in bond distances) is represented by the cycloocta-triene-bicyclooctadiene interconversion (Fig. 7b) which takes place readily at room temperature. This kind of rapid isomerization is referred to as valence bond isomerism or valance tautomerism. Two other examples of rapidly interconverting isomers are shown in **Fig. 8**. The first is chlorocyclohexane which exists in an equatorial and an axial conformation rapidly interconverted by reversal (flipping) of the cyclohexane chair. The barrier to this interconversion is about 10 kcal/mol (42 kJ/mol), which is so low that the two conformational isomers (sometimes called conformers) can be isolated only at temperatures as low as $-150°C$ ($-238°F$). However, the two isomers can be seen separately in nuclear magnetic resonance (NMR) spectra below about $-65°C$ or $-85°F$; (the exact temperature depends on the frequency of the instrument and the nucleus observed). The axial and equatorial conformations of chlorocyclohexane are thus diastereomeric. In 1,2-dichloroethane (Fig. 8), three stereoisomeric conformational minima (two enantiomeric gauche conformations and the achiral anti conformation that is diastereomeric to the two others) can be discerned, but the barrier to rotation is so low (about 3 kcal/mol or 13 kJ/mol) that isolation is out of the question, and even many physical techniques (such as electron diffraction, NMR spectroscopy, and dipole moment measurement) yield only weighted average values of the physical properties of the three conformations. However, the gauche isomers and antiisomers can be seen distinctly in the infrared spectrum of the substance and are thus clearly different species.

Ultimately there is the question of what will happen if the barrier to interconversion becomes even lower. A point must come where it is no longer possible to speak of two distinct

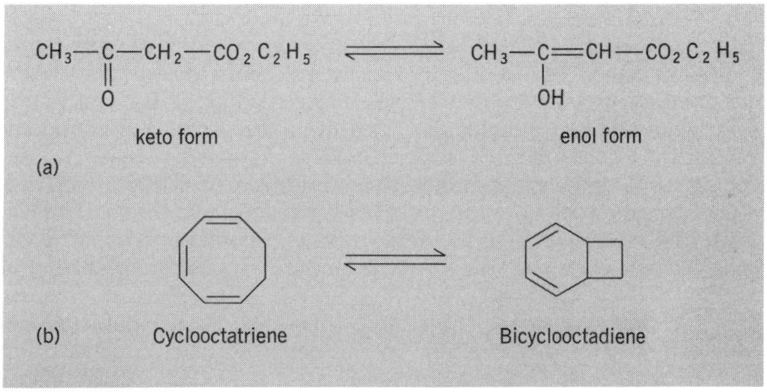

Fig. 7. Easily interconverted isomers. (a) Ethyl acetoacetate. (b) Cyclooctatriene and bicyclooctadiene.

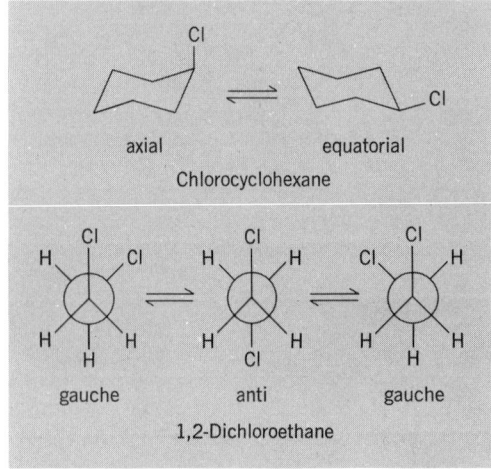

Fig. 8. Conformational isomers.

isomeric molecules but where only a single molecule is deemed to exist. This happens when there is no longer an operational way of demonstrating the existence of two separate energy minima. It has been suggested that there are not two isomeric molecules when the barrier (Fig. 6) is lower than the product of the gas constant R on the absolute temperature T (about 0.6 kcal/mol or 2.5 kJ/mol at room temperature), since under those circumstances the molecule traverses the barrier in a single molecular vibration, but this limit is clearly somewhat arbitrary. In any case, it is clear that the differentiation between two distinct isomeric molecules and two energy states of a single molecule is not a sharp one in those instances where the energy barrier between isomers is very low.

Bibliography. E. L. Eliel, On the concept of isomerism, *Israel J. Chem*, 15(1-2):7–11, 1977; J. G. Nourse, Applications of artificial intelligence for chemical inference, *J. Amer. Chem. Soc.*, 101(5): 1210–1216, 1979; J. F. Stoddart, in D. Barton and W. D. Ollis (eds.), *Comprehensive Organic Chemistry*, vol. 1, pp. 13–15, 24–26, 1979; P. Uzzell, *Aspects of Isomerism*, 1971.

MOLECULAR STRUCTURE AND SPECTRA
ROBERT S. MULLIKEN AND U. FANO

Until the advent of quantum theory, ideas about the structure of molecules evolved gradually from analysis and interpretation of the facts of chemistry. Chemists developed the concept of molecules as built from atoms in definite proportions, and identified and constructed (synthesized) a great variety of molecules. Later, when the structure of atoms as built from nuclei and electrons began to be understood with the help of quantum theory, a beginning was made in seeing why atoms can combine in definite ways to form molecules; also, infrared spectra began to be used to obtain information about the dimensions and the nuclear motions (vibrations) in molecules. However, a fundamental understanding of chemical binding and molecular structure became possible only by application of the present form of quantum theory, called quantum mechanics. This theory makes it possible to obtain from the spectra of molecules a great deal of information about the nature of molecules in their normal as well as excited states, and about dissociation energies and other characteristics of molecules. For an important aspect of molecular structure which is treated separately SEE CHEMICAL BONDING.

Molecular sizes. The size of a molecule varies approximately in proportion to the numbers and sizes of the atoms in the molecule. Simplest are diatomic molecules. These may be thought of as built of two spherical atoms of radii r and r', flattened where they are joined. The

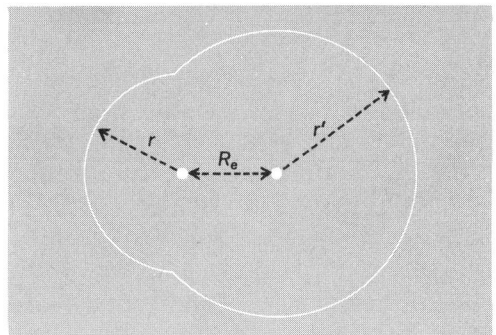

Fig. 1. Diatomic molecule with nuclei at distance R_e apart, built from atoms of radii r and r'.

equilibrium value R_e of the distance R between their nuclei is then smaller than the sum of the atomic radii (**Fig. 1**). However, the nuclei of atoms in two different molecules cannot normally approach more closely than a distance $r + r'$; r and r' are called the van der Waals radii of the atoms. The smallest molecule is hydrogen (H_2), with two electrons whose negative charges equal the positive charges of the two nuclei. Here r is about 0.12 nanometer giving $r + r' = 0.24$ nm, but R_e is only 0.074 nm. In HCl $r = 0.12$ nm for H and 0.18 nm for Cl, but R_e is only 0.127 nm.

To describe a polyatomic molecule, one must specify not merely its size but also its shape or configuration. For example, carbon dioxide (CO_2) is a linear symmetrical molecule, the O—C—O angle being 180°. The H—O—H angle in the nonlinear water (H_2O) molecule is 105°. Many molecules which are essential for life contain thousands or even millions of atoms. Proteins are often coiled or twisted and cross-linked in curious ways which are important for their biological functioning.

Dipole moments. Most molecules have an electric dipole moment. In atoms, the electron cloud surrounds the nucleus so symmetrically that its electrical center coincides with the nucleus, giving zero dipole moment; in a molecule, however, these coincidences are disturbed, and a dipole moment usually results.

Thus, when the atoms of HCl come together, there is some shifting of the H-atom electron toward the Cl. A complete shift would give H^+Cl^-, which would constitute an electric dipole of magnitude eR_e, where e is the electronic charge. But in fact the dipole moment is only 0.17 eR_e. This is because the actual electronic shift is only fractional. SEE ELECTRONEGATIVITY.

Although in molecules such as H_2, N_2, and CO_2 partial shifts of electronic charge from the original atoms do take place, these necessarily occur so symmetrically that no dipole moment results. Many larger molecules also have zero dipole moments by virtue of high symmetry. Examples are methane (CH_4), uranium hexafluoride (UF_6), and benzene (C_6H_6).

In general, the dipole moment of a neutral molecule is defined as the vector sum of quantities $+Q\mathbf{S}$ for the nuclei and $-e\mathbf{s}$ for the electrons. Here Q is the charge on any nucleus and $\mathbf{S}$ its vector distance from any fixed point in the molecule; $\mathbf{s}$ is the average vector distance of any electron from the same point. To calculate a dipole moment with these definitions, quantum mechanics must be used.

However, a study of what is known experimentally about molecular dipole moments has led to useful semiempirical generalizations. Bond moments and group moments have been obtained for various types of chemical bonds and of chemical groups or radicals. By adding these vectorially, the actual dipole moment of a large molecule can often be reproduced fairly accurately. In CH_4 or CO_2, one can assume a moment for each C—H or C═O bond, even though these cancel out vectorially to give a zero resultant. In the linear molecule OCS, the unequal moments of the C═O and C═S bonds give a nonzero resultant. Because of the zero moment of CH_4, the CH bond moment and the CH_3 group moment must be equal and opposite. In CH_3Cl, the total moment can be thought of as the vector sum of the H_3C group moment and the C—Cl bond moment.

When molecules vibrate, their dipole moments usually vary. **Figure 2** shows how the dipole moment μ in a diatomic molecule may vary with R; the quantity previously discussed is μ_e,

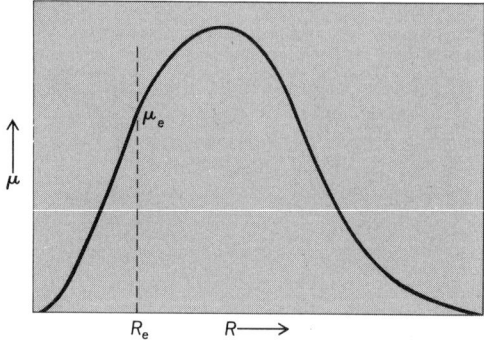

Fig. 2. Electric dipole moment μ of typical diatomic molecule as function of internuclear distance R; μ_e is the dipole moment at the equilibrium distance R_e.

the value of μ at R_e. When μ_e is zero because of symmetry, it remains zero for symmetrical vibrations but, in polyatomic molecules, varies during unsymmetrical vibrations.

Molecules may possess magnetic as well as electric dipole moments.

Molecular polarizability. In the preceding consideration of dipole moments, the discussion has been in terms of atoms and molecules free from external forces. An atom field pulls the electrons of an atom or molecule toward it and pushes the nuclei away, or vice versa. This action creates a small induced dipole moment, whose magnitude per unit strength of the field is called the polarizability.

Molecular polarizabilities can be expressed as sums of atomic polarizabilities, plus corrections depending on the types of bonds present. Polarizabilities increase rather rapidly in such series as F, Cl, Br, I, and also from HF to HI, or F_2 to I_2.

Molecular polarizabilities can also be expressed approximately as sums of bond polarizabilities. These polarizabilities are anisotropic, being greater along bonds than perpendicular to bonds.

Molecular energy levels. The stationary states of motion of nuclei and electrons in a molecule, or of electrons in an atom, are restricted by quantum mechanics to special forms with definite energies. (Nonstationary states, which vary in the course of time, are constructed by mixing stationary states of different energies.) The state of lowest energy is called the ground state; all others are excited states. In analogy to water levels, the energies of the stationary states are called energy levels. Excited states exist only momentarily, following an electrical or other stimulus. SEE QUANTUM CHEMISTRY.

Energy levels are either discrete or continuous. The levels of a self-contained atom or molecule are restricted to special, sharply defined values (discrete levels). When an atom or molecule is ionized, that is, when one of its electrons has enough energy to escape completely, the energy can take on any value exceeding the minimum escape energy. This range of energies is called a continuous level or an ionization continuum. Molecules also have dissociation continua, which are discussed below.

Excitation of an atom consists of a change in the state of motion of its electrons. Electronic excitation of molecules can also occur, but alternatively or additionally, molecules can be excited to discrete states of vibration and rotation.

In a diatomic vibration, R varies periodically above and below R_e. The possible vibration energies E_v are given by Eq. (1), where $c\omega_e$ is just the small-amplitude vibration frequency, and h

$$E_v = hc\omega_e[(v + 1/2) - x_e(v + 1/2)^2] + \ldots \quad (1)$$

is Planck's constant (6.62×10^{-34} joule·s); x_e is a small quantity which is nearly always positive. The vibrational quantum number v can take whole number values 0, 1, 2, etc. The $+ \ldots$ in Eq. (1) indicates small correction terms. The zero-point vibration energy $1/2 hc\omega_e(1 - 1/2 x_e)$ present in the ground vibration state ($v = 0$) is a characteristic manifestation of quantum phenomena.

The value of $c\omega_e$ depends on the masses m_1 and m_2 of the atoms and the force constant k, as shown in Eq. (2). The frequency $c\omega_e$ (c = speed of light) is written in this manner for reasons

$$c\omega_e = \sqrt{k[(1/m_1) + (1/m_2)]} \qquad (2)$$

of convenience in spectroscopic work, where the factor c is usually dropped.

The quantities R_e, k, and the dissociation energy D are the most important properties of a potential curve, which shows how the energy of attraction $U(R)$ of the atoms varies with R; k is d^2U/dR^2 taken at R_e. The $U(R)$ curve and vibrational levels for the ground electronic state of H_2 are shown in **Fig. 3**. Similar curves, but with other R_e, k, and D values, exist for other electronic

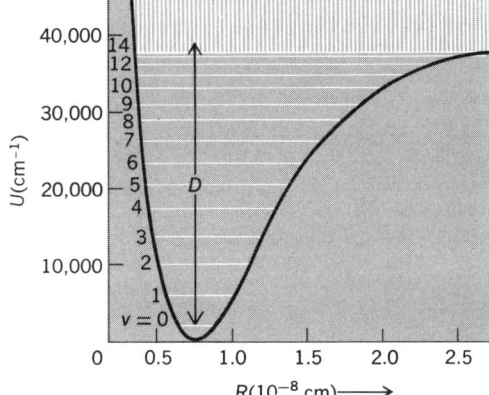

Fig. 3. $U(R)$ curve of ground electronic state of H_2 with vibrational levels and dissociation continuum. D indicates the dissociation energy. Maximum v here is 14. (*After G. Herzberg, Molecular Spectra and Molecular Structure*, vol. 1, 2d. ed., Van Nostrand, 1950)

states and other molecules. Molecules have also repulsive electronic states, whose $U(R)$ curves rise steadily with decreasing R. These are often important for spectroscopy and in atomic collisions. For stable (attractive) $U(R)$ curves, the vibrational levels decrease in spacing as v increases, until finally, as the spacing approaches zero, a maximum v is reached; in Fig. 3 this is 14. After a small gap, a dissociation continuum of energy levels then sets in. Here the atoms have enough mutual kinetic energy to fly apart. For repulsive states, there is only a dissociation continuum, with no vibrational levels. **Figure 4** illustrates how strongly vibration level spacings can vary:

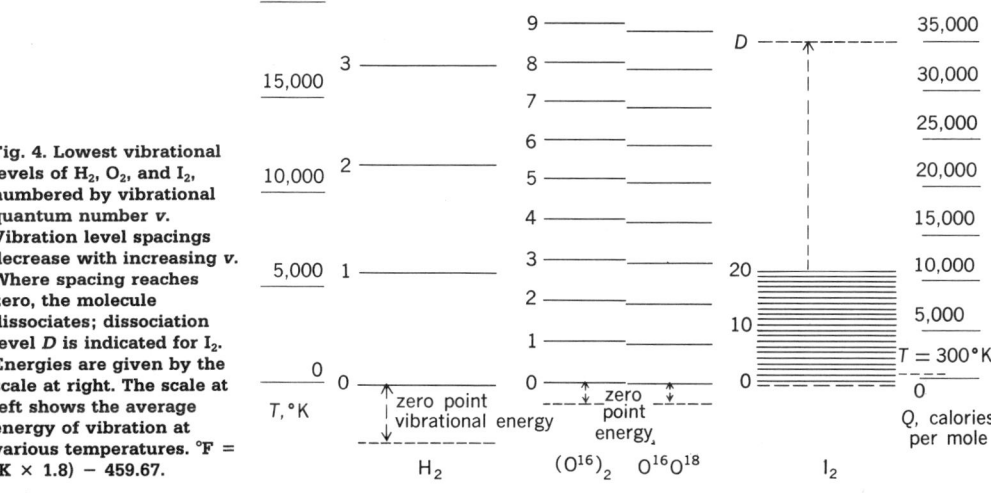

Fig. 4. Lowest vibrational levels of H_2, O_2, and I_2, numbered by vibrational quantum number v. Vibration level spacings decrease with increasing v. Where spacing reaches zero, the molecule dissociates; dissociation level D is indicated for I_2. Energies are given by the scale at right. The scale at left shows the average energy of vibration at various temperatures. °F = (K × 1.8) − 459.67.

both k and $1/m$, and therefore $c\omega_e$, decrease from H_2 to O_2 to I_2. Figure 4 likewise illustrates the effect of mass in isotopic molecules.

The total energy of any molecule can be written as Eq. (3). Both the electronic energy E_{el}

$$E = E_{el} + E_v + (E_r + E_{fs} + E_{hfs} + E_{ext}) \tag{3}$$

and vibration energy E_v can be discrete or continuous. The quantities E_r, E_{fs}, and E_{hfs} denote rotational, fine-structure, and hyperfine-structure energies. The last two appear as small or minute splittings of the rotation levels. The spacings ΔE of adjacent discrete levels of each type are usually in the order given in notation (4).

$$\Delta E_{el} \gg \Delta E_v \gg \Delta E_r \gg \Delta E_{fs} \gg \Delta E_{hfs} \tag{4}$$

The fine structures of rotational levels differ strongly for different types of electronic states. The simplest diatomic electronic states are called $^1\Sigma$ states, and include $^1\Sigma^+$ and $^1\Sigma^-$ types for heteropolar and $^1\Sigma_g^+$, $^1\Sigma_u^+$, $^1\Sigma_g^-$, and $^1\Sigma_u^-$ for homopolar molecules. Most even-electron diatomic and linear polyatomic molecule ground states are $^1\Sigma^+$ states ($^1\Sigma_g^+$ if homopolar). The rotational levels of $^1\Sigma$ states have no fine structure; hyperfine structure, because of interaction with nuclear spins, is usually on too small a scale to detect by optical spectroscopy, to which the present article is limited. The E_{ext} term in Eq. (3) refers to additional fine structure which appears on subjecting molecules to external magnetic fields (Zeeman effect) or electric fields (Stark effect).

The rotational levels of any $^1\Sigma$ state are given by Eq. (5). The quantity B_v is related to the

$$E_r = hcB_vJ(J + 1) + \ldots \tag{5}$$

moment of inertia I [$I = m_1m_2R^2/(m_1 + m_2)$], and to v, by Eq. (6). The rotational quantum number

$$B_v = (h/8\pi^2c)\overline{(1/I)}_v = B_e - \alpha_e(v + 1/2) + \ldots = B_O - \alpha_e v + \ldots \tag{6}$$

J can have any whole number value from 0 up, and corresponds to an angular momentum $(h/2\pi)\sqrt{J(J + 1)}$. The averaging of $1/I$ in Eq. (6) normally results in a slow decrease of B with increasing v (α_e is usually a small positive quantity). The quantity B_e refers to a hypothetical nonvibrating molecule ($R = R_e$).

Figure 5 illustrates how enormously rotational level spacings can vary because of differences in m and R_e (both are much greater for I_2 than H_2). The effect of mass for isotopic molecules is illustrated for O_2. Comparison with Fig. 4 illustrates the relation $\Delta E_v \gg \Delta E_r$ mentioned earlier.

Polyatomic molecules have much more complicated patterns of vibrational and (usually) of rotational energy levels than diatomic molecules. The number of normal modes (independent forms) of vibration for a molecule with n atoms is $3n - 6$ for nonlinear molecules, and $3n - 5$ for linear molecules. Each normal mode is a cooperative vibration of some or all the atoms moving

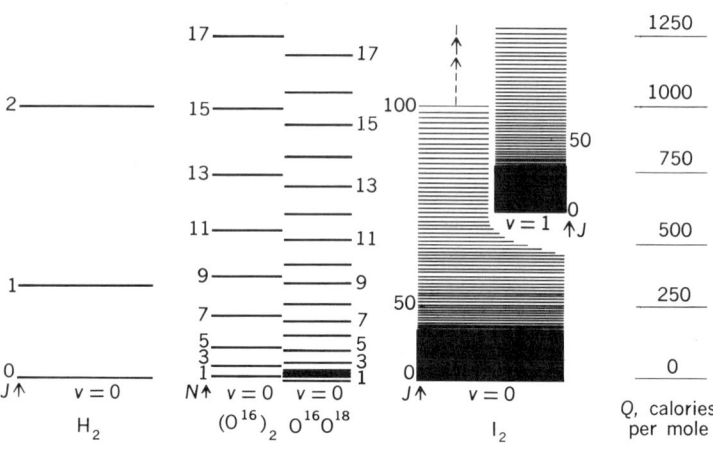

Fig. 5. Lowest rotational levels of H_2, O_2, and I_2. For H_2 and I_2, J is the rotational quantum number, according to Eq. (5) in the text O_2 is in a Hund's case b triplet state, and the rotational levels are designated by N, where the total angular momentum $J = N + 1$, N, and $N - 1$. This narrow spin tripling is indicated for the $N = 1$ level of $O^{16}O^{18}$ only. Energies are given by the scale shown at right.

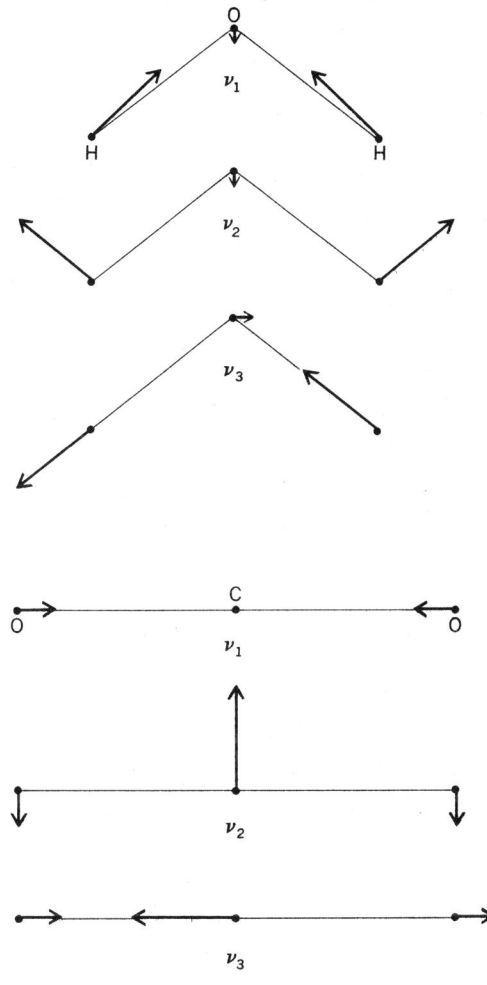

Fig. 6. Normal vibration modes of H_2O and CO_2. Synchronized displacements of atoms occur in proportion to lengths of the arrows. Diagram corresponds to snapshot taken at one phase of vibration.

with the same frequency, characteristic of the mode. Sometimes two or even three modes are so related in form that their frequencies are identical. These are called degenerate vibrations.

Figure 6 depicts the normal modes of H_2O and CO_2. They are labeled by symbols which also denote their frequencies. The arrows indicate the directions of motion of the atoms during one phase of vibration. The CO_2 frequency ν_2 is twofold degenerate: there are two independent modes corresponding to motion in either of two planes at right angles. The other two CO_2 modes, and all three H_2O modes, are nondegenerate.

Molecular spectra. The frequencies $c\nu$ of electromagnetic spectra obey the Einstein-Bohr equation, Eq. (7). The quantities ν, in waves per centimeter, or wave numbers (cm^{-1}), will

$$hc\nu = E' - E'' \qquad (7)$$

hereafter be called frequencies, as is usual in spectroscopy, although properly only the $c\nu$ are frequencies. Molecular emission spectra accompany jumps in energy from higher to lower levels; absorption spectra accompany transitions from lower to higher levels. Both E' and E'' can be either discrete or continuous levels. If both are discrete, they give a spectrum of discrete frequencies; otherwise, they give a continuous spectrum. Discrete spectra are the main type considered here. Discrete frequencies are usually called spectrum lines because of their appearance when recorded by an optical spectrograph. The wealth and precision of spectroscopic observations have been increased by orders of magnitude by the advent of laser sources and techniques.

Besides its frequency, the intensity and width of a spectrum line are important. Intensities vary over wide ranges. In the extreme case of nearly zero intensity for a spectroscopic transition, the transition is called forbidden. Only a small minority of all pairs of levels yield allowed transitions. These are governed by selection rules derivable from quantum theory.

Under disturbing influences, however, some lines are seen, weakly, which violate these rules. Further, the usual selection rules are electric dipole rules, and additional transitions become very weakly allowed if magnetic dipole, electric quadrupole, and other selection rules are also considered. The following discussion is confined to spectra which obey the electric dipole rules.

Molecular spectra can be classified as fine-structure or low-frequency spectra, rotation spectra, vibration-rotation spectra, and electronic spectra. Low-frequency spectra are discussed elsewhere.

Pure rotation spectra. Transitions between energy levels differing only in rotational state give rise to pure rotation spectra. For diatomic molecules in $^1\Sigma$ states, Eq. (5), the relation is given by Eq. (8). The transitions obey the selection rule $\Delta J = 1$ (ΔJ means $J' - J''$). Putting $J' = J'' + 1$,

$$hc\nu = E'_r - E''_r = hcB_v[J'(J' + 1) - J''(J'' + 1)] + \ldots \quad (8)$$

Eq. (9) is obtained. Equation (9) represents a sequence of lines spaced almost equidistantly ($2B_v$,

$$\nu = 2B_v(J'' + 1) + \ldots \quad (9)$$

$4B_v$, $6B_v$, ...), and lying in the far infrared or (for small B or low J'') the microwave region. Their intensities are proportional to μ_e^2, where μ_e is the electric moment at R_e (Fig. 2); hence homopolar molecules (H_2, N_2, and so on) show no pure rotation spectra. The intensities are proportional also to the lower-state (v'', J'') level population and to ν (for absorption) or ν^4 (for emission).

Pure rotation spectra of linear polyatomic molecules are like those of diatomic molecules. Polyatomic molecules having $\mu_e = 0$, whatever their shape (examples are CO_2, CH_4, C_6H_6), have no pure rotation spectra. In other cases, the spectra can be obtained using $hc\nu = E'_r - E''_r$ with appropriate E_r expressions and selection rules.

Vibration-rotation bands. Spectra involving only vibrational and rotational state changes lie mainly in the infrared. For a $^1\Sigma$ diatomic state, using Eqs. (1), (5), and (7), Eq. (10) is obtained, with ν_0 defined in Eq. (11). In Eq. (10) B' and B'' mean B_v for v' and v'', respectively. Each band

$$\nu = \nu_0(v',v'') + [B'J'(J' + 1) - B''J''(J'' + 1)] + \ldots \quad (10)$$

$$\nu_0 = \omega_e(1 - x_e)(v' - v'') - x_e\omega_e(v'^2 - v''^2) \quad (11)$$

consists of two sets of rotational lines, one on each side of its ν_0. Each line corresponds to a particular rotational transition conforming to $\Delta J = \pm 1$. The two series (branches) have frequencies defined in Eq. (12) for R or positive branch ($J' = J'' + 1$), and in Eq. (13), for P or negative

$$\nu = \nu_0 + 2B''(J'' + 1) + (B' - B'')(J'' + 1)(J'' + 2) \quad (12)$$

$$\nu = \nu_0 - 2B''J + (B' - B'')J''(J'' - 1) \quad (13)$$

branch ($J' = J'' - 1$). Both can be represented by a single equation, Eq. (14), by letting $M =$

$$\nu = \nu_0 + (B' + B'')M + (B' - B'')M^2 + \ldots \quad (14)$$

$J'' + 1$ for the R and $M = -J''$ for the P branch. Neglecting the term in M^2, Eq. (14) represents a series of equidistant lines with one missing ($M = 0$) at ν_0. **Figure 7** shows how the line positions are related to the upper (v') and lower (v'') sets of rotational levels.

Since $B' - B''$ is a small negative quantity [see Eq. (6), noting that $v' > v''$], the M^2 term makes the P line spacing increase and the R line spacing decrease slowly as M increases. This is shown, exaggerated, in Fig. 7. At some large M value, the R branch turns back on itself, but usually the lines have become weak before this value is reached.

The relative intensities of band lines depend primarily on the initial rotational distribution of molecules. More precisely, Eq. (15) holds. Here B_{in}, J_{in}, and n in ν_n are B', J', and 4, respectively,

$$\text{Intensity} = C(v',v'')\nu^n(J' + J'' + 1)e^{-B_{in}J_{in}(J_{in} + 1)hc/kT} \quad (15)$$

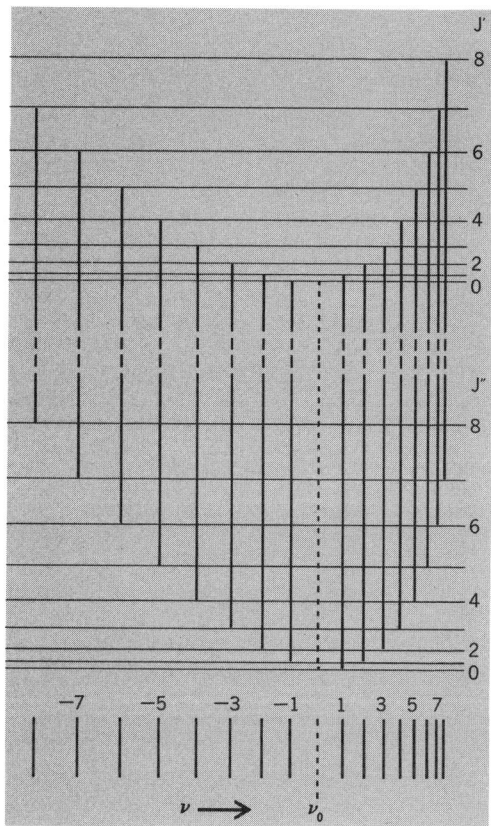

Fig. 7. Relation of band lines (lower part) [see Eqs. (8) and (9)] to rotational levels [see Eq. (6)] for a vibration-rotation band or an electronic band. In the former case, the upper and lower sets of rotational levels belong to two vibrational levels of a $^1\Sigma$ electronic state. In the latter case, they belong to two different $^1\Sigma$ states. Positive M values, R branch; negative M values, P branch.

for an emission band, and B'', J'', and 1, respectively, for an absorption band. **Figure 8** shows diagrammatically how the values of B and T affect the appearance of a typical absorption band ($B' = B''$ has been assumed for simplicity in Fig. 8). **Figure 9** shows the appearance of an actual HCl band. The weaker HCl^{37} lines are at slightly lower frequencies than the HCl^{35} lines, mainly because ω_e is smaller [see Eqs. (2) and (11)].

The factor $C(v',v'')$ is largest by far for fundamental bands ($\Delta v = 1$), and falls rapidly with increasing Δv in the overtone bands or harmonics ($\Delta v = v' - v''$). For fundamental bands, C depends on the slope of the $\mu(R)$ curve (Fig. 2), being approximately proportional to $(d\mu/dR)^2$ taken at R_e. For overtone bands, C depends on the detailed shapes of both the $\mu(R)$ and $U(R)$ curves. Fundamental or overtone bands arising from $v'' > 0$ are called hot bands.

Vibration-rotation absorption bands of liquids and solutions are widely used in chemical analysis. Here the rotational structure is blurred out, and only an "envelope" is seen. For many purposes, it is sufficient to know empirically the spectrum of each molecule which may be present. Also, groups of atoms which recur in many molecules often have nearly constant frequencies, of use for identification and in determining molecular structure.

Electronic band spectra. These are the most general type of molecular spectra. The characteristic feature is a change of electronic state. From Eqs. (3) and (7), neglecting fine structure, Eqs. (16) and (17) are obtained. Diatomic electronic spectra are often observed in emission,

$$\nu = \frac{(E'_{el} - E''_{el}) + (E'_v - E''_v) + (E'_r - E''_r)}{hc} \quad (16) \qquad \nu = \nu_{el} + \nu_v + \nu_r = \nu_0 + \nu_r \quad (17)$$

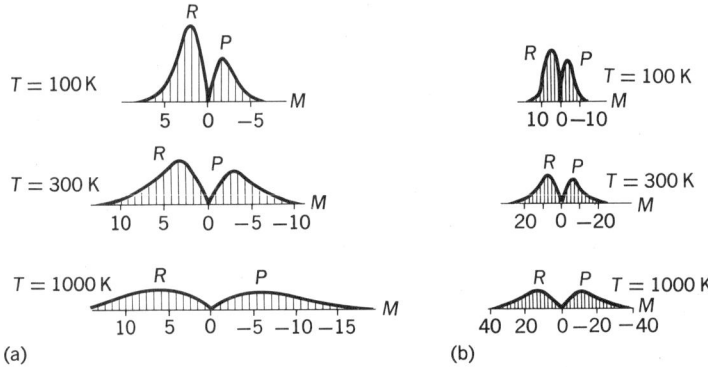

Fig. 8. Intensity distribution at several temperatures for a diatomic absorption band. Line positions are based on Eq. (9) assuming $B' = B''$ for simplicity; frequency increases toward the left (opposite to Fig. 7). (a) and (b) correspond respectively to B values of HCl ($B = 10.44$ cm^{-1}) and of 2 cm^{-1} (approximately the value for CO, for which $B = 1.93$ cm^{-1}). 100 K = $-280°$F; 300 K = 80°F; 1000 K = 1340°F. (*After G. Herzberg, Molecular Spectra and Molecular Structure, vol. 1, 2d ed., Van Nostrand, 1950*)

while the electronic spectra of polyatomic molecules are usually absorption spectra. Depending mainly on the magnitude of ν_{el}, electronic spectra occur in the infrared, visible, ultraviolet, or vacuum ultraviolet.

For any one electronic transition, the spectrum consists typically of many bands. These lie in general at frequencies both above and below ν_{el}, since ν_v can be positive or negative. They constitute a band system. Each band consists of numerous rotational lines arranged in two or more branches and lying on both sides of a central position ν_0.

For diatomic molecules, ν_0 depends on a single v' and v'' and, using Eq. (1) for each electronic state, is given by Eq. (18). Since ω_e and $x_e\omega_e$ are now different (often strongly) in the upper and lower states, $\nu_0(v',v'')$ cannot be reduced to as simple an expression as the corresponding Eq. (11) for vibration-rotation bands. Eq. (18) is more convenient when rewritten as Eq. (19), where

$$\nu_0(v',v'') = \nu_{el} + [\omega'_e(v' + \tfrac{1}{2}) - x'_e\omega'_e(v' + \tfrac{1}{2})^2 + \ldots]$$
$$- [\omega''_e(v'' + \tfrac{1}{2}) - x''_e\omega''_e(v'' + \tfrac{1}{2})^2 + \ldots] \quad (18)$$

$$\nu_0(v',v'') = \nu_{00} + (\omega'_0 v' - a' v'^2) - (\omega''_0 v'' - a'' v''^2) + \ldots \quad (19)$$

Eqs. (20) apply. The relative intensities of the bands depend on (1) the initial distribution of mol-

$$\nu_{00} = \nu_{el} + \tfrac{1}{2}(\omega'_e - \omega''_e) - \tfrac{1}{4}(x'_e\omega'_e - x''_e\omega''_e) \qquad \omega'_0 = \omega'_e(1 - x'_e) \quad a' = x'_e\omega'_e, \text{ etc.} \quad (20)$$

ecules among vibrational levels, and (2) the relative transition probabilities from any initial to various final vibrational levels.

The simplest example is the absorption spectrum of a cool gas of low molecular weight, for which all molecules initially have $v'' = 0$. The spectrum then consists of one "v' progression," a single series of bands with various values of v'; the frequencies are given by $\nu = \nu_{00} + \omega'_0 v' - a'v'^2$. For a hot or a heavy gas, additional weaker v' progressions with $v'' > 0$ also appear.

In emission spectra, the initial population usually ranges over a number of v' values, from each of which transitions occur to a number of v'' values, so that the system contains many bands on both sides of ν_{00}. In the special case of fluorescence spectra, the molecule is excited to various v' values by absorbing light; it then emits light belonging to the same (or sometimes another) electronic transition. From each v', it can descend not only to the original v'' but also to various other, mainly larger, values. Hence fluorescence bands lie mainly at lower frequencies than the absorption bands used to excite them. *See* FLUORESCENCE.

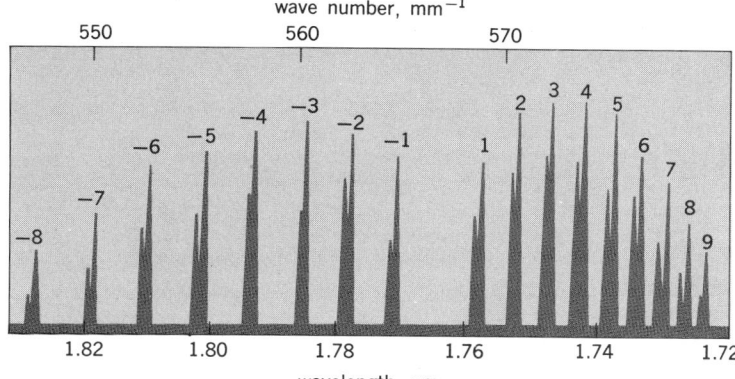

Fig. 9. First harmonic (2,0) vibration-rotation band of HCl in absorption. *R* branch to right, *P* branch to left, showing intensity distribution. The stronger lines are ^{35}HCl; the weaker companions, at lower frequencies, are ^{37}HCl. (*After C. F. Meyer and A. A. Levin, Phys. Rev., 34:44, 1929*)

Relative transition probabilities are governed by the Franck-Condon principle. This takes note of the very great rapidity of electronic motions as compared with those of the far more massive nuclei, and concludes that during the extremely brief time for an electronic transition, the nuclei tend to remain unchanged in their positions and momenta. It is applicable to both polyatomic and diatomic spectra. Consider first a diatomic molecule starting from the $v'' = 0$ level of a ground state $U(R)$ curve like the lower curves in **Fig. 10**. A vertical line drawn from the bottom point A ($v'' = 0$ if zero-point vibration is neglected) to point B on any one of the upper curves corresponds to an electronic absorption transition in which the nuclei have not moved.

In the case of Fig. 10a, point B corresponds to $v' = 0$, and the conclusion is that this is the most probable v' for $v'' = 0$. In the case of Fig. 10b, point B corresponds to an excited molecule at the inner turning point of a vibration with a v' of possibly 3 or 4, in a typical case. One then concludes (with J. Franck) that the strongest absorption bands for $v'' = 0$ have $v' = $ 3.0 and 4.0. To obtain more exact information, a quantum-mechanical calculation (first carried out by E. U. Condon) is necessary.

In the case of Fig. 10c, point B corresponds to an energy level in the dissociation continuum above the asymptote of the upper $U(R)$ curve. According to the Franck-Condon principle, the absorption spectrum will have maximum intensity in a continuous range of frequencies, with $hc\nu$ about equal to the energy difference AB. The quantum-mechanical calculation shows that the actual spectrum will extend with appreciable intensity over a range of both higher and lower frequencies than this, including, on the lower-frequency side, a number of high-v' bands. Actual examples of such spectra (a long v' progression followed by a strong continuum) are the far-ultraviolet Schumann-Runge bands of oxygen and the visible bands of iodine. By measuring the frequency at which the continuum begins, one obtains an exact value of the dissociation energy

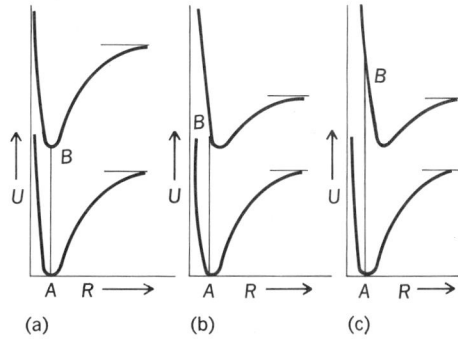

Fig. 10. Diatomic $U(R)$ curves for three cases to explain the vibrational intensity distribution according to the Franck-Condon principle. The asymptote of each curve for large R corresponds to dissociation into atoms, with one or both atoms excited in the case of the upper curves. Starting in each case from the bottom of the lower curve (essentially $v'' = 0$), the most probable transition in absorption is (a) to $v' = 0$, (b) to $v' = 3$ or 4, and (c) to the dissociation continuum, as shown by vertical lines. (*After G. Herzberg, Molecular Spectra and Molecular Structure, vol. 1, 2d ed., Van Nostrand, 1950*)

of these molecules. In so doing, any excitation energy of the atomic dissociation products to which the upper $U(R)$ curve leads is subtracted.

The Franck-Condon method is useful in understanding intensity distributions and structure in emission as well as absorption band systems. For diatomic spectra, various patterns of intensity as functions of v' and v'' occur, depending largely on the R_e values of the two $U(R)$ curves and, of course, also on the initial distribution among v' levels. Sometimes the upper-state $U(R)$ curve is stable (has a minimum) but the lower state is repulsive. Continuous emission spectra then occur, with the atoms flying apart on reaching the lower state. The H_2 molecule shows such a spectrum, as do rare gas molecules such as He_2 and Kr_2, which are stable only in excited or ionized states.

Molecular electronic states. Before discussing the structures of electronic bands, one must consider the nature of molecular electronic states. Each electronic state has orbital and spin characteristics. The spin quantum number S has a whole-number value if the number of electrons is even, a half-integral value if it is odd. Electronic states with $S = 0$ are called singlet states, all others multiplet states. The orbital characteristics differ sharply for linear (including diatomic) and nonlinear molecules.

For linear molecules only, there is a quantum number Λ such that $\pm \Lambda h/2\pi$ is the component of angular momentum around the line of nuclear centers. Linear-molecule electronic states can be discussed under three headings: (1) singlet states; (2) multiplet states with strong spin coupling (Hund's case a); and (3) multiplet states with weak spin coupling (Hund's case b). Strictly speaking, actual multiplet states are intermediate between cases a and b, or between these and certain other cases called c and d. The discussion to follow is largely restricted to singlet electronic states.

Singlet states with $\Lambda = 0$ include $^1\Sigma^+$ and $^1\Sigma^-$ states: states with $\Lambda = 1, 2, \ldots$ are called $^1\Pi$, $^1\Delta$, and so on. In linear molecules with a center of symmetry (H_2, CO_2 and so on), one must further distinguish even and odd (g and u) states: $^1\Sigma^+_g$, $^1\Sigma^+_u$, $^1\Sigma^-_g$, $^1\Sigma^-_u$, $^1\Pi_u$, $^1\Pi_g$, $^1\Delta_g$, $^1\Delta_u$, etc. The rotational levels of singlet states obey the symmetric rotor formula, Eq. (21). Here J is

$$E_r = hc[BJ(J+1) - \Lambda^2] + \ldots \tag{21}$$

restricted to integral values equal to or greater than Λ.

For $\Lambda > 0$, each rotational level is a narrow doublet (Λ-doubling). Corresponding fine structure [see E_{fs} in Eq. (3)] can usually be detected in electronic bands, but (for ground states only) it can be much more accurately studied in low-frequency spectra. Hyperfine structure [see E_{hfs} in Eq. (3)] is usually on too small a scale to be detected in electronic band lines, but has been found in a few cases. Hyperfine structure is best studied in low-frequency spectra.

Electronic band structures. The simplest electronic bands occur for transitions between singlet electronic states. The possible types of electronic transitions are limited by the selection rule $\Delta \Lambda = 0, \pm 1$. The structures of $^1\Sigma - ^1\Sigma$ bands are essentially the same as for the $^1\Sigma$-state vibration-rotation bands described earlier. Equations (12) to (15) and Fig. 7, also Fig. 8, for the intensities in absorption are still applicable if Eq. (18) instead of Eq. (11) is used for v_0, and it is recognized that B' and B'' now belong to two different electronic states.

The quantity $B' - B''$ in Eq. (14), instead of always being a small negative quantity, may now be either positive or negative, and $(B' - B'')/(B' + B'')$ is often fairly large (although it can also be nearly zero). As a result, it is usual in electronic bands to find heads. A head is a position of maximum or minimum frequency; by using Eq. (14) to obtain $dv/dM = 0$, one finds $M_{head} = (B' + B'')/2(B' - B'')$. Then, on inserting M_{head} into Eq. (14), one obtains $v_{head} = v_0 - (B' + B'')^2/4(B' - B'')$. [Since $(B' + B'')/2(B' - B'')$ is not usually a whole number, the actual M_{head} is the nearest whole-number M to that calculated.] According to whether $B' - B''$ is negative or positive, the positive (R) or the negative (P) branch forms the head. Figure 7, if continued to somewhat larger M values, illustrates the formation of an R-branch head at a calculated M of 10.5; the actual head is formed by the two coincident lines $M = 10$ and 11.

Although homopolar molecules (H_2, N_2, and so on) have no pure rotation or vibration-rotation spectra, they do have electronic spectra. For homonuclear homopolar molecules, the band lines show alternating intensities. The lines in each branch are alternately stronger and weaker as M increases, this effect being superposed on the otherwise smoothly varying intensity distribution. The alternation ratio depends on the nuclear spin I and has been, in several cases, the means of determining I. When $I = 0$, alternate lines are completely missing. Heteronuclear molecules, even if homopolar (for example, HD or $^{16}O^{18}O$) do not show alternating intensities.

Polyatomic electronic spectra. These differ from diatomic electronic spectra because several initial and final vibration quantum numbers are involved, and because the rotational structure (except for linear molecules) is usually much more complicated. However, the detailed structures of the electronic spectra of a number of simple molecules and radicals in the vapor state in emission and in absorption have been studied. Nevertheless, for the most part, the spectra of polyatomic molecules are examined as absorption spectra in solution. The rotational structure is then completely blurred out, but the vibrational structure can be seen.

The Franck-Condon principle is here a useful guide. One of its corollaries, which amounts almost to a selection rule, is that only totally symmetric vibrations (vibrations during which the equilibrium symmetry of the molecule is preserved) undergo quantum number changes. This greatly simplifies the vibrational structure, especially of absorption spectra where most molecules are initially mainly in the $v''=0$ state of all vibrations. One finds then mostly v' progressions of one or a very few totally symmetric vibrations, and combinations of these.

Rather often, polyatomic band systems do not even show obvious vibrational structure. This can happen for any of several reasons: The upper state may involve dissociation; in CH_3I, for example, the first ultraviolet absorption region yields $CH_3 + I$; there may be so many low-frequency, upper-state vibrations that the spectrum looks continuous; or there may be a combination of these and other reasons. Such continuous or pseudocontinuous band systems are often loosely referred to as bands. For complicated molecules, the spectra of several different electronic transitions often overlap strongly so that it is difficult even to separate one system from another. SEE INTERMOLECULAR FORCES; MOLECULAR WEIGHT; VALENCE.

Bibliography. C. N. Banwell, *Fundamentals of Molecular Spectroscopy*, 3d ed., 1983; E. F. Brittain et al., *Introduction to Molecular Spectroscopy: Theory and Experiment*, 1978; P. R. Bunker, *Molecular Symmetry and Spectroscopy*, 1979; W. H. Flygare, *Molecular Structure and Dynamics*, 1978; I. N. Levine, *Molecular Spectroscopy*, 1975; W. G. Richards and P. R. Scott, *Structure and Spectra of Molecules*, 1985; J. Steinfeld, *Molecules and Radiation: An Introduction to Modern Molecular Spectroscopy*, 1978; reprint 1985.

STEREOCHEMISTRY
SAMUEL H. WILEN

The study of the three-dimensional arrangement of atoms or groups within molecules and the properties which follow from such arrangement. Molecules that have identical molecular structures (that is, the same kind, number, and sequential arrangement of atoms) but differ in the relative spatial arrangement of component parts are stereoisomers. Inorganic and organic compounds exhibit stereoisomerism. Examples are structures (I) to (VIII). In the example, pairs of stereoisomers are related by permutation (transposition) of bonded atoms or groups. In structures (III) and (IV), a single permutation of Cl and H about atom C_2 produces two stereoisomers. Atom C_2 may be called a stereogenic atom (or a stereocenter).

Significance. Stereochemistry has played a significant role in the historical development of theories of molecular structure. The optical activity of substances such as sugar and turpentine was discovered by J. B. Biot in 1813, and L. Pasteur was the first to separate or resolve enantiomers from one another (1848). While Pasteur recognized that enantiomers must differ in symmetry, it was not until 1874 that J. H. van't Hoff and J. A. LeBel were able to relate symmetry, structure, and properties into the hypothesis of the tetrahedral carbon which formed the cornerstone of modern stereochemistry. The foundation of inorganic stereochemistry rests on the coordination theory of A. Werner (1893).

Since the latter part of the nineteenth century, stereochemical studies have been prominent in the evolution of both inorganic and organic chemistry. These studies have concerned themselves with the determination of configuration, the interconversion of diastereomers and the assessment of their energy differences, and conformational analysis.

Methods for the analysis and separation of stereoisomers have been devised, and stereochemical principles have been applied to the elucidation of reaction mechanisms, to the development of asymmetric syntheses, and to attempts to understand biological processes at the molecular level.

Symmetry. The nature and the number of stereoisomers of a molecule are determined by the permutation of atoms or groups (called ligands) at stereocenters as well as by the symmetry of the molecule. The symmetry elements to be considered are: planes of symmetry, axes of symmetry, centers of symmetry, and rotation-reflection (or alternating) axes of symmetry. Two types of stereoisomers are known. Those such as (VII) and (VIII), which are devoid of reflection symmetry—which cannot be superposed on their image in a mirror—are called enantiomers. All other stereoisomers, such as the pairs (I)–(II), (III)–(IV), and (V)–(VI), are called diastereomers. The configuration of a stereoisomer designates the relative position of the atoms associated with a specific structure. The structures of stereoisomers (I) and (II) differ only in configuration. The same is true for (III) and (IV), (V) and (VII), and (VII) and (VIII).

Enantiomers are related to one another as a left hand is to a right hand. Such structures are said to be chiral (Greek *cheir* = hand) or dissymmetric. Unlike diastereomers, which differ in most physical and chemical properties, enantiomers have identical properties other than the sign (+ or −) of their optical activity. This identical behavior exists toward all agents and processes which are themselves achiral. Chiral agents act differently toward enantiomers, however. Biological specificity toward enantiomers is dramatic; for example, the enantiomers of the amino acid leucine have different tastes: one is bitter, while the other is sweet.

Configuration. In order to understand chemical processes involving stereoisomers, it is necessary to know the configurations of reactants and products. The configuration of stereoisomers (I) and (II) are designated cis (chlorines on the same side) and trans (chlorines on opposite sides), respectively; (III) and (V) are trans; (IV) and (VI) are cis. Configurations of diastereomers can often be determined from their physical properties. For example, the isomer of $C_2H_2Cl_2$ with zero dipole moment is (III), while (IV) has a finite dipole moment.

The absolute configuration of chiral molecules such as (VII) and (VIII) is the actual order of atoms about the stereogenic atom (tetrahedral carbon atom) which defines the specific enantiomer. While the three-dimensional picture of a molecule defines its stereochemistry just as the picture of a hand defines its handedness (right or left), it is not convenient or useful to rely on full three-dimensional representations in most cases, and a shorthand notation to designate configuration is widely accepted. Groups and atoms which surround the stereocenter are assigned priorities related by unambiguous rules to atomic number, for example, Br > Cl > F > H, and to substitution pattern. The molecule is viewed from the tetrahedral face opposite the carbon substituent with the lowest priority, that is, from the face opposite hydrogen in the case of (VII). The

(VII)   (VIIa)   (VIIIa)

counterclockwise descent (VIIa) from high- to low-priority substituent (Br followed by Cl followed by F) is designated S (Latin *sinister* = left). Enantiomer (VIII) exhibits a similar but clockwise order (VIIIa) designated R (Latin *rectus* = right). When the order of groups immediately bound to the stereocenter is equivocal, the priority is determined by atoms or groups attached to the atoms whose priorities are equivocal.

The same priorities may be used in the designation of the configurations of diastereomers such as (III) and (IV). When the atoms or groups of highest priority lie on opposite sides of the double bond, as in (III), the stereoisomer is designated E (German *entgegen* = opposite). Isomer (IV) in which the reference atoms are on the same side of the double bond are termed Z (German *zusammen* = together). This system is less ambiguous than the cis-trans system which is still applicable in simple cases.

Enantiomers are characterized by the sign of their optical activity at a given wavelength, temperature, and specified solvent. The assignment of configuration to specific enantiomers is carried out by chemical transformations of known stereochemistry, by spectroscopic means such as circular dichroism and optical rotatory dispersion, and by diffraction of single crystals employing anomalous x-ray scattering. From such studies it is known that the lactic acid isomer that is experimentally found to be levorotatory, that is, whose optical activity is (−)- or counterclockwise-rotating, has structure (IX) corresponding to the R configuration (IXa). Configuration has no simple connection to the sign of the optical rotation.

Two-dimensional (planar) projections of three-dimensional structures are very commonly used to represent configurations. Projection formulas compress the tetrahedral geometry into the plane of the paper. In (IX) the horizontal atoms or groups are forward of the plane (wedges) of the stereogenic atom, and the vertical groups are behind (solid or broken lines). Tipping formula (IX) forward places the COOH group in the plane of the paper and the H atom behind, corresponding to structure (IXa). The planar projection of (IX) may be written as (IXb). To avoid confusion in their use, the following conventions apply to planar projections: they must not be rotated in this plane through any but integral multiples of 180°; an odd number of exchanges of any pair of substituents is equivalent to transformation into its mirror image, and an even number of exchanges leaves the configuration unchanged.

The relative configurations of stereoisomers applicable to diastereomers are illustrated by the tartaric acids (X) and (XI). Structure (X) represents (+)-tartaric acid (both stereocenters * have R configurations), and (XI) represents *meso*-tartaric acid, the diastereomer which is optically inactive (the two stereocenters, one R and one S, are mirror images of one another. Older configurational symbolism (D and L) involving intermolecular comparisons to reference substances such as glyceraldehyde and serine persists, but it is giving way to the R-S designations.

Stereoregular polymers with defined configurations at stereogenic atoms or at carbon-carbon double bond stereocenters differ substantially in properties from their diastereomers. Polymeric carbohydrates such as starch and cellulose and hydrocarbons such as rubber and gutta-percha exemplify natural diastereomer pairs. Synthetic polystyrene diastereomers (XII) [isotactic] and (XIII) [syndiotactic] differ only in the configuration at stereogenic atoms. The high-melting,

(XII)

(XIII)

fiber-forming isomer is isotactic polystyrene (XII).

Configurational studies are powerful probes in the elucidation of reaction mechanisms. Replacement of substituent atoms or groups in a reaction is often attended by a change in configuration at the stereocenter (Walden inversion), or it may result in the loss of optical activity (racemization). In some cases retention of configuration prevails. Such results provide evidence for the geometry of transition states or for the intervention of intermediates (ions or radicals).

Resolution. Most chiral substances occur in nature as only one enantiomer. Few natural products are represented in nature by both enantiomers, and few also exist as mixtures with equal proportions of enantiomers called racemates.

On the other hand, synthesis of the chiral substances in the laboratory or in industry in the absence of chiral agents or conditions results in racemates which exhibit no measurable optical activity, since the activity of one enantiomer cancels that of the other.

The separation of one or both enantiomers from a racemate, called resolution, requires the intervention of an optically active reagent, catalyst, or other chiral influence. The most common resolution procedure involves the reversible conversion of an enantiomer mixture into a pair of diastereomers, with as widely different physical properties as practicable, by reaction with a single pure enantiomer. For example, a ($\pm$)-amine reacts with a (+)-acid to give two diastereomeric salts: (+)-RNH$_3^\oplus$(+)-R'COO$^\ominus$ and (−)-R-NH$_3^\oplus$(+)-R'COO$^\ominus$.

These diastereomeric salts may be separated by fractional crystallization and may then be converted to the individual enantiomeric amines by cleavage with a strong base. The (+)-acid resolving agent may be recovered and reused. Covalent diastereomers may also be separated by crystallization or increasingly by gas, liquid, or thin-layer chromatography.

Kinetic resolution takes advantage of differences in rates of reaction of enantiomers with chiral reagents. Enzymes are chiral catalysts which can preferentially catalyze reactions of one enantiomer. Enzymatic resolutions play a major role in the syntheses of substances of biochemical importance. Chromatographic resolution on chiral stationary phases, which bind or dissolve one enantiomer more strongly than the other, constitutes another useful type of resolution.

In resolution by preferential crystallization or entrainment, one enantiomer preferentially crystallizes upon seeding of supersaturated solutions with seed crystals of that enantiomer. Though relatively few racemates are amenable to this type of resolution, it is nevertheless of considerable importance on an industrial scale. SEE CRYSTALLIZATION.

Conformational analysis. Molecules are not rigid collections of atoms. Torsional stereoisomerism arises as a consequence of the rotation (torsion) of atoms about bonds within molecules. This gives rise to stereoisomers called conformers which are interconvertible by rotation about carbon-to-carbon single bonds. The existence of two extreme forms of ethane (in terms of energy), (XIV) [staggered] and (XV) [eclipsed], was predicted by K. S. Pitzer in 1936 and

(XIV)   (XV)

later experimentally verified. An infinite number of forms (confirmations) intermediate between the equilibrium (low-energy) conformer (XIV) and the higher-energy conformation (XV) is possible. At room temperature most ethane molecules resemble (XIV). Though their presence is experimentally readily demonstrated, separation of conformers is normally not possible since the small energy difference between them [3 kcal/mole; 12.6 kilojoules/mole in the case of (XIV), and (XV) makes their interconversion facile]. The ease of interconversion explains why some enantiomeric structures, such as the *gauche* conformers (large groups, CH_3, 60° apart) of *n*-butane (XVI), are incapable of resolution.

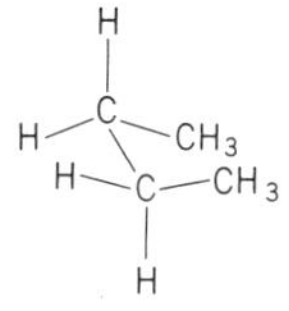

(XVI)

The relationship between the physical and chemical properties of substances and their preferred conformations (conformational analysis) has been the subject of many studies since 1950. Cyclohexane (XVII) exists overwhelmingly in the form of conformer (XVIII; the chair form), which

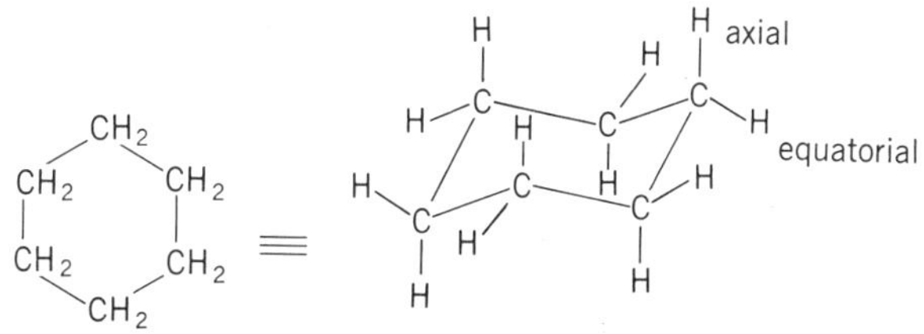

(XVII)   (XVIII)

is the structural subunit from which the carbon atom lattice of diamond is constructed. Substituents on six-membered rings are designated as either axial or equatorial, as in (XVIII). Conformational analysis of synthetic hydrocarbon polymers such as polypropylene and biopolymers such as DNA has led to an understanding of their properties. Much of the biological activity of enzymes, for example, is made possible by the specific conformations adopted by these macromolecules.

Stereocontrolled synthesis. An understanding of the stereochemical consequences of chemical reactions and the determination of configuration of complex natural products has permitted the total synthesis of many stereochemically complex substances since around 1940. For example, cholesterol and gibberellic acid each contain eight stereogenic carbon atoms, which makes possible, in principle, $2^8 = 256$ stereoisomers (128 racemates). Disregard for stereochemical consequences during synthesis would lead to extremely complex mixtures requiring repeated separation and to very low yields.

Living organisms unerringly synthesize (biosynthesis) just one of these isomers in fully stereoselective processes. Laboratory and commercial syntheses of substances which have desirable biological properties have increasingly been devised so as to mimic biosyntheses by taking advantage of stereocontrolled reactions. These are of two types: stereospecific reactions and stereoselective reactions. Stereospecific reactions are those in which different stereoisomers are transformed into stereochemically different products. For example, Z-2-butene (XIX) [cis] is epoxidized to the *cis*-epoxide (XX) only. The diastereomeric E-2-butene [trans] forms only the *trans*-epoxide.

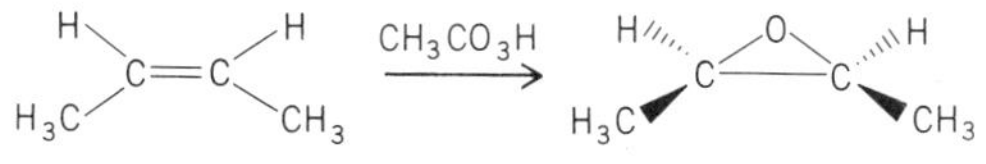

(XIX)   (XX)

Stereoselective reactions are those in which a single reactant gives rise to a mixture of diastereomers in which one predominates. Those stereoselective reactions in which new stereocenters are formed in unequal amounts are called asymmetric syntheses. Transformations of optically active compounds into more complex substances which incorporate additional stereocenters are also asymmetric syntheses.

Syntheses of natural products which originate in achiral starting materials require at least one resolution step if nonracemic products are desired. Alternatively, a total synthesis may well originate in a single enantiomer of a compound having but one or two stereogenic atoms. Additional stereocenters are introduced during the synthesis through stereospecific and especially through stereoselective steps. Separation of diastereomers at various stages may be required.

Bibliography. R. Bentley, *Molecular Asymmetry in Biology*, 1969–1970; E. L. Eliel, *Elements of Stereochemistry*, 1969; E. L. Eliel, *Stereochemistry of Carbon Compounds*, 1962; K. Mislow, *Introduction to Stereochemistry*, 1965; K. Mislow and J. Siegel, Stereomerism and local chirality, *J. Amer. Chem. Soc.*, 106:3319, 1984; C. Tamm, *Stereochemistry*, 1982; F. Vogtle and E. Weber (eds.), *Stereochemistry*, 1984.

ELECTRON AFFINITY
W. C. LINEBERGER

The amount of energy release when an electron at rest is captured by a species M, producing the negative ion M^-. The electron affinity of a species M can also be thought of as the ionization potential of the negative ion M^-. Stated in terms of a chemical equation, the electron affinity of a species M is equal to the exothermicity of the reaction $e + M \rightarrow M^-$, where the negative ion M^- is left in its lowest electronic, vibrational, and rotational state.

If the electron affinity of M is negative, the M^- ion is unstable with respect to decomposition into $M + e$. Most atoms have positive electron affinities, even though there is no net Coulomb attraction between the electron and the atom until the electron is close enough to be "a part of the atom." The simple rules of chemical valency provide a qualitative guide to the magnitude of electron affinities. Thus the noble gases, which have a filled outer electronic shell and are chemically inert, are not capable of binding an additional electron to form a negative ion. The largest electron affinities are possessed by the halogens, atoms which require only one additional electron to fill the valence shell.

The major exception to this concept is that multiply charged negative ions—for example, O^{2-}, one of many multiply charged negative ions which are stable in solution—are not stable in the gas phase. The ability to place more than one additional electron in the valence shell of a neutral atom or molecule appears to come from the medium; the solvent shell surrounding the ion in liquid solutions and the amorphous or crystalline region surrounding the ion in solids.

Experimental methods. Accurate ionization potentials of the elements were known for a number of years before comparable data for electron affinities of the elements became available. In order to determine the ionization potential of an element, one can simply make a vapor of the element, place it in an optical spectrometer, and look for the onset of absorption corresponding to the photoionization process $h\nu + A \rightarrow A^+ + e$. The energy of the photon corresponding to the threshold wavelength for this process is the ionization energy of the species A. The analogous method for determination of an electron affinity is through observation of the threshold of the very similar photodetachment reaction $h\nu + A^- \rightarrow A + e$. Again, the threshold wavelength for this process corresponds to the electron affinity of the species A. Unfortunately, it has not proved possible to produce sufficiently large densities of negative ions to be able to observe directly the threshold for the photodetachment process in a photoabsorption measurement; consequently, determination of accurate electron affinities has lagged far behind the determination of accurate ionization potentials.

The major experimental advances which enabled accurate electron affinity determinations have been the development of ion-beam techniques and the availability of intense light sources in the form of lasers. In experiments to determine electron affinities, negative ions are formed in an electrical discharge, extracted through an aperture into a high-vacuum region, formed into a negative ion beam, mass-analyzed, and intersected by an intense laser beam. The laser-beam–negative-ion-beam intersection takes place in a high-vacuum region where no collisions are likely. The occurrence of a photodetachment event is determined by detection of the photodetached electron.

Two experimental methods evolved which produce accurate electron affinities. In the first

Periodic table showing the best values for the electron affinities of the elements. All values are reported in electronvolts. The value <0 implies that the negative ion is unstable with respect to decomposition to an electron and a neutral atom. The solid bar below represents the relative uncertainty in the electron affinity. (*After H. Hotop and W. C. Lineberger, Binding energies in atomic negative ions, J. Phys. Chem. Ref. Data, 4:539–576, 1975*)

1 H 0.7542							2 He <0
3 Li 0.620	4 Be <0	5 B 0.282	6 C 1.268	7 N ≤0	8 O 1.462	9 F 3.399	10 Ne <0
11 Na 0.546	12 Mg <0	13 Al 0.442	14 Si 1.385	15 P 0.743	16 S 2.0772	17 Cl 3.615	18 Ar <0
19 K 0.5012	20 Ca <0	31 Ga 0.3	32 Ge 1.2	33 As 0.80	34 Se 2.0206	35 Br 3.364	36 Kr <0
37 Rb 0.4860	38 Sr <0	49 In 0.3	50 Sn 1.25	51 Sb 1.05	52 Te 1.9708	53 I 3.061	54 Xe <0
55 Cs 0.4715	56 Ba <0	81 Tl 0.3	82 Pb 0.349	83 Bi 0.947	84 Po 1.9	85 At 2.8	86 Rn <0

21 Sc <0	22 Ti 0.2	23 V 0.526	24 Cr 0.667	25 Mn <0	26 Fe 0.164	27 Co 0.667	28 Ni 1.157	29 Cu 1.226	30 Zn <0
39 Y ≈0	40 Zr 0.429	41 Nb 0.886	42 Mo 0.747	43 Tc 0.7	44 Ru 1.1	45 Rh 1.138	46 Pd 0.558	47 Ag 1.303	48 Cd <0
57 La 0.5	72 Hf <0	73 Ta 0.323	74 W 0.816	75 Re 0.15	76 Os 1.1	77 Ir 1.566	78 Pt 2.128	79 Au 2.3086	80 Hg <0

method the laser is a tunable laser, and one searches for the wavelength corresponding to the threshold for the photodetachment process. In this case the electron affinity is given directly by the threshold wavelength, and is in principle determinable to accuracies of 10^{-6} eV. In the second type of experiment, called photoelectron spectroscopy, a fixed-frequency laser (of known photon energy) is employed, and electrostatic fields are used to determine the kinetic energy of the ejected electron. From simple energy conservation arguments, the electron affinity is then given by the photon energy less the kinetic energy of the ejected electron. This latter technique is quite general, but is limited in accuracy by the resolution of the electron energy analyzer (typically 10^{-2} eV).

Periodic trends. These laser photodetachment studies dramatically improved knowledge of the electron affinities of the elements. The **illustration** is a periodic table showing values of the electron affinities of the elements. Most of the data shown here were obtained by using laser photodetachment methods. The periodic trends in electron affinities and the qualitative effects described earlier are immediately apparent. In addition, a number of more subtle trends are observable. For example, while one expects that filled-shell species such as the rare gases will not be capable of binding an additional electron, the illustration shows that half-filled shells (for example, N and P) also exhibit small or negative electron affinities. Again, this effect is the result of the fact that a half-filled valence shell is spherically symmetric and behaves somewhat as though it were a filled shell. A similar situation is also seen for half-filled d-shells, as in the transition metals.

These same techniques have provided a number of accurate electron affinity determinations for molecules and free radicals, and new insight into the structural and chemical properties of ions in the gas phase.

Bibliography. R. R. Corderman and W. C. Lineberger, Negative ion spectroscopy, *Annu. Rev. Phys. Chem.*, 30:347–376, 1979; H. Hotop and W. C. Lineberger, Binding energies in atomic negative ions, *J. Phys. Chem. Ref. Data*, 4:539–576, 1975; B. K. Janousek and J. I. Brauman, Electron affinities, in M. T. Bowers (ed.), *Gas Phase Ion Chemistry*, 1980; H. S. W. Massey, *Negative Ions*, 3d ed., 1976.

ELECTRONEGATIVITY

A. Louis Allred

Electronegativity, according to L. Pauling, is "the power of an atom in a molecule to attract electrons to itself." With the concept of electronegativity, a vast number of observations of chemical and physical properties have been either correlated or predicted. Quantitative definitions and scales of electronegativity have been based not on electron distribution itself but on properties which were assumed to reflect electronegativity.

The electronegativity of an element depends upon its valence state and thus is not an invariant atomic property. As an example, the electron-withdrawing ability of an sp^n hybrid orbital centered on carbon and directed toward hydrogen increases as the percentage of s character in the orbital increases in the series ethane < ethylene < acetylene. Thus, according to this concept of orbital electronegativity, each element exhibits a range of electronegativity values. In the following paragraphs, a few scales of electronegativity will be discussed.

The original scale, proposed by Pauling in 1932, is based upon the difference Δ between the energy of the A—B bond in the compound AB_n and the mean of the energies of the homopolar bonds A—A and B—B, as in Eq. (1). The A—B bond energy exceeds the arithmetic means of the

$$\Delta = E(A\text{—}B) - \frac{E(A\text{—}A) + E(B\text{—}B)}{2} \tag{1}$$

A—A and B—B bonds to an increasing extent as the elements A and B diverge in electron-attracting ability. The difference in electronegativities of bonded atoms is proportional to the square root of the energy difference Δ, as in Eq. (2). The proportionality factor 0.208 converts the

$$\chi_A - \chi_B = 0.208\sqrt{\Delta} \tag{2}$$

units of energy from kilocalories to electronvolts. After the electronegativity of one element is arbitrarily assigned, other values of electronegativity can be calculated from thermochemical data. Values for selected elements in common oxidation states are presented in the **table**. Electronegativity increases with increasing oxidation state. For example, $\chi_{Sn(II)} = 1.80$ and $\chi_{Sn(IV)} = 1.96$.

R. S. Mulliken proposed that the electronegativity of an element is given by the average of the valence-state ionization potential and electron affinity: $\chi_M = (IP_V + EA_V)/2$. The quantities IP_V and EA_V are not observable properties of the ground state of an atom, but are energies for a hypothetical state of the isolated atom having the same electronic configuration (hybridization, electron-electron interaction, and so forth) as the atom in the molecule. The Mulliken approach has a sound theoretical basis, is consistent with Pauling's original definition, and gives orbital electronegativities, not invariant atomic electronegativities. Valence-state ionization potentials

Average electronegativities from thermochemical data

Element	Value	Element	Value
H	2.20	Al	1.61
Li	0.98	Ga	1.81
Na	0.93	In	1.78
K	0.82	Tl	2.04
Rb	0.82	C	2.55
Cs	0.79	Si	1.90
Be	1.57	Ge	2.01
Mg	1.31	Sn	1.96
Ca	1.00	Pb	2.33
Sr	0.95	N	3.04
Ba	0.89	P	2.19
Sc	1.36	As	2.18
Ti	1.54	Sb	2.05
V	1.63	Bi	2.02
Cr	1.66	O	3.44
Mn	1.55	S	2.58
Fe	1.83	Se	2.55
Co	1.88	F	3.98
Ni	1.91	Cl	3.16
Cu	1.90	Br	2.96
Zn	1.65	I	2.66
B	2.04		

and electron affinities have been calculated from the equations $IP_V = IP_g + P^+ - P^0$ and $EA_V = EA_g + P^0 - P^-$, where IP_g and EA_g are ground-state potentials and affinities, respectively, and P^+, P^0, and P^- are promotion energies of the positive ion, atom, and negative ion, respectively. The calculation of d-orbital electronegativity by the Mulliken method has not been accomplished for nontransition elements due to the lack of spectroscopic data. Since electronegativity is a sensitive function of d-orbital hybridization and since the extent of d-orbital participation generally cannot be ascertained quantitatively, the calculations of electronegativities for the heavier elements are limited.

The energy of an ion relative to the neutral atom can be expressed as a power series $E = aq + bq^2 + cq^3 + dq^4$, where q is the formal oxidation state or ionic charge for a particular state of ionization. Z. R. P. Iczkowaski and J. L. Margrave defined the electronegativity of a neutral atom as the derivative, $\chi_{IM} = (dE/dq)_{q=0}$. As a fairly good approximation, the last two terms in the above power series can be dropped, giving Eq. (3). The units of χ_{IM} are energy/electron and

$$\chi_{IM} = \frac{dE}{dq} = \frac{d(aq + bq^2)}{dq} = a + 2bq \qquad (3)$$

the magnitudes are the same, in accordance with theory, as those of Mulliken if E is evaluated only from the electron affinity and the first ionization potential. The quantity a is the Mulliken electronegativity for a neutral atom, and electronegativity is shown by Eq. (3) to increase linearly with increasing positive charge. By using IP_V and EA_V values, H. H. Jaffé and coworkers calculated the electronegativities of certain vacant, singly occupied, and doubly occupied orbitals.

Electronegativity was defined by A. L. Allred and E. G. Rochow as the force of attraction between a nucleus and an electron from a bonded atom. The electrostatic force was calculated simply from the effective nuclear charge and the atomic radius, as in Eq. (4).

$$\chi = \frac{0.359 Z_{\text{eff}}}{r^2} + 0.744 \qquad (4)$$

A quantum-defect electronegativity scale has been developed from potentials based on atomic spectral data, and a nonempirical scale has been calculated by an ab initio method using floating gaussian orbitals.

Other methods for calculating electronegativities utilize such observables as bond-stretching force constants, electrostatic potentials, spectra, and covalent radii. The fact that the various scales of electronegativity have different dimensions (energy$^{1/2}$, energy/electron, force, potential, and so forth) or no dimension reflects the widespread results of differences in electron-attracting ability. The measurement of electronegativities involves observations of properties dependent upon electron distribution. Close agreement of electronegativity values obtained from measurements of several diverse properties lends confidence and utility to the concept.

VALENCE
Robert G. Parr

A term commonly used by chemists to characterize the combining power of an element for other elements, as measured by the number of bonds to other atoms which one atom of the given element forms upon chemical combination The term also has come to signify the theory of all the physical and chemical properties of molecules that specially depend on molecular electronic structure.

Thus, in water, H_2O (I), the valence of each hydrogen atom is 1; the valence of oxygen, 2. In methane, CH_4 (II), the valence of hydrogen again is 1; of carbon, 4. In NaCl and CCl_4, the valence

```
         H                    H
         |                    |
         O—H              H—C—H
        (I)                   |
                              H
                             (II)
```

of chlorine is 1, and in CH_2 the valence of carbon is 2.

Much more is known about a water molecule than that it contains two hydrogen atoms and one oxygen atom. Each OH distance is 9.57 nanometers and the HOH bond angle is 104°27'. The oxygen and hydrogen ends of the molecule are negatively and positively charged, giving it a dipole moment 1.84×10^{-18} electrostatic unit (esu). The molecule absorbs infrared light strongly but is transparent to visible light. Scientists are striving for an understanding of these properties and many more in terms of the fundamental theory of valence. SEE BOND ANGLE AND DISTANCE.

Here the term valency is used as a synonym for valence.

Combining power of an element. By the 1920s the most important facts about atoms had been established experimentally. A neutral atom of atomic number Z comprises a massive nucleus of charge $+Ze$, and Z very light electrons each of charge $-e$, where $e = 4.80 \times 10^{-10}$ esu; most of the space within the atom is empty. Atomic nuclei are immutable through ordinary chemical changes; when one molecule of H_2 combines with one molecule of Cl_2 to give two molecules of HCl, the four nuclei (two hydrogen nuclei, or protons, of charge $+1e$ and two chlorine nuclei of charge $+17e$) are unchanged. It is redistribution of electrons between atoms which constitutes chemical combination. This is what valences of atoms control, and this is what a theory of valence must explain.

Atomic structure. To understand molecule formation, then, one first must understand the electronic structure of atoms. According to Neils Bohr, electrons in an atom move in orbits much like the orbits of planets about a sun, held to the nucleus by electrical attractions for it, prevented from falling into it by centrifugal forces. A special quantum effect is operative at the atomic level, however, which possesses no analogy in the motions of planets; not all orbits are possible for an electron, but only those for which the angular momentum of the electron as it moves about the nucleus is an integral multiple of $h/2\pi$, where $h = 6.63 \times 10^{-34}$ J · s is Planck's constant, and for which the energy is similarly quantized. Furthermore, not more than two electrons can move in one orbit at once. SEE QUANTUM CHEMISTRY.

When the consequences of these ideas are worked out, there actually emerges the periodic classification of the elements. To cover just part of the periodic table, occupation of orbits by electrons in the lighter atoms are shown in the **table**, where the symbol $2p$ stands for three distinct orbits of the same energy and shape but differently oriented in space. The lowest energy orbit is $1s$, forming the K shell. Next in energy are $2s$ and $2p$, making up the L shell. The $3s$ state is still higher, in the M shell. The chemically inert gases helium, He, and neon, Ne, are characterized by closed shells of 2, and $2 + 8 = 10$ electrons, respectively. The next inert gas is argon, Ar, with a closed shell of $2 + 8 + 8 = 18$ electrons, followed by krypton, Kr, with $2 + 8 + 18 + 8 = 36$ electrons, and the others.

Electron configuration of some atoms

Atom	Z	K shell	L shell		M shell		
		1s	2s	2p	3s	3p	3d
H	1	1	0	0	0	0	0
He	2	2	0	0	0	0	0
Li	3	2	1	0	0	0	0
Be	4	2	2	0	0	0	0
B	5	2	2	1	0	0	0
C	6	2	2	2	0	0	0
N	7	2	2	3	0	0	0
O	8	2	2	4	0	0	0
F	9	2	2	5	0	0	0
Ne	10	2	2	6	0	0	0
Na	11	2	2	6	1	0	0
Mg	12	2	2	6	2	0	0
Al	13	2	2	6	2	1	0

Rule of eight. Most of the simple facts of valence (though certainly not all) follow from the postulate that atoms combine in such a way as to seek closed-shell or inert-gas structures (rule of eight) by the transfer of electrons between them or the sharing of a pair of electrons between them. Following G. N. Lewis, many molecular structures may be obtained by inspection using these rules. Letting a dot represent an electron, one may write:

$$\overset{..}{\underset{..}{\text{O}}}\text{:H} \quad\quad \text{H}:\overset{\text{H}}{\underset{\text{H}}{\text{C}}}:\text{H} \quad\quad [\text{Na}]^+ \quad [:\overset{..}{\underset{..}{\text{F}}}:]^-$$

In these electron-dot symbols, the electrons in the K shell are not included for atoms after He, nor are the electrons in the K and L shells for atoms following Ne.

Hydrogen has a valence of 1, because one more electron will give a hydrogen atom an inert gas structure. Carbon can form four bonds because four more electrons give it the neon electronic structure.

Bond types. The bond between two atoms is covalent if one electron in the bonding electron pair comes from each atom, as in H:H or the CH bonds in CH_4. It is coordinate covalent if both electrons come from one atom, as the boron-nitrogen bond in the compound

$$\begin{array}{c} :\overset{..}{\underset{..}{\text{F}}}:\text{H} \\ :\overset{..}{\underset{..}{\text{F}}}:\text{B}:\text{N}:\text{H} \\ :\overset{..}{\underset{..}{\text{F}}}:\text{H} \end{array}$$

If there is complete transfer of electrons from one atom to another the bond is electrovalent or ionic, as in sodium fluoride, NaF. Bonds intermediate in type are possible; the bond in hydrogen fluoride, HF is between covalent and ionic. An ionic bond X^+Y^- will be more stable the less the ionization potential of X and the greater the affinity of Y for electrons, that is, when X is a metallic element from the lower left corner of the periodic table and Y is a nonmetallic element from the upper right corner. Bond type can be inferred from both chemical and physical evidence. SEE CHEMICAL BONDING; ELECTRONEGATIVITY.

Bonds, involving one or three electrons are known, but they are rare; H_2^+ and HeH are examples. Multiple bonds between atoms are common and important; examples are the carbon-carbon bond and the carbon-oxygen bonds in ethylene and carbon dioxide

$$\begin{array}{ccc} H_2C::CH_2 & \text{or} & H_2C=CH_2 \\ O::C::O & \text{or} & O=C=O \end{array}$$

For discussion of a bond of special importance in biology SEE HYDROGEN BOND.

Valence electrons are the electrons of an atom that can participate in chemical binding, for example, for H and He the 1s electrons, for Li through Ne the 2s and 2p electrons, and for Na the 3s electrons.

Oxidation-reduction. As generally used and here defined, the word valence is ambiguous. Before a value can be assigned to the valence of an atom in a molecule, the electronic structure of the molecule must be exactly known, and this structure must be describable simply in terms of simple bonds. In practice neither of these conditions is ever precisely fulfilled. A term not so ambiguous is oxidation number or valence number. Oxidation numbers are useful for the balancing of oxidation-reduction equations, but they are not related simply to ordinary valences. Thus the valence of carbon in CH_4, $CHCl_3$, and CCl_4 is 4; oxidation numbers of carbon in these three substances are -4, $+2$, and $+4$. SEE OXIDATION-REDUCTION.

Quantum theory of valence. The above theory of valence is inadequate in at least three ways. First, it fails to account for many experimental facts, such as why the six C—C bonds in the molecule benzene, C_6H_6, are physically and chemically equivalent, what the electronic structures of the boron hydrides are, why the H—H bond is much stronger than the C—C bond, why CO_2 is a linear molecule but H_2O nonlinear, and what principles govern the rates of chemical combination. Second, the explanations that are offered are not physically satisfying. The stability conferred upon a molecule by the sharing of a pair of electrons by two atoms is established, but what is the real origin of this stability? And third, the theory is not comprehensive or quantitative enough to allow correlation and prediction of the many different properties of molecules. Dozens of properties of molecules can be measured, many to a high degree of accuracy. A theory should ultimately, and quantitatively, account for all of these. SEE MOLECULAR STRUCTURE AND SPECTRA.

The quantum theory of valence does not have these faults. It is based on the new precise laws of physics for the atomic domain which were formulated in the 1920s by E. Schrödinger and others, the discipline called quantum mechanics. The quantum ideas of M. Planck and N. Bohr require modification to take care of experimental observations that electrons and other particles at times act like waves. Like waves, they interfere when they are on top of one another in a manner that can be precisely calculated. According to nineteenth-century physics, an electron moving about a proton would collapse onto it. In the Bohr theory this collapse is prevented by a special quantum hypothesis; in the new mechanics it is prevented by elementary energy considerations. It would be favored by the attractive potential energy of the particle pair, but it turns out to be catastrophic for their kinetic energy. Instead of collapse a compromise is reached; the electron, or wave, is smudged out over a region about the nucleus which defines the atomic size.

The pattern of the periodic table comes out as before. The orbits of Bohr are replaced by new entities, orbitals, which represent not the paths of the electrons but the amplitudes of the electron waves at different points in space. Furthermore, electrons are treated as if they were spinning, but only in two possible ways. The rule that generates the periodic table then is that in an atom no two electrons can occupy the same atomic orbital with the same spin.

In a chemical bond again there is interplay of kinetic and potential energies. An electron pair will tend to be shared by two atoms instead of being located on one of them if that situation is energetically favorable. The region between nuclei is more favorable for the potential energy of electron-nuclear attraction than other regions the same distance from just one nucleus. Moving in this restricted region is not as favorable for the kinetic energy as moving on individual atoms, but the potential energy predominates when a bond is formed. The normal covalent bond may be described as two electrons occupying one molecular orbital, rather than two distinct atomic orbitals, with opposite spins because the exclusion principle is still operative. SEE MOLECULAR ORBITAL THEORY.

When a detailed examination is made of these effects with the new theory, the stabilities of actual molecules and other of their properties can be quantitatively accounted for. In particular, if two atoms approach which have low-energy atomic orbitals which overlap each other in space, and if two electrons are available, the conditions are favorable for forming, with evolution of heat, a chemical bond. It follows that the valence of an atom is given by the number of unpaired electrons it possesses, an old basic rule of valence.

The greater the overlap between two atomic orbitals, the stronger the bond that can be formed with them (criterion of maximum overlapping). This condition may be regarded as determining the shapes of molecules. Two or more orbitals of comparable energy, as 2s and 2p orbitals, can be combined (hybridized) to give orbitals concentrated along certain directions in space, and these are the orbitals that participate in directed bond formation. In the carbon atom, for instance, the four electrons in the 2s and 2p subshells are potential valence electrons. The two 2s electrons are paired, however, so that to make four bonds possible one of these must be promoted to a vacant 2p orbital. Four bonds than are possible, in various directions. Four equivalent bonds can be formed, tetrahedrally directed, as in CH_4. Three bonds in a plane and one other less strong one can be formed, as in CH_2=CH_2. In this manner Linus Pauling and others have accounted for a multitude of phenomena in stereochemistry.

The peculiar binding in benzene and other aromatic molecules has been explained, together with its consequences for chemical reactivity. The principles governing reaction rates have been formulated and applied.

Research in valence theory through the 1930s and 1940s led to understanding of a great deal of chemistry, and contributed toward acceptance of the language of modern physics as a proper language for chemistry. However, considerable research in this field continues. New substances with new types of bonds are being synthesized constantly. New physical methods for studying molecules are constantly revealing more intimate details of molecular structure which demand explanation (for example, the new techniques of magnetic resonance). SEE CHELATIONS; CHEMICAL DYNAMICS; CONJUGATION AND HYPERCONJUGATION; CRYSTAL FIELD THEORY; STEREOCHEMISTRY.

Bibliography. A. L. Companion, *Chemical Bonding*, 2d ed., 1979; H. Gray, *Electrons and Chemical Bonding*, 1964; R. McSweeney, *Coulson's Valence*, 3d ed., 1980; L. Pauling, *The Nature of the*

Chemical Bond and the Structure of Molecules and Crystals, 3d ed., 1960; L. Pauling and P. Pauling, *Chemistry*, 1975; A. N. Stranges, *Electrons and Valence: Development of the Theory*, 1982.

INTERMOLECULAR FORCES
GEORGE E. EWING

Attractive or repulsive interactions that occur between all atoms and molecules. These forces, which become significant at molecular separations of about 1 nanometer or less, are much weaker than forces associated with chemical bonds or electrostatic interactions of charged particles. They are important, however, since they are responsible for many of the physical properties of solids, liquids, and pressurized gases. Intermolecular forces also determine to an important extent the three-dimensional arrangement of biological molecules, polymers, and even smaller molecules.

Description. A simple description of intermolecular forces can begin with the example of two interacting argon atoms. The atoms are electrically neutral and do not undergo chemical bond formation.

Figure 1 shows the potential energy of two argon atoms as a function of their separation. At distances of about 1 nm or greater this energy is essentially zero and the atoms exert no forces on each other. (The force is the negative gradient, or slope, of the potential energy.) Between 0.4 and 0.8 nm the potential energy decreases and the atoms experience forces of attraction. For distances less than 0.3 nm the potential energy rises sharply as the atoms repel each other. At a distance of 0.38 nm the forces of attraction and repulsion balance each other. The potential energy (and corresponding intermolecular forces) between other pairs of atoms exhibits the same general shape as shown in Fig. 1, although the quantitative values of energy and separation are somewhat different. For intermolecular forces between molecules the relative orientations as well as distances are important and the description is more complex. In general, for either atoms or molecules at separations of 0.3 nm or less, the intermolecular forces are repulsive. At longer range, usually greater than 0.3 nm, the intermolecular forces are attractive. And at some intermediate distance, usually 0.3–0.4 nm (which depends on orientation in the case of molecules), the intermolecular forces of attraction and repulsion just balance.

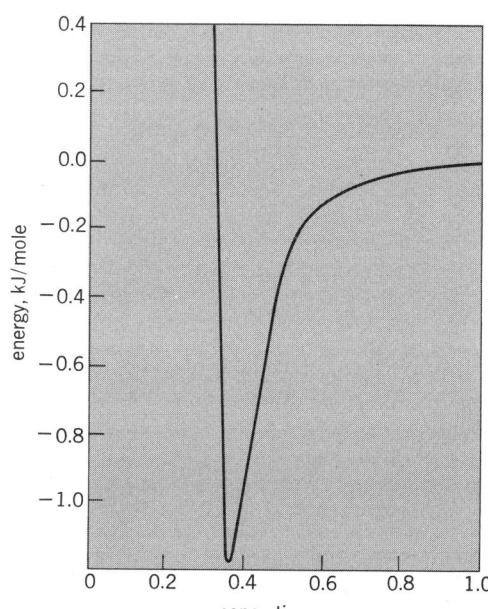

Fig. 1. The intermolecular potential energy of two argon atoms. (*After J. M. Parson, P. E. Siska, and Y. T. Lee, Intermolecular potentials from cross-beam differential elastic scattering measurements, IV:Ar + Ar, J. Chem, Phys., 56:1511–1516, 1972*)

Origin. The origin of intermolecular forces is again most simply discussed by considering two interacting atoms. Quantum mechanics indicates that the rapid motion of the electrons causes instantaneous fluctuations in the charge density around the nucleus. For atoms far apart the electrons in one atom move independently of electrons in the other atom, and on the average the charge distribution is symmetric as shown in **Fig. 2**a. At distances where attractive forces become important, the average charge distribution is still symmetric. However, an instantaneous fluctuation in the electron distribution in one atom can now affect its neighbor nearby. A charge separation in one atom occurs when the electron cloud shifts toward one side of the atom, barring its nucleus to a slight extent. In the other atom the electrons have moved in concert toward this barred nucleus, and an electrostatic attraction is set up. This is illustrated schematically in an exaggerated fashion in Fig. 2b. At another instant the electron clouds may shift in the opposite direction, and the other atom has its nuclear charge partially exposed to the electrons of its neighbor. The electron motions in both atoms are correlated so that an attractive electrostatic force is maintained while the averaged motions assure a symmetric distribution about each atom. These attractive forces are often called London or dispersion forces.

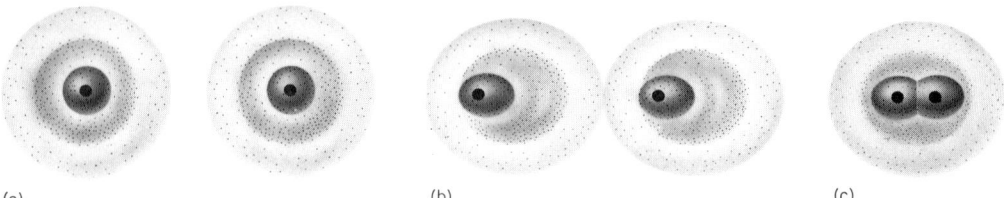

(a)   (b)   (c)

Fig. 2. Schematic diagram of intermolecular interaction. (a) There is no interaction between the atoms that are 1 nm or more apart. (b) For atoms separated by about 0.8 nm or less, dispersion forces which are attractive result from correlated fluctuations of the electron charge distribution in the atoms. (Distribution shown is greatly exaggerated.) (c) For the atoms closer together, 0.3 nm or less, exchange forces which are repulsive cause a permanent distortion of the electron charge distribution. (Distribution shown is greatly exaggerated.)

At small separations the electron clouds can overlap, and repulsive forces are set up. These are called Pauli or exchange forces and are also explained by quantum mechanics. They are essentially a consequence of the reluctance of electrons to be confined into the same small region of space. Atoms or molecules brought close together will respond to exchange forces by a permanent distortion of their electron distribution as shown in Fig. 2c.

All atoms and molecules experience dispersion and exchange forces, which thus are a common component of intermolecular forces. Neutral molecules, in addition, may interact with each other because they possess permanent electrical polarity expressed as a dipole, quadrupole, or higher multipole moments. The electrostatic forces associated with these interacting multipole moments depend on the orientation of the molecules and may be either attractive or repulsive. The corresponding energies are usually somewhat less than dispersion or exchange energies. The dispersion, exchange, and permanent multipole electrostatic forces taken together are usually called van der Waals forces. Energies associated with the formation of hydrogen bonds (that is, between two HF or H_2O molecules) are somewhat larger than van der Waals energies.

Interactions considerably stronger than those just discussed sometimes occur between atoms or molecules. The energies of chemical bond formation are hundreds of times greater than that shown by the intermolecular potential well of Fig. 1. Electrostatic interactions between charged particles are likewise relatively strong. These interactions are usually not classified as intermolecular forces. *See* CHEMICAL BONDING.

Occurrence. Intermolecular forces are responsible for many of the bulk properties of matter in all its phases. A realistic description of the relationship among pressure, volume, and temperature of a gas must include the effects of attractive and repulsive forces between molecules. At increased pressures and sufficiently low temperatures the attractive forces between molecules in the gas will cause it to liquefy. The viscosity, surface tension, and diffusion of liquids are examples of physical properties which are a consequence of intermolecular forces. Repulsive

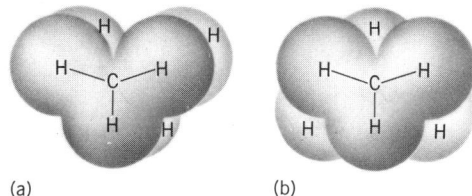

Fig. 3. Intermolecular forces and ethane. (*a*) Eclipsed configuration. (*b*) Staggered configuration.

forces prevent the molecules from approaching one another too closely and account for the high compressibility of liquids. Intermolecular forces between near and distant neighbors dictate the ordered molecular arrangements in crystalline solids. These forces also account for the elasticity of solids. A detailed accounting of the intermolecular forces in the condensed phase is complex since it must include the interactions of each molecule with many of its neighbors. Nevertheless, the energy of each pair of atom interactions is approximately described by an intermolecular potential of the sort shown in Fig. 1.

Intermolecular forces are also important between atoms within a molecule. Even for a molecule as small as ethane, CH_3CH_3, they direct important details in the molecular structure. Chemical bonds dictate the arrangement of the hydrogen atoms about each carbon atom as well as the distance between carbon atoms. However, intermolecular forces mold the final structure, which keeps the hydrogen atoms on one CH_3 group staggered with respect to those on the other CH_3 group. Thus the staggered rather than the eclipsed configuration for ethane as shown in **Fig. 3** is the most stable three-dimensional structure. In an analogous way for larger molecules, proteins, and other biological molecules, the complex spatial arrangement assumed is determined in part by the balance of attractive and repulsive intermolecular forces between atoms that are chemically bonded within the molecule.

Atoms and molecules may be held to a solid surface by intermolecular or van der Waals forces. This weak bonding, called physisorption, has many important applications. The trapping out of molecules from the gas phase onto cooled surfaces is the basis of pumps for producing vacuums. Undesirable odors or colors in food or water may sometimes be removed by use of filters which capture by physisorption the offending contamination. The selective adsorption of molecules by surfaces is a useful method for separation of mixtures of molecules. SEE ADSORPTION.

Study methods. The importance of intermolecular forces has been responsible for their extensive study. In the early 1970s most of the information on intermolecular forces was inferred from the study of matter in bulk. Measurements of the viscosity of gases, or crystal structure of solids, for example, were used to estimate the quantitative nature of the intermolecular interactions that must produce these physical properties. However, it has since been found that studies of individual molecular interactions yield the information more directly.

In molecular beam experiments, low-density streams of atoms or molecules are directed so that individual particles collide. The way in which the molecules rebound as a result of their collision is determined by their initial velocities which can be controlled. Intermolecular forces can be extracted from the experimental data. The intermolecular potential energy curve shown in Fig. 1 was obtained from studies of the collision dynamics of argon atoms. Mappings of the potential energy surfaces of other atoms and molecules are being obtained by this technique.

Another approach is to study van der Waals molecules. In these experiments clusters of atoms or molecules are formed at low temperatures in the gas phase because of their intermolecular attraction for each other. Clusters of two or three atoms or molecules are called van der Waals molecules. For example, gaseous argon at the temperature of the boiling liquid ($-186°C$ or $-303°F$) contains about 98% Ar atoms, and the remaining 2% are Ar_2 van der Waals molecules. the ultraviolet spectrum of the gas at low temperatures reveals features due to Ar_2 which can be used to characterize the bond strength and the intermolecular forces between the argon atoms in the van der Waals molecule.

Spectroscopy of van der Waals molecules formed by clusters of chemically bonded molecules has also revealed much about intermolecular forces which depend on the orientation of the molecules within the cluster. Gaseous H_2, O_2, or HF contains small concentrations of $(H_2)_2$, $(O_2)_2$, or $(HF)_2$. The structures of these van der Waals molecules are shown in **Fig. 4**. The chemical

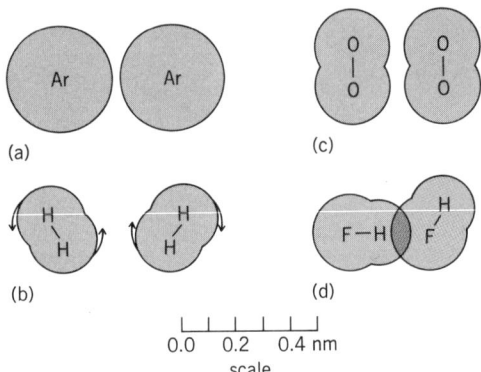

Fig. 4. Structures of some van de Waals molecules. (a) Argon. (b) Hydrogen. (c) Oxygen. (d) Hydrogen fluoride. (*After G. Ewing, Structure and properties of van der Waals molecules, Accounts Chem. Res., 8:185–192, 1975*)

bonds in H_2, O_2, or HF are about 0.1 nm long and not affected by the formation of the 0.3–0.4 nm intermolecular bond of the van der Waals molecule. In $(H_2)_2$ the intermolecular forces do not depend much on the orientation of either H_2, and as a consequence each H_2 molecule, while weakly bound to its neighbor, rotates freely within the cluster. The arrows shown in Fig. 4 are meant to represent this freedom of internal rotation. The $(O_2)_2$ van der Waals molecule appears to reside in a rectangular configuration, while $(HF)_2$ exhibits a bent structure characteristic of hydrogen bond formation. While chemical bonds produce rigid molecules with well-defined geometries, intermolecular forces maintain rather floppy structures of the van der Waals molecules. Internal motions in $(O_2)_2$ or $(HF)_2$ produce considerable distortions of the static structure representations in Fig. 4. The structures of several dozen van der Waals molecules are now known. The determination of properties of this new class of compounds promises to provide a deeper insight into the nature of intermolecular forces. SEE MOLECULAR STRUCTURE AND SPECTRA.

Theoretical approaches to intermolecular interactions have taken two directions. Detailed quantum-mechanical calculations have been performed on the interactions of very simple systems, for example, two He atoms. These calculations seek to determine the wave functions, importance of the correlated motions of the electrons, and the precise nature of the energy of the interaction. This theoretical approach then seeks a deeper understanding of the quantum-mechanical origin of intermolcular forces. A more pragmatic approach uses the electron distribution of the isolated molecule from previous calculations. This distribution is treated as an "electron gas" with an associated electric field. It is the response of the interacting molecules to these electric fields that is responsible for intermolecular forces. Calculations of the electron gas model appear to produce reliable intermolecular energies for both interacting atoms and molecules, with a modest amount of computational effort. SEE SOLUTION; VALENCE.

Bibliography. G. P. Arrighini, *Intermolecular Forces and Their Evaluation by Perturbation Theory*, 1982; B. L. Blaney and G. E. Ewing, Van der Waals molecules, *Annu. Rev. Phys. Chem.*, 27:553–586, 1976; S. T. Ceyer and G. A. Somojai, Surface scattering, *Annu. Rev. Phys. Chem.*, 28:477–499, 1977; J. O. Hirschfelder, C. F. Curtiss, and R. B. Bird, *Molecular Theory of Gases and Liquids*, 1964; T. Kihara, *Intermolecular Forces*, 1978; Y. S. Kim and R. G. Gordon, Unified theory for intermolecular forces between closed shell atoms and ions, *J. Chem. Phys.* 61:1–16, 1974; G. C. Maitland et al., *Intermolecular Forces: Their Origin and Determination*, 1982; B. Pullman (ed.), *Intermolecular Forces*, 1981.

FREE RADICAL
JAMES R. BOLTON

Any molecule or atom which possesses one unpaired electron. This definition does not include transition-metal ions or molecules with more than one unpaired electron, such as O_2. Most chemists accept this definition; however, spectroscopists have adopted a looser definition, in which they define any transient species (atom, molecule, or ion) in the gas phase to be a free radical. Free radicals can be chemically very reactive (for example, the methyl radical, $CH_3\cdot$) or they can be very stable entities (for example, nitric oxide, NO). Often, a free radical is designated by a dot (·). The modifier free probably arose because organic chemists often used to refer to a substituent (for example, hydroxyl or carboxyl) as a radical. However, this usage of the term radical is no longer prevalent, and increasingly the terms radical and free radical are used synonymously. The term biradical is used for a molecule which contains two unpaired electrons so that the two are clearly confined to distinct parts of the molecule and do not interact significantly. That is, a biradical corresponds to two free radicals which are chemically bonded to each other.

Free radicals were frequently, and incorrectly, postulated throughout the nineteenth century. Avogadro's hypothesis was not taken too seriously by the early organic chemists, and substances such as C_2H_6 and C_4H_{10} were frequently described as CH_3 and C_2H_5, respectively. However, by the end of the century this situation had been cleared up, and the impossibility of the independent existence of free radicals appeared to be well established. This situation was completely upset by Moses Gomberg's discovery of the triphenyl methyl radical during an attempt to synthesize hexaphenylethane. He found that the solution, which was expected to be colorless, was yellow, the color deepened when the solution was heated, molecular-weight determinations gave values lower than expected, and the new species was very reactive toward O_2, I_2, and NO at room temperature. This evidence provided strong support for the hypothesis that hexaphenylethane in solution is dissociated according to reaction (1). In spite of this evidence, Gomberg's

$$(C_6H_5)_3C\text{---}C(C_6H_5)_3 \rightleftharpoons 2\ (C_6H_5)_3C\cdot \qquad (1)$$

claim met with years of doubt and denial. Since then, many similar radicals have been discovered, and such entities are now freely postulated in organic reaction mechanisms.

In 1929, the first evidence for very reactive free radicals was obtained by F. Paneth. He observed that when tetramethyllead $Pb(CH_3)_4$ vapor, with a carrier gas, was passed very rapidly, over a heated tube, a lead mirror was formed some distance downstream from the point of heating, and ethane was detected as the principal gaseous product. If the tube was then heated upstream of the mirror with tetramethyl lead passing through, the old mirror was observed to disappear. This was interpreted in terms of indirect evidence that methyl radicals ($CH_3\cdot$) were present as transient free radicals in the gas phase.

Free radicals can be grouped into three major classes: atoms (for example, H·, F·, and Cl·), inorganic radicals (for example, OH·, CN·, NO·, and $ClO_3\cdot$), and organic radicals (for example, $CH_3\cdot$, $C_2H_5\cdot$, and $C_6H_6^-$). Such radicals are of great importance because they have been shown to appear as intermediates in both thermal and photochemical reactions. Free radicals are also known to initiate and propagate polymerization and combustion reactions.

Production. In general, free radicals are formed by the homolytic rupture of a bond in a stable molecule with the production of two fragments, each with an unpaired electron. The resulting free radicals may participate in further reactions or may combine to reform the original compound. In the gas phase, equilibria such as reaction (2) may be established, especially at

$$R_1\text{---}R_2 \rightleftharpoons R_1\cdot + R_2\cdot \qquad (2)$$

elevated temperatures. Because in most cases recombination occurs at nearly every collision of R_1 and R_2, the composition of the reaction mixture under ordinary circumstances indicates only a minute amount of decomposition into radicals. Even though radical partial pressures may be less than 10^{-6} torr (1.33×10^{-4} pascal) and their lifetimes less than 1 millisecond, these radicals do play an important role in reaction kinetics. The transitory existence of such radicals has been confirmed by spectroscopic study. SEE CHEMICAL DYNAMICS.

There are many ways by which free radicals may be generated, including thermal decomposition, electric discharge, microwave discharge, photochemical decomposition, electrolysis at an electrode such as mercury or platinum, rapid mixing of two reactants, and radiolysis with gamma-ray, x-ray, or electron irradiation. In thermal methods, a stable molecule is decomposed at an elevated temperature. In exceptional circumstances, the dissociation into radicals at equilibrium may be considerable. Thus, hydrogen atoms may be produced by heating hydrogen gas (H_2) to a very high temperature, as in reaction (3). At 1900 K (2960°F), this equilibrium corresponds to

$$H_2 \rightleftharpoons 2H\cdot \qquad (3)$$

1% dissociation into atoms when the pressure is 1 atmosphere (100 kilopascals). In a few cases, substances in solution are considerably dissociated into radicals at room temperature. It is then possible to obtain radicals of apparent long life in high concentrations. Two examples are hexaphenylethane [reaction (1)], which in benzene solution at 5°C (41°F) and a concentration of 2–3% is about 3% dissociated into triphenylmethyl radicals; and hexa-(p-biphenyl)ethane [$(C_6H_5-C_6H_4)_3C-C(CH_6H_5-C_6H_4)_3$], which is virtually 100% dissociated under similar circumstances. Usually, however, thermal decompositions are irreversible under the conditions of the reaction. Most gaseous organic substances decompose wholly or in part by a mechanism involving an initial split into radicals. For example, in reaction (4), ethane splits into two methyl radicals.

$$C_2H_6 \rightarrow 2\ CH_3\cdot \qquad (4)$$

Free radicals may also be produced by passing a gas through an electrical discharge at high speed. Atoms, such as hydrogen, oxygen, or nitrogen, are often produced in this manner and then subjected to further chemical reactions which often produce other radicals. A gentler method is to use a microwave discharge.

Photochemical methods of producing free radicals are very common and have been studied extensively. When subjected to ultraviolet light with wavelengths less than 200 nanometers, most gaseous organic molecules decompose photochemically into free radicals. Certain classes of molecules (for example, carbonyls) will react photochemically with ultraviolet light from 200 to 350 nanometers' wavelength, while a few molecules, such as NO_2 and Cl_2, will decompose with visible light. The photolysis of acetone is a good example of photochemical decomposition. When photolyzed in the range 250–310 nm, acetone splits, as in reaction (5). This reaction has been one of

$$CH_3COCH_3 + h\nu \rightarrow CH_3\cdot + CH_3CO\cdot \qquad (5)$$

the most frequently used sources of methyl and acetyl radicals in gas-phase photochemistry. Chlorine decomposes on irradiation with visible light to give chlorine atoms [reaction (6)]. Many reac-

$$Cl_2 + h\nu \rightarrow 2\ Cl\cdot \qquad (6)$$

tions of chlorine atoms have been investigated by this method. $NO_2\cdot$ decomposes via reaction (7)

$$NO_2\cdot + h\nu \rightarrow NO\cdot + O \qquad (7)$$

with visible and near-ultraviolet light. This is a reaction of considerable importance in the chemistry of air pollution. A photochemical production of free radicals via an electron transfer reaction from a chlorophyll species is the primary photochemical reaction of photosynthesis by plants and certain bacteria.

Electrolysis often produces radicals at one of the electrodes. For instance, the electrolysis of anthracene in dimethylformamide, in the absence of oxygen and water, leads to the generation of the anthracene negative ion at the cathode. *See Electrolysis.*

Rapid mixing of two suitable reactants can lead to moderate steady-state concentrations of unstable intermediates. For example, rapid mixing of aqueous solutions of titanium trichloride ($TiCl_3$), and of hydrogen peroxide (H_2O_2), results in the formation of the $OH\cdot$ radical, presumably by reaction (8).

$$Ti^{3+} + H_2O_2 \rightarrow Ti^{4+} + OH\cdot + OH^- \qquad (8)$$

High-energy radiation such as gamma rays and x-rays, as well as high-energy particles given off by radioactive nuclei, can cause extensive disruption of molecules leading to radical and ion production. Such systems are important in cancer research and in research on the mechanisms of radiation damage. The use of pulsed electron beams in pulsed radiolysis has proved particularly valuable in these mechanism studies.

Detection and estimation. The earliest methods of detection involved the chemical properties of the radicals, such as the Paneth method for the detection of methyl radicals. Later and more reliable methods used absorption spectroscopy, mass spectrometry, and especially electron paramagnetic resonance spectroscopy.

One chemical method that has proved effective uses chemical traps. For example, if iodine is added to a gas-phase photochemical system, rapid reactions (9) and (10) occur. Under appro-

$$R\cdot + I_2 \rightarrow RI + I\cdot \qquad (9) \qquad R\cdot + I\cdot \rightarrow RI \qquad (10)$$

priate conditions, all radicals may be removed from the system and trapped as iodides. In solution, nitrones such as 5,5′-dimethyl-1-pyrroline-1-oxide (DMPD) have proved to be effective spin traps via reaction (11). The resulting spin adduct is usually a stable nitroxide free radical which may be

$$\begin{array}{c}\text{H}_3\text{C} \\ \text{CH}_3\end{array}\!\!\!\underset{\underset{\text{O}^-}{|}}{\overset{+}{\text{N}}}\!\!\!=\!\text{H} \quad + R \quad \longrightarrow \quad \begin{array}{c}\text{H}_3\text{C} \\ \text{CH}_3\end{array}\!\!\!\underset{\underset{\text{O}}{|}}{\text{N}}\!\!\!\begin{array}{c}\text{H} \\ \text{R}\end{array} \qquad (11)$$

detected by electron paramagnetic resonance spectroscopy. In this manner, both quantitative and qualitative information can be obtained about the nature of the reactive radical R·. However, valid interpretation of results requires further information.

Absorption spectroscopy is a simple method of detecting free radicals in the gas phase and has been used extensively for many years. For example, the following transient free radicals (under the spectroscopists' definition) have been identified spectroscopically: OH, NH, NH_2, CN, CF_2, CF, C_2, CH, CHO, C_3, CH_3O, C_2H_5O, NCO, NCS, HNO, PH_2, and CH_3; many others have been identified with varying degrees of certainty. The flash-photolysis technique, by which large amounts of light energy may be absorbed by a gas in a short period of time, has greatly advanced the detection of radicals by absorption spectroscopy. In an interesting variation of the method, an organic compound frozen in a glassy solvent may be photolyzed at very low temperatures so that the radicals formed are trapped indefinitely. These frozen intermediates may then be studied at will by using ultraviolet, visible, or infrared spectroscopy.

An ingenious method for the detection of gas-phase radicals has been developed using mass spectrometry. A rapid-flow system is used in which radicals are produced thermally or photochemically. The gas stream is sampled through a leak into the mass spectrometer. An electron energy is selected so that radicals are ionized, but stable molecules are not dissociated. Any radical ions detected in the mass spectrometer must come from radicals already present in the gas mixture, and a direct and unequivocal detection of radicals is possible. Under favorable circumstances, the method can be used for the quantitative estimation of radical concentrations.

Electron paramagnetic resonance spectroscopy is by far the best technique available for the detection and characterization of free radicals in solids, liquids, or gases. The technique is very sensitive (10^{-9} moles per liter free radicals can be detected under favorable conditions) and, furthermore, the technique is specific to molecules which contain one or more unpaired electrons. Generally, each radical has a characteristic electron paramagnetic resonance spectrum which can be used both for identification and for deducing structural information about the radical. For example, the methyl radical $CH_3\cdot$ was shown to have a planar structure by analyzing its electron paramagnetic resonance spectrum; whereas the corresponding $CF_3\cdot$ radical was shown to have a pyramidal structure. Analysis of the electron paramagnetic resonance spectra of radicals such as the anthracene negative ion yields information concerning the distribution of the unpaired electron over the molecule.

Free-radical mechanisms. In well-investigated photochemical reactions, it is frequently possible to establish beyond any reasonable doubt a multistep mechanism involving free-radical intermediates by using some of the analytical techniques. A typical example is the thermal decomposition of ethane, reactions (12)–(17). General mechanisms of this type, the Rice mecha-

$$C_2H_6 \rightarrow 2\ CH_3\cdot \qquad (12)$$

$$CH_3\cdot + C_2H_6 \rightarrow CH_4 + C_2H_5\cdot \quad (13) \qquad C_2H_5\cdot \rightarrow C_2H_4 + H\cdot \quad (14) \qquad H\cdot + C_2H_6 \rightarrow C_2H_5\cdot + H_2 \quad (15)$$

$$H\cdot + C_2H_5\cdot \rightarrow C_2H_6 \quad (16) \qquad 2\ C_2H_5\cdot \rightarrow C_4H_{10} \quad (17a) \qquad 2\ C_2H_5\cdot \rightarrow C_2H_4 + C_2H_6 \quad (17b)$$

nisms, have been well-established in many cases by detailed investigation of the kinetic behavior. The Rice mechanism is an example of a chain reaction; reaction (12) initiates the chain, (13) to (15) propagate it, and (16) and (17) terminate it. Similar mechanisms are involved in a large number of other thermal and photochemical reactions. A large body of information has become available concerning the individual reaction steps. With the aid of computer simulation, it is possible to predict the behavior of very complex reactions over a wide range of conditions. This approach has been particularly valuable in the study of gas-phase chemistry in the upper atmosphere and in polluted air. SEE CHAIN REACTION.

MOLECULE
ROBERT S. MULLIKEN

A molecule may be thought of either as a structure built of atoms bound together by chemical forces or as a structure in which two or more nuclei are maintained in some definite geometrical configuration by attractive forces from a surrounding swarm of negative electrons. Besides chemically stable molecules, short-lived molecular fragments called free radicals can be observed under special circumstances, for example, at high temperatures, in electrical discharges, in chemical reactions, and even (but in small quantities) frozen into ordinary solid substances under some conditions. Free radicals are really just highly active molecules. SEE CHEMICAL BONDING; FREE RADICAL; MOLECULAR STRUCTURE AND SPECTRA.

ION
THOMAS C. WADDINGTON

An atom, or group of atoms, which by loss or gain of one or more electrons has acquired an electric charge. If the ion is formed from an atom of hydrogen or an atom of a metal, it is usually positively charged; if the ion is formed from an atom of a nonmetal or from a group of atoms, it is usually negatively charged. The number of electronic charges carried by an ion is called its electrovalence. The charges are denoted by superscripts which give their sign and number; for example, a sodium ion, which carries one positive charge, is denoted by Na^+; a sulfate ion, which carries two negative charges, by SO_4^{2-}. SEE CHEMICAL BONDING.

Salts are usually composed of orderly arrangements of ions which are not free to move easily in the solid. However, when the salt is fused or dissolved in water, the ions become free, and when an electric field is applied to the salt in solution, the positively charged cations move to the cathode and the negatively charged anions move to the anode. At the electrodes the ions lose their electric charge. This process is called electrolysis. SEE ELECTROLYSIS; VALENCE.

GAS
C. F. CURTISS AND J. O. HIRSCHFELDER

A phase of matter characterized by relatively low density, high fluidity, and lack of rigidity. A gas expands readily to fill any containing vessel. Usually a small change of pressure or temperature produces a large change in the volume of the gas. The equation of state describes the relation between the pressure, volume, and temperature of the gas. In contrast to a crystal, the molecules in a gas have no long-range order.

At sufficiently high temperatures and sufficiently low pressures, all substances obey the ideal gas, or perfect gas, equation of state, shown as Eq. (1), where p is the pressure, T is the

$$pv = RT \qquad (1)$$

absolute temperature, v is the molar volume, and R is the gas constant. Absolute temperature T expressed on the Kelvin scale is related to temperature t expressed on the Celsius scale as in Eq. (2).

$$T = t + 273.16 \qquad (2)$$

The gas constant is $R = 82.0567$ cm^3-atm/(mole)(K) $= 82.0544$ ml-atm/(mole)(K).
The molar volume is the molecular weight divided by the gas density.

Empirical equations of state. At lower temperatures and higher pressures, the equation of state of a real gas deviates from that of a perfect gas. Various empirical relations have been proposed to explain the behavior of real gases. The equations of J. van der Waals (1899), Eq. (3),

$$\left(p + \frac{a}{v^2}\right)(v - b) = RT \tag{3}$$

of P. E. M. Berthelot (1907), Eq. (4), and F. Dieterici (1899), Eq. (5), are frequently used. In these

$$\left(p + \frac{a}{Tv^2}\right)(v - b) = RT \tag{4} \qquad pe^{a/vRT}(v - b) = RT \tag{5}$$

equations, a and b are constants characteristic of the particular substance under considerations. In a qualitative sense, b is the excluded volume due to the finite size of the molecules and roughly equal to four times the volume of 1 mole of molecules. The constant a represents the effect of the forces of attraction between the molecules. In particular, the internal energy of a van der Waals gas is $-a/v$. None of these relations gives a good representation of the compressibility of real gases over a wide range of temperature and pressure. However, they reproduce qualitatively the leading features of experimental pressure-volume-temperature surfaces.

Schematic isotherms of a real gas, or curves showing the pressure as a function of the volume for fixed values of the temperature, are shown in **Fig. 1**. Here T_1 is a very high tempera-

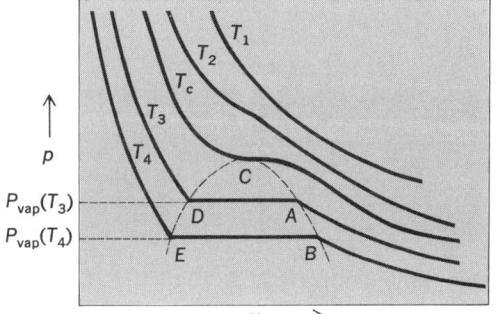

Fig. 1. Schematic isotherms of a real gas. C is the critical point. Points A and B give the volume of gas in equilibrium with the liquid phase at their respective vapor pressures. Similarly, D and E are the volumes of liquid in equilibrium with the gas phase.

ture and its isotherm deviates only slightly from that of a perfect gas; T_2 is a somewhat lower temperature where the deviations from the perfect gas equation are quite large; and T_c is the critical temperature. The critical temperature is the highest temperature at which a liquid can exist. That is, at temperatures equal to or greater than the critical temperature, the gas phase is the only phase that can exist (at equilibrium) regardless of the pressure. Along the isotherm for T_c lies the critical point, C, which is characterized by zero first and second partial derivatives of the pressure with respect to the volume. This is expressed as Eq. (6). At temperatures lower than the

$$(\partial p/\partial v)_c = (\partial^2 p/\partial v^2)_c = 0 \tag{6}$$

critical, such as T_3 or T_4, the equilibrium isotherms have a discontinuous slope at the vapor pressure. At pressures less than the vapor pressure, the substance is gaseous; at pressures greater than the vapor pressure, the substance is liquid; at the vapor pressure, the gas and liquid phases (separated by an interface) exist in equilibrium.

Along one of the isotherms of the empirical equations of state discussed above, the first and second derivatives of the pressure with respect to the volume are zero. The location of this critical point in terms of the constants a and b is shown in the following; p_c and v_c are the pressure and volume at the critical temperature.

	van der Waals	Berthelot	Dieterici
p_c	$\dfrac{a}{27b^2}$	$\left(\dfrac{aR}{216b^3}\right)^{1/2}$	$\dfrac{a}{4e^2b^2}$
v_c	$3b$	$3b$	$2b$
T_c	$\dfrac{8a}{27Rb}$	$\left(\dfrac{8a}{27Rb}\right)^{1/2}$	$\dfrac{a}{4Rb}$
$\dfrac{p_c v_c}{RT_c}$	0.3750	0.3750	0.2706

Some typical values of $p_c v_c/RT_c$ for real gases are as follows: 0.30 for the noble gases, 0.27 for most of the hydrocarbons, 0.243 for ammonia, and 0.232 for water. The van der Waals and Berthelot equations of state, Eqs. (3) and (4), cannot quantitatively reproduce the critical behavior of real gases because no substance has a value of $p_c v_c/RT_c$ as large as 0.375. The Dieterici equation, Eq. (5), gives a good representation of the critical region for the light hydrocarbons but does not represent well the noble gases or water.

At temperatures lower than the critical point, the analytical equations of state, such as the van der Waals, Berthelot, or Dieterici equations, give S-shaped isotherms as shown in **Fig. 2**. From thermodynamic considerations, the vapor pressure is determined by the requirement that the cross-hatched area *DEO* be equal to the cross-hatched area *AOB*. Under equilibrium conditions, the portion of this isotherm lying between *A* and *D* cannot occur. However, if a gas is suddenly compressed, points along the segment *AB* may be realized for a short period until enough condensation nuclei form to create the liquid phase. Similarly, if a liquid is suddenly overexpanded, points along *DE* may occur for a short time. For low temperatures, the point *E* may represent a negative pressure corresponding to the tensile strength of the liquid. However, the simple analytical equations of state cannot be used for a quantitative estimate of these transient phenomena. Actually, it is easy to show that the van der Waals, Berthelot, and Dieterici equations give poor representations of the liquid phase since the volume of most liquids (near their freezing point) is considerably less than the constant *b*.

Principle of corresponding states. In the early studies, it was observed that the equations of state of many substances are qualitatively similar and can be correlated by a simple scaling of the variables. To describe this result, the reduced or dimensionless variables, indicated by a subscript *r*, are defined by dividing each variable by its value at the critical point. These variables are given in Eqs. (7)–(9).

$$p_r = p/p_c \quad (7) \qquad T_r = T/T_c \quad (8) \qquad v_r = v/v_c \quad (9)$$

In its most elementary form, the principle of corresponding states asserts that the reduced pressure, p_r, is the same function of the reduced volume and temperature, v_r and T_r, for all sub-

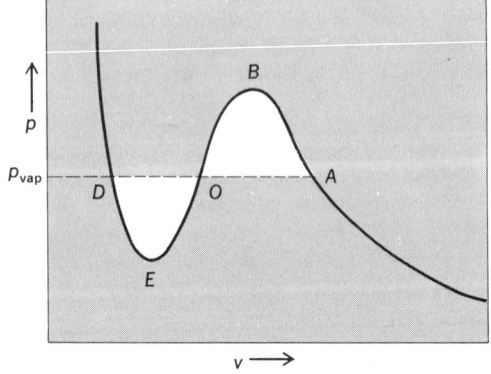

Fig. 2. Schematic low-temperature isotherm as given by van der Waals, Berthelot, or Dieterici equations of state. Here the line *DOA* corresponds to the vapor pressure. The point *A* gives the volume of the gas in equilibrium with the liquid phase, and *D* gives the volume of the liquid. The segment of the curve *DE* represents overexpansion of the liquid. The segment *AB* corresponds to supersaturation of the vapor. However, the segment *EOB* could not be attained experimentally.

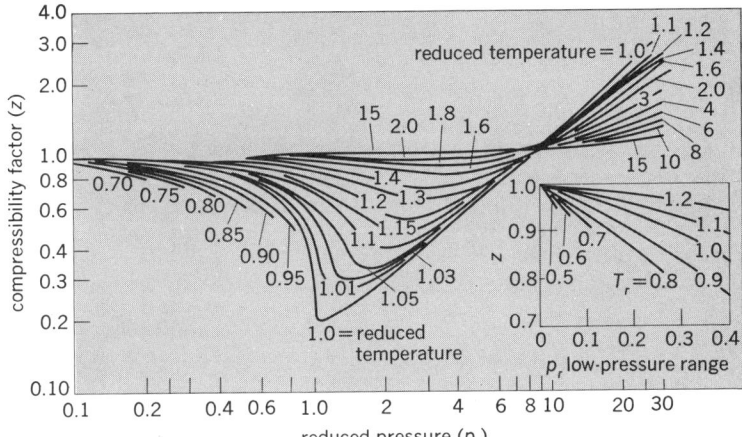

Fig. 3. The compressibility factor pv/RT as a function of the reduced pressure $p_r = p/p_c$ and reduced temperature $T_r = T/T_c$. (After O. A. Hougen, K. M. Watson, and R. A. Ragatz, *Chemical Process Principles*, pt. 2, Wiley, 1959)

stances. An immediate consequence of this statement is the statement that the compressibility factor, expressed as Eq. (10), is a universal function of the reduced pressure and the reduced

$$z = pv/RT \qquad (10)$$

temperature. This principle is the basis of the generalized compressibility chart shown in **Fig. 3**. This chart was derived from data on the equation-of-state behavior of a number of common gases.

It follows directly from the principle of corresponding states that the compressibility factor at the critical point z_c should be a universal constant. It is found experimentally that this constant varies somewhat from one substance to another. On this account, empirical tables were developed of the compressibility factor and other thermodynamic properties of gases as functions of the reduced pressure and reduced temperature for a range of values of z_c. Such generalized corresponding-states treatments are very useful in predicting the behavior of a substance on the basis of scant experimental data.

Theoretical considerations. The equation-of-state behavior of a substance is closely related to the manner in which the constituent molecules interact. Through statistical mechanical considerations, it is possible to obtain some information about this relationship. If the molecules are spherically symmetrical, the force acting between a pair of molecules depends only on r, the distance between them. It is then convenient to describe this interaction by means of the intermolecular potential $\varphi(r)$ defined so that the force is the negative of the derivative of $\varphi(r)$ with respect to r. *See* INTERMOLECULAR FORCES.

Two theoretical approaches to the equation of state have been developed. In one of these approaches, the pressure is expressed in terms of the partition function Z and the total volume V of the container in the manner of Eq. (11). Here k is the Boltzmann constant or the gas constant

$$p = kT\,(\partial \ln Z/\partial V) \qquad (11)$$

divided by the Avogadro number N_0, $k = R/N_0$. For a gas made up of spherical molecules or atoms with no internal structure, the partition function is given as Eq. (12). In this expression, φ_{ij}

$$Z = \frac{1}{N!}\left(\frac{2\pi mkT}{h^2}\right)^{3N/2} \times \int \exp\left(-\sum_{i>j}\frac{\varphi_{ij}}{kT}\right) dv_1\, dv_2 \cdots dv_N \qquad (12)$$

is the energy of interaction of molecules i and j and the summation is over all pairs of molecules, h is Planck's constant, N is the total number of molecules, and the integration is over the three cartesian coordinates of each of the N molecules. The expression for the partition function may easily be generalized to include the effects of the structure of the molecules and the effects of quantum mechanics.

In another theoretical approach to the equation of state, the pressure may be written as Eq. (13), where $g(r)$ is the radial distribution function. This function is defined by the statement

$$p = \frac{NkT}{V} - \frac{2\pi N^2}{3V^2} \int g(r) \frac{d\varphi}{dr} r^3 \, dr \qquad (13)$$

that $2\pi(N^2/V)g(r)r^2 \, dr$ is the number of pairs of molecules in the gas for which the separation distance lies between r and $r + dr$. The radial distribution function may be determined experimentally by the scattering of x-rays. Theoretical expressions for $g(r)$ are being developed.

The compressibility factor $z = pV/NkT$ may be considered as a function of the temperature, T, and the molar volume, v. In the virial form of the equation of state, z is expressed as a series expansion in inverse powers of v, as in Eq. (14). The coefficients $B(T)$, $C(T)$, . . . , which

$$z = 1 + B(T)/v + C(T)/v^2 + \cdots \qquad (14)$$

are functions of the temperature, are referred to as the second, third, . . . , virial coefficients. This expansion is an important method of representing the deviations from ideal gas behavior. From statistical mechanics, the virial coefficients can be expressed in terms of the intermolecular potential. In particular, the second virial coefficient is Eq. (15). If the intermolecular potential is

$$B(T) = 2\pi N_0 \int (1 - e^{-\varphi/kT}) r^2 \, dr \qquad (15)$$

known, Eq. (15) provides a convenient method of predicting the first-order deviation of the gas from perfect gas behavior. The relation has often been used in the reverse manner to obtain information about the intermolecular potential. Often $\varphi(r)$ is expressed in the Lennard-Jones (6–12) form, Eq. (16), where ϵ and σ are constants characteristic of a particular substance. Values of

$$\varphi(r) = 4\epsilon \left[\left(\frac{\sigma}{r}\right)^{12} - \left(\frac{\sigma}{r}\right)^6 \right] \qquad (16)$$

these constants for many substances have been tabulated. In terms of these constants, the second virial coefficient has the form of Eq. (17), where $B^*(kT/\epsilon)$ is a universal function. If all substances

$$B(T) = (2/3)\pi N_0 \sigma^3 B^*(kT/\epsilon) \qquad (17)$$

obeyed this Lennard-Jones (6–12) potential, the simple form of the law of corresponding states would be rigorously correct. See CHEMICAL THERMODYNAMICS.

Bibliography. I. B. Cohen (ed.), *Laws of Gases*, 1981; J. O. Hirshfelder et al., *Molecular Theory of Gases and Liquids*, 1964; R. Holub and P. Vonka, *The Chemical Equilibrium of Gaseous Systems*, 1976; R. Mohilla and B. Ferencz, *Chemical Process Dynamics*, 1982.

LIQUID
NORMAN H. NACHTRIEB

A state of matter intermediate between that of crystalline solids and gases. Macroscopically, liquids are distinguished from crystalline solids in their capacity to flow under the action of extremely small shear stresses and to conform to the shape of a confining vessel. Liquids differ from gases in possessing a free surface and in lacking the capacity to expand without limit. On the scale of molecular dimensions liquids lack the long-range order that characterizes the crystalline state, but nevertheless they possess a degree of structural regularity that extends over distances of a few molecular diameters. In this respect, liquids are wholly unlike gases, whose molecular organization is completely random.

Thermodynamic relations. The thermodynamic conditions under which a substance may exist indefinitely in the liquid state are described by its phase diagram, shown schematically in the **illustration.** The area designated by L depicts those pressures and temperatures for which the liquid state is energetically the lowest and therefore the stable state. The areas denoted by S and V similarly indicate those pressures and temperatures for which only the solid or vapor phase may exist. The connecting lines OC, OB, and OA define pressures and temperatures for which

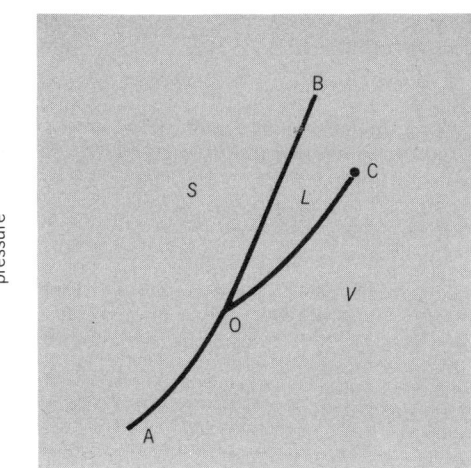

Phase diagram of a pure substance.

the liquid and its vapor, the solid and its liquid, and the solid and its vapor, respectively, may coexist in equilibrium. They are usually termed phase boundary or phase coexistence lines. The intersection of the three lines at O defines a triple point which, for the three states of matter under discussion, is the unique pressure and temperature at which they may coexist at equilibrium. Other triple points exist in the phase diagram of a substance that possesses two or more crystalline modifications, but the one depicted in the figure is the only triple point for the coexistence of the vapor, liquid, and solid. Line OA has its origin at the absolute zero of temperature and OB, the melting line, has no upper limit. The liquid-vapor pressure line OC is different from OB, however, in that it terminates at a precisely reproducible point C, called the critical point. Above the critical temperature no pressure, however large, will liquefy a gas. *See* TRIPLE POINT.

Along any of the coexistence curves the relationship between pressure and temperature is given by the Clausius-Clapeyron equation: $dP/dT = \Delta H/T\Delta V$, where ΔV is the difference in molar volume of the corresponding phases (gas-liquid, gas-solid, or liquid-solid) and ΔH is the molar heat of transition at the temperature in question. By means of this equation the change in the melting point of the solid or the boiling point of the liquid as a function of pressure may be calculated. When a liquid in equilibrium with its vapor is heated in a closed vessel, its vapor pressure and temperature increase along the line OC. ΔH and ΔV both decrease and become zero at the critical point, where all distinction between the two phases vanishes. *See* PHASE EQUILIBRIUM.

Transport properties. Liquids possess important transport properties, notably their capacity to transmit heat (thermal conductivity), to transfer momentum under shear stresses (viscosity), and to attain a state of homogeneous composition when mixed with other miscible liquids (diffusion). These nonequilibrium properties of liquids are well understood in macroscopic terms and are exploited in large-scale engineering and chemical-process operations. Thus, the rate of flow of heat across a layer of liquid is given by $\dot{Q} = \kappa\, dT/dx$, where $\dot{Q}$ is the heat flux, dT/dx is the thermal gradient, and κ is the coefficient of thermal conductivity. Similarly, the shearing of one liquid layer against another is resisted by a force equal to the momentum transfer: $F = \dot{p} = \eta\, dv/dx$, where dv/dx is the velocity gradient and η is the coefficient of viscosity. Likewise, the rate of transport of matter under nonconvective conditions is governed by the gradient of concentration of the diffusing species: $J = -D\, dC/dx$, where J is the matter flux and D is the coefficient of diffusion. Each of these transport coefficients depends upon temperature, pressure, and composition and may be determined experimentally. An a priori calculation of κ, η, and D is a very difficult problem, however, and only approximate theories exist.

Theoretical explanations. In fact, although a great deal of effort has been expended, there still exists no satisfactory theory of the liquid state. Even so commonplace a phenomenon as the melting of ice has no adequate theoretical explanation. The reason for this state of affairs lies in the tremendous structural and dynamical complexity of the liquid state. To understand this, it is useful to compare the structural and kinetic properties of liquids with those of crystalline solids on the one hand and with those of gases on the other.

In crystals, atoms or molecules occupy well-defined positions on a three-dimensional lattice, oscillating about them with small amplitudes; their kinetic energy is entirely distributed among these quantized vibrational states up to the melting point. This nearly perfect spatial order is revealed by diffraction techniques, which utilize the coherent scattering of x-rays or particles have wavelengths comparable with interatomic spacings. The structural and dynamical properties are sufficiently tractable mathematically so that the theory of solids is quite well understood.

The theory of gases is also simple, but for quite a different reason. No vestige of positional regularity of atoms remains in gases, and their energy resides entirely in high-speed translational motion. Except for collisions, which deflect their motions into new straight-line trajectories, atoms in gases do not interact with one another; vibrational modes in monatomic gases are absent.

Liquids, by contrast, lie intermediate between gases and crystals from both a structural and dynamic point of view. Kinetic energy is partitioned among translational and vibrational modes, and diffraction studies reveal a degree of short-range order that extends over several molecular diameters. Moreover, this "structure" is continually changing under the influence of vibrational and translational displacements of atoms. Physical reality may be attributed to this short-range structure, nevertheless, in the sense that a time average over the huge number of possible configurations of atoms may show that a fairly definite number of neighboring atoms lie close to any arbitrary atom. At a somewhat greater distance from this reference atom, the density of neighbors oscillates above and below the average density of atoms in the liquid as a whole.

Information about the degree of local order is contained in the radial distribution function, a mathematical property which may be deduced from diffraction measurements. This is the starting point for a theory of the liquid state, and although research efforts have yielded partial successes, prodigious mathematical difficulties lie in the path of an entirely satisfactory solution. SEE BOILING POINT; MELTING POINT; VISCOSITY.

Bibliography. J. P. Hansen and I. R. McDonald, *The Theory of Simple Liquids*, 1977; P. Kruus, *Liquids and Solutions: Structure and Dynamics*, 1977; G. W. Rothchild, *Dynamics of Molecular Liquids*, 1983; D. Tabor, *Gases, Liquids, and Solids*, 2d ed., 1980; H. N. Temperley and D. H. Trevena, *Liquids and Their Properties: A Molecular and Microscopic Treatise*, 1978.

AMORPHOUS SOLID
Brian G. Bagley

A rigid material whose structure lacks crystalline periodicity; that is, the pattern of its constituent atoms or molecules does not repeat periodically in three dimensions. In the present terminology amorphous and noncrystalline are synonymous. A solid is distinguished from its other amorphous counterparts (liquids and gases) by its viscosity: a material is considered solid (rigid) if its shear viscosity exceeds $10^{14.6}$ poise ($10^{13.6}$ pascal·second). SEE VISCOSITY.

Preparation. Techniques commonly used to prepare amorphous solids include vapor deposition, electrodeposition, anodization, evaporation of a solvent (gel, glue), and chemical reaction (often oxidation) of a crystalline solid. None of these techniques involves the liquid state of the material. A distinctive class of amorphous solids consists of glasses, which are defined as amorphous solids obtained by cooling of the melt. Upon continued cooling below the crystalline melting point, a liquid either crystallizes with a discontinuous change in volume, viscosity, entropy, and internal energy, or (if the crystallization kinetics are slow enough and the quenching rate is fast enough) forms a glass with a continuous change in these properties. The glass transition temperature is defined as the temperature at which the fluid becomes solid (that is, the viscosity = $10^{14.6}$ poise = $10^{13.6}$ Pa·s) and is generally marked by a change in the thermal expansion coefficient and heat capacity. [Silicon dioxide (SiO_2) and germanium dioxide (GeO_2) are exceptions.] It is intuitively appealing to consider a glass to be both structurally and thermodynamically related to its liquid; such a connection is more tenuous for amorphous solids prepared by the other techniques.

Types of solids. Oxide glasses, generally the silicates, are the most familiar amorphous solids. However, as a state of matter, amorphous solids are much more widespread than just the oxide glasses. There are both organic (for example, polyethylene and some hard candies) and inorganic (for example, the silicates) amorphous solids. Examples of glass formers exist for each of the bonding types: covalent [As_2S_3], ionic [$KNO_3-Ca(NO_3)_2$], metallic [Pd_4Si], van der Waals [o-terphenyl], and hydrogen [$KHSO_4$]. Glasses can be prepared which span a broad range of physical properties. Dielectrics (for example, SiO_2) have very low electrical conductivity and are optically transparent, hard, and brittle. Semiconductors (for example, As_2SeTe_2) have intermediate electrical conductivities and are optically opaque and brittle. Metallic glasses (for example, Pd_4Si) have high electrical and thermal conductivities, have metallic luster, and are ductile and strong.

Bibliography. N. March et al., (eds.), *Amorphous Solids and the Liquid State*, 1985; N. F. Mott and E. A. Davis, *Electronic Processes in Non-Crystalline Materials*, 2d ed., 1979; R. Zallen, *The Physics of Amorphous Solids*, 1983.

LIQUID CRYSTALS
GLENN H. BROWN

A state of matter that mixes the properties of both the liquid and solid states. Liquid crystals may be described as condensed fluid states with spontaneous anisotropy. They are categorized in two ways: thermotropic liquid crystals, prepared by heating the substance, and lyotropic liquid crystals, prepared by mixing two or more components, one of which is rather polar in character (for example, water). Thermotropic liquid crystals are divided, according to structural characteristics, into two classes, nematic and smectic. Nematics are further subdivided into ordinary and cholesterics.

When the solid which forms a liquid crystal is heated it undergoes transformation into a turbid system that is both fluid and birefringent. The consistency of the fluid varies with different compounds from a paste to a free-flowing liquid. When the turbid system is heated, it is converted into an isotropic liquid (optical properties are the same regardless of the direction of the measurement). These changes in phases can be represented schematically as follows:

$$\text{Solid} \underset{\text{cool}}{\overset{\text{heat}}{\rightleftarrows}} \text{liquid crystal} \underset{\text{cool}}{\overset{\text{heat}}{\rightleftarrows}} \text{liquid}$$

On cooling the system, the process reverses and goes from isotropic liquid to liquid crystal and finally to the solid. *See* LIQUID.

Lyotropic liquid crystals often have an amphiphilic component, a compound with a polar head attached to a long hydrophobic tail. Sodium stearate and lecithin (a phospholipid) are typical examples of amphiphiles. Starting with a solid amphiphile and adding water, the lamellar structure (molecular packing in layers) is formed. By step-wise addition of water, the molecular packing may take on a cubic structure, then hexagonal, then micellar, followed by true solution. The process is reversed by withdrawing water. Thousands of compounds will form liquid crystals on heating, and still more will do so if two or more components are mixed. A few representative compounds are listed in the **table**.

Classification and structure. Conventionally, matter is considered to have only three states: solid, liquid, and gas. In the solid state, the molecules or atoms show small vibrations about rigidly fixed lattice positions, but they cannot rotate. A liquid will take the shape of its container and will bound itself at the top with its own free surface. The liquid state is characterized by relatively unhindered rotation and no long-range order. The space in a system constituting a gas is sparsely occupied. The molecules are free to occupy the entire volume of their container.

Liquid crystals are a state of matter that combines a kind of long-range order (in the sense of a solid) with the ability to form droplets and to pour (in the sense of waterlike liquids). They also exhibit properties of their own such as the ability to form monocrystals with the application of a normal magnetic or electric field; an optical activity of a magnitude without parallel in either solids or liquids; and a temperature sensitivity which results in a color change in certain liquid crystals. Thermotropic liquid crystals are either nematic or smectic.

Name, formula, and liquid crystalline range of selected compounds

THERMOTROPIC LIQUID CRYSTALS

1. Nematic liquid crystals

$H_3C-O-\langle\text{ring}\rangle-\underset{H}{\overset{}{C}}=N-\langle\text{ring}\rangle-C_4H_9\text{-}n$ 70–117°F (21–47°C)

p-Methoxybenzylidene-p'-butylaniline (MBBA)

$H_{11}C_5-\langle\text{ring}\rangle-\langle\text{ring}\rangle-CN$ 75–95°F (24–35°C)

4-Cyano-4'-n-pentyl-biphenyl

$H_{11}C_5-\langle\text{ring}\rangle-\langle\text{ring}\rangle-\langle\text{ring}\rangle-CN$ 268–464°F (131–240°C)

4-Cyano-4"-n-pentyl-n-terphenyl

$H_3C-O-\langle\text{ring}\rangle-N\overset{O\uparrow}{=}N-\langle\text{ring}\rangle-O-CH_3$ 243–279°F (117–137°C)

p-Azoxyanisole (PAA)

2. Cholesteric esters

Cholesteryl nonanoate   293–354°F (145–179°C)

3. Noncholesteryl, chiral-type compound

$H_3C-O-\langle\text{ring}\rangle-\underset{H}{\overset{}{C}}=N-\langle\text{ring}\rangle-\underset{H}{\overset{}{C}}=\underset{H}{\overset{}{C}}-\underset{O}{\overset{}{C}}-O-CH_2-\underset{H}{\overset{CH_3}{\overset{|}{C}}}-C_2H_5$ 169–257°F (76–125°C)

(−)-2-Methylbutyl-p-(p'-methoxy-benzylideneamino) cinnamate

Nematic structure. Nematic liquid crystals are subdivided into the ordinary nematic and the cholesteric-nematic. The molecules in the ordinary nematic structure maintain a parallel or nearly parallel arrangement to each other along the long molecular axes (**illus.** a). They are mobile in three directions and can rotate about one axis. This structure is one-dimensional.

When the nematic structure is heated, it is generally transformed into the isotropic liquid (illus. b). The nematic structure is the highest-temperature mesophase in thermotropic liquid crystals. The energy required to deform a nematic liquid crystal is so small that even the slightest perturbation caused by a dust particle can distort the structure considerably.

In the cholesteric-nematic structure (illus. c), the direction of the long axis of the molecule in a given layer is slightly displaced from the direction of the molecular axes of the molecules in

Name, formula, and liquid crystalline range of selected compounds (cont.)

THERMOTROPIC LIQUID CRYSTALS (cont.)

4. Smectic A

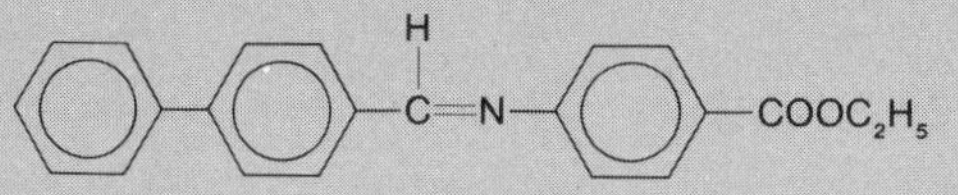

Ethyl p(p'-phenylbenzalamino) benzoate — 250–268°F (121–131°C)

5. Smectic B

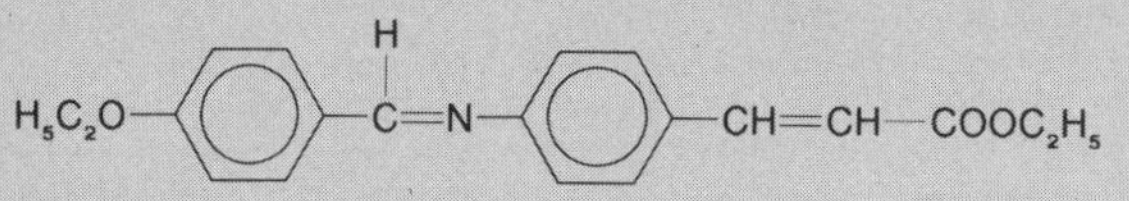

Ethyl p-ethoxybenzal-p'-aminocinnamate — 171–241°F (77–116°C)

6. Smectic C

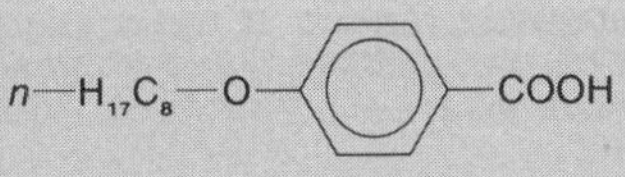

$n-H_{17}C_8-O-\bigcirc-COOH$

p-n-Octyloxybenzoic acid — 226–297°F (108–147°C)

LYOTROPIC LIQUID CRYSTALLINE COMPOUNDS

1. Sodium stearate

$CH_3(CH_2)_{16}COO^-Na^+$

2. α-Lecithin

$CH_2-O-CO-(CH_2)_{16}-CH_3$
$CH-O-C-O(CH_2)_{16}-CH_3$
$CH_2-O-\overset{O^-}{\underset{\underset{O}{\parallel}}{P}}-O-CH_2-CH_2-N^+(CH_3)_3$

(Stearic acid derivative)

---

an adjacent layer. If a twist is applied to this molecular packing, a helical structure is formed. The helix has a pitch which is temperature-sensitive. The helical structure serves as a diffraction grating for visible light. Chiral compounds show the cholesteric-nematic structure (twisted nematic), for example, the cholesteric esters.

Smectic structure. The term smectic covers all thermotropic liquid crystals that are not nematics. At least seven smectic structures have been described. Indications are that two more can be added, making a total of nine from smectic A(S_A) to smectic I (S_I). The alphabetic subscripts only indicate the order in which the smectic structures were first recognized and identified. In most smectic structure, molecules are arranged in strata. The molecules (except in smectic D) are arranged in layers with their long axes parallel to each other. They can move in two directions

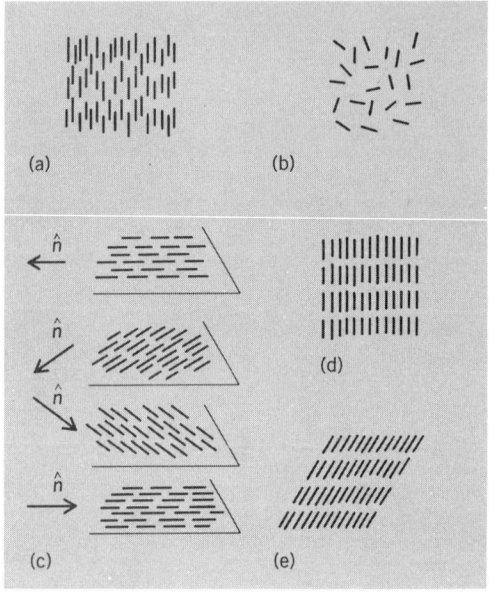

Schematic representation of the molecular arrangement in the (a) ordinary nematic liquid crystal; (b) isotropic liquid; (c) cholesteric-nematic liquid crystal; (d) smectic A liquid crystal; and (e) smectic C liquid crystal. (*After G. H. Brown and J. J. Wolken, Liquid Crystals and Biological Structures, Academic Press, 1979*)

in the plane and can rotate about one axis. Those within layers can be in neat rows or randomly distributed.

Smectic liquid crystals may have structured or unstructured strata. Structured smectic liquid crystals have long-range order in the arrangement of molecules in layers to form a regular two-dimensional lattice. The most common of the structured liquid crystals is smectic B. Molecular layers are in well-defined order, and the arrangement of the molecules within the strata is also well ordered. The long axes of the molecules lie perpendicular to the plane of the layers. In the smectic A (illus. d) structure, molecules are also packed in strata, but the molecules in a stratum are randomly arranged. The long axes of the molecules in the smectic A structure lie perpendicular to the plane of the layers. Molecular packing in a smectic C (illus. e) is the same as that in smectic A, except the molecules in the stratum are tilted at an angle to the plane of the stratum.

There is also a unique kind of liquid crystal known as the smectic D which is isotropic, but nevertheless shows three-dimensional order in the molecular packing of the structure.

Applications. Liquid crystals have many applications. They are used as displays in digital wristwatches, calculators, panel meters, and industrial products. They can be used to record, store, and display images which can be projected onto a large screen. They also have potential use as television displays.

Bibliography. G. H. Brown (ed.), *Advances in Liquid Crystals*, vols, 1–6, 1975–1983; J. D. Margerum and L. J. Miller, Electro-optical applications of liquid crystals, *J. Coll. Int. Sci.*, 58: 559–580, 1977; F. D. Saeva (ed.), *Liquid Crystals; The Fourth State of Matter*, 1979; A. Skoulios, Amphiphiles: Organization et diagrammes des phases, *Ann. Phys.*, 3:421–450, 1978.

COLLOID
Egon Matijević

A state of matter characterized by large specific surface areas, that is, large surfaces per unit volume or unit mass. The term colloid refers to any matter, regardless of chemical composition, structure (crystalline or amorphous), geometric form, or degree of condensation (solid, liquid, or gas), as long as at least one of the dimensions is less than ~ 1 micrometer but larger than ~ 1

Types of colloid dispersions

Medium	Dispersed matter	Technical name	Examples
Gas	Liquid	Aerosol	Fog, sprays
	Solid	Aerosol	Smoke, atmospheric or interstellar dust
Liquid	Gas	Foam	Head on beer, lather
	Liquid	Emulsion	Milk, cosmetic lotions
	Solid	Sol	Paints, muddy water
Solid	Gas	Solid foam	Foam rubber
	Liquid	Solid emulsion	Opal
	Solid	Solid sol	Steel

nanometer. Thus, it is possible to distinguish films (for example, oil slick), fibers (spider web), or colloidal particles (fog) if one, two, or three dimensions, respectively, are within the submicrometer range.

A colloid consists of dispersed matter in a given medium. In the case of finely subdivided particles, classification of a number of systems is possible, as given in the **table**. In addition to the colloids listed in the table, there are systems that do not fit into any of the listed categories. Among these are gels, which consist of a network-type internal structure loaded with larger or smaller amounts of fluid. Some gels may have the consistency of a solid, while others are truly elastic bodies that can reversibly deform. Another colloid system that may occur is termed coacervate, and is identified as a liquid phase separated on coagulation of hydrophilic colloids, such as proteins. *See* GEL.

It is customary to distinguish between hydrophobic and hydrophilic colloids. The former are assumed to be solvent-repellent, while the latter are solvent-attractant (dispersed matter is said to be solvated). In reality there are various degrees of hydrophilicity for which the degree of solvation cannot be determined quantitatively.

Properties. Certain properties of matter are greatly enhanced in the colloidal state due to the large specific surface area. Thus, finely dispersed particles are excellent adsorbents; that is, they can bind various molecules or ions on their surfaces. This property may be used for removal of toxic gases from the atmosphere (in gas masks), for elimination of soluble contaminants in purification of water, or decolorization of sugar, to give just a few examples. *See* ADSORPTION.

Colloids are too small to be seen by the naked eye or in optical microscopes. However, they can be observed and photographed in transmission or scanning electron microscopes. Owing to their small size, they cannot be separated from the medium (liquid or gas) by simple filtration or normal centrifugation. Special membranes with exceedingly small pores, known as ultrafilters, can be used for collection of such finely dispersed particles. The ultracentrifuge, which spins at very high velocities, can also be employed to promote colloid settling.

Colloids show characteristic optical properties. They strongly scatter light, causing turbidity such as in fog, milk, or muddy water. Scattering of light (recognized by the Tyndall beam) can be used for the observation of tiny particles in the ultramicroscope. Colloidal state of silica is also responsible for iridescence, the beautiful effect observed with opals. *See* TYNDALL EFFECT.

Preparation. Since the characteristic dimensions of colloids fall between those of simple ions or molecules and those of coarse systems, there are in principle two sets of techniques available for their preparation: dispersion and condensation. In dispersion methods the starting materials consist of coarse units which are broken down into finely dispersed particles, drawn into

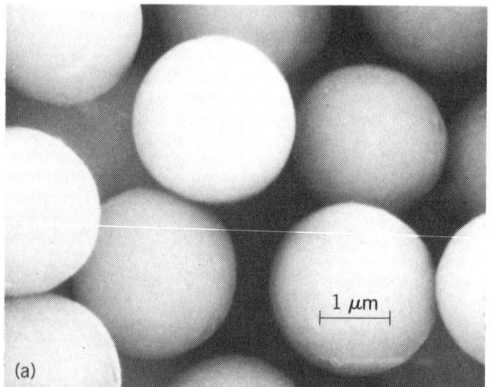

Examples of colloid particles. (a) Zinc sulfide (sphalerite). (b) Cadmium carbonate; at right is an enlargement of the boxed section in the left photograph (*Courtesy of Egon Matijević*).

fibers, or flattened into films. For example, colloid mills grind solids to colloid sizes, nebulizers can produce finely dispersed droplets from bulk liquids, and blenders are used to prepare emulsions from two immiscible liquids (such as oil and water). SEE EMULSION.

In condensation methods, ions or molecules are aggregated to give colloidal particles, fibers, or films. Thus, insoluble monolayer films can be developed by spreading onto the surface of water a long-chain fatty acid (for example, stearic acid) from a solution in an organic liquid (such as benzene or ethyl ether). Colloidal aggregates of detergents (micelles) form by dissolving the surface-active material in an aqueous solution in amounts that exceed the critical micelle concentration. SEE MICELLE; MONOMOLECULAR FILM.

The most common procedure to prepare sols is by homogeneous precipitation of electrolytes. Thus, if aqueous silver nitrate and potassium bromide solutions are mixed in proper concentrations, colloidal dispersions of silver bromide will form, which may remain stable for a long time. Major efforts have focused on preparation of monodispersed sols, which consist of colloidal particles uniform in size, shape, and composition (see **illus.**).

Dispersion stability. Dispersions of colloids (sols) are inherently unstable because they represent systems of high free energy. Particles may remain in suspended state as long as they are small enough for gravity to be compensated by the kinetic energy (brownian movement). Consequently, particle aggregation on collision must be prevented, which can be achieved by various means.

The most common cause for sol stability is electrostatic repulsion, which results from the charge on colloidal particles. The latter can be due to adsorption of excess ions (such as of bromide ions on silver bromide particles), surface acid or base reactions (for example, removal or addition of protons on the surface of metal hydroxides), or other interfacial chemical reactions. Alternately, sol stability can be induced by adsorption of polymers or by solvation. Stability of colloid dispersions that occur naturally may represent an undesirable state, as is the case with muddy waters whose turbidity is due to the presence of tiny particles (clays, iron oxides, and so forth). When needed, stability can be artificially induced, as in the production of paints.

Coagulation or flocculation. Processes by which a large number of finely dispersed particles is aggregated into larger units, known as coagula or flocs, are known as coagulation or flocculation; the distinction between the two is rather subtle. The concepts most commonly refer to colloidal solids in liquids (sols), although analogous principles can be employed in explaining aggregation of solid particles or liquid droplets in gases (aerosols).

Coagulation of stable sols can be accomplished by various additives (electrolytes, surfactants, polyelectrolytes, or other colloids) whose efficiency depends on their charge and adsorptivity, as well as on the nature of suspended particles to be treated. A common example of applications of coagulation is in water purification in which particulate colloid contaminants must be aggregated in order to be removed. Coagulation is also used to describe solidification of polymers (for example, proteins), which can be achieved by heating, as exemplified by boiling an egg. SEE EMULSION.

Bibliography. P. C. Hiemenz, *Principles of Colloid and Surface Chemistry*, 1977; J. N. Israelachvili, *Intermolecular and Surface Forces*, 1985; R. D. Vold and M. J. Vold, *Colloid and Interface Chemistry*, 1983.

COLLOIDAL CRYSTALS
NOEL A. CLARK

Periodic arrays of suspended colloidal particles. Common colloidal suspensions (colloids) such as milk, blood, or latex are polydisperse; that is, the suspended particles have a distribution of sizes and shapes. However, suspensions of particles of identical size, shape, and interaction, the so-called monodisperse colloids, do occur. In such suspensions, a new phenomenon that is not found in polydisperse systems, colloidal crystallization, appears: under appropriate conditions, the particles can spontaneously arrange themselves into spatially periodic structures. This ordering is analogous to that of identical atoms or molecules into periodic arrays to form atomic or molecular crystals. However, colloidal crystals are distinguished from molecular crystals, such as those formed by very large protein molecules, in that the individual particles do not have precisely identical internal atomic or molecular arrangements. On the other hand, they are distinguished from periodic stackings of macroscopic objects like cannonballs in that the periodic ordering is spontaneously adopted by the system through the thermal agitation (brownian motion) of the particles. These conditions limit the sizes of particles which can form colloidal crystals to the range from about 0.01 to about 5 micrometers.

The most spectacular evidence for colloidal crystallization is the existence of naturally occurring opals. The ideal opal structure is a periodic close-packed three-dimensional array of silica microspheres with hydrated silica filling the spaces not occupied by particles. Opals are the fossilized remains of an earlier colloidal crystal suspension. Another important class of naturally occurring colloidal crystals are found in concentrated suspensions of nearly spherical virus particles, such as *Tipula* iridescent virus and tomato bushy stunt virus. Colloidal crystals can also be made from the synthetic monodisperse colloids, suspensions of plastic (organic polymer) microspheres. Such suspensions have become important systems for the study of colloidal crystals, by virtue of the controllability of the particle size and interaction.

Crystal structure and properties. In colloidal crystals of spherical or nearly spherical particles, the crystal structures adopted are identical to those found in crystals formed by spherical

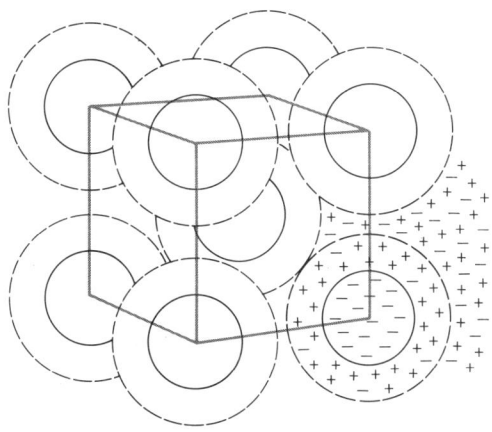

Fig. 1. Arrangement of particles and distribution of electrical charges in a charged stabilized body-centered cubic colloidal crystal.

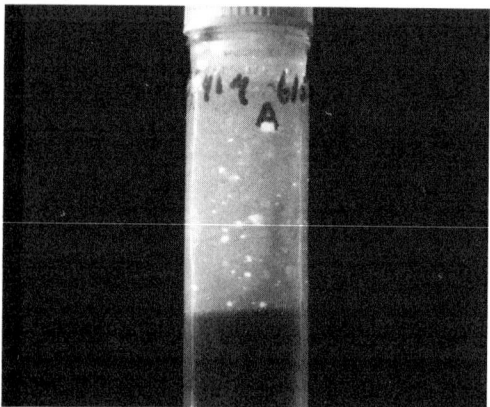

Fig. 2. Reflection of white light from a polycrystalline colloidal crystal suspension. The crystals are body-centered cubic, with a particle diameter of 0.1 μm and interparticle spacing of 1.5 μm.

atoms, such as the noble gases and alkali metals. The phase behavior is also quite similar. Consider, for example, a suspension in water of polymer microspheres. These particles are readily prepared with acid surface groups which dissociate in water, leaving the particle with a large negative charge (as large as several thousand electron charges). This charge attracts positively charged small ions, which form clouds around the particles (**Fig. 1**). The spheres will repel each other if they come close enough so that their ion clouds overlap, much the same as noble gas atoms repel if their electron clouds overlap. At low concentration, the particles rarely encounter one another, executing independent brownian motion and forming the colloidal gas phase. As the concentration is increased to where the average interparticle separation is comparable to the overlap separation, the particles become strongly confined by their neighbors and form the colloidal liquid phase. If the concentration is further increased, an abrupt freezing transition to the ordered colloidal crystal phase will occur. The colloidal crystal structure found near the freezing transition is the body-centered cubic (Fig. 1), as is often the case in atomic systems. At higher concentrations, the more densely packed, in fact closest-packed, face-centered cubic and hexagonal close-packed structures are found. *See* Gas; Liquid.

Despite these structural similarities, atomic and colloidal crystals are characterized by vastly different length, stiffness, and time scales. First, particles in atomic crystals are separated by less than a nanometer compared to micrometers in colloidal crystals. This means that colloidal crystal structure can be determined by visible-light techniques, as opposed to x-rays, which are required in the atomic crystal case. The spectacular iridescent light reflection by opal and other colloidal crystals (**Fig. 2**) is a direct result of the periodic lattice structure. Second, colloidal atomic crystals are about 10^8 times weaker than atomic crystals. Finally, significant atomic motions, such as the relative vibration of adjacent particles, occur in 10^{-12} s, whereas the equivalent time in a colloidal system is 10^{-2} s. Hence, it is possible to watch in detail through a microscope the

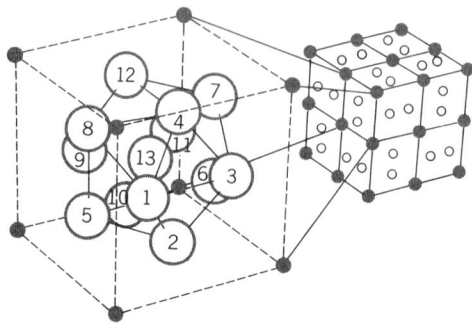

Fig. 3. Crystal structure found in a binary mixture of polystyrene microspheres of two different sizes: 550 nm in diameter (open circles) and 270 nm in diameter (closed circles). The colloidal alloy has the crystal structure found in atomic binary alloys such as $NaZn_{13}$. (*After S. Hachisu and S. Yoshimura, Optical demonstration of crystalline superstructures in binary mixtures of latex globules, Nature 283:188–189, 1980*).

motions of particles in a colloidal lattice. The distinctive properties of colloidal crystals thus provide a new avenue to the understanding of many phenomena found in atomic systems, such as lattice vibrations (sound waves), crystal dislocation and other defects, the response of crystals to large stresses (plastic flow), and crystallization itself. SEE OPALESCENCE.

Complex systems. In addition to the simple colloidal crystals found in bulk (three-dimensional) suspensions of spherical particles, there are a number of more exotic colloidal systems. Colloidal alloys can be formed in suspensions containing particles of several different sizes. Binary mixtures form mixed crystals, sometimes with approximately 100 particles per unit cell, analogous to the alloy structures found in binary atomic metal mixtures (**Fig. 3**). Other systems include colloidal glasses, which are binary or multicomponent suspensions that exhibit shear rigidity without periodic structure; two-dimensional colloidal crystals, which are obtained by trapping a monolayer or several layers of particles; and colloidal liquid crystals, which are found in suspensions of interacting nonspherical (rod- or disk-shaped) particles such as tobacco mosaic virus. SEE AMORPHOUS SOLID; COLLOID; LIQUID CRYSTALS.

Bibliography. N. A. Clark, A. J. Hurd, and B. J. Ackerson, Single colloidal crystals, *Nature*, 281:57–60, 1979; D. J. Darragh, A. J. Gaskin, and J. V. Sanders, Opals, *Sci. Amer.*, 234(4):84–95, April 1976; R. M. Fitch (ed.), *Polymer Colloids*, vols. 1 and 2, 1978; S. Hachisu and S. Yoshimura, Optical demonstration of crystalline superstructures in binary mixtures of latex globules, *Nature*, 283:188–189, 1980; P. Pieranski, Colloidal crystals, *Contemp. Phys.*, 24:25–73, 1983.

MICELLE
J. K. THOMAS

A colloidal aggregate of a unique number (50 to 100) of amphipathic molecules, which occurs at a well-defined concentration called the critical micelle concentration. In polar media such as water, the hydrophobic part of the amphiphiles forming the micelle tends to locate away from the polar phase while the polar parts of the molecule (head groups) tend to locate at the polar micelle solvent interface.

A micelle may take several forms, depending on the conditions and composition of the system, such as distorted spheres, disks, or rods (**illus**. *a*, *b*, and *c*). The dimensions of the particle are derived from those of the amphiphile, for example, a sphere radius of about 2 nanometers or a rod cross-sectional radius of 2 nanometers. Frequently the polar head group is a salt which ionizes in polar media. The micelle may have about 30% of the amphiphiles in the ionized state, giving rise to a highly charged particle surrounded by a cloud of counterions (ions with a charge opposite to that of the micelle ions), and also counterions bound into the micelle surface. Micelles are formed in nonpolar media such as benzene, where the amphiphiles cluster around small water droplets in the system, forming an assembly known as a reversed micelle (illus. *d*).

When the surfactant contains two long alkyl chains, a vesicle structure rather than a micelle is formed because of geometric restraints. This entity consists of two closed concentric spherical layers of surfactant which enclose an internal volume of water. Such a structure is similar to that of a biological membrane; micelles and vesicles mimic biological systems both in structure and in many kinetic processes.

Micellar systems have the unique property of being able to solubilize both hydrophobic and hydrophilic compounds. They are used extensively in industry for detergency and as solubilizing agents. A strong catalytic action is often associated with these systems and is attributed to the clustering of reactants in the micelle, thereby creating high local reactant concentration, and also to the strongly charged surface which influences the transition state of a reaction. SEE SURFACE-ACTIVE AGENT.

The locally high concentration of guest molecules or reactants in a micelle surface leads to enhanced rates of reaction over those observed when the reactants are dispersed in the bulk phase. This catalytic effect can be as large as 1000-fold and can, in some cases, be used selectively. Photolysis of dialkyl ketones in micelles leads to radicals whose local concentrations are high. Only a portion of these radicals can react in micelles, those that have experienced spin relaxation due to their particular isotopic constituency. The unreacted portion with a different

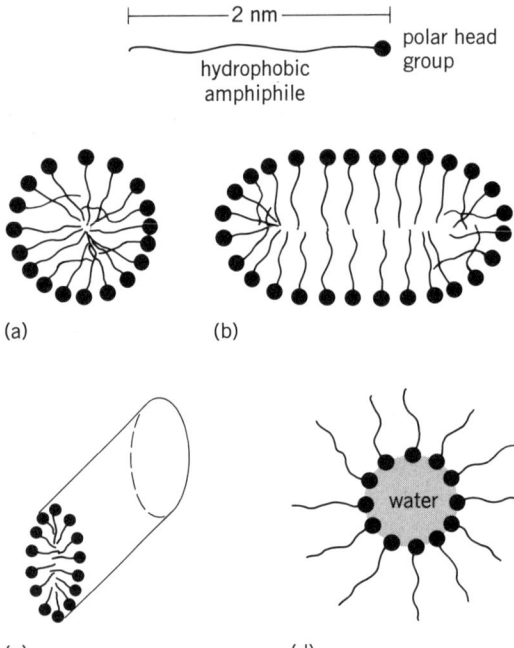

Form of an amphiphile and several forms of micelle: (**a**) spherical, (**b**) disk, (**c**) rod, and (**d**) reversed.

isotopic constituency diffuse into the bulk phase and give alternative products. The technique has been used successfully to produce isotopic enrichment. *See* CATALYSIS; FREE RADICAL.

Micelles play an important role in photo-induced reactions, in particular photo-induced electron transfer reactions. Anionic micelles such as sodium dodecyl sulfate strongly promote photoionization of several molecules (for example, phenothiazine, aminopyrene, or tetramethyl benzidine) located at the micelle interface. The ionization threshold is reduced by more than 3.0 eV [>70 kilocalories (300 kilojoules)/mole], an effect not observed in homogeneous solution. The electron of the guest molecule is transferred to the aqueous phase and observed as a hydrated electron, while the cation is stablized by the micelle. Electron transfer between two guest molecules, giving two radical ions, often occurs most efficiently on micellar systems. The high local concentration and the interface promote electron transfer and lead to efficient reaction. The charged micelle interface stabilizes the ion of opposite charge, while repelling into the bulk water phase the ion of like charge. This latter effect of the micelle surface leads both to efficient charge separation and to long-lived radical ions. The ions can be used to generate useful chemistry via subsequent reaction with other components of the system. In particular, the presence of catalysts such as colloidal platinum can lead to water breakdown, the radical anion donating an electron to the platinum which leads to the formation of hydrogen. These types of micelle-mediated processes are of use in the storage of solar energy. *See* PHOTOCHEMISTRY.

Bibliography. J. H. Fendler, *Membrane Mimetic Chemistry*, 1982; J. K. Thomas, *The Chemistry of Excitation at Interfaces*, Amer. Chem. Soc. Monog. Ser. 181, 1984; N. Turro, A. Braun, and M. Grätzel, Photophysical and photochemical processes in micellar systems, *Angewandte Chem.*, 19:675–696, 1980.

GEL
W. O. MILLIGAN

A two-phase colloidal system consisting of a solid and a liquid. Gels behave as elastic solids and retain their characteristic shape, whereas sols (colloidal dispersions) possess the shape of the container. Commonly, gels have a low solid content, for example, 2–5% for ferric oxide, and as little as 0.1% for coagulated blood. Gels include jellies or transparent elastic gels rich in liquid, and gelatinous precipitates which are believed to consist of minute particles of jelly. Gels or jellies which have dried until apparently solid may be called zerogels, and sometimes will swell or redisperse to form a sol when treated with a suitable solvent. The lyophobic gels (usually inorganic) may be prepared by double decomposition reactions employing high reagent concentrations. Rapid mixing favors the formation of clear, transparent gels. Other methods include slow coagulation of sols by electrolytes and concentration of sols by slow evaporation of the dispersion medium.

The lyophilic gels (usually organic), such as gelatin, agar-agar, and certain soaps, may be prepared by cooling a sol prepared at an elevated temperature or by allowing air-dry gels to swell in a solvent. The setting or gelation of a sol is characterized by (1) time of set, (2) gelation temperature, (3) critical concentration of setting, and (4) rate of viscosity increase.

Setting of a sol to a gel occurs with negligible heat effects; this phenomenon is designated as the isothermal sol-gel transformation. Gels such as clays, which contain platelike particles, may be transformed into sols by shaking; these sols revert to gels on standing. This phenomenon is called thixotropy. Rheopexy refers to the converse process, in which gelation occurs more rapidly when the sol is stirred or vibrated.

The classical explanations of gel structure were the so-called honeycomb and brush-heap theories. In the former, the continuous phase was the solid, with the liquid contained in the connected holes or pores. In the latter theory, the liquid was the continuous phase, and the pores or capillaries were the interstices between the solid particles. The brush-heap theory appears to be correct for many common gels. For example, one sample of silica gel, as viewed in the electron microscope, consisted of small primary particles about 10 nanometers in diameter. The continuous pore structure (the interstices between the loosely packed primary particles) had a most frequent pore radius of 3–3.5 nm. These considerations do not preclude the possibility that some types of gels, such as sintered or partially sintered gels, may possess the honeycomb structure. SEE COLLOID.

EMULSION
GEORGE S. MILL AND W. O. MILLIGAN

A dispersion of one liquid in a second immiscible liquid. Since the majority of emulsions contain water as one of the phases, it is customary to classify emulsions into two types: the oil-in-water (O/W) type consisting of droplets of oil dispersed in water, and the water-in-oil (W/O) type in which the phases are reversed. The continuous liquid is referred to as the dispersion medium, and the liquid which is in the form of droplets is called the disperse phase.

A stable emulsion consisting of two pure liquids cannot be prepared; to achieve stability, a third component, an emulsifying agent, must be present. Generally, the introduction of an emulsifying agent will lower the interfacial tension of the two phases.

Classification. A large number of emulsifying agents are known; they can be classified broadly into several groups. The largest group is that of the soaps, detergents, and other compounds whose basic structure is a paraffin chain terminating in a polar group. Water-soluble soaps (for example, sodium or potassium soaps) are effective in stabilizing oil-in-water emulsions; in this case, the paraffin chains of the soap molecules are concentrated in the oil droplets, and their polar groups are oriented toward the continuous water medium. Dissociation of the polar groups leaves the oil droplets with a charge situated at the interface, and the counter ions form a diffuse double layer in the water, thus preventing coalescence. Water-insoluble soaps (for example, calcium soaps) are effective in stabilizing water-in-oil emulsions. In this case the stabilizing action is believed to be similar to that exhibited by certain solid powders. For a powder to act as an emulsifier, it must be wetted more by one phase than by the other. Whichever phase shows the greater wetting power will become the dispersion medium, because such powders congregate at the interfaces and present the greater portions of their surfaces to the liquid which wets them preferentially. For example, precipitated sulfur, which is wetted preferentially by water, stabilizes oil-in-water emulsions; lamp-black, which is wetted preferentially by oil, stabilizes water-in-oil emulsions. Many naturally occurring emulsions, such as milk or rubber latex, are stabilized by proteins. Egg yolk proteins stabilize mayonnaise and salad dressing. In these and similar types of emulsions, stability results from the formation of a protective coating of the protein around each droplet of the disperse phase. Certain hydrophilic colloids such as gum arabic or gelatin also stabilize water-in-oil emulsions by a similar mode of action.

Geometrically, the maximum amount of one liquid which can be dispersed in another in the form of spheres of equal size is about 74% of the total available space, independent of the diameter of the spheres. However, considerably more concentrated emulsions of either type can be obtained because droplets are not necessarily uniform in size, and the emulsifying agent permits distortion of the droplets without coalescence. The creams used in cosmetics are examples of high-concentration emulsions.

Properties. Various methods are available for determining the type of a particular emulsion, but recognition of the following three characteristics is usually sufficient for most purposes. (1) The electrical conductivity of an oil-in-water emulsion is much greater than that of a water-in-oil emulsion. (2) A water-soluble dye such as methyl orange will color an oil-in-water system easily, but will not color a water-in-oil system. For an oil-soluble dye such as fuchsin, the reverse is true. (3) An emulsion will mix perfectly with more of its continuous phase when this is added in pure form.

Emulsions may be prepared readily by shaking together the two liquids or by adding one phase drop by drop to the other phase with some form of agitation, such as irradiation by ultrasonic waves of high intensity. In industry, emulsification is accomplished by means of emulsifying machines. In a typical machine, a mixture of the two liquids containing an emulsifying agent is forced through a narrow slit between a rapidly rotating rotor and a stator. The preparation of stable emulsions must be controlled carefully, since emulsions are sensitive to such variations as the mode of agitation, the nature and amount of the emulsifying agent, and temperature changes.

The breaking of emulsions is necessary in many industrial operations, for example, in the separation of water-in-oil emulsions in the petroleum industry and in product recovery from emulsions produced by the steam distillation of organic liquids. Emulsions may be broken by (1) addition of multivalent ions of charge opposite to the emulsion droplet, (2) chemical action (addition of acids to emulsions stabilized by soaps), (3) freezing, (4) heating, (5) aging, (6) centrifuging, (7) application of high-potential alternating electric fields, and (8) treatment with ultrasonic waves of low intensity. SEE COLLOID.

SOLUTION
JOEL H. HILDEBRAND

A homogeneous mixture of two or more components whose properties vary continuously with varying proportions of the components. A liquid solution can be distinguished experimentally from a pure liquid by the fact that during transfers into other single phases at equilibrium (freezing and vaporizing at constant pressure) the temperature and other properties vary continuously, whereas those of a pure liquid remain constant.

Gases, unless highly compressed, are mutually soluble in all proportions.

A solid solution is, similarly, a single phase whose composition and other properties vary continuously with changing composition of the liquid phase with which it is in equilibrium. SEE SOLID SOLUTION.

Types of intermolecular force. The extent to which substances can form solutions depends upon the kind and strength of the attractive forces between the several molecular species involved. It is necessary to consider the attractive forces exerted by molecules of the following types: (1) nonpolar molecules; (2) polar molecules, that is, those containing electric dipoles; (3) ions; and (4) metallic atoms.

London forces. The theory of attraction between nonpolar molecules, developed by F. London in 1930, is based upon the quantum-mechanical interaction between pairs of electron systems. For two molecules with electrons having frequencies v_1 and v_2, polarizabilities α_1 and α_2, and separated by the distance r between centers, the attraction potential is shown as Eq. (1), where h is the Planck constant. For molecules of the same species, this reduces to Eq. (2). The

$$\epsilon_{12} = -\frac{3\alpha_1\alpha_2 h}{2r^6} \cdot \frac{\nu_{0,1}\nu_{0,2}}{\nu_{0,1} + \nu_{0,2}} \quad (1) \qquad \epsilon_{11} = -\frac{3\alpha_1^2}{4r^6} h\nu_0 \quad (2)$$

frequency ν_0 is that corresponding to $h\nu_0$, the zero-point energy, of the molecule in its unperturbed state. The perturbation by another molecule is related to its perturbation by light of varying frequencies, as seen in the variation of refractive index n with the frequency of light, that is, the dispersion. For this reason London designated these forces as dispersion forces. It is equally appropriate to speak of London forces by analogy with the nearly equivalent term, van der Waals forces. In the case of gases the dipersion n_ν is related to the frequency as in Eq. (3), where C is a

$$n_\nu - 1 = \frac{C}{\nu_0^2 - \nu^2} \quad (3)$$

constant. The polarizability α can be determined from the refractive index with the aid of the Lorentz-Lorenz formula. As a substitute for zero-point energy, London proposed the ionization potential.

The model upon which these relations are based is much simpler than the polyatomic molecules in most solutions of interest. In these the potential field is not central and radial; the interaction is between the electrons in the peripheral bonds. A striking example is octamethylcyclotetrasiloxane, whose core of alternating silicon and oxygen atoms is so buried within the eight methyl groups that it behaves toward other molecules essentially as an aliphatic hydrocarbon. The normal paraffins themselves are not symmetrical spherically. Moreover, the electrons in the molecules have many frequencies.

Although attempts have been made to extend London's basic concept to take account of such complexities, only the general implications of the concept as to the characteristics of the London forces are necessary here. These forces are (1) of very short range; (2) additive and nonspecific; (3) temperature independent; (4) operative between molecules of all types, whether nonpolar, polar, or metallic; (5) dependent in magnitude upon the number and "looseness" of the electrons; and (6) ordinarily less than average between molecules of different species. This last property can be seen by comparing ϵ_{12} with ϵ_{11} and ϵ_{22} as given by Eqs. (1) and (2). Eliminating the α terms one obtains Eq. (4). Most component pairs differ much less in ionization potential

$$\epsilon_{12} = \frac{(\nu_1\nu_2)^{1/2}}{(\nu_1 + \nu_2)/2} \cdot (\epsilon_{11}\epsilon_{22})^{1/2} \quad (4)$$

than in polarizability, and the factor representing frequencies in this expression is not far from unity. Thus, relation (5) can be written. This means that the interaction potential between unlike

$$\epsilon_{12} \approx (\epsilon_{11}\epsilon_{22})^{1/2} \quad (5)$$

molecules is less than the arithmetic mean of the like potentials.

The pair potentials can be integrated over all the molecules in the pure components as well as the solution to obtain approximate attraction constants, a's, corresponding to the attraction constant of the van der Waals equation. M. P. E. Berthelot proposed the relation $a_{12} = (a_{11}a_{22})^{1/2}$ between the attraction constants of like and unlike species. J. Hildebrand and H. M. Carter found the Berthelot relation to be valid within 1% for seven liquid pairs, for example, $a = 31.21$ for liquid CCl_4 and 64.79 for $SnBr_4$. The calculated geometric mean is 46.46, and a_{12}, observed, is 46.86.

The consequence of this geometric mean relation is that the cohesion in a mixture of two liquids having different cohesion is less than their average. This results in expansion in volume, absorption of heat, and vapor pressures greater than additive upon mixing.

This geometric mean relation is usually adhered to rather well in cases of unlike molecules whose outer electrons are of similar types, such as (1) "N-electrons," nonbonded, as in halogenated paraffins and halogens; (2) π-electrons, as in olefins and aromatics; (3) bonding electrons only, as in H_2, CH_4, and other aliphatics; and (4) fluorochemicals. But deviations are found between molecules whose outer electrons are of different types. Illustrations will be found below in the sections on regular solutions and on solubility of gases.

Dipole interaction. This is the attraction between molecules containing permanent electric dipoles; it includes both the London forces and an electrostatic interaction of the dipoles. The latter depends upon the dipole moments of the molecules; it is temperature dependent because thermal agitation opposes the antiparallel orientation in which the interaction is greatest. Its magnitude depends also upon the geometry of the molecules because it is related to the distance of approach of the dipoles, not the molecular centers; the dipoles of some molecules are buried more deeply than those of others. This is the case with chloroform, which has solvent properties similar to those of carbon tetrachloride, except in a few specific cases. J. G. Kirkwood has expressed the degree of interaction between the dipoles of pure liquids by a g factor. For pyridine, the dipole moment μ is 2.20×10^{-18} cgs units, the dielectric constant ϵ is 12.5, and the dipole interaction g is 0.9. For water, $\mu = 1.84 \times 10^{18}$, $\epsilon = 78.5$, and $g = 2.7$; for ethyl alcohol, $\mu = 2.80 \times 10^{-18}$, $\epsilon = 24.6$, and $g = 3.0$.

It is the g factor, not the dipole moment or the dielectric constant, that is most significant for understanding solubility relations. Furthermore, in the case of molecules having more than one polar bond, it is the separate polar bonds, not their vector sum of the overall dipole moment, that determine solubility relations. The three isomeric dinitrobenzenes all affect the vapor pressure of benzene virtually to the same extent, even though their dipole moments are quite different.

The substances with the largest g-factors are those that form hydrogen bonds. These have exceptionally high boiling points and are poor solvents for nonpolar substances. These liquids resist penetration by nonpolar molecules. The best-known pairs of incompletely miscible liquids are composed of a nonpolar liquid and water.

Electron donor-acceptor interaction. In the modern theory of generalized acids and bases, initiated by G. N. Lewis, a base is a substance having electrons that may be "accepted" into the vacant orbitals of other molecules, termed acids. This acceptance of electrons takes place reversibly and with little or no activation energy. Typical bases or donors are pyridine, acetone, ether, alkyl bromides, alkyl iodides, alkyl sulfides, iodide ion, thiocyanate ion, and aromatic hydrocarbons. Typical acids are the pure and mixed halogens, sulfur dioxide and trioxide, boron trichloride and trifluoride, aluminum halides, and stannic chloride. R. S. Mulliken and his co-workers have pointed out the close relationship between base strength and ionization potential and elaborated a theory of charge transfer complexes. H. A. Benesi and Hildebrand discovered the strong absorption in the ultraviolet characteristic of such complexes. They found that the basic strength increases in the order benzene < toluene < xylene < mesitylene. R. L. Scott determined that acid strength increases in the order $Cl_2 < Br_2 < I_2 < BrI < BrCl < ICl$.

This type of interaction is specific and saturating, and it reduces the escaping tendencies of the components. It corresponds to $\epsilon_{12} > (\epsilon_{11}\epsilon_{22})^{1/2}$. In cases where it is weak it may reduce but not overcome the opposite effect of unequal London forces.

Ion-ion interaction. The ions in a solid or liquid salt attract and repel electrostatically according to Coulomb's law, but there is also a London force component, and large ions are polarized by smaller ones. This last effect is illustrated by solid silver bromide, which is colored although its ions in aqueous solutions are colorless and whose crystals have the sodium chloride structure. The evidence is that in both the solid and the fused salt the electron cloud of the bromide ion is distorted equally by each of its six neighboring silver ions. S*ee* F*used-salt phase equilibria*.

Ion-dipole interaction. In order to dissolve a solid salt, its lattice energy must be supplanted by the ion-ion action of another salt already in the liquid state or by the predominantly electrostatic attraction of a polar solvent or by the specific chemical interaction represented by complex ions.

Ideal solution. It is profitable to deal with actual solutions in terms of their departure from a simple idealized model—a mixture of components having the same attractive fields, which mix without change in volume or heat content. This is analogous to an ideal gas mixture, which is formed with no heat of mixing and in which the total pressure is the sum of the partial pressures. In such a solution the escaping tendency of the individual molecules is the same, whether they are surrounded by similar or by different molecules. Therefore, the combined escaping tendency of all the molecules of species 1, f_1, is given by Eq. (6), where x_1 is the mole fraction of

$$f_1 = f_1^0 x_1 \qquad (6)$$

species 1 and f_1^0 is the escaping tendency of the molecules from the pure liquid. For a binary mixture, $x_1 + x_2 = 1$. If the gas imperfections of vapors are disregarded, vapor pressures may be substituted for fugacities to give Raoult's law (1886) in its usual form, $p_1 = p_1^0 x_1$. The total pressure is $P = p_1 + p_2 = p_1^0 x_1 + p_2^0 x_2$.

A more sophisticated derivation than the foregoing requires one to postulate molecules of the same size and shape. The gross structure of the solution containing n_1 plus n_2 molecules of two components is identical with those of the pure liquid components. The number of configurations of the components in the mixture within this structure is $(n_1 + n_2)!/n_1!n_2!$, and the configurational entropy of mixing is, by aid by Stirling's formula, Eq. (7). The substitution of a number

$$\Delta S = k \left[n_1 \ln \frac{n_1 + n_2}{n_1} + n_2 \ln \frac{n_1 + n_2}{n_2} \right] \tag{7}$$

of moles of each, N_1 and N_2, gives Eq. (8). The partial derivative $(\partial \Delta S / \partial N_1)_{N_2}$ represents the partial

$$\Delta S = R \left[N_1 \ln \frac{N_1 + N_2}{N_1} + N_2 \ln \frac{N_1 + N_2}{N_2} \right] \tag{8}$$

molal entropy of transfer of component 1 from pure liquid into an ideal solution of mole fraction x_1, Eq. (9). Because the model postulates no change in enthalpy, Eq. (10) applies, and $f_1/f_1^0 = x_1$,

$$\overline{S}_1 - S_1^0 = -R \ln x_1 \tag{9} \qquad T(\overline{S}_1 - S_1^0) = -RT \ln (f_4/f_1^0) \tag{10}$$

which is Raoult's law, where f is fugacity. SEE FUGACITY.

But to arrive at this conclusion one must assume identical structures in the solution and the pure liquid components. This is very far from the case in solutions of high polymers in ordinary solvents, even though, as with polystyrene in benzene, the heat of mixing is practically zero.

Moderate difference in molal volume between components of high symmetry has little effect, as might be expected from the fact that the radius varies only with the cube root of the volume.

As a foundation for dealing with actual solutions in terms of the deviations of their properties from those of the model, it is necessary to derive other equivalent thermodynamic relationships.

Solubility of a crystalline solid. The fugacity of a crystalline substance f^s at temperature T is less than that of its supercooled liquid f_0 to an extent depending upon its melting point T_m and heat of fusion ΔH^F, as given by Eq. (11). If ΔH^F is assumed constant, this gives upon integration Eq. (12). If the heat capacities of the solid and liquid forms are known, the variation of

$$\frac{d \ln (f^s/f^0)}{dT} = \frac{\Delta H^F}{RT^2} \tag{11} \qquad \ln \frac{f^s}{f^0} = -\frac{\Delta H^F}{R} \left(\frac{1}{T} - \frac{1}{T_m} \right) \tag{12}$$

ΔH^F with temperature can be taken into account. This is hardly necessary for the present purpose because the deviations from ideal solubility involve factors that are more uncertain than this.

Molecular weight measurements. If a solid dissolves to form an ideal solution, its heat of solution is the same as its heat of fusion ΔH^F and $f_1^s/f_1^0 = x_1$. Therefore, Eq. (13) is formed. This

$$-\ln x_1 = \frac{\Delta H^F}{R} \left(\frac{T_m - T}{T_m T} \right) \tag{13}$$

is the approximate equation for solubility of a solid that forms an ideal solution. It can be transformed into one much used for determining the molal weight of a solute by the depression of the freezing point of the solvent, here component 1. For $-\ln x_1$, one can write $\ln (1 + N_2/N_1)$. Expanding in powers of N_2/N_1 gives Eq. (14). When $N_2 \ll N_1$, the higher powers may be neglected

$$\ln \left(1 + \frac{N_2}{N_1} \right) = \frac{N_2}{N_1} \left[1 - \frac{1}{2} \frac{N_2}{N_1} + \frac{1}{3} \left(\frac{N_2}{N_1} \right)^2 - \cdots \right] \tag{14}$$

to give Eq. (15), where $\Delta T = T_m - T$. By measuring ΔT for a known weight of solute in N_1 moles

$$\frac{N_2}{N_1} \approx \frac{\Delta H_1^F}{RT_m^2} \Delta T \tag{15}$$

of solvent, the molal weight of the solute can be calculated. Because of the approximations made and the fact that even good solvents for the solid in question are seldom ideal, the resulting molal weights are not very exact unless extrapolated to $x_2 = 0$ from a series of values.

The lowering of the vapor pressure of a solvent upon the addition of a nonvolatile solute (component 2) may be offset by raising the temperature to restore the pressure of the solvent. These changes are related as in Eq. (16). This relation is far less useful than that for the freezing

$$x_2 = \frac{\Delta H_1^{vap}}{RT_b^2} \Delta T \tag{16}$$

point depression because the heat of vaporization is much greater than the heat of fusion; therefore, the elevation of the boiling point is much smaller than the depression of the freezing point and also is harder to determine.

Osmotic pressure. One mole of a solvent can be removed from a large quantity of a solution in which its mole fraction is x_1 in two reversible, and hence equivalent, ways. If it is distilled from the solution into pure liquid, the gain in (Gibbs) free energy is $\Delta G_1 = RT \ln (f_1^0/f_1)$. If it is pressed out through a semipermeable membrane against the hydrostatic pressure difference, osmotic pressure ΔP, the gain in free energy is $\Delta P \overline{V}_1$, where $\overline{V}_1$ is the partial molal volume of the solvent. In an ideal solution this is the molal volume. Equating the free energy of these processes gives Eq. (17). Expanding as before and neglecting the higher powers gives $P\overline{V}_1 \approx$

$$\Delta P \overline{V}_1 = RT \ln \frac{f_1^0}{f_1} = RT \ln \frac{N_1 + N_2}{N_1} \tag{17}$$

$(N_2/N_1)RT$ or $PV = RT$, where $V = N_2\overline{V}_1/N_1$, the volume containing 1 mole of solute.

This is the van't Hoff law for osmotic pressure, put forth in 1887. The theoretical basis of Raoult's law, discovered at almost the same time, was not yet appreciated. The formal correspondence between the van't Hoff law and the perfect gas law seemed to lend unique significance to osmotic pressure and elevated the van't Hoff law to the status of an ideal solution law. It is a limiting law only and not valid at high concentrations; it neglects the specific nature of the solvent. The solvent is regarded as furnishing space for a quasi-gas solute. Thus the law cannot cover molecular states in solutions of finite concentration.

The determination of osmotic pressure offers a valuable means for determining molal weights of high polymers in solution, where high weight concentrations correspond to mole fractions so low as to have only minute effects upon the vapor pressure and the freezing point of the solvent. For example, consider a solution of 0.001 mole of solute in 1 mole of benzene; $T_m = 279$ K and $\Delta H_1^F = 2370$ cal/mole, ΔT by Eq. (15) would be only 0.065°, but ΔP by Eq. (17) would be 194 mm. The latter is large enough to be measured with some precision.

Nonideal solutions. Unlikeness of the components of a binary mixture leads, as explained earlier, to fugacities in excess of ideal values. The excess is largest when the molecules of one species are surrounded mostly by those of the other, as shown in the **illustration**. They

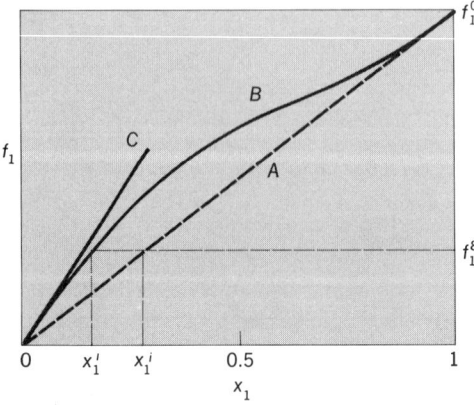

Fugacity and mole fraction. Line A, ideal. Line B, typical nonideal. Line C, Henry's law.

approach Raoult's law at the upper end and Henry's law, $p_1 = kx_1$ (or $f_1 = k_1 x_1$), at the lower end, where k is an experimentally determined constant.

An important relation between the two components is given by the Gibbs-Duhem equation, Eq. (18). If Raoult's law holds in the limit when $x_1 = 1$, since $(\partial \ln f_1)/(\partial \ln x_1) = 1$, then also

$$\left(\frac{\partial \ln f_1}{\partial \ln x_1}\right)_T = \left(\frac{\partial \ln f_2}{\partial \ln x_2}\right)_T \tag{18}$$

$(\partial \ln f_2)/(\partial \ln x_2) = 1$. Integrating gives $\ln f_2 = \ln x_2 + \ln k_2$ or $f_2 = kx_2$, where k is a constant of integration that cannot be evaluated unless Raoult's law holds for component 1 over the whole range. In that case it also holds for component 2.

The activity in the case of nonelectrolytes is defined as $a_1 = f_1/f_1^0$ and so on. In an ideal solution, $a_1 = x_1$ and $a_2 = x_2$. The activity coefficient is $\gamma_1 = a_1/x_1$ and so on. Alternate, equivalent forms of the Gibbs-Duhem equation include Eq. (19), and $N_1 d\bar{G}_1 + N_2 d\bar{G}_2 = 0$, where $\bar{F}$

$$\frac{\partial \ln a_1}{\partial \ln x_1} = \frac{\partial \ln a_2}{\partial \ln x_2} \tag{19}$$

denotes partial molal Gibbs free energies.

If one component is a crystalling solid, its activity a_s is less than that of the liquid, which is 1, as given by Eq. (12); its maximum solubilities would be x_1' in the illustration, if in an ideal solvent, and x_1', if in a real solution.

Regular solutions. There are many mixtures of nonpolar components in which, except in the immediate neighborhood of a critical mixing temperature, thermal agitation suffices to neutralize tendencies to segregate and yields virtually complete randomness of mixing, with a close approach to ideal entropy of mixing, Eqs. (8) and (9).

The enthalpy of mixing can be calculated as the difference between the potential energy of the mixture and the sum of the potential energies of the liquid components. The potential energy of a mole of liquid may be related to the potential between a pair of molecules $\epsilon(r)$. The lattice energy of a crystal is obtained by summation of $\epsilon(r)$ over all of the lattice distances; that of a liquid is obtained by integration over the continuous distribution function $\rho(r)$. The expression for a pure liquid is Eq. (20). Here N_{Av} is the Avogadro number and V is the molal volume of the

$$\Delta E_{\text{vap}} = -\frac{2\pi N_{Av}^2}{V} \int \rho(r)\epsilon(r) r^2 dr \tag{20}$$

liquid. The corresponding expression for the potential of the mixture of N_1 and N_2 moles of the pure components involves their relative sizes. With certain simplifying assumptions, including the geometric mean for $\epsilon_{12}(r)$, Eq. (21a) was obtained for the energy of mixing N_1 and N_2 moles of

$$\Delta E_M = \frac{N_1 V_1 N_2 V_2}{N_1 V_1 + N_2 V_2}\left[\left(\frac{\Delta E_1}{V_1}\right)^{1/2} - \left(\frac{\Delta E_2}{V_2}\right)^{1/2}\right]^2 \tag{21a} \qquad \bar{E} - E_2^0 = V_2 \phi_1^2 (\delta_1 - \delta_2)^2 \tag{21b}$$

two nonpolar liquids. The corresponding partial molal energy of transferring pure liquid to solution, for component 2, is Eq. (21b). Here ϕ_1 denotes volume fraction, neglecting expansion on mixing, and the δ's are $(\Delta E_{\text{vap}}/V)^{1/2}$, designated solubility parameters. Because energy and enthalpy are virtually identical for liquids, Eq. (21b) may be combined with the entropy of transfer as given by Eq. (9) to give the free energy of transfer, as in Eqs. (22a) and (22b).

$$\bar{G}_2 - G_2^0 = \bar{H}_2 - H_2^0 - T(\bar{S}_2 - S_2^0) \tag{22a}$$

$$RT \ln a_2 = V_2 \phi_1^2(\delta_1 - \delta_2)^2 + RT \ln x_2 \quad \text{or} \quad RT \ln \gamma_2 = V_2 \phi_1^2(\delta_1 - \delta_2)^2 \tag{22b}$$

The quantity in square brackets in Eq. (21a) is the excess of the arithmetic mean, $\frac{1}{2}[\Delta E/V_1) + (\Delta E_2/V_2)]$, over the geometric mean, $[(\Delta E_1/V_1)(\Delta E_2/V_2)]^{1/2}$. As mentioned earlier, the geometric mean assumption is amazingly valid for many species of similar electronic structure but fails for others.

Representative values of solubility parameters at 77°F (25°C) are given in the **table**. Parameters for substances solid at 77°F (25°C) have been calculated by Eq. (14a) from solubilities of these substances in solvents whose parameters are well determined. With paraffins the solubility data give concordant δ-values a little greater than $(\Delta E_{\text{vap}}/V)^{1/2}$.

Solubility parameters and molal volumes, 77°F (25°C)			
Liquid	Formula	Molal volume, ml	Solubility parameter
Perfluoroheptane	C_7F_{16}	225	5.9
Perfluorotributylamine	$(C_4F_9)_3N$	360	6.0
Perfluoromethylcyclohexane	$c\text{-}C_6F_{11}CF_3$	196	6.1
n-Heptane	$n\text{-}C_7H_{16}$	147	7.4
Silicon tetrachloride	$SiCl_4$	115	7.6
Cyclohexane	C_4H_{12}	109	8.2
Carbon tetrachloride	CCl_4	97	8.6
Chloroform	$CHCl_3$	81	9.3
Benzene	C_6H_6	89	9.2
Carbon disulfide	CS_2	60	10.0
Bromine	Br_2	52	11.5
Iodine	I_2	59	14.1

Bibliography. *Annual Review of Physical Chemistry*, annually; J. H. Hildebrand, J. M. Prausnitz, and R. L. Scott, *Regular and Related Solutions: Solubility of Solids, Liquids and Gases*, 1970; J. N. Murrell and E. A. Boucher, *Properties of Liquids and Solutions*, 1982.

SOLID SOLUTION
PAUL B. MOORE

A crystalline state in which at least one atomic position may accommodate more than one atomic species. It is easiest to conceptualize the phenomenon by defining structure type, mineral species, and end-member composition. The structure type is a design of distinct-symmetry independent atomic positions possessing an asymmetric unit of atomic positions which must be specified along with the unit cell properties, including axial parameters, crystal system, and space group. The greater effort in crystal structure analysis is establishing unit cell properties and the loci of the atoms in the asymmetric unit, some of whose parameters may have up to three degrees of freedom. Mineral species is defined on the basis of structure type and end-member composition, where each end member in principle is a distinct species.

In the prevalent olivine group of minerals, the two important end members are fayalite, $Fe_2^{2+}(SiO_4)$, and forsterite, $Mg_2^{2+}(SiO_4)$. Both belong to the same space group and structure type. The general formula for the silicate olivines can be written $M_2^{2+}(SiO_4)$, where M stands for the divalent cations octahedrally coordinated by the oxide anions. When Fe^{2+} and Mg^{2+} can mix at any ratio over these M sites, a perfect solid solution exists. Another example is the series calcite-magnesite, $CaCO_3$–$MgCO_3$. Both end members belong to the same structure type, and with increasing temperature, mixing of Ca and Mg increases over the large cation site. At some temperature, mixing of the two cationic species can be complete, as in the olivines. Below this temperature the Ca^{2+} and Mg^{2+} cations segregate within the crystal and can form a stable ordered species, dolomite, $CaMg(CO_3)_2$, plus an exsolved phase such as calcite.

Ordered arrangements. If an ordered arrangement appears, one of these phenomena may take place. The space group may remain unchanged if the crystal symmetry is not violated through ordering; the space group may be reduced to a subgroup due to the partitioning of two cations into separate positions, which destroys some of the symmetry elements of the space group; the segregation may lead to an unmixing of the structure into two separate phases which usually bear a crystallographic or epitaxial relation with each other. The calcite-magnesite series is an excellent example. Both end members possess the space group $R\bar{3}c$; dolomite possesses the subgroup $R\bar{3}$. Samples can be found with calcite exsolved from a magnesian calcite, a dolomite, or a calcian magnesite. Formerly at crystallization temperature, both Ca^{2+} and Mg^{2+} were completely mixed at some higher temperature. The composition of the magnesian calcite and the exsolved calcite can be used as a geothermometer to estimate the initial temperature of crystallization in the system $CaO-MgO-CO_2$.

Omission solid solution. This type of solid solution can also exist. In the hypothetical composition $\Box X_3(TO_4)_2$, $\Box$ is a symbol for a vacancy or an empty atomic position in the structure. Imagine the series $\Box X_3(PO_4)_2-XX_3(PO_4)_2[=X_4(PO_4)_2]$. The series $\Box-X$ is also possible, and frequently structures and compositions are found where the filling of the $\Box$ may be only partial; like the join calcite-magnesite, the fraction of occupancy is temperature-dependent. Such models can be also used as geothermometers, providing the thermochemistry of the system is known in detail. One remarkable example is triphylite-sarcopside, $LiFe(PO_4)-Fe_3(PO_4)_2$. Large crystals of triphylite can be found with exsolved sarcopside. Sarcopside possesses a subgroup of the triphylite space group. Triphylite is an ordered olivine structure where Li^+ and Fe^{2+} are segregated into two independent atomic positions in the crystal. It is believed that there is exsolution from the series $Li_2Fe_2(PO_4)_2\Box FeFe_2(PO_4)_2$, where at some higher temperature the crystal was completely mixed, that is, $(Li,Fe,\Box)_4(PO_4)_2$. Charge balance restricts the admissible ratios for $\Box^0:Li^+:Fe^{2+}$. *See* Thermochemistry.

Occurrence. In mineralogy and metallurgy, solid solution is a frequent and important phenomenon. In metallurgy, examples are the systems Au-Ag-Hg and Fe-Ni. In mineralogy, the most important examples are the systems $CaO-MgO-CO_2$, $Na_2O-CaO-Al_2O_3-SiO_2$, and $MgO-Al_2O_3-SiO_2$. The feldspar series anorthite-albite, $CaAl_2Si_2O_8-NaAlSi_3O_8$, is renowned for a great range of unmixing phenomena at subsolidus temperatures and an astonishing diversity in group-subgroup relations in their crystal structures, even though the framework topology of their structures remains unchanged.

Solid solution is a statistical phenomenon, and implies only that the sites which accommodate distinct ionic species are energetically similar. Since x-ray diffraction experiments (the most important tool) essentially average the atomic populations over their sites according to the space group extinctions, and since about 10^{15} unit cells are involved in the experiment, it is not implied that atoms hop from site to site.

Some general observations have been made about those cations and anions which can form solid solution. Most important are similarity in their ionic radii, and similarity in their electronic structure. Important examples of ions which can form solid solutions are $Fe^{2+}-Mg^{2+}$, $Fe^{2+}-Mn^{2+}$, $Mn^{2+}-Ca^{2+}$, $(OH)^--F^-$, $Mg^{2+}-Al^{3+}$, and $Al^{3+}-Si^{4+}$. Coordination numbers must be the same in these cases.

Many admired gemstones are results of solid solution. Pure corundum, $\alpha-Al_2O_3$, is colorless. If doped with small amounts of Cr^{3+}, it is red (ruby), or with $Fe^{2+}Ti^{4+}$, it is blue (sapphire). Tourmaline when pure is $NaMg_3Al_6(BO_3)_3(Si_6O_{18})(OH)_4$. Since no chromophores are present, it is colorless. But small amounts of Fe^{3+} color it green (elbaite); $Fe^{2+}Fe^{3+}$, blue (indicolite); and Mn^{3+}, pink to red (rubellite). Typical chromophores are ions of the first transition series metals. Another well-known example is the soft, platy mineral vivianite, $Fe_3^{2+}(H_2O)_8(PO_4)_2$. When pure it is pale green, but when exposed to air some Fe^{2+} is oxidized to Fe^{3+}, leading to a blue vivianite. Extensive oxidation renders it nearly opaque and black, due to homonuclear valence transfer absorption, $Fe^{2+} + Fe^{3+} \rightleftharpoons Fe^{3+} + Fe^{2+}$. In such a phenomenon, an increase of mixed valences over an equivalent (or structurally adjacent) set of sites increases the absorption of light and the darkness of the color. In these mixed-valence phenomena it is believed that the ions do not jump from one site to another, but that electrons do. In such relations the cations could be written $Fe_x^{3+}Fe_{1-x}^{2+}$, and values should be ascribed to x. Calorimetric heats of mixing have been obtained on important geochemical solid solutions such as the feldspars, pyroxenes, and garnets.

CRYSTALLIZATION
William R. Wilcox

The formation of a solid from a solution, melt, vapor, or a different solid phase. Crystallization from solution is an important industrial operation because of the large number of materials marketed as crystalline particles. Fractional crystallization is one of the most widely used methods of separating and purifying chemicals. This article discusses crystallization of substances from solutions and melts. Not discussed is biological crystallization, involving formation of teeth and bone, otoconia in the inner ear, renal calculi (kidney stones), biliary calculi (gallstones), crystals in some forms of arthritis, and dental plaque. For solubility and other relationships between solid and liquid phases SEE PHASE EQUILIBRIUM; SOLUTION.

Solutions. In order for crystals to nucleate and grow, the solution must be supersaturated; that is, the solute must be present in solution at a concentration above its solubility. Different methods may be used for creating a supersaturated solution from one which is initially undersaturated. The possible methods depend on how the solubility varies with temperature. Two examples of solubility behavior are shown in the **illustration**. Either evaporation of water or cooling may be used to crystallize potassium nitrate (KNO_3), while only evaporation would be effective for NaCl. An alternative is to add a solvent such as ethanol which greatly lowers the solubility of the salt, or to add a reactant which produces an insoluble product. This causes a rapid crystallization perhaps more properly known as precipitation.

Nucleation. The formation of new crystals is called nucleation. At the extremely high supersaturations produced by addition of a reactant or a lower-solubility solvent, this nucleation may take place in the bulk of the solution in the absence of any solid surface. This is known as homogeneous nucleation. At more moderate supersaturations new crystals form on solid particles or surfaces already present in the solution (dust, motes, nucleation catalysts, and so on). This is called heterogeneous nucleation. When solutions are well agitated, nucleation is primarily secondary, that is, from crystals already present. Probably this is usually due to minute pieces breaking off the crystals by impact with other crystals, with the impeller, or with the walls of the vessel.

Crystal size. Generally, large crystals are considered more desirable than small crystals, probably in the belief that they are more pure. However, crystals sometimes trap (occlude) more solvent as inclusions when they grow larger. Thus there may be an optimal size. The size distribution of crystals is influenced primarily by the supersaturation, the amount of agitation, and the growth time. Generally, the nucleation rate increases faster with increasing supersaturation than does the growth rate. Thus, lower supersaturations, gentle stirring, and long times usually favor large crystals. Low supersaturations require slow evaporation and cooling rates.

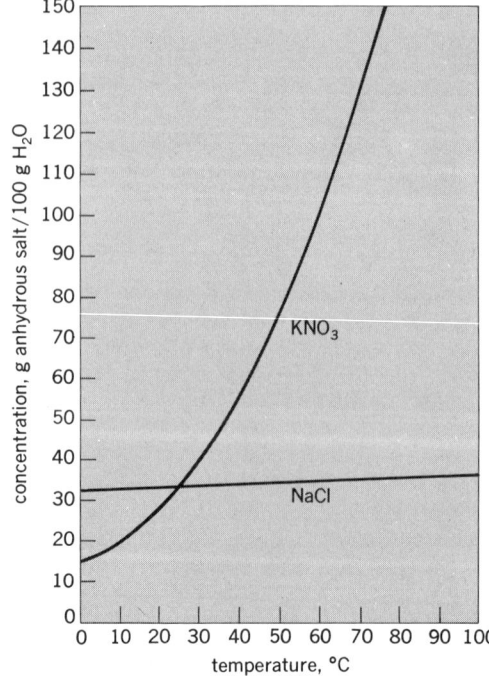

Temperature-solubility curves for two salts. °F = (°C × 1.8) + 32.

Crystal habit. Often the habit (shape) of crystals is also an important commercial characteristic. In growth from solutions, crystals usually display facets along well-defined crystallographic planes, determined by growth kinetics rather than by equilibrium considerations. The slowest-growing facets are the ones that survive and are seen. While habit depends somewhat on the supersaturation during growth, very dramatic changes in habit are usually brought about by additives to the solution. Strong habit modifiers are usually incorporated into the crystal, sometimes preferentially. Thus the additive may finally be an impurity in the crystals.

Fractional crystallization. In fractional crystallization it is desired to separate several solutes present in the same solution. This is generally done by picking crystallization temperatures and solvents such that only one solute is supersaturated and crystallizes out. By changing conditions, other solutes may be crystallized subsequently. Occasionally the solution may be supersaturated with respect to more than one solute, and yet only one may crystallize out because the others do not nucleate. Preferential nucleation inhibitors for the other solutes or seeding with crystals of the desired solute may be helpful. Since solid solubilities are frequently very small, it is often possible to achieve almost complete separation in one step. But for optimal separation it is necessary to remove the impure mother liquor from the crystals. Rinsing of solution trapped between crystals is more effective for large-faceted crystals. Internally occluded solution cannot be removed by rinsing. Even high temperatures may not burst these inclusions from the crystal. Repeated crystallizations are necessary to achieve desired purities when many inclusions are present or when the solid solubility of other solutes is significant.

Melts. If a solid is melted without adding a solvent, it is called a melt, even though it may be a mixture of many substances. That is the only real distinction between a melt and a solution. Crystallization of a melt is often called solidification, particularly if the process is controlled by heat transfer so as to produce a relatively sharp boundary between the solid and the melt. It is then possible to slowly solidify the melt and bring about a separation, as indicated by the phase diagram. The melt may be stirred to enhance the separation. The resulting solid is usually cut into sections, and the purest portion is subjected to additional fractional solidifications. Alternatively the melt may be poured off after part of it has solidified. Zone melting was invented to permit multiple fractional solidifications without the necessity for handling between each step. Fractionation by the above techniques appears to be limited to purification of small batches of materials already fairly pure, say above 95%.

Fractional crystallization from the melt is also being used for large-scale commercial separation and purification of organic chemicals. It has also been tested for desalination of water. Rather than imposing a sharp temperature gradient, as in the above processes, the melt is relatively isothermal. Small, discrete crystals are formed and forced to move countercurrent to the melt. At the end from which the crystals exit, all or part of the crystals are melted. This melt then flows countercurrent to the crystals, thereby rinsing them of the adhering mother liquor.

Bibliography. S. J. Jancic and P. A. Grootscholten, *Industrial Crystallization*, 1984; J. Nývlt, *Industrial Crystallization: The Present State of the Art*, 2d ed., 1983; A. D. Randolph and M. A. Larson, *Theory of Particulate Processes: Analysis and Techniques of Continuous Crystallization*, 1971.

AVOGADRO'S NUMBER
Ira N. Levine

The number of elementary entities in one mole of a substance. A mole is defined as an amount of a substance that contains as many elementary entities as there are atoms in exactly 12 g of ^{12}C; the elementary entities must be specified and may be atoms, molecules, ions, electrons, other particles, or specified groups of such particles. Experiments give 6.02×10^{23} as the value of Avogadro's number. Thus, a mole of ^{12}C atoms has 6.02×10^{23} carbon atoms, a mole of water molecules contains 6.02×10^{23} H$_2$O molecules, a mole of electrons contains 6.02×10^{23} electrons, and so forth.

Significance. The atomic weight (relative atomic mass) of ^{12}C is exactly 12, by definition. Consider 12 g of ^{12}C (which is one mole and contains Avogadro's number of atoms) compared with 4 g of He, whose atomic weight is 4. The 12 g to 4 g ratio of the masses of the two samples is the same as the 12 to 4 ratio of the masses of the atoms of ^{12}C and He. Therefore the two samples must contain the same number of atoms, and 4 g of He contains Avogadro's number of atoms. The same argument holds for any element. Thus, for an element with atomic weight x, a sample with mass x grams contains Avogadro's number of atoms. Similarly, for a substance with molecular weight y, a sample whose mass is y grams must contain Avogadro's number of molecules. For example, 18 g of water contains 6.02×10^{23} H$_2$O molecules. SEE ATOMIC WEIGHT; RELATIVE ATOMIC MASS.

Avogadro's number is a dimensionless number. The Avogadro constant is defined as Avogadro's number divided by the unit "mole." The Avogadro constant is usually symbolized by N_A, N_0, or L. Since N_A gives the number of molecules per mole, $N_A = N/n$, where N is the number of molecules present in n moles of a substance.

Avogadro's number relates the mass of a mole of a substance to the mass of a single molecule. For example, for H$_2$O (whose molecular weight is 18) the mass of one mole is 18 g and the mass of one molecule is $(18 \text{ g})/(6.02 \times 10^{23}) \approx 3 \times 10^{-23}$ g. The mass m of one molecule of a substance with molar mass M is $m = M/N_A$.

The Avogadro constant N_A is related to other fundamental physical constants. The Faraday constant F is the absolute value of the charge on one mole of electrons. Therefore $F = N_A e$, where e is the absolute value of the charge on one electron. Also, $R = N_A k$, where R is the gas constant and k is Boltzmann's constant. SEE GAS CONSTANT.

Widespread use of the mole concept began only around 1900. The nineteenth-century concept most closely related to Avogadro's number is the number of molecules per unit volume in a gas at 0°C and 1 atm. [The ideal-gas law $PV = nRT = (N/N_A)RT$ gives $N/V = N_A P/RT$, so N/V, the number of gas molecules per unit volume, is proportional to the Avogadro constant N_A at fixed pressure P and temperature T.] Avogadro hypothesized in 1811 that at a fixed temperature and pressure the number of molecules per unit volume is the same for different gases, but he had no way of estimating this number.

Determination. The first estimate of the number of molecules per unit volume in a gas was given by J. Loschmidt in 1867. Using a kinetic-theory-of-gases equation for the gas viscosity and estimating NV_{molecule} (where V_{molecule} is the volume of a single molecule) from the volume of the liquid formed by condensing the gas, Loschmidt gave a value of N/V which is one-thirtieth the correct value.

The number of molecules per cubic centimeter in an ideal gas at 0°C and 1 atm is called Loschmidt's number in English-speaking countries. (However, in Germany, Loschmidt's number means the number of molecules per mole.)

In 1900 M. Planck, using the hypothesis of energy quantization, derived the law for blackbody radiation. Planck's law contains Boltzmann's constant k and Planck's constant h. By fitting experimental data to his law, Planck found values for h and k, and from k (which equals R/N_A) he calculated Avogadro's number as 6.18×10^{23}. Using $F = N_A e$, he calculated $e = 4.69 \times 10^{-10}$ statC. Thus Planck had obtained the first reasonably accurate values of h, k, N_A, and e.

In 1906 A. Einstein used his molecular theory of diffusion and data for aqueous sugar solutions to estimate $N_A = 4 \times 10^{23}$ mol^{-1}; Einstein subsequently found an error in his work and recalculated this estimate as 6.56×10^{23} in 1911.

J. B. Perrin in 1908 observed the distribution in the Earth's gravitational field of colloidal gamboge particles of uniform size. According to the Boltzmann distribution law, the ratio n_2/n_1 of numbers of particles at heights h_2 and h_1 in the gravitational field is given by Eq. (1).

$$n_2/n_1 = \exp[-(E_2 - E_1)/kT] \tag{1}$$

The potential energies of the particles are $E_1 = mgh_1$ and $E_2 = mgh_2$, where m is the mass of a particle (corrected to allow for the buoyancy of the suspending fluid) and g is the gravitational acceleration. Observation of n_2/n_1 allows calculation of Boltzmann's constant k, and use of $R = N_A k$ then gives N_A. Perrin found $N_A = 7.0 \times 10^{23}$ mol^{-1}.

From 1908 to 1917, R. A. Millikan and H. Fletcher determined the electron charge by observing the motions of charged oil drops. Using their final result for e and the value of the Faraday constant known from electrolysis experiments, Millikan found $N_A = F/e = 6.06 \times 10^{23}$ mol^{-1} in 1917.

A very accurate way to find N_A uses x-rays to measure the lattice spacing in a crystal. The crystal's density ρ equals the mass of a unit cell divided by the volume of a unit cell: $\rho = m_{cell}/V_{cell}$. The mass of one formula unit is M/N_A, where M is the molar mass; hence $m_{cell} = ZM/N_A$, where Z is the number of formula units per unit cell. If the unit cell is cubic, then $V_{cell} = a^3$, where a is the unit-cell edge length. The crystal density is then given by Eq. (2). Z is readily found from x-ray diffraction study of the crystal.

$$\rho = MZ/a^3 N_A \qquad (2)$$

In the period 1930–1970, several determinations of N_A were done by using Eq. (2) with a determined from the diffraction pattern produced by x-rays of known wavelength. The wavelength was determined by diffracting the x-rays with a grating having closely ruled lines of known spacing.

The accuracy of the x-ray crystal-density method was improved substantially by R. Deslattes and coworkers in 1974 by using a combination of x-ray and optical interferometry to determine the lattice spacing in a very pure silicon (Si) crystal. These workers also did very accurate measurements of the density and atomic weight of silicon. P. Becker and coworkers in 1981 repeated the interferometry measurement of the Si lattice spacing and obtained a slightly different result from that of Deslattes. When the Becker determination of a is combined with the Deslattes group's determination of M and ρ, the value $N_A = 6.02213 \times 10^{23}$ mol^{-1} is found.

Bibliography. R. D. Deslattes, The Avogadro constant, *Annu. Rev. Phys. Chem.*, 31, 435–461, 1980; D. Kolb, The mole, *J. Chem. Educ.*, 55, 728–732, 1978.

GAS CONSTANT
Thomas C. Waddington

Boyle's law and Charles' law may be combined into a single expression, pV/T = a constant, showing how the volume V of a given mass of gas depends upon its temperature T and the pressure p. If the mass of gas chosen is 1 g-mole, then the constant, known as the gas constant, is written as R. Hence, $pV = RT$ for 1 g-mole. *See Gas*.

The numerical value of the gas constant R is obtained by dividing the molar or gram-molecular volume of a perfect gas at the melting point of ice and a pressure of 760 mm of mercury (101,325 pascals) at the same temperature (the standard atmosphere) by the absolute temperature at the ice point, $R = V_0/T_0$. The best values available for R, in various units, are given below.

$$R = 8.31441 \pm 0.00026 \text{ joules/(kelvin)(mole)}$$
$$= 1.98719 \pm 0.00006 \text{ cal/(kelvin)(mole)}$$
$$= 8.20568 \pm 0.00026 \times 10^{-2} \text{ liter-atm/(kelvin)(mole)}$$

The kinetic theory of gases relates R to the specific heats of a perfect monatomic gas at constant pressure C_p and constant volume C_v. C_p is equal to $5/2R$ and C_v to $3/2R$. Maxwell's law of the equipartition of energy shows that any degree of freedom of a system possesses an energy of $1/2RT$ per mole, making a specific heat contribution of $1/2R$. A perfect monatomic gas has three degrees of freedom, three independent directions of molecular motion, and a specific heat C_v of $3/2R$. In a perfect diatomic gas molecules rotate in two independent directions; their specific heat C_v is $5/2R$.

MASS
Leo Nedelsky

The quantitative or numerical measure of a body's inertia, that is, of its resistance to being accelerated.

Before this rather abstract definition is developed, it is useful to consider a description of mass that, although less general, is more easily grasped intuitively. Isaac Newton said that the mass of a body is the measure of the quantity of matter the body contains. This description is useful for comparing the masses of samples of a particular type of matter, say, sugar, because it correctly suggests that the mass of a sample is the basic factor in determining the extent to which the sample possesses the fundamental unchangeable properties peculiar to that type of matter. Thus a given mass of sugar, that is, a quantity of sugar that exhibits a given measure of inertia, contains a definite number of molecules and will therefore sweeten a definite number of cups of coffee or supply a definite number of calories if it is burned or eaten. Twice the mass of sugar will contain twice as many molecules.

These properties are inherent or inalienable. Their extent can be changed only by adding more sugar to the sample or taking some away, that is, only by changing mass; it will not be changed by taking the sample of sugar into a spaceship or to the Moon. Other quantities associated with a given sample of sugar are nonpermanent. Thus its volume can be changed by pressure and its weight changed by taking it to a different altitude. Conversely, equal volumes of sugar may contain different masses and thus different numbers of molecules. In contrast to volume or weight of a body, mass is a measure of something that is fundamental and permanent. (However, see the subsequent discussion of mass and energy.)

Gravitation and inertia. The preceding discussion shows that the mass of a body could be defined in terms of one of the properties of the material of the body, for example, in the case of sugar, in terms of the amount of energy released when the sugar is burned in a specified way. Because, however, it is often necessary to compare masses of such dissimilar bodies as a sample of sugar, a sample of air, an electron, and the Moon, it is necessary to define mass in terms of a property that not only is inherent and permanent but is also universal in that it is possessed by all known forms of matter. All matter possesses two properties, gravitation and inertia. The property of gravitation is that every material body attracts every other material body. The property of inertia is that every material body resists any attempt to change its motion. A body's motion is said to change if the body is accelerated, that is, if it increases or decreases its speed or changes the direction of its motion. Because of its inertia a body cannot be accelerated unless a force is exerted on it. The greater the inertia of a body, the less will be the acceleration produced by a given force.

As indicated in the beginning of this article, the present definition of mass is in terms of inertia. The masses of two bodies are compared by applying equal forces to the bodies and measuring their accelerations. For example, the two bodies may be allowed to collide. According to Newton's third law; each body will then experience an equally strong force. If there are no external forces, and if a_1 and a_2 are the measured accelerations of the two bodies, the ratio of the masses of the two bodies is by definition given by Eq. (1).

$$\frac{m_1}{m_2} = \frac{a_2}{a_1} \tag{1}$$

This equation gives only ratios of masses; it is therefore necessary to designate the mass of some one body as the standard mass to which the masses of all other bodies can be compared. The body that has been chosen for this purpose is a cylinder of platinum-iridium alloy. It is known as the international standard of mass; its mass is called 1 kilogram (kg), and it is kept at the International Bureau of Weights and Measures near Paris, France. Replicas of the standard mass, kept at various national laboratories, are periodically compared with this standard.

It would be possible to define mass in terms of the property of gravitation. This gravitational mass of a body—to distinguish it from the inertial mass already defined—could be determined from the force of attraction exerted by the body on some standard body at a specified distance. It could also be determined (by Newton's third law, which must give the same result) from the force of attraction exerted on the body by the standard body. For example, to compare the gravitational masses of two bodies, the forces with which the Earth attracts them at any one point on the Earth's surface could be compared. The force of the Earth's attraction on a body is called the weight of the body. Thus the gravitational masses of bodies can be compared by simply weighing the bodies.

The ease and precision of weighing contrasts strongly and favorably with the difficult and necessarily imprecise experimental determination of the inertial mass of a body that requires measurement of acceleration. It is therefore extremely fortunate that R. von Eötvös in the nineteenth century proved experimentally with great precision that the inertial masses of such bodies as could be tested in the laboratory are equal to their gravitational masses. Astronomical observations support this finding. Finally, Albert Einstein, in his general theory of relativity, presented strong theoretical arguments that the two definitions of mass—in terms of inertia and in terms of gravitation—are identical.

At present, mass is defined in terms of inertia but is measured by weighing. The result is a combination of the more useful of two definitions and of the simpler of two experimental determinations. The relation between the weight W and the mass m of a body is very simple. In a locality where the acceleration of gravity is g, $W = mg$ (in appropriate units).

Other properties. Besides inertia and gravitational attraction, mass has two other properties that point up the genius of Newton in abstracting, from the infinity of possible observations and descriptions, a concept that is as simple as it is basic. The first of these properties is that mass is linearly additive; for example, the total mass of a 1-kg body and a 3-kg body is 4 kg. (For exceptions, see the subsequent discussion of mass and energy.) As a consequence of this experimental fact, it is easy to define masses that are multiples or fractions of the standard kilogram mass. Thus 1 gram is defined as a mass, a thousand of which add up to 1 kg; 1 pound, as a mass of 0.45359237 kg. This awkward conversion factor is used in order to make the current definition of a 1-lb mass agree closely with the old definition, which was made in terms of a standard pound, given by the mass of a block of platinum. Another useful unit of mass is the slug, which is defined as the mass of a body whose inertia is such that a force of 1 lb, which is the weight of a 1-lb mass at a certain specified locality, gives the body an acceleration of 1 ft/s^2. One slug is equal to 32.174 lb (14.59 kg).

The second property of mass is that it is conserved; that is, it can neither be created nor be destroyed; a loss of mass by a system of bodies is always accompanied by an equal gain by some other system.

Mass and energy. Einstein's special theory of relativity predicts that the inertia of a body should increase if the energy of the body increases. This prediction has been conclusively verified experimentally. It follows that the mass of a body will increase if, for example, the body gains speed (addition of kinetic energy), or its temperature rises (addition of heat energy), or the body is compressed (addition of elastic energy).

The increase Δm of a body's mass (in grams), if its energy is increased by ΔE (in ergs), is given by $\Delta m = \Delta E/c^2$, where $c = 3 \times 10^{10}$ cm/s is the velocity of light in vacuum. Thus the mass of a body increases by about 1.1×10^{-21} g for each erg of energy added to it. It follows that changes in mass are observable only if energy changes are extremely large. Nevertheless, experimental confirmation of Einstein's relation between mass and energy is excellent. For example, the mass m of an electron moving with a speed v that is near the speed of light has been experimentally found to be greater than its mass m_0 when at rest, by observing that a given force produces smaller acceleration of the rapidly moving electron than of an electron at rest. The results agree with Einstein's prediction defined in Eq. (2).

$$m = \frac{m_0}{\sqrt{1 - (v^2/c^2)}} \qquad (2)$$

Another example is given by neutrons and protons which when combined to form a stable atomic nucleus have, because of tremendous forces of mutual attraction, much lower energy than when they are free and separated. The mass of such a nucleus is measurably smaller than the sum of the masses of its constituent particles. On the other hand, in the explosion of dynamite, the mass of the products of explosion after they have come to rest is smaller than the original mass of the dynamite by less than one part in 10^{10}. Even in the explosion of a plutonium bomb, the "loss" of mass of plutonium is a small fraction of 1%.

The mass of a body can be changed not only by cutting a part of the body off, for example, but by merely changing the body's energy; thus Newton's description of mass as the measure of the quantity of matter is not useful when extremely high energies are involved, and may even be misleading. Indeed, mass must be assigned to all forms of energy and therefore to such nonmaterial entities as light. Although so-called particles of light, called photons, have zero rest mass, they have energy when they are—as they always are—in motion, and thus must have mass, equal to their energy divided by c^2. The mass of one of the more massive known photons, the photon of a gamma ray of wavelength 2×10^{-11} cm, is approximately the same as the mass of an electron. This means that such a photon will have just as much inertia as an electron and will weigh as much in the same location.

Even with this broader, ultra-material concept of mass, mass is conserved. That is, if a system loses some of its mass by losing either some matter or some energy, another system must gain just as much mass. Matter can be destroyed or created by converting it into the energy of light, or vice versa, but the total mass remains unchanged. Mass, however, is no longer simply additive. Two 1-kg blocks, when brought close together, will have less gravitational potential energy than when far apart, and the mass of the combination will be less than 2 kg. The difference, however, is too small to be detectable now or in the foreseeable future.

Bibliography. F. Bueche, *Principles of Physics*, 4th ed., 1982; A. Einstein and L. Infeld, *The Evolution of Physics*, 1938, reprint 1960; L. B. Macurdy, Standards of mass, *Phys. Today*, 4(4):7–11, 1951.

DENSITY
Leo Nedelsky

The mass per unit volume of a material. The term is applicable to mixtures and pure substances and to matter in the solid, liquid, gaseous, or plasma state. Density of all matter depends on temperature; the density of a mixture may depend on its composition, and the density of a gas on its pressure. Common units of density are grams per cubic centimeter, and slugs or pounds per cubic foot. The specific gravity of a material is defined as the ratio of its density to the density of some standard material, such as water at a specified temperature, for example, 60°F (15.6°C), or, for gases the basis may be air at standard temperature and pressure. Another related concept is weight density, which is defined as the weight of a unit volume of the material. *See* Mass.

RELATIVE ATOMIC MASS
Thomas C. Waddington

The ratio of the average mass per atom of the natural nuclidic composition of an element to $\frac{1}{12}$ of the mass of an atom of nuclide ^{12}C. For example, $\mu(\text{Cl}) = 35.453$. Relative atomic mass replaces the concept of atomic weight. It is also known as relative nuclidic mass. *See* Atomic weight.

ATOMIC WEIGHT
Cornelius P. Browne

A measure of the average mass of the atoms of a chemical element. Most elements consist of a mixture of isotopes, which are atoms having the same chemical properties but different masses. Although the mixture of isotopes may depend on the origin of the sample, especially when some of the material is the product of radioactive decay, these variations are small in most natural materials. The assumption that all atoms of a given isotope have the same mass is a fundamental part of atomic theory. It is confirmed by the existence of chemical combining weights, that is, the fact that the ratio of the weights of elements which combine to form compounds is very nearly the ratio of small whole numbers. These ratios are indeed the ratios of the numbers of atoms in the molecules. This verifies Avogadro's hypothesis that a mass of any material having a weight equal to its molecular weight contains the same number of molecules. This number is called the Avogadro number.

Relative weights. The relative weight of elements may be determined by measuring the chemical combining weights. For example, if it is found that 1 g of carbon combines with 1.332 g of oxygen to form CO, then the ratio of the atomic weight of oxygen to that of carbon is 1.332 (approximately 16/12). A second method of determining atomic weights is based on the fact that equal volumes of ideal gases at the same temperature and pressure contain equal numbers of molecules (Avogadro's law). By measuring the ratio of the densities of two real gases at a series of decreasing pressures and extrapolating to zero density, the ratio of molecular weights may be determined.

Relative atomic weights may be used to form a table of chemical atomic weights if it is agreed to assign a value for the weight of one element. For example, chemists used to choose a value of exactly 16 for the average atomic weight of naturally occurring oxygen. Carbon then had a weight of about 12, hydrogen a weight of about 1, and so forth. To obtain the mass of a single atom from the relative weights, it is necessary to determine the Avogadro number. There are many methods of doing this, one of the most accurate being the use of x-ray crystallography to measure the distance between atoms in a crystal lattice. This spacing, together with the overall dimensions of the crystal, gives the number of atoms in the crystal, whereas the weight of the crystal gives the fraction of an atomic weight. The ratio of the two is the Avogadro number. SEE AVOGADRO'S NUMBER.

Unified scale. To the physicist, the mass of an atom of a given isotope is the significant quantity. Since it is now possible to measure the masses of individual atoms, and in order to remove the uncertainties of isotopic composition, a Unified Scale of Atomic Masses has been adopted by international agreement. The chemical and physical scales are unified by defining the mass of the abundant isotope of carbon to be exactly 12 u, where u is the symbol for the unified atomic mass unit. On this scale all atoms have mass values that are nearly integers. The difference, called the mass deficit, represents the difference in mass between the sum of the masses of the constituent protons, neutrons, and electrons and the mass of the atom. This difference represents the binding energy of the nucleus.

Accurate mass values. Results from two methods of accurate measurements of individual atomic masses are combined to give the best values. These methods are mass spectroscopy and nuclear reaction energy measurements.

Mass spectroscopy. In mass spectroscopy a beam of ionized atoms is sent through a combination of electric and magnetic fields. For an ion of mass m and charge q moving perpendicular to an electric field of strength E, with velocity v, the radius r of the path is given by Eq. (1). In a magnetic field of strength B the radius r' is given by Eq. (2).

$$r = \frac{m}{q}\frac{v^2}{E} \qquad (1) \qquad\qquad r' = \frac{m}{q}\frac{v}{B} \qquad (2)$$

By proper choice of the pattern of E and B fields, ions of a given m/q ratio may be focused along a given line on a detector. In practice, rather than relying on absolute measurements of the fields and radii, the small differences in position of ions of nearly equal m/q are measured. Thus mass differences between ions of the unknown mass and known mass are determined. For example, the masses of $^{12}C^1H_4^+$ and $^{16}O^+$ are both nearly 16 u. Given the mass of carbon (12 u) and the mass of hydrogen, a very precise value for the mass of ^{16}O may then be found.

Nuclear reaction energies. Differences in atomic masses are measured very accurately by using the conversion of mass to energy in a nuclear reaction. From special relativity theory $E = (\Delta m)c^2$, where E is the energy released or absorbed in the reaction, c is a fundamental constant (the velocity of light), and Δm is the difference between the total mass before and after the reaction. The energy may appear in the form of kinetic energy of the recoiling product masses or as electromagnetic energy in the form of gamma rays or x-rays, or a combination of these forms. For example, if a hydrogen atom ^1H and a neutron n combine to form a deuterium atom ^2H, some 2.2 MeV of energy is released as a gamma ray. This energy may be very accurately measured by diffraction techniques or with solid-state detectors. Again, if ^{12}C atoms are bombarded by high-speed ^2H ions, the reaction $^{12}C + \,^2H \rightarrow \,^{13}C + \,^1H$ occurs. The kinetic energies of the ^2H ions before the reaction and of the ^1H ions produced may be precisely measured by deflections in magnetic fields. From these and the direction of the ^1H ions, the energy of the ^{13}C recoil may be deduced and thus the net change in kinetic energy. From this, Δm is calculated.

Mass tables. When the mass differences from enough different reactions and mass spectroscopy ion pairs are available, a set of equations having the masses as unknowns may be written. There are actually many more known differences than masses, so the equations may be solved for all the unknown masses. All the nuclear reaction and mass spectroscopic data are combined to give the Table of Atomic Masses. There are thousands of input values, and great computing power is required. The table is periodically revised as new measurements are made and new unstable nuclear species are discovered. At present the masses of most stable atoms are known to 1 part in 10^7, and the masses of a great many artificially produced unstable atoms are known to 1 part in 10^6 or better. SEE MOLECULAR WEIGHT.

Bibliography. R. Eisberg and R. Resnick, *Quantum Physics*, 2d ed., 1985; D. C. Peaslee, *Elements of Atomic Physics*, 1955; A. Septier (ed.), *Focusing of Charged Particles*, vol. 2, 1967; A. H. Wapstra and K. Bos, The 1977 Atomic Mass Evaluation, Atomic Mass Tables, *Atomic Nucl. Data Tables*, 19:177–214, 1977.

RELATIVE MOLECULAR MASS
THOMAS C. WADDINGTON

The ratio of the average mass per formula unit of the natural nuclidic composition of a substance to $1/12$ of the mass of an atom of nuclide ^{12}C. For example, $\mu(KCl) = 74.555$. Relative molecular mass replaces the concept of molecular weight. SEE MOLECULAR WEIGHT.

MOLECULAR WEIGHT
C. DENISE CALDWELL

The sum of the atomic weights of all atoms making up a molecule. Actually, what is meant by molecular weight is molecular mass. The use of this expression is historical, however, and will be maintained. The atomic weight is the mass, in atomic mass units, of an atom. It is approximately equal to the total number of nucleons, protons and neutrons, composing the nucleus. Since 1961 the official definition of the atomic mass unit (amu) has been that it is 1/12 the mass of the carbon-12 isotope, which is assigned the value 12.000 exactly.

The microscopic atomic weight is connected to the fundamental macroscopic mass unit, the kilogram, through the definition of the mole (mol). The kilogram is defined as the mass of a platinum-iridium bar kept at the International Bureau of Weights and Measures in Paris, France. A mole is an amount of substance containing Avogadro's number, N_A, approximately 6.023×10^{23}, of molecules or atoms. Molecule, in this definition, is understood to be the smallest unit making up the characteristic compound. Originally, the mole was interpreted as that number of particles whose total mass in grams was numerically equivalent to the atomic or molecular weight in atomic mass units, referred to as gram-atomic or gram-molecular weight. This is how the above value for N_A was calculated. As the ability to make measurements of the absolute masses of single atoms and molecules has improved, however, modern metrology is tending to alter its approach and define Avogadro's number as an exact quantity, thereby changing slightly the definition of the atomic mass unit and removing the need to define atomic weight with respect to a particular isotopic species. The latest and most accurate value for Avogadro's number is $6.0220978(63) \times 10^{23} \text{ mol}^{-1}$. SEE AVOGADRO'S NUMBER.

Mass spectrometry. The best values for microscopic atomic and molecular masses are derived from measurements utilizing mass spectrometry, an experimental technique employed both for mass detection and for absolute mass determinations based on deflection of charged particles in electric or magnetic fields. Of the two methods, magnetic deflection yields the most accurate results and is the one most often used. A magnetic deflection apparatus which is used for mass determination is called a mass spectrometer. While particular types may vary as to the details of construction, the principle of operation for all is based on the fact that a charged particle moving in a magnetic field will follow a curved path, the curvature of which depends on the mass and charge of the species.

For a spectrometer employing deflection in a constant magnetic field with induction $\vec{B}$, the curvature can be easily calculated. Let $\vec{v}_0$ be the initial velocity with which the particle enters the magnetic field region, m the mass of the particle, and q the charge. The magnetic force does not alter the magnitude of the particle's velocity; the sole action is to cause the particle to execute a circular motion in the field. If $\vec{v}_0$ is perpendicular to $\vec{B}$, so that $\vec{v}_0$ is equal to $\vec{v}_\perp$, the component of the velocity perpendicular to $\vec{B}$, the plane of the orbit will be perpendicular to the region in which $\vec{B}$ is constant. The radius of curvature of the circle r is then given by Eq. (1). Thus, the particle's mass is given by Eq. (2).

$$r = mv_\perp/qB \qquad (1) \qquad\qquad m = qrB/v_\perp \qquad (2)$$

Mass spectroscopic measurements are made by first ionizing the atom or small molecule by using bombardment with high-energy electrons or absorption of ultraviolet light, then accelerating it across an electric potential difference V to produce a initial velocity given by Eq. (3), and

$$v = (2qV/m)^{1/2} \qquad (3)$$

finally allowing it to be deflected in the magnetic field and strike a detector. With q known, m is extracted by measuring the radius of curvature r of the circle of motion, at least in principle. In practice, r and $\vec{B}$ for a particular apparatus will be kept fixed, and $\vec{v}$ will be chosen by varying the accelerating voltage in such a way that the path followed by the particle will deposit it in the detector. In general, q will be equal to the charge on the electron, as most ionization will produce only the singly ionized species. As the mass of the electron is so much smaller than the mass of the proton, approximately 1/1836, the loss of one electron will not appreciably alter the mass of the atom or molecule.

Mass spectrometry has made possible determinations of absolute masses of virtually all the elements as well as many small molecules. The drawback to this technique is that the species to be measured must be in the vapor phase in order to execute motion in the fields. Its primary advantage is that it, taken together with Avogadro's number, puts the atomic mass scale on an absolute level.

Importance of determination. As the masses of all the atomic species are now known so well, masses of molecules can be determined once the composition of the molecule has been ascertained. Alternatively, if the molecular weight of the molecule is known and enough additional information about composition is available, such as the basic atomic constituents, it is possible to begin to assemble structural information about the molecule. Thus, the determination of the molecular weight is one of the first steps in the analysis of an unknown species. Given the increasing emphasis on the study of biologically important molecules, particular attention has been focused on the determination of molecular weights of larger and larger units. There are a number of methods available, and the particular one chosen will depend on the size and physical state of the molecule. All processes are physical macroscopic measurements and determine the molecular weight directly. Connection to the absolute mass scale is straightforward by using Avogadro's number, although, for extremely large molecules, this connection is often unnecessary or impossible, as the accuracy of the measurements is not that good. The main function of molecular weight determination of large molecules is elucidation of structure.

Weighing of gases. Although molecular weight determination for gases has been largely taken over by the mass spectroscopic analysis, it is still possible to determine molecular weights by direct weighing of gases. This method has some historical value in that much of present-day knowledge about chemical reactions was originally derived on the bases of observations concerning gases. For any real gas, if the pressure is low enough, the relationship between the pressure P, volume V, temperature T, and number of moles n of that gas can be expressed through the ideal gas law in the form of Eq. (4). Here R is the universal gas constant, equal to 8.31441 J/

$$R = \lim_{P \to 0} (PV/nT) \qquad (4)$$

mol K. The number of moles of the gas is related to the molecular weight by Avogadro's number, as discussed above. Thus, by measuring the pressure and weight of a known volume of gas, it is possible to derive the molecular weight M of the gas from Eq. (5). Here ρ is the mass density,

$$M = RT \lim_{P \to 0} (\rho/P) \qquad (5)$$

equal to the total mass of the gas as measured in the experiment divided by the volume V. S<small>EE</small> G<small>AS</small>.

Colligative properties of solutions. Molecular weight determination of materials which are solid or liquid at room temperature is best achieved by taking advantage of one of the colligative properties of solutions, boiling-point elevation, freezing-point lowering, or osmotic pressure, which depend on the number of particles in solution, not on the nature of the particle. The choice of which to use will depend on a number of properties of the substance, the most important of which will be the size. All require that the molecule be small enough to dissolve in the solution but large enough not to participate in the phase change or pass through a semipermeable membrane. Freezing-point lowering is an excellent method for determining molecular weights of smaller organic molecules, and osmometry, as the osmotic pressure determination is called, for determining molecular weights of larger organic molecules, particularly polymeric species. Boiling-point elevation is used less frequently.

The basis of all the methods involving colligative properties of solutions is that the chemical potentials of all phases must be the same. (Chemical potential is the partial change in energy of a system as matter is transferred into or out of it. For two systems in contact at equilibrium, the chemical potentials for each must be equal.) SEE CHEMICAL EQUILIBRIUM; CHEMICAL THERMODYNAMICS.

Phase change methods. For the two methods involving a phase change, analysis requires that it be possible to form a solution of the molecule in some known solvent which is dilute enough so that the solution may be considered ideal but concentrated enough that a measurable raising of the boiling point or lowering of the freezing point will be produced. It must also be possible to make the assumption that there are no solute particles in either the vapor or the solid phase. The balance of chemical potentials then demands that the solvent remain in solution rather than freeze out or vaporize. Under these restricted conditions of a dilute, ideal solution, the extent ΔT of freezing-point lowering or boiling-point elevation is related to the concentration of solute, and alternatively its molecular weight M through Eq. (6). Here R is the univeral gas constant, T_0 the

$$\Delta T = \frac{RT_0^2 M}{1000 \Delta H} c \qquad (6)$$

temperature at which freezing or boiling would occur in the pure solvent, ΔH the enthalpy of fusion or vaporization of the solvent (the amount of energy which must be supplied or removed at constant pressure to cause 1 mole to change phase), and c the molal concentration of the solute. (1 molal equals 1 mole of solute per 1 kg of solvent.) Because of the restrictions on the solutions which must be formed and for which the above expression is valid, combined with the accuracy with which temperature differences can be measured, these two methods are best employed for molecules with molecular weights less than 10,000.

Osmometry. The basis for the application of the osmotic pressure method (osmometry) is that the chemical potentials of a solution and a pure solvent separated from each other by a semipermeable membrane must be the same. The membrane must be coarse enough to allow solvent molecules to pass through, but fine enough to prohibit passage of solute. In an attempt to equalize the chemical potentials on either side of the membrane, solvent will flow from the side containing only solvent to the solution. The resulting change in chemical potential causes an increase in pressure on the side containing the solution, which will continue until the pressure produced prohibits further flow of solvent, that is, the chemical potentials are equal. This pressure Π is a real physical pressure which can be measured and is related to the concentration in the solution through the van't Hoff equation (7). Here R is again the universal gas constant, T the

$$\Pi = RTc \qquad (7)$$

temperature, and c the concentration in number of moles per given amount of solvent. Because of restrictions on passage through the membrane, osmometry can be used to determine molecular weights from about 1000 to about 30,000. For larger molecules other methods are necessary. SEE OSMOSIS; SOLUTION.

Sedimentation. Of all techniques available for determination of molecular weights of large molecules, sedimentation, or the settling of large molecules out of solution under the action of an external force, is perhaps the most common. This process may take a number of varied forms, but its successful utilization for weight determination depends on the fact that the solute molecules in a solution are never at rest but are free to move among the solvent molecules. The practical application of sedimentation involves producing a directed motion of these particles. This is done in a high-speed centrifuge. A particle of mass m when rotated in a circle experiences a force which is equal to $m\omega^2 r$, where ω is the angular velocity and r the distance of the particle from the center of rotation. In the absence of any other forces, the particle would tend to move away from the force center and to the edge of the container, where it would be collected as soon as it can move no further. However, as the molecules move away from the center of the rotation, a concentration gradient is set up. This causes a diffusion of the particles in the opposite direction to the gradient, and a tendency for the particles to remain distributed throughout the solvent. In addition, there is also a buoyant force on the particles produced by collisions with solvent molecules which tends to maintain the particles in solution. At equilibrium the external centrifugal force driving the sedimentation must balance the total forces produced by the buoyant effect and the concentration gradient. If V is the volume of the solute molecule, ρ_M its density, and ρ_0 the density of the solvent, this balancing of forces leads to Eq. (8). Here D is the diffusion coefficient,

$$N_A V \rho_M (1 - \rho_M^{-1} \rho_0) \omega^2 r = RTr/D \qquad (8)$$

N_A Avogadro's number, R the universal gas constant, and T the temperature. The molecular weight M is related to the mass density by Eq. (9). At equilibrium the rate of sedimentation (rate

$$M = N_A V \rho_M \qquad (9)$$

of deposit of particles across an area in the direction of the rotation axis) will equal the rate of diffusion across that same boundary. Combining this requirement with the force balance yields Eq. (10) for the molecular weight. Successful application of this method requires determination of

$$M = \frac{2RT}{(1 - \rho_M^{-1} \rho_0) \omega^2} \frac{\ln(c_2/c_1)}{r_2^2 - r_1^2} \qquad (10)$$

the concentrations c_1 and c_2 at two values of the radius r_1 and r_2. This is carried out by optical refractometry. Light when passing from one medium to another will be refracted so as to follow a different path from the original one. If the second medium happens to be a solution, the angle of refraction will depend on the concentration of the solution. A modern centrifuge has containers for the solution which are constructed of quartz to provide for this measurement while the solution is being centrifuged.

A variant on the sedimentation equilibrium method described above involves dissolving the molecule in a concentrated salt solution and centrifuging the resulting mixture. The buoyant force produced by the dense salt solution will force the molecular species to be concentrated in that region in which its density is equal to that of the concentrated solution, $\rho_M = \rho_0$. Continuous motion of the molecules, however, prevents that the concentration be a perfect line, but some distribution about the central point r_0 occurs. From the width σ of this distribution about the center and the density gradient $\Delta\rho/\Delta r$ of the salt solution, the molecular weight can be extracted according to Eq. (11).

$$M = \frac{2RT}{\rho_M^{-1}(\Delta\rho/\Delta r)\omega^2 r_0 \sigma^2} \qquad (11)$$

Light scattering. Another measurement from which molecular weights can be obtained is based on the scattering of light from the molecule. A beam of light falling on a molecule will induce in the molecule a dipole moment which in its turn will radiate. The interference between the radiated beam and the incoming beam produces an angular dependence of the scattered radiation which depends on the molecular weight of the molecule. This occurs whether the molecule is free or in solution. While the theory for this effect is complicated and varies according to the size of the molecule, the general result for molecules whose size is considerably less than that of the wavelength λ of the radiation (less than $\lambda/50$) is given by Eq. (12). Here $I(\theta)$ is the intensity

$$I(\theta)/I_0 = \text{constant} (1 + \cos^2 \theta) Mc \qquad (12)$$

of radiation at angle θ, I_0 the intensity of the incoming beam, M the molecular weight, and c the concentration in grams per cubic centimeter of the molecule. If the molecules are much larger than $\lambda/50$ (about 9 nanometers for visible light), this relationship in this simple form is no longer valid, but the method is still viable with appropriate adjustments to the theory. In fact, it can be used in its extended version even for large aggregates.

Sample purity. One requirement which all these methods demand is purity of the sample if the resulting weight is to be very accurate. Impurities result in a molecular weight which is not absolute for the species but is the average weight of the species plus the impurities. Depending on whether or not the influence of the impurities is due to their number (for low-molecular-weight impurities) or their weight (high-weight impurities), the average is either a number average M_n or a weight average M_w. The two average molecular weights are defined by Eqs. (13) and (14),

$$M_n = \frac{\Sigma_i(n_i M_i)}{\Sigma_i n_i} \quad (13) \qquad M_w = \frac{\Sigma_i(n_i M_i^2)}{\Sigma_i(n_i M_i)} \quad (14)$$

where n_i is the number of particles of species i with molecular weight M_i.

Measurement of auxiliary parameters. All the methods in use in biophysics require the knowledge of parameters other than the one actually being measured in the experiment. In most cases this will be the volume of the molecule, or ρ_M^{-1}. For these methods to give accurate results, this number must be obtained by other means. One of the best, but often one of the hardest to apply, is x-ray crystallography, the same technique used to determine Avogadro's number.

Other determination methods. Other methods of molecular weight determination include end-group labeling, useful for proteins; viscosity, useful for relative molecular weight determination; gel electrophoresis, also useful for relative measurements; and electron microscopy, useful if the molecule is large enough to be well resolved in the microscope and again only for relative measurements. For a truly accurate molecular weight determination, a combination of these techniques may be necessary. SEE ELECTROPHORESIS; VISCOSITY.

GRAM-EQUIVALENT WEIGHT
THOMAS C. WADDINGTON

The equivalent weight of an element or compound, expressed in grams (g), on a scale in which ^{12}C (the isotope of carbon of nuclear mass 12) has an equivalent weight of 3 g in those compounds in which its formal valency is 4. This replaces the older definition in which the fixed point on the scale was the equivalent weight of oxygen, 8.000 g. Because variable valency is much more common in carbon than in oxygen, it is now necessary to specify the formal valency in defining the equivalent weight concept. The gram-equivalent weight of an electrolyte is usually the weight in grams which is associated with 1 faraday of electricity. For example, the gram-equivalent weight of sodium chloride, Na^+Cl^-, is equal to the gram-molecular weight; that of calcium chloride, $Ca^{2+}Cl_2^-$, is one-half the gram-molecular weight; that of aluminum sulfate, $Al_2^{3+}(SO_4)_3^{2-}$, is one-sixth the gram-molecular weight. This concept, together with that of equivalent weight, tends to be abandoned, and relations are expressed in terms of balanced stoichiometric chemical equations and relative numbers of moles reacting. SEE ELECTROLYSIS; ELECTROLYTE; GRAM-MOLECULAR WEIGHT; VALENCE.

GRAM-MOLECULAR WEIGHT
THOMAS C. WADDINGTON

The molecular weight of an element or compound expressed in grams (g), that is, the molecular weight on a scale on which the atomic weight of the ^{12}C isotope of carbon is taken as 12 exactly. This replaces the earlier scale on which the atomic weight of oxygen was taken as 16.00 g. In the International System of Units, gram-molecular weight is replaced by the mole.

The ratio of the gram-molecular weights of any two elements or compounds must be identical with the ratio of the absolute weights of their individual molecules. Therefore, the gram-molecular weights of all elements or compounds contain the same number of molecules. This number, called the Avogadro number, N, is 6.022×10^{23}.

Since they contain the same number of molecules, the gram-molecular weights of all gases occupy the same volume at the same temperature and pressure. At 0°C and 1 atm (100 kilopascals) pressure this volume, called the gram-molecular volume, is 22.4 liters. SEE AVOGADRO'S NUMBER; GAS; MOLECULAR WEIGHT; RELATIVE MOLECULAR MASS.

BOILING POINT
ROBERT L. SCOTT

The temperature at which the transition from the liquid to the gaseous phase occurs. For pure substances at a fixed pressure, the boiling or vaporization process occurs at a single temperature; as heat is added, the temperature remains constant until all the liquid has boiled.

The normal boiling point is defined as the boiling point at a total applied pressure of 1 atm, that is, the temperature at which the vapor pressure of the liquid equals 1 atm (760 mmHg or 100 kilopascals).

The pressure dependence of the boiling point (expressed as absolute temperature T_b) is given by the Clapeyron equation, which is shown below. In the equation ΔV_v and ΔH_v are the

$$\frac{dT_b}{dP} = \frac{T_b \, \Delta V_v}{\Delta H_v}$$

volume change and the heat absorbed during the vaporization process, and P is the pressure exerted on the liquid. All are necessarily positive, so the boiling point increases with applied pressure. For substances boiling in the region of room temperature, the rate of change of boiling point with temperature is approximately 0.04°C/mmHg [0.0003°C per pascal or 1.83°F per inch of mercury] (where the pressure is approximately 1 atm or 100 kPa). For example, for water at 212°F (100°C) and 1 atm pressure, $dT_b/dP = 0.0369$°C/mmHg [0.00028°C per pascal or 1.687°F per inch of mercury].

The boiling point cannot be raised indefinitely, however. As the pressure is increased, the density of the gas phase increases until it finally becomes indistinguishable from the liquid phase with which it is in equilibrium; this is the critical temperature, above which no distinct liquid phase exists. Helium has the lowest normal boiling point (4.2 K or -452°F) of any substance, and tungsten carbide has one of the highest (6300 K or 10,880°F).

The boiling point is a rough measure of the intermolecular potential energy of the system, although the heat of vaporization is a better measure. A liquid whose molecules are held together weakly boils at a low temperature; substances with stronger intermolecular forces boil at higher temperatures.

For solutions of two or more components, the boiling process normally occurs over a range of temperatures. A distinction is made between the boiling point (or bubble point), the temperature at which the first bubbles of vapor appear, and the condensation point (or dew point), the temperature at which the last trace of liquid disappears, or alternatively at which the first droplets of liquid appear. Measurement of the boiling point of a solution and the difference between it and the boiling point of the pure solvent provides a convenient method of determining the molecular weight of a dissolved nonvolatile solute. SEE PHASE EQUILIBRIUM.

The well-known process by which liquids evaporate into air at temperatures far below their normal boiling points (for example, the evaporation of water at room temperature) can be regarded as an extreme case of "boiling" in a system of several components. At equilibrium the liquid phase (essentially one substance since the solubility of the air is small) coexists with a gas phase, which is a mixture of air with the vapor of the liquid at a partial pressure almost equivalent to its vapor pressure. For this mixture the dew point is virtually the same as the boiling point of the pure liquid, whereas the bubble point is a temperature as low as the freezing point of the liquid (below which the solid can also evaporate). Similar principles are involved in steam distillation.

Precision measurement of boiling points is known as ebulliometry. The principal experimental difficulty is to be sure that true equilibrium is reached and that the liquid is not superheated into a metastable state. SEE LIQUID; SOLUTION.

MELTING POINT
ROBERT L. SCOTT

The temperature at which a solid changes to a liquid. For pure substances, the melting or fusion process occurs at a single temperature, the temperature rise with addition of heat being arrested until melting is complete. The direct transition from solid phase to gas phase is not properly called melting, but preferably, sublimation.

Melting points reported in the literature, unless specifically stated otherwise, have been measured under an applied pressure of 1 atm (100 kilopascals), usually 1 atm of air. (The solubility of air in the liquid is a complicating factor in precision measurements.) The Clapeyron equation for the pressure dependence of the absolute melting temperature T_m is the equation below, where

$$\frac{dT_m}{dP} = \frac{T_m \Delta V_f}{\Delta H_f}$$

ΔV_f is the change in volume, ΔH_f the heat absorbed during the fusion process, and P the applied pressure. Upon melting, all substances absorb heat, and most substances expand; consequently an increase in pressure normally raises the melting point. A few substances, of which water is the most notable example, contract upon melting; thus, the application of pressure to ice at 0°C (32°F) causes it to melt. Large changes in pressure are required to produce significant shifts in the melting point; a pressure of 10 atm (1 megapascal) lowers the melting point of ice by only 0.075°C (0.135°F).

A sufficient decrease in temperature at ordinary pressures causes all pure substances except helium to freeze to solids; the lowest normal melting point is that of hydrogen, at 14 K (−434°F) and one of the highest is that of rhenium, at 3700 K. Liquid helium can be transformed into a solid only by applying a pressure in excess of 25 atm (2500 kPa).

For solutions of two or more components, the melting process normally occurs over a range of temperatures, and a distinction is made between the melting point, the temperature at which the first trace of liquid appears, and the freezing point, the higher temperature at which the last trace of solid disappears, or equivalently, if one is cooling rather than heating, the temperature at which the first trace of solid appears. Measurement of the freezing point of a solution and the difference between it and the freezing point of the pure solvent provides a convenient method of determining the molecular weight of a dissolved solute, because the freezing point of a solution is lower than that of a pure solvent. SEE PHASE EQUILIBRIUM; SOLUTION; TRIPLE POINT.

TRANSITION POINT
ROBERT L. SCOTT

The point at which a substance changes from one state of aggregation to another. This general definition would include the melting point (transition from solid to liquid), boiling point (liquid to gas), or sublimation point (solid to gas); but in practice the term transition point is usually restricted to the transition from one solid phase to another, that is, the temperature (for a fixed pressure, usually 1 atm or 100 kilopascals) at which a substance changes from one crystal structure to another.

Some typical examples of transition points are:

$$\beta = \text{Fe} \xrightarrow[(1664.3°F)]{\text{at 1180 K}} \gamma\text{FE}$$
(body-centered cubic) (face-centered cubic)

$$S_8 \xrightarrow[(253.1°F)]{\text{at 369 K}} S_8$$
(rhombic) (monoclinic)

$$CCl_4 \xrightarrow[(117.8°F)]{225.5K} CCl_4$$
(monoclinic) (tetragonal)

$$\text{NH}_4\text{NO}_3 \text{ (β-rhombic)} \xrightarrow[\text{(90.2°F)}]{\text{at 305.3 K}} \text{NH}_4\text{NO}_3 \text{ (α-rhombic)}$$

$$\text{NH}_4\text{NO}_3 \text{ (α-rhombic)} \xrightarrow[\text{(183.6°F)}]{\text{at 357.4 K}} \text{NH}_4\text{NO}_3 \text{ (trigonal)}$$

Another kind of transition point is the culmination of a gradual change (for example, the loss of ferromagnetism in iron or nickel) at the lambda point, or Curie point. This behavior is typical of second-order transitions. SEE BOILING POINT; MELTING POINT; PHASE EQUILIBRIUM; TRIPLE POINT.

TRIPLE POINT
Robert L. Scott

A particular temperature and pressure at which three different phases of one substance can coexist in equilibrium. In common usage these three phases are normally solid, liquid, and gas, although triple points can also occur with two solid phases and one liquid phase, with two solid phases and one gas phase, or with three solid phases.

According to the Gibbs phase rule, a three-phase situation in a one-component system has no degrees of freedom (that is, it is invariant). Consequently, a triple point occurs at a unique temperature and pressure, because any change in either variable will result in the disappearance of at least one of the three phases. SEE PHASE EQUILIBRIUM.

Triple points are shown in the **illustration** of part of the phase diagram for water. Point A is the well-known triple point for Ice I (the ordinary low-pressure solid form) + liquid + water + water vapor at 0.01°C (273.16 K) and a pressure of 0.00603 atm (4.58 mmHg or 611 pascals). In 1954 the thermodynamic temperature scale (the absolute of Kelvin scale) was redefined by setting this triple-point temperature for water equal to exactly 273.16 K. Point B, at 251.1 K (− 7.6°F) and 2047 atm (207.4 megapascals) pressure, is the triple point for liquid water + Ice I + Ice III; and point C, at 238.4 K (−31°F) and 2100 atm (212.8 MPa) pressure, is the triple point for Ice I + Ice II + Ice III. At least four other triple points are known at higher pressures, involving other crystalline forms of ice.

For most substances the solid-liquid-vapor triple point has a pressure less than 1 atm (about 100 kilopascals); such substances then have a liquid-vapor transition at 1 atm (normal boiling point). However, if this triple point has a pressure above 1 atm, the substance passes directly from solid to vapor at 1 atm. SEE SUBLIMATION.

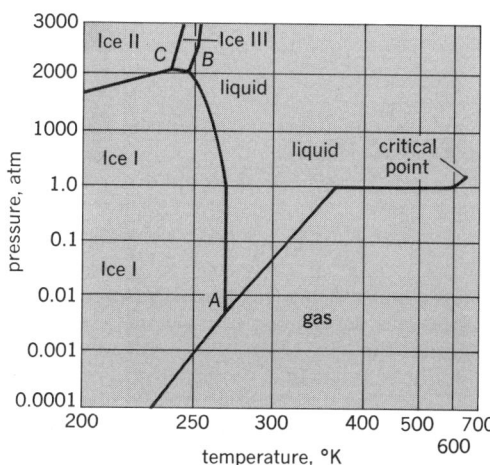

Phase diagram for water, showing gas, liquid, and several solid (ice) phases; triple points at A, B, and C. The pressure scale changes at 1 atm from logarithmic scale at low pressure to linear at high pressure.
1 atm = 100 kPa; °F = (K × 1.8) − 459.67.

For a two-component system, the invariant point in a phase diagram is a quadruple point at which four phases coexist. The three-phase situation is then represented by a line in the three-dimensional pressure-temperature-composition diagram. SEE BOILING POINT; MELTING POINT; TRANSITION POINT.

ISOELECTRIC POINT
W. O. MILLIGAN

The pH value of the dispersion medium of a colloidal suspension at which the colloidal particles do not move in an electric field; that is, the particles are electrophoretically inert (see **illus**.). The isoelectric point is often employed to characterize colloidal material such as the proteins. However, a range of values is usually necessary, since the isoelectric point varies detectably with (1) the size of the particles, (2) the purity, and (3) the concentration of other than hydrogen ions.

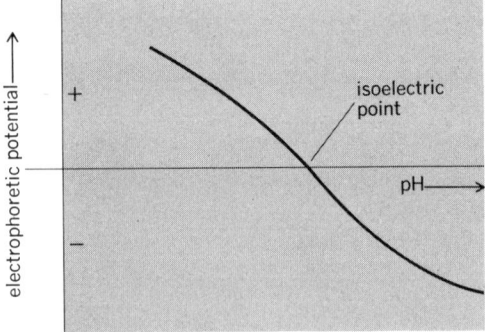

Graph showing mobility of a sample colloidal suspension as a function of pH. Mobility ceases at the isoelectric point.

If the pH value of an extrinsic sol (stability attributed primarily to electrical charge) is adjusted toward the isoelectric point, coagulation will occur at or near the isoelectric point. Intrinsic sols (stability attributed primarily to solvation) may be carried to the isoelectric point without coagulation, but such sols will be in a region of minimum stability, so that a minimum concentration of desolvating agent will cause coagulation. Likewise, viscosity changes often reach a minimum at or near the isoelectric point. SEE COLLOID; ELECTROPHORESIS; ION SELECTIVE MEMBRANES AND ELECTRODES; PH.

SPECIFIC HEAT
HAROLD CHRISTIAN WEBER

The ratio of the amount of heat required to raise unit mass of a material 1 degree in temperature to the amount of heat required to raise the same mass of a reference substance 1 degree in temperature. Both measurements are made at a reference temperature and in nearly all cases at either constant volume or constant pressure. Water is usually the reference substance. Because the heat capacity of water is nearly unity, the value of specific heat for a material is nearly equal to its heat capacity. Specific heat, as defined here, is a ratio without units, although it is often defined differently. For clarity it is recommended that thermodynamic discussion be carried out in terms of heat capacity instead of specific heat. Also, it is desirable to define heat units in electrical terms. SEE CHEMICAL THERMODYNAMICS.

ELECTROCHEMISTRY

Electrochemistry	**242**
Electrolyte	**243**
Electrolytic conductance	**244**
Electrochemical series	**247**
Electrode	**249**
Electrode potential	**249**
Electrochemical techniques	**252**
Electrochemical process	**256**
Electrolysis	**270**
Electrokinetic phenomena	**271**
Electrophoresis	**274**
Streaming potential	**278**

ELECTROCHEMISTRY
Herbert A. Laitinen

The science dealing with the chemical changes accompanying the passage of an electric current, or the reverse process in which a chemical reaction is used as the source of energy to produce an electric current, as in a battery. Electric conduction occurs through the motion of charged particles. The charged particles may be electrons (as in metals or semiconductors) or ions, which are electrically charged atoms, molecules, or molecular aggregates. Ionic conduction in electrolytes (liquid solutions, molten salts, and certain ionically conductive solids) is a phase of electrochemistry. Conduction in metals, semiconductors, and gases is generally considered a portion of physics. Other aspects of electrochemistry are described below. *See* ELECTROLYTIC CONDUCTANCE.

Galvanic cells. These are better known as electric batteries. Many chemical reactions can be arranged to produce electrical energy by physically separating the reaction into two half-reactions, one supplying electrons to an electrode forming the negative terminal of the cell, and the other removing the electrons from the positive terminal. In the lead storage battery, for example, electrons are supplied to the negative terminal by the half-reaction shown in (1), which

$$Pb + SO_4^{2-} \rightarrow PbSO_4 + 2e^- \tag{1}$$

represents the oxidation of metallic lead to form lead sulfate. At the postive terminal, lead dioxide is reduced to lead sulfate by the half-reaction shown in (2).

$$2e^- + PbO_2 + 4H^+ + SO_4^{2-} \rightarrow PbSO_4 + H_2O \tag{2}$$

The electrons flowing in the external circuit from the negative to the positive terminal constitute the desired electric current. Charging of the lead storage battery by forcing a current to flow in the reverse direction results in the reversal of both half-reactions, and the storage of electric energy in the form of lead and lead dioxide. Such a cell is called a secondary cell, in contrast to a primary cell, such as the Lechanché cell or familiar dry cell, which is not designed to be recharged. In the Lechanché cell, electric energy is produced by the oxidation of zinc and the reduction of manganese dioxide at a carbon electrode. The electrolyte is a moist mixture of zinc chloride, ammonium chloride, and powdered carbon. The fuel cell is designed for the continuous production of electric current through the consumption of oxidant and reductant at separate electrodes. The most common fuel cell is the hydrogen-oxygen (or air) cell with alkaline electrolyte. Many other systems, including hydrocarbon-oxygen, carbon monoxide–oxygen, lithium-chlorine, and sodium-sulfur have been proposed. *See* BATTERY; DRY CELL; FUEL CELL; STORAGE BATTERY.

Electrodeposition. The most important type of chemical reaction brought about by the passage of electric current is the deposition of a metal at a cathode from a solution of its ions. Electroplating of many metals, such as silver, cadmium, nickel, and chromium is used for protective and decorative coatings. Electroforming is a variety of electrodeposition in which an article to be reproduced is rendered conductive by spraying with a thin metallic coating, then electroplated with a metallic deposit that is stripped from its substrate and filled with backing to reproduce the original article. Electrowinning is used for the commercial production of active metals, such as aluminum, magnesium, and sodium, from molten salts and others, such as copper, manganese, and antimony, from aqueous solution. Electrorefining is commonly used to purify metals such as silver, lead, and copper. The impure metal is used as the anode, and purified metal is deposited at the cathode.

Electrolytic processes. Many electrode reactions other than metal deposition are of commercial or scientific use. Electrolysis of brine to yield chlorine at the anode, hydrogen at the cathode, and sodium hydroxide in the electrolyte is an important industrial process. Many organic compounds can be prepared electrolytically. *See* ELECTROLYSIS.

Electrothermics. While not strictly electrochemical, electrothermics is generally recognized as a part of the field. It includes high-temperature processes involving electric arc or resistance furnaces.

Electroanalytical chemistry. Many electrochemical measurements are useful for analytical purposes. Electrodes that are commonly used for analytical purposes through measurement of their potentials include the glass electrode for pH measurements, and ion-selective electrodes for certain ions, such as sodium or potassium ion (special glass compositions), calcium ion (liquid membrane), and fluoride ion (doped lanthanum fluoride single crystals). Polarography involves the use of a dropping mercury electrode as one elctrode of an electrolytic cell. Qualitative analysis is carried out by measurement of characteristic potentials (half-wave potentials) for electrode processes, and quantitative analysis by measurement of diffusion-controlled currents. Coulometry involves the application of Faraday's law for analytical purposes.

Several methods involving electrolysis during short periods of electrolysis permit the application of diffusion theory (in the absence of convection) to calculate mass transport rates. These include chronopotentiometry (measurement of potential-time transients under constant current conditions), and linear sweep and cyclic voltammetry (measurement of currents with linear voltage scan). Several titration methods involve electrochemical measurements, for example, conductometric, potentiometric, and amperometric titrations. S*ee* C*oulometric analysis;* E*lectrode;* E*lectrodeposition analysis;* I*on-selective membranes and electrodes;* P*olarographic analysis.*

Electrode kinetics. Studies of kinetics of electrode processes are valuable not only for the understanding of mechanisms of electrode reactions, but also for the study of homogeneous reactions occurring in solutions either preceding or following the charge-transfer step. Studies of this kind are made by pulse or transient techniques, or by steady-state methods involving dynamic systems, such as rotating disc electrodes for flowing solutions. S*ee* E*lectrochemical techniques.*

Miscellaneous phenomena. Electrochemical transport of ions through synthetic or natural membranes is important for processes, such as desalination of water and electrodialysis. In biological systems, the transmittal of nerve impulses and the generation of electrical signals, such as brain waves, are basically of electrochemical origin. A set of related phenomena can be grouped together under electrokinetic behavior, including the motion of colloidal particles in an electric field (electrophoresis), the motion of the liquid phase relative to the stationary solid under the influence of a potential gradient (electroosmosis), and the inverse generation of a potential gradient caused by a flowing liquid (streaming potential). Alternating-current phenomena, such as dielectric behavior, double-layer charging, and faradaic rectification, may also be included in a general definition of electrochemistry. Corrosion and passivation of metals are electrochemical in nature. S*ee* W*et cell.*

Bibliography. J. O. Bockris et al. (eds.), *Modern Aspects of Electrochemistry*, vol. 17, 1986; D. Rand and A. M. Bond, *Electrochemistry: The Interfacing Science*, 1984.

ELECTROLYTE

T*homas* C. W*addington*

A chemical compound which when fused or dissolved in certain solvents, usually water, will conduct an electric current. The passage of the current is always accompanied by decomposition of the electrolyte, called electrolysis, which takes place at the electrodes. All electrolytes in the fused state or in solution give rise to ions which conduct the electric current. The phenomena of electrolysis are summarized by Faraday's laws of electrolysis. All acids, bases, and salts are electrolytes. S*ee* E*lectrode;* E*lectrolysis;* I*on.*

Electrolytes are divided into strong and weak electrolytes. Strong electrolytes usually contain a stable ionic bond and are wholly ionized in solution and usually in the crystalline state. Weak electrolytes are only partially ionized in solution. Metallic hydroxides and salts are usually strong electrolytes, for example, potassium hydroxide and sodium chloride. A weak electrolyte contains a covalent bond which on dissolving in a solvent such as water may be transformed into an ionic bond. A solution of a weak electrolyte contains both the ionic and the covalent forms in equilibrium; for example, acetic acid in water consists of a mixture of undissociated molecules, CH_3COOH, and of the ions CH_3COO^- and H^+. S*ee* C*hemical bonding;* I*onic equilibrium.*

ELECTROLYTIC CONDUCTANCE

Herbert A. Laitinen

The transport of electric charges, under electric potential differences, by particles of atomic or larger size. This phenomenon is distinguished from metallic conductance, which is due to the movement of electrons. The charged particles that carry the electricity are called ions.

Positively charged ions are termed cations; the sodium ion, Na^+, is an example. The negatively charged chloride ion, Cl^-, is typical of anions. The negative charges are identical with those of electrons or integral multiples thereof. The unit positive charges have the same magnitude as those of electrons but are of opposite sign. Colloidal particles, which may have relatively large weights, may be ions, and may carry many positive or negative charges. Electrolytic conductors may be solids, liquids or gases. Semiconductors have properties that are intermediate between the metallic and electrolytic types.

Measurement. Conductances are usually reported as specific conductances κ, which are the reciprocals of the resistances of cubes of the materials, 1 cm in each dimension, placed between electrodes 1 cm square, on opposite sides. These units are sometimes called mhos, that is, ohms spelled backward. Conductances of solutions are usually measured by Friedrich Kohlrausch's method, in which a Wheatstone bridge is employed. Such a bridge is shown diagrammatically in **Fig. 1**. The resistances R_3 and R_4 (usually of the same value) form two arms of the bridge. Resistance R_2 is adjustable, and the remaining arm is the cell holding the electrolytic conductor, or as is usually stated, solution of electrolyte. Direct current and the usual galvanometers cannot be used because of an apparent failure of Ohm's law. Passage of direct current produces chemical reactions and a back electromotive force (emf) is generated by the galvanic action of the products. By using an alternative current, the electrochemical reactions occurring when the current is briefly passed in one direction may be reversed when the direction of the current is changed. When a small alternating-current input signal is used, practically all the electric charge passed during each half cycle is stored in the electric double layer, which acts as a capacitor. The electrodes are usually made of platinum and are platinized, that is, coated with finely divided platinum. The surface area, and hence the electrode capacitance, is thereby greatly increased. By making measurements at several frequencies and extrapolating to infinite frequency, the effect of electrode reactions can be eliminated. For less exact measurements, a fixed frequency of 60–1000 Hz is commonly used.

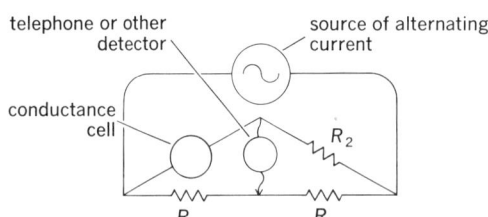

Fig. 1. Wheatstone bridge circuit for the measurement of electrolytic conductance. R_3 and R_4 are fixed resistances; R_2 is a variable resistance.

To determine the conductance C, that is, the reciprocal resistance $1/R$ of the cell of Fig. 1, the resistance R_2 of the bridge is adjusted until a minimum of sound is heard in the telephone.

Greater sensitivity may be obtained by electronic amplification of the off-balance signal and by using a "tuning eye" or an oscilloscope to detect the point of balance. When the bridge is in balance, the conductance is given by the relation $C = R_4/R_2R_3$. From this the specific conductance κ may be obtained from the equation $κ = KC$, in which K is the cell constant. Occasionally, this constant can be computed from the dimensions of the cell. Usually, however, it is determined by using a solution whose κ value is accurately known from measurements in such a cell, or by comparison with the specific conductance of mercury.

For precision work, care must be taken to avoid errors due to electrical reactances. This has been done in specially designed bridges. A typical, properly designed conductance cell is

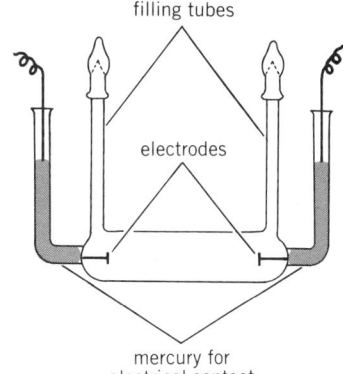

Fig. 2. Diagram of conductance cell.

mercury for electrical contact

shown in **Fig. 2**. The cell is filled with solution through the center tubes. Electrical contact is made with the electrodes by platinum wires sealed through the glass wall. These connect the mercury in the outside tubes, which are widely spread to avoid errors due to electrical capacity.

Equivalent conductance. Although many substances and mixtures show electrolytic conductance, the greater part of the research on the subject has been on aqueous solutions of salts, acids, and bases. There have been considerable data accumulated for solutions of such electrolytes in nonaqueous solvents, such as alcohols. The data are usually given in terms of equivalent conductance Λ, which id defined by Eq. (1), in which κ is the specific conductance

$$\Lambda = \frac{1000\kappa}{c} \qquad (1)$$

and c is the concentration in equivalents per liter. Values of Λ change with the concentration and, in general, increase as the solutions measured are made more dilute, that is, as c is decreased. A plot of values of the equivalent conductance Λ against $\sqrt{c}$ for some typical electrolytes is shown in **Fig. 3**. Svante Arrhenius, who was the first to assume that electrolytic conductance is due to freely moving charged ions, explained the decrease of Λ with increasing c by assuming

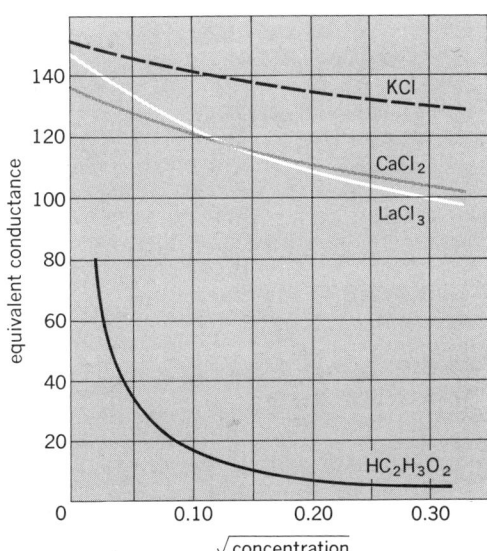

Fig. 3. Equivalent conductance at infinite dilution for some typical electrolytes.

that the number of ionic carriers gets smaller as the concentration increases, and he computed a degree of dissociation α by the formula shown in Eq. (2). The term Λ_0 is obtained by determing

$$\alpha = \frac{\Lambda}{\Lambda_0} \tag{2}$$

Λ at a series of low concentrations and extrapolating to a limiting value, termed the equivalent conductance at infinite dilution. Though Eq. (2) has been shown by later work to give nearly the right values of α for certain poorly conducting solutions, it is now considered to be much in error for the so-called strong electrolytes. These include most salts, such as potassium chloride, KCl, and sodium sulfate, Na_2SO_4, and inorganic acids and bases, such as hydrochloric acid, HCl, and sodium hydroxide, NaOH. For an electrolyte which yields two types of ion, it can be shown that Eq. (3) holds, in which F is the faraday and U^+ and U^- are the mobilities, or speeds, under unit

$$\Lambda = F\alpha \, (U^+ + U^-) \tag{3}$$

potential difference of the positive and negative ions, respectively. For Eq. (2) to hold, these mobilities must be constant from the equivalent concentration c at which Λ is measured to infinite dilution. This requires that the transference numbers of the ions be constant in the same range, which is seldom the case.

Since the advent of the Debye-Hückel theory of interionic attractions, strong electrolytes have been considered to be completely dissociated, that is, the term α of Eq. (3) is equal to unity for these substances. The decreases observed in the values of the equivalent conductances Λ, with increases in concentration, are assumed to be due to reductions in the values of the ionic mobilities U^+ and U^-. According to the theory of P. Debye and E. Hückel, the ion possesses an ionic atmosphere distributed with radial symmetry around the ion as center. This is due to the fact that interionic attractions and repulsions, together with thermal vibrations, tend to produce a slight preponderance of negative ions around a positive ion, and vice versa. The presence of this atmosphere leads to the lowering of ionic mobilities with increasing ion concentrations. The adaptation of the Debye-Hückel theory for conducting solutions is due to Lars Onsager. His equation for very dilute uni-univalent electrolytes, such as sodium chloride, is shown in Eq. (4), in which θ

$$\Lambda = \Lambda_0 - (\theta \Lambda_0 + \sigma) \sqrt{c} \tag{4}$$

and σ are given for uni-univalent electrolytes by Eqs. (5) and (6), in which D is the dielectric

$$\theta = \frac{8.16 \times 10^5}{(DT)^{3/2}} \tag{5} \qquad \sigma = \frac{8.28}{\eta \, (DT)^{1/2}} \tag{6}$$

constant at the absolute temperature T and η is the viscosity.

Equation (4) yields accurate values of the data for strong electrolytes up to concentrations of about 0.001 M, above which there are small deviations. Modifications of Eq. (4) for solutions of salts of higher valence types, such as calcium chloride, $CaCl_2$, and lanthanum chloride, $LaCl_3$, are available and have also been found to agree with the data for dilute solutions.

Onsager, in his derivation of Eq. (4), treated the ions as point charges. Later Raymond Fuoss and Onsager extended the theory to include the radii of the ions and also the effects of higher concentrations. The ion sizes obtained from conductance data agree closely with those calculated from activity measurements. *See* ACTIVITY.

Equation (4), or its empirical and theoretical extensions, can be used to obtain values of Λ_0 from the data on equivalent conductances of dilute solutions. Some typical figures for limiting equivalent conductances Λ_0 of typical strong electrolytes in aqueous solution at 25°C (77°F) are listed below

HCl	426.16	NH_4Cl	149.86	$MgCl_2$	129.40
NaOH	247.8	KNO_3	144.96	NaI	126.94
KBr	151.9	$CaCl_2$	135.84	NaCl	126.45
KI	150.38	Na_2SO_4	129.9	LiCl	115.03

Ion conductances. Values of the limiting equivalent conductance Λ_0 may be assigned to each of the ions of an electrolyte. Thus, for potassium chloride, $\Lambda_{0,KCl} = \lambda_{0,K^+} + \lambda_{0,Cl^-}$, and the value of λ_{0,Cl^-} is the same whether it is derived from measurements on HCl, NaCl, or KCl solutions. This additive relation is known as Kohlrausch's law of the independent mobility of ions. However, it is necessary to obtain the value of λ_0 for at least one ion constituent independently in order to establish the ion conductances of the other ions. The relation used is shown in Eqs. (7), in which

$$t_0^+ \Lambda_0 = \lambda_0^+ \quad \text{or} \quad t_0^- \Lambda_0 = \lambda_0^- \qquad (7)$$

Λ_0 is the limiting equivalent conductance of an electrolyte and t_0^+ and t_0^- are the limiting transference numbers of the positive and negative ion constituents, respectively.

The same value of λ_{0,Cl^-}, within 0.02%, is obtained from precision conductance and transference measurements on solutions of hydrogen, lithium, sodium, and potassium chlorides. Values of the limiting ionic conductance at 25°C (77°F) are given below for some ions.

H^+	349.82	$1/2 Ba^{2+}$	63.64
OH^-	198	Ag^+	61.92
SO_4^{2-}	79.8	$1/2 Ca^{2+}$	59.50
Br^-	78.4	$1/2 Mg^{2+}$	53.06
I^-	76.8	Na^+	50.11
Cl^-	76.34	HCO_3^-	44.48
K^+	73.52	$CH_3CO_2^-$	40.9
NH_4^+	73.4	$CH_2ClCO_2^-$	39.7
NO_3^-	71.44	Li^+	38.69

Nonaqueous systems. In addition to the study of water solutions of electrolytes, considerable study has been given to electrolytes in nonaqueous and mixed solvents. In general, the same principles as those outlined above apply to the interpretation of the results. However, fewer of the electrolytes are completely dissociated, and the degrees of dissociation of the weaker acids and bases are lower. This is due to the fact that, in general, the dielectric constants of nonaqueous solvents are smaller than those of water, so that the attractions between positive and negative ions are greater.

It will be observed that in this article discussion is confined to quite dilute solutions of electrolytes. For concentrated solutions few generalizations of any value can be given.

Molten salts exhibit a wide range of conductivities, depending upon their structures. Salts of alkali and alkaline earth metals usually are largely ionic in character and are highly conductive in the molten state, whereas heavy metal salts may be essentially covalent and exhibit little or no conductivity. Thus the conductivities of liquid $AsCl_3$ and $BiCl_3$ near their melting points are approximately 10^{-6} and 0.44 ohm^{-1} cm^{-1}, respectively, reflecting the more ionic structure of $BiCl_3$.

If, instead of using quite low potentials in the measurement of electrolytic conductances, voltages of the order of 100,000 are employed, the conductances observed are no longer constant but tend to increase with the potential used. Under these conditions Ohm's law evidently is not valid. This increase of conductance with high potentials is called the Wien effect. This effect is in accord with the interionic attraction theory. When the velocity of the ions becomes sufficiently great, the ion atmospheres do not have time to form to their full extent, so that both the electrophoretic and time of relaxation effects exert less influence on the conductance. However, a large Wien effect is also found for weak acids and bases. It would appear that the high potentials produce, temporarily, additional ionization of these substances. This explanation has been proposed and discussed theoretically by Onsager. If very high frequencies are used in the measurements, an increase in the conductance, termed the Debye-Falkenhagen effect, is observed. This can also be explained by the interionic attraction theory. SEE ELECTROCHEMISTRY.

ELECTROCHEMICAL SERIES
EUGENE G. ROCHOW

A series in which the metals are listed in the order of their chemical reactivity, the most active at the top and the less reactive or more "noble" metals at the bottom. In a broader sense such an activity series need not be limited to the metals but may be carried on through the electronegative (nonmetallic) elements as well. See the **table** for a list of common elements.

Electrochemical series of the elements*

Element	Symbol	Element	Symbol
Lithium	Li	Zinc	Zn
Potassium	K	Chromium	Cr
Rubidium	Rb	Gallium	Ga
Cesium	Cs	Iron	Fe
Radium	Ra	Cadmium	Cd
Barium	Ba	Indium	In
Strontium	Sr	Thallium	Tl
Calcium	Ca	Cobalt	Co
Sodium	Na	Nickel	Ni
Lanthanum	La	Molybdenum	Mo
Cerium	Ce	Tin	Sn
Magnesium	Mg	Lead	Pb
Scandium	Sc	Germanium	Ge
Plutonium	Pu	Tungsten	W
Thorium	Th	**Hydrogen**	**H**
Beryllium	Be		
Uranium	U	Copper	Cu
Hafnium	Hf	Mercury	Hg
Aluminum	Al	Silver	Ag
Titanium	Ti	Gold	Au
Zirconium	Zr	Rhodium	Rh
Manganese	Mn	Platinum	Pt
Vanadium	V	Palladium	Pd
Niobium	Nb	Bromine	Br
Boron	B	Chlorine	Cl
Silicon	Si	Oxygen	O
Tantalum	Ta	Fluorine	F

*According to standard oxidation potentials $E°$ at 25°C or 77°F.

The electrochemical series as it applies to metals was first established by laboratory experiments in which the purpose was to determine which metals would displace others from solutions of their salts. Thus a clean strip of zinc immersed in a solution of copper sulfate is soon found to be covered by a deposit of copper, while zinc in turn goes into solution from the strip as zinc ions. By definition, then, zinc is a more reactive metal than copper, since it will displace copper from a solution of Cu^{2+} ions. The reaction is readily seen to be an oxidation-reduction transfer of electrons, which can be summarized by the equation below. Similarly, copper will displace silver from

$$Cu^{2+} + Zn = Cu + Zn^{2+}$$

a solution containing Ag^+ ions, depositing crystals of metallic silver and coloring the solution with Cu^{2+} ions. From these observations an activity series may be set up in the order Zn, Cu, Ag. By exhaustive experiments with other metals it becomes possible to draw up a complete list in the order of chemical activity, in which the metals at the top of the list are those which are found to give up their electrons most readily (that is, are the most electropositive elements). Such a list is shown in the table, where lithium exhibits the most reactivity as a metal.

The ease with which an isolated atom of an element gives up an electron, known as the first ionization potential, is a precise physical quantity which can be measured by electrical experiments on gases or vapors at low pressure. The replacement experiments which determine the order of the electrochemical series take place in a very different environment, since they involve solid phases and also aqueous solutions with their consequent hydration effects. Moreover, it might well be expected that displacement reactions in solution would depend upon the concentrations of the reagents used, and also upon the presence or absence of other dissolved substances.

To obtain a more accurate and reproducible activity series, it is best to turn to the more exact quantity called electrode potential, or oxidation-reduction potential, which is defined as the voltage developed by a sample of pure metal immersed in a solution of one of its salts (at unit activity and at 25°C or 77°F) versus a hydrogen electrode immersed in hydrochloric or sulfuric acid of equivalent concentration. For details about this measurement SEE ELECTRODE POTENTIAL.

It is evident that by confining the experimental conditions to a standard concentration and temperature the hydration and concentration effects can be kept quite constant, making possible a more exact listing of metals according to their activity. Hence any present-day electrochemical series must rely on the measurements of oxidation potential and should be in agreement with the accepted values determined from such electrochemical cells. Such reliance has the further advantage that the series need not then be confined to metals but may be extended to the nonmetals, or electronegative elements. As before, those metals which will liberate hydrogen from dilute acids (such as hydrochloric or sulfuric) will stand above hydrogen in the series, while those metals and nonmetallic elements which will not liberate hydrogen from such dilute acids will stand below hydrogen in the list. Since the oxidation potentials are also related to the equilibrium constants for reversible reactions, it becomes possible to calculate oxidation potentials from other information when direct experiments are inconvenient, as in the case of the alkali metals versus aqueous solutions of their salts. SEE ELECTROCHEMISTRY; ELECTRONEGATIVITY; OXIDATION-REDUCTION.

Bibliography. F. A. Cotton and G. Wilkinson, *Advanced Inorganic Chemistry: A Comprehensive Text*, 4th ed., 1980; C. A. Hampel (ed.), *Encyclopedia of Electrochemistry*, 1964, reprint 1972.

ELECTRODE
WALTER J. HAMER

An electrical conductor through which an electric current enters or leaves a conducting medium, whether it be an electrolytic solution, solid, molten mass, gas, or vacuum. For electrolytic solutions, many solids, and molten masses, an electrode is an electric conductor at the surface of which a change occurs from conduction by electrons to conduction by ions. For gases and vacuum, the electrodes merely serve to conduct electricity to and from the medium. SEE ELECTRODE POTENTIAL; ELECTRODEPOSITION ANALYSIS; ELECTROLYSIS.

ELECTRODE POTENTIAL
RICHARD GLICKSMAN

The potential which a metal or gas electrode takes up relative to a solution of ions.

The various electrodes encountered in electrochemical work may be grouped into seven types: (1) metal–metal ion, (2) amalgam, (3) nonmetal nongas, (4) gas, (5) metal–insoluble salt, (6) metal–insoluble oxide, and (7) oxidation-reduction. Any of these electrodes may be combined with any other to give a cell, the emf of which is equal to the algebraic sum of the potentials of the two electrodes. Metal–metal ion electrodes and gas electrodes are discussed.

Metal–metal ion electrodes. When a metal is immersed in an electrolyte, an equilibrium tends to be established in which a steady difference of electric potential exists across the region of the interface between metal and solution. This equilibrium electrode potential results from the ionization of the atoms of the metal until the displacement of electric charges produced thereby exactly balances the tendency for additional metallic atoms to ionize. If the metal electrode M has a valence of n, reaction (1) occurs, where e^- indicates an electron and M^{n+} an ion

$$M \rightleftharpoons M^{n+} + ne^- \tag{1}$$

in solution. Examples of metal–metal ion electrodes are zinc, copper, and sodium electrodes.

Because the potential of such an electrode changes with the concentration of ions, it is necessary to adopt some standard concentration at which to compare the potentials of various electrodes. The standard electrode potential $E°$, expressed in volts, is defined as the potential of an element immersed in a solution of its ions at unit activity, that is, the effective concentration of 1 mole/1000 g of water. The electrode potential E at other concentrations is given by Eq. (2),

$$E_M = E_M^\circ - \frac{RT}{nF} \ln a_{M^{n+}} \qquad (2)$$

where T is the absolute temperature, F is the faraday (96,487 coulombs), R the gas constant, and $a_{M^{n+}}$ the effective concentration (activity) of M^{n+} ions in the solution. At 25°C (77°F) Eq. (2) can be written as Eq. (3).

$$E_M = E_M^\circ - \frac{0.05916}{n} \log a_{M^{n+}} \qquad (3)$$

Because the single electrode potential E involves the activity of an individual ionic species, it has no strict thermodynamic significance. This difficulty is overcome by defining the standard hydrogen electrode as an arbitrary zero of potential. Electrode potentials based on this zero are thus said to refer to the hydrogen scale. Such a potential is actually the electromotive force (emf) of a cell obtained by combining the given electrode with a standard hydrogen electrode. See ELECTROCHEMICAL SERIES.

Gas electrodes. A gas electrode is formed by partially immersing an inert metal (usually platinized platinum) in a solution of the ions of the gas. The gas must establish a reversible equilibrium with the ions in solution in the presence of the metal. The metal wire or foil helps establish an equilibrium between the gas and its ions and serves as the electric contact for the electrode.

The potential of such an electrode is determined by the pressure of the gas and the activity of its ions in solution. Thus, for the chlorine electrode, the reaction is shown in (4), and the equation for the electrode potential E_{Cl_2} is shown in Eq. (5), where P_{Cl_2} is the pressure of chlorine

$$\tfrac{1}{2}Cl_2(g) + e^- \rightleftharpoons Cl^- \qquad (4) \qquad\qquad E_{Cl_2} = E_{Cl_2}^\circ - \frac{RT}{F} \ln \frac{a_{Cl^-}}{P_{Cl_2}^{1/2}} \qquad (5)$$

in atmospheres, a_{Cl^-} the effective concentration (activity) of chloride ions in the solution, and $E_{Cl_2}^\circ$ the standard electrode potential for the chlorine electrode, which is equal to 1.3583 V at 25°C (77°F). The standard electrode potential of a gas electrode is defined as the potential of the electrode when the gases involved in the reaction are at a fugacity of 1 atm (100 kilopascals), that is, an effective pressure of 1 atm and all dissolved substances are at an effective concentration (activity) of 1 molal, that is, 1 mole/1000 g of water. See ACTIVITY; FUGACITY.

The most important gas electrode is the hydrogen electrode, which is reversible to hydrogen ions. The reaction for this electrode is shown in (6), and the electrode potential is given by Eq. (7). But $E_{H_2}^\circ$, the standard electrode potential of hydrogen, is the reference of all emf measure-

$$\tfrac{1}{2}H_2(g) \rightleftharpoons H^+ + e^- \qquad (6) \qquad\qquad E_{H_2} = E_{H_2}^\circ - \frac{RT}{F} \ln \frac{a_{H^+}}{P_{H_2}^{1/2}} \qquad (7)$$

ments and is taken by definition to be zero at all temperatures. Thus Eq. (7) becomes Eq. (8).

$$E_{H_2} = -\frac{RT}{F} \ln \frac{a_{H^+}}{P_{H_2}^{1/2}} \qquad (8)$$

Another gas electrode which has received considerable attention is the oxygen electrode, whose potential depends on the activity of hydroxyl ions. However, unlike the hydrogen and chlorine electrodes, the oxygen electrode cannot be made reversible because no suitable electrode material has been found which can catalyze the establishment of the equilibrium between oxygen and hydroxyl ions. This equilibrium is shown in (9). The standard potential of the oxygen electrode

$$\tfrac{1}{2}O_2 + H_2O + 2e^- \rightleftharpoons 2OH^- \qquad (9)$$

cannot be determined directly from emf measurements because of the irreversible behavior of this electrode. It is possible, however, to derive the value in an indirect manner, and it has been found to be +0.401 V.

Measurements. In order to measure the potential of any electrode, it is necessary, in principle, to combine the electrode with a reference electrode such as the hydrogen electrode,

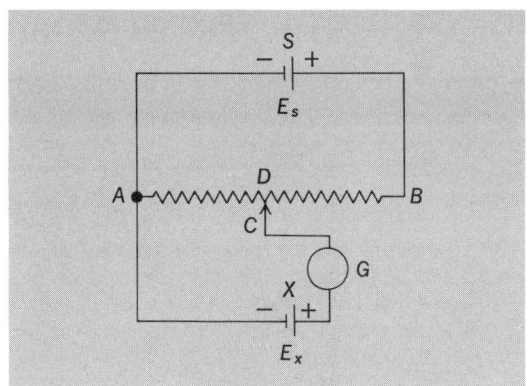

Poggendorff compensation method for measuring emf.

which has an arbitrarily assigned potential, and to measure the total voltage across the two half-cells. The potential of the reference electrode is then subtracted to give the required electrode potential. For various reasons, such as the difficulty in setting up a hydrogen gas electrode, several subsidiary reference electrodes, whose potentials are known on the hydrogen scale, have been devised. The most common of these is the calomel electrode, which consists of mercury in contact with a solution of potassium chloride saturated with mercurous chloride.

A simple voltmeter cannot be used alone for measuring the emf of a small cell because the operation of the voltmeter draws some current, which causes chemical changes at the electrodes and produces a different voltage. To avoid these difficulties, the emf is measured by balancing against the cell a known voltage under conditions such that practically no current flows.

The constructional details of a potentiometer, as the apparatus for measuring emf is called, is shown schematically in the **illustration**. In this diagram S is a cell of know emf E_S, whose potential is impressed across a uniform resistance AB. Connected with S in such a way that the two emfs oppose each other is the source X of unknown potential E_X. To find E_X the sliding contact C is moved along AB until a position D is found at which the galvanometer G gives no deflection. From E_S and the distances AB and AD, the unknown emf E_X is found as follows. Since E_S is impressed across the full length AB, E_S for any given current passing through the resistance must be proportional to AB. Again, because E_X is impressed only across the distance AD, it must be proportional to this length. Consequently, dividing E_X by E_S gives Eqs. (10) and (11).

$$\frac{E_X}{E_S} = \frac{AD}{AB} \qquad (10) \qquad\qquad E_X = \left(\frac{AD}{AB}\right) E_S \qquad (11)$$

At the present time the majority of emf measurements are made by means of special potentiometers, available commercially, which operate on the foregoing principle. In these potentiometers, the conductor AB consists of a number of resistance coils with a movable contact, together with a slide wire for fine adjustment. A standard cell is used for calibration purposes, and the emf of the cell being studied can then be read with an accuracy of 0.1 mV or better.

The standard cell that is widely employed for emf measurements is some form of the Weston cell. It is highly reproducible, its emf remains constant over long periods of time, and it has a small temperature coefficient. In order to retain constancy of emf during use, only very minute currents should be drawn from the cell, as is actually done if the potentiometer is operated properly. SEE ELECTROCHEMISTRY; ELECTRONEGATIVITY; OXIDATION-REDUCTION.

Bibliography. M. S. Antelmon, *The Encyclopedia of Chemical Electrode Potentials*, 1982; J. Bockris et al., *An Introduction to Electrochemical Science*, 1974; B. E. Conway, *Theory and Principles of Electrode Processes*, 1965; D. R. Crow, *Principles and Applications of Electrochemistry*, 2d ed., 1979; I. N. Levine, *Physical Chemistry*, 2d ed., 1983.

ELECTROCHEMICAL TECHNIQUES
Sam P. Perone

Experimental methods developed to study the physical and chemical phenomena associated with electron transfer at the interface of an electrode and solution. The objective is to obtain either analytical or fundamental information regarding electroactive species in solution. Fundamental electrode characteristics may be investigated also.

The physical and chemical phenomena important in electrode processes generally occur very close to the electrode surface (usually within a few micrometers). Mass transfer of species involved in an electrode process to and from the bulk of solution is one important aspect. Inclusion of a large excess of inert electrolyte in most electrochemical systems eliminates electrical migration as an important means of mass transfer for electroactive species, and only convection and diffusion are considered.

Important chemical aspects of electrode processes include the oxidation or reduction occurring as a result of electron transfer, and coupled chemical reactions. Coupled reactions are initiated by production or depletion of the primary products or reactants at the electrode surface.

The primary experimental variables involved in electrochemical techniques are the potential E, the current I, and the time t. Either the potential or current at the working electrode is controlled and the other observed as a function of time. The many ways in which either may be controlled give rise to the wide variety of controlled-potential or controlled-current techniques.

The general scheme for electron transfer at an electrode in solution is shown in reaction (1), where O is the number of electrons transferred. When k_f and k_b, the rate constants for the

$$O + ne^- \underset{k_b}{\overset{k_f}{\rightleftharpoons}} R \tag{1}$$

forward and back reaction, respectively, are very fast, the system is called reversible and the Nernst equation, Eq. (2), holds. In this relation E is the electrode potential, $E^{o\prime}$ is the formal

$$E = E^{o\prime} + (0.059/n) \log (C_O/C_R) \tag{2}$$

standard potential for the redox couple, and C_O and C_R are concentrations at the electrode surface. In the following discussions, only reversible reduction processes will be considered, although oxidations are equally applicable. SEE ELECTROCHEMICAL PROCESS; ELECTRODE; ELECTRODE POTENTIAL; ELECTROLYSIS.

Controlled potential. Two methods have been developed in this area and are described.

Constant potential with convection. The electrode potential is held constant or varied slowly with time (about 5 mV/s) as the solution is stirred or the electrode rotated at a constant rate. The current is measured as a function of potential, and it increases sharply whenever the potential passes through the region of $E^{o\prime}$ for the particular electroactive species involved. Between reduction steps, current plateaus are established, the heights of which are proportional to the concentrations of each electroactive species. This is plotted in **Fig. 1**a, where I_L is the limiting or plateau current which is proportional to the bulk concentration of electroactive species C^*. The point designated $E_{1/2}$ is the half-wave potential, about equal to the formal standard potential $E^{o\prime}$. In Fig. 1b an alternating current potential is superimposed on the direct current one of Fig. 1a. E_p is peak voltage and I_p is alternating current.

Potentiostatic chronoamperometry. This technique consists of maintaining a constant potential without convection. When a reducing potential is imposed instantaneously on a stationary working electrode in quiescent solution, current will rise sharply and then decay as the electroactive species in the electrode vicinity is depleted by electrolysis. The magnitude of the current is proportional to the bulk concentration of electroactive species and is related to the electrode potential through the Nernst equation, Eq. (2). If the potential is sufficiently beyond $E^{o\prime}$, the Nernst equation demands complete conversion to the reduced form. Thus, under these conditions the current is diffusion-controlled and decays with $1/t^{1/2}$. This is shown by Eq. (3) where F is the

$$I = nFAC^*(D/\pi t)^{1/2} \tag{3}$$

faraday, A is the electrode area, D is the diffusion coefficient for the electroactive species, and C^* is the bulk concentration of electroactive species.

One important variation of potentiostatic chronoamperometry is to apply a double (or cyclic)

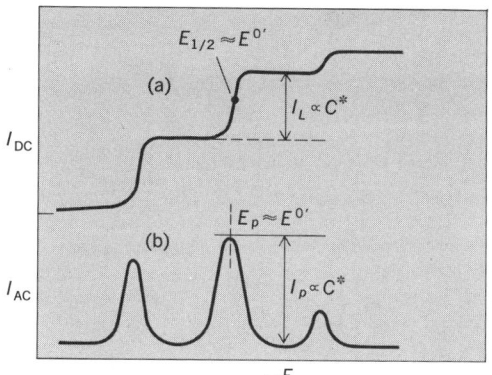

Fig. 1. Plots of the variation of current with constant applied potential in reduction processes. (a) Direct current versus voltage at a rotated working electrode, where three successive reductions of electroactive species occur. (b) Alternating current versus direct-current voltage applied at a working electrode in an alternating-current polarographic experiment.

potential step. During the initial step, electrolysis occurs, depleting the oxidized form O, but producing the reduced form R in the immediate vicinity of the electrode. If the potential is instantaneously switchd back to the initial value, species R will be reoxidized, and an inverted I-t curve will be obtained with well-defined diffusion-controlled decay characteristics (**Fig. 2**). If species R is chemically reactive in the solution (coupled chemical reaction), its contribution to the reverse current will be diminished not only because of mass transfer limitations, but also because of chemical decay. The ratio of reverse to forward currents (i_a/i_c) will be smaller for larger chemical rates and a specific switching time τ. Thus, within electronic limitations, one can adjust τ to fit the reactivity of the chemical system in order to evaluate the kinetics.

Variable potential. Several procedures that have been developed in this area are described.

Linearly varying potential (LVP). The term chronoamperometry with LVP is applied here. When the potential of a stationary electrode in quiescent solution is varied in a linear fashion, as in conventional polarography, but without restrictions on the sweep rate, a peak-shaped stationary electrode polarogram is obtained. The generalized system and a current-voltage curve are given in **Fig. 3**. Peak current I_p is proportional to bulk concentration of electroactive species, and the peak potential E_p is related to $E^{o\prime}$. Analytically, the approach is applicable over the range of 10^{-3} to 10^{-6} M electroactive species.

A variation, called cyclic voltammetry, involves application of a triangular potential sweep, allowing one to sweep back through the potential region just covered. A reverse current I_r is

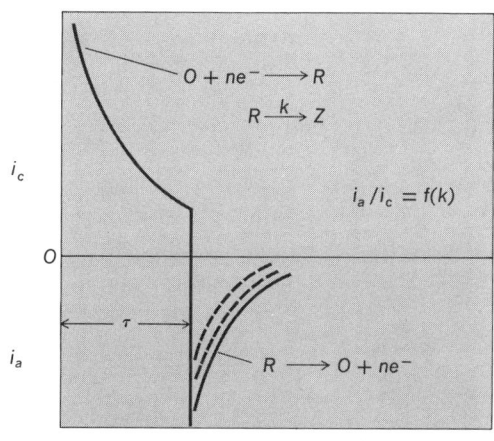

Fig. 2. Current-time behavior for double potential-step experiment at a stationary working electrode. Broken lines in reverse current-time curve represent cases where the rate of chemical reaction is fast compared to the time scale of the experiment.

(a)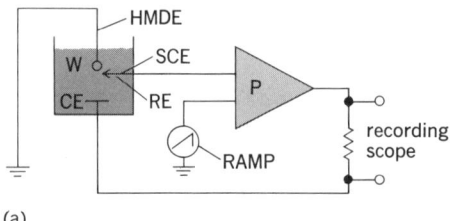

(b)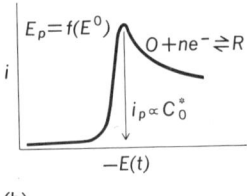

Fig. 3. Stationary electrode polarography. (a) Generalized instrumentation and (b) the current-voltage curve obtained on the scope. HMDE = hanging mercury drop electrode; SCE = saturated calomel reference electrode; CE = auxiliary counter electrode; P = potentiostate; W = working electrode; RE = reference electrode; and RAMP = linearly increasing voltage signal.

obtained with characteristics related to the chemical reactivity of the initial electrode product. This is analogous to cyclic potential-step chronoamperometry. Current-voltage curves are displayed in **Fig. 4**.

An enhancement in analytical sensitivity is obtained by another variation called stripping analysis. Here one applies a constant reducing potential for a period of time (about 1 to 60 min), sufficient to collect substantial amounts of the reduced species as an amalgam or deposit at the working electrode. Then the concentrated product is stripped by applying a linear oxidizing potential sweep. The stripping current is considerably larger than would be obtained from a direct measurement. The stripping current is proportional to the amount accumulated, which in turn is proportional to bulk concentration and prior electrolysis time. Sensitivity to 10^{-9} M is achieved.

Alternating potential. Three variations of this method are outlined below.

Alternating-current polarography. This involves an approach identical to conventional polarography at a dropping mercury electrode, except that a small-amplitude (about 1–15 mV), sinusoidal alternating voltage (frequency about 10–100 Hz) is superimposed on the direct-current controlling potential. Only the ac component of the electrolysis current is measured. A plot of I_{ac} versus E_{dc} results in a derivative-type curve with symmetrical triangular peaks (Fig. 1b). The pri-

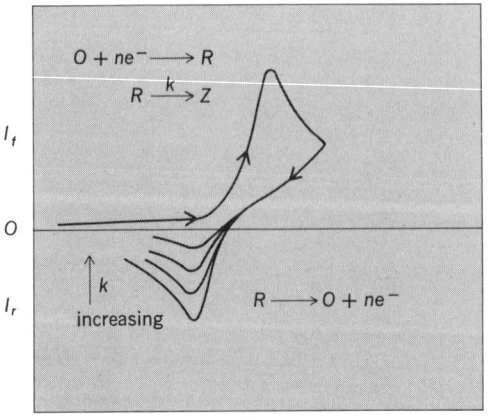

Fig. 4. Current-voltage curves in cyclic voltammetry.

mary analytical advantage is that the interference from succeeding or preceding reductions is considerably reduced. Sensitivity to 10^{-6} M is possible.

Square-wave polarography. This also involves the basic polarographic approach, except that a small-amplitude (about 1–25 mV) square wave (frequency about 225 Hz) is imposed on the controlling dc potential. The current measured is an alternating component obtained by measuring the current at times near the end of each half-cycle. This avoids the large capacitive background current associated with instantaneous electrode potential changes and allows greater sensitivity. Analyses over the range of 10^{-3} to 10^{-7} M are possible.

Pulse polarography. This is a variation of square-wave polarography, where a single small potential step is imposed on the controlling dc potential during the life of each mercury drop. The change in current with each step is measured and plotted as a function of the dc potential. Like ac and square-wave polarography, a derivative-type curve is obtained. Sensitivity is about the same as square-wave polarography.

Controlled current. Several analytical methods based on controlled current have evolved.

Chronopotentiometry. A constant current is imposed at the working electrode, and its potential is monitored with time. The electrode must assume a potential which will cause electrolysis sufficient to maintain the imposed current. Thus, the electrode adopts first the potential of the most easily reduced species. As this species is depleted near the electrode, the potential shifts to that of the next most easily reduced species. This continues until reduction of solvent or electrolyte occurs. Characteristic curves are plotted in **Fig. 5**. The transition time τ is related to the bulk concentration of the electroactive species giving rise to the transition. For the first transition, C^* is proportional to $\tau^{1/2}$. However, for the reduction of a second species the transition time τ_2 depends on the first transition time. Thus, the term $[(\tau_1 + \tau_2)^{1/2} - \tau_1^{1/2}]$ is proportional to the bulk concentration of the second reducible substance. Further steps are similarly complicated.

Cyclic chronopotentiometry. This involves following a normal chronopotentiometric experiment with an instantaneous current reversal. The product is reoxidized and a reverse transition is seen. Results obtained are analogous to the cycle voltammetric and double potential-step experiments. The ratio of the reverse transition time τ_r to the forward transition time τ_f reflects the kinetics of a succeeding chemical reaction of species R. For negligible chemical reactivity $\tau_r/\tau_f = 1/3$.

Coulostatic analysis. This does not involve controlled current, but the application of a very short, large pulse of current to the electrode. This pulse serves to charge up the capacitive electrode–solution interface to some new potential. The cell circuit is then opened, and the return of the working electrode potential to its initial value is monitored. The open-circuit condition requires that the current necessary to discharge the electrode interface comes from electrolysis of electroactive species in solution. The change in electrode potential ΔE versus $t^{1/2}$ results in a straight-line plot, the slope of which is proportional to concentration, as in Eq. (4), where C is the

$$\pm \Delta E = \frac{2nFD^{1/2}C^*}{\pi^{1/2}C}t^{1/2} \tag{4}$$

electrical capacitance associated with the electrode-solution interface.

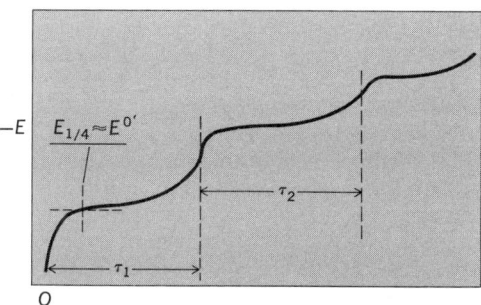

Fig. 5. Potential-time curves for chronopotentiometric experiments. Curve represents successive reduction of three different electroactive species in solution.

Thin-layer electrochemistry. Electrochemical techniques can be applied in cells where only a 10–100-μm-thick solution is electrolyzed. The principal advantage is that, for experiments lasting more than about 1 s, mass transfer limitations can be ignored. Thus, correlations between observed parameters and system characteristics are more straightforward. For example, for a chronopotentiometric experiment the transition time represents complete depletion of the electroactive species. Thus, C^* is proportional to τ for each of a sequence of reduction steps. Also, if a linear potential sweep is applied, a symmetrical current peak is obtained, the height or area of which is related to the concentration of electroactive species.

Bibliography. J. O'M. Bockris et al. (eds.), *Modern Aspects of Electrochemistry*, vol. 17, 1986; B. E. Conway et al. (eds.), *Modern Aspects of Electrochemistry*, vol. 16, 1985; P. Delahay, *New Instrumental Methods in Electrochemistry: Theory, Instrumentation, and Applications to Analytical and Physical Chemistry*, 1954, reprint 1980.

ELECTROCHEMICAL PROCESS
CHARLES L. MANTELL

The principles of electrochemistry may be adapted for use in the preparation of commercially important quantities of certain substances, both inorganic and organic in nature.

INORGANIC PROCESSES

Inorganic chemical processes can be classified as electrolytic, electrothermic, and miscellaneous processes including electric discharge through gases and separation by electrical means. In electrolytic processes, chemical and electrical energy are interchanged. Current passed through an electrolytic cell causes chemical reactions at the electrodes. Voltaic cells convert chemicals into electricity. Electrothermic processes use electricity to attain the necessary temperature for reaction. For related information SEE ELECTROCHEMISTRY; ELECTROLYSIS; ELECTROLYTIC CONDUCTANCE.

Voltaic cells are used for the intermittent production of small amounts of electricity. When the chemicals involved are exhausted and must be replaced, the unit is called a primary cell. A special case of the primary cell is the fuel cell in which the fuel and an oxidizer are fed continuously to the cell, converted to electricity, and the products removed. If exhausted components can be revived by passing electricity backward through the unit, it is called a secondary cell, storage battery, or accumulator. Cells may be connected in parallel or in series to form a battery.

For a discussion of the theory and description of commercial primary, secondary, and fuel cells SEE BATTERY; FUEL CELL.

Electrolysis in aqueous solutions. The electrolysis of water to form hydrogen and oxygen, according to the reaction $2H_2O \rightarrow 2H_2 + O_2$, may be considered as the simplest process for aqueous electrolytes. It does not compete with hydrogen from propane or from natural gas and with oxygen from liquid air, except in small installations. While simplicity, high hydrogen purity requirement, and lower capital cost (in small plants) have justified electrolytic plants, severely rising energy costs limited such applications. The electrolyte is 18–30% NaOH or KOH, the latter having a lower resistance but higher cost. The cathode is steel and the anode is nickel-plated steel separated by a diaphragm, usually of asbestos (**Fig. 1**). A cell voltage of 2.0–2.5 V is the summation of the decomposition voltage of water (1.23 V at room temperature) and the oxygen and hydrogen overvoltages, plus the *IR* drop through the electrolyte, electrode contacts, and bus bars. The raw material is distilled or demineralized water. There are monopolar cells operating up to 20,000 A, bipolar or filter-press cells using 2000–5000 A, and 150 cells in series at 300–400 V overall. Cells are available which operate at 600 psi (4.1 megapascals). A. T. Kuh described 11 cells using 25–30% KOH and indicated designs operating at pressures up to 200 atm (20 MPa).

Heavy water, or deuterium oxide, used in moderating nuclear reactors is also a by-product of the electrolysis of water. Protium (^{1}H) is preferentially discharged, so that the electrolyte becomes richer in deuterium. Electrolysis must be combined with catalytic exchange or distillation processes when used in primary or earlier stages of concentration; it is also used in final stages to produce 99–100% concentration.

Metallurgical applications. Protective or decorative coatings on a base metal such as steel are obtained by electroplating. Plating may also be used to replace worn metal or to provide

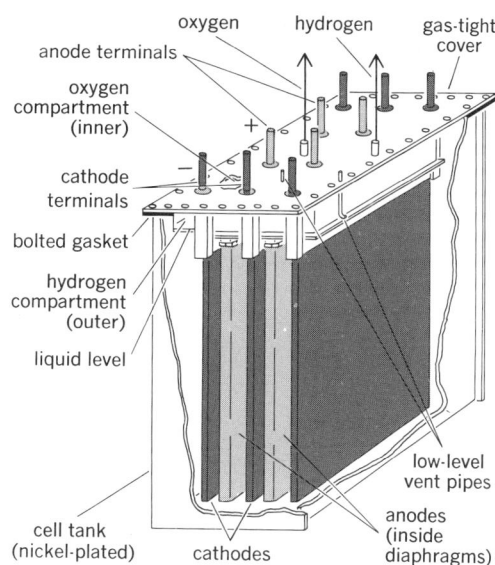

Fig. 1. Diagram of Stuart hydrogen-oxygen cell. (*Electrolyser Corp. Ltd., Toronto, Canada*)

a wear-resistant surface. The final surface may require several layers of different metals or even layers of the same metal deposited under varying conditions. The metals plated are: copper, cadmium, chromium, cobalt, gold, iron, lead, nickel, the platinum metals, silver, tin, and zinc, and many alloys. Electrogalvanizing is preferred over hot dipping for applying zinc to steel. Tin plate for containers is electrolytic. Factors affecting the resulting plate include pretreatment and cleaning of the metal surface, current density, concentration of metal ions, agitation, temperature, conductance of solution, pH, and addition agents.

Electroforming is a method of forming or reproducing articles by electrodeposition. In contrast to electroplating, the product is removed from the base surface or mold. A nonconducting surface can be made conductive by metallizing or with graphite powder. A low-melting-point metal mold can be used, or a mold can be plated with a metal from which the final metal can be removed. The electrodeposits may be up to 0.5 in. (13 mm) thick, and after completion are removed from the mold. Phonograph records, electrotype, textile replicas, and brass instrument bells are common products of electroforming, which is also widely used in the automotive and electronic industries.

Electrodeposition of metal powders is used to produce particles in the 1–1000-micrometer range for use in powder metallurgy and metallic pigments. Powdered iron made electrolytically makes stronger parts than other types, and electrolytic powdered copper makes parts that are easier to machine and have improved wear.

In the anodizing of aluminum articles, a coating approximately 0.001 in. (25 μm) thick is applied, which can be dyed or made impervious. The article to be anodized is cleaned and made anodic in sulfuric acid solution.

Electrolytic polishing of metals is accomplished by making the article anodic in an electrolyte of mixed acids, such as phosphoric-chromic acids or phosphoric-sulfuric acids. High points on the surface apparently dissolve, to give a unique polish not achievable by mechanical polishing.

Electrolytic machining of metals is accomplished by making the metal part anodic in a suitable electrolyte. Metal dissolves at the anode and hydrogen discharges at the cathode. The cathode is contoured as a negative of the shape to be developed in the workpiece. By circulating the electrolyte under very high pressure and using high current densities (30–2000 A/in.^{2} or 5–300 A/cm^2), practical machining rates are achieved. Very hard and very thin metal surfaces can be machined without changing the heat treatment.

Electrorefining is a process for purifying metals and recovering their impurities, which at times are more valuable than the original metal. Copper from its ore or scrap is purified of volatilizable impurities, cast into an anode, dissolved in an electrolyte in a cell, and deposited at a cathode as a very pure, highly conductive metal for electrical engineering purposes (**Fig. 2**). Gold, silver platinum, selenium, and tellurium are recovered as by-products. Nickel is freed of copper by using a diaphragm between anode and cathode, with cobalt and the platinum metals as by-products. Lead is freed from bismuth; tin from lead, antimony, and bismuth with silver as a by-

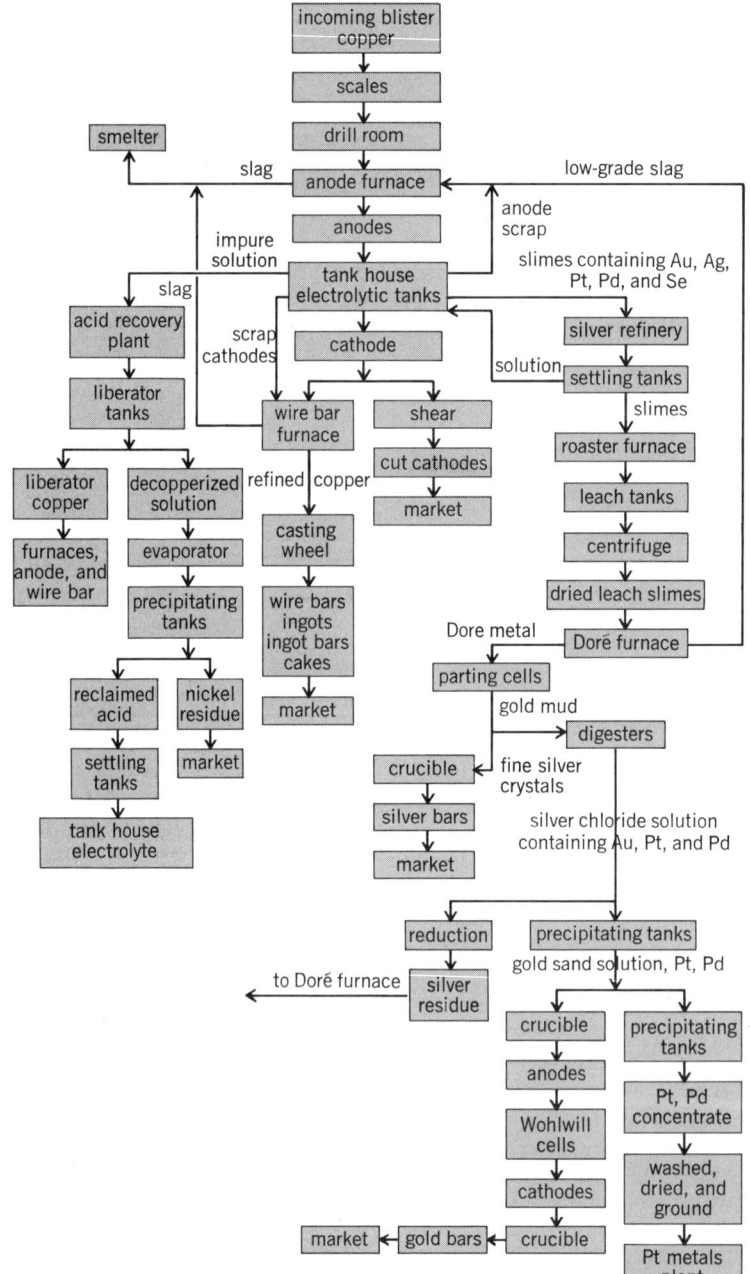

Fig. 2. Flowsheet for a copper plant and the refinery by-products resulting from a product of 99.97% conductor copper. (*Ontario Refining Co., International Nickel Co. of Canada, Ltd.*)

product; and zinc from copper and lead, with cadmium as a by-product, to make commercial a 99.99% pure metal for die casting. The last operation is usually referred to as electrowinning. This has rapidly replaced pyrometallurgy or fire processing because it can eliminate sulfur dioxide atmospheric pollution and contamination by particulates.

Electrowinning, sometimes termed aqueous electrometallurgy, involves processing of metallic ores, usually of very low metal content but large in volume, by leaching solutions, usually

Table 1. Electrolytics for various individual metals

Metal	Source	Method	Electrolyte	Application
Antimony	Crude metal	Refining	Sulfates	Alloys, batteries, chemicals
Antimony	Antimony ores	Winning	Sulfides	
Bismuth	Lead refining slimes	Refining	Chlorides	Alloying agent, medicine
Cadmium	Zinc residues and slimes	Winning	Sulfates	Electroplating, alloys
Chromium	Chromium ores	Winning	Sulfates	High-temperature, alloys
Cobalt	Complex cobalt ores	Winning	Sulfates	Alloying agent, electronics
Cobalt	Complex nickel ores	Refining	Sulfates and chlorides	Alloying agent, electronics
Copper	Copper ores, complex ores	Winning	Sulfates	Electrical conductors, wire, brass alloys
Copper	Crude metal, secondary materials, waste	Refining	Sulfates	Electrical conductors, wire, brass alloys
Copper powder	Refined metal	Plating	Sulfates	Powders for powder metallurgy, oilless bearings
Copper sheet	Purified electrolyte	Plating	Sulfates	Sheet for printed circuits, electronics
Gallium	Sodium aluminate liquors	Winning	Caustic	Low-melting-point metals and alloys
Gold	Copper refining slimes	Refining	Chlorides	Jewelry, dentistry, plating
Indium	Residues, wastes, ores	Refining	Chlorides	Silver alloys, jewelry, television
Iron	Low-carbon steel	Refining	Sulfates	Powder metallurgy
Lead	Crude lead	Refining	Fluosilicates	Separation of bismuth etc., alloys, fittings
Manganese	Manganese ores	Winning	Sulfates	Stainless steels, carbon-free metal, alloys
Nickel	Crude metal, nickel matte	Refining	Sulfate-chloride	Plating, alloys, stainless steel
Silver	Copper refinery slimes, lead residues, crude metal, ore	Refining	Nitrates	Jewelry, electrical applications, alloys
Silver	Silver alloys, photographic wastes	Winning	Nitrates	Jewelry, electrical applications, alloys
Solder	Waste and crudes	Refining	Fluosilicate	Metal joining, electrinic components
Tin	Crude metal	Refining	Cresol-sulfonates	Tin plate, solder, alloys
Zinc	Zinc ores, complex ores	Winning	Sulfates	Die-casting alloys, battery cups, brass, galvanizing
Zinc	Ores and residues	Winning	Caustic	Chemicals, paint

Table 2. Anodes, cathodes, and diaphragms used on commercial processes

Metal	Method	Anode	Anolyte	Catholyte	Diaphragm	Cathode	Voltage
Antimony	Winning	Insoluble	Same		No	Steel	2.5–3
Cadmium	Winning	Insoluble	Same		No	Aluminum	4.0
Chromium	Winning	Insoluble	More acid	Acid	Yes	Hastelloy	4.2
Cobalt	Refining	Soluble	Same		No	Cobalt-stainless steel	2.5
Copper	Refining	Soluble	Same		No	Copper	0.2–0.3
Copper	Winning	Insoluble	Same		No	Copper	2–2.1
Gold	Refining	Soluble	Same		No	Gold	0.5–2.8
Lead	Refining	Soluble	Same		No	Lead	0.35–0.45
Manganese	Winning	Insoluble	Acid	Alkaline	Yes	Stainless steel	
Nickel	Refining	Soluble	Same	Pure	Yes	Nickel	2.4
Silver	Refining	Soluble	Same		Yes	Stainless steel or carbon	1.3–5.4
Tin	Refining	Soluble	Same		No	Tin	0.3
Zinc	Winning	Insoluble	Same		No	Aluminum	3.25–3.7

sulfates, to obtain metal-containing electrolytes which can be processed with insoluble anodes and metal cathodes. Examples are found in cadmium, copper, cobalt, manganese, and zinc.

Electrorefining, electrowinning, and electroforming are summarized as to the individual metals in **Table 1**, while **Table 2** gives anodes, cathodes, and diaphragms for commercially successful operation. **Table 3** lists energy consumption of aqueous electrochemical operations. *See* Electrometallurgy.

Electrolytic corrosion of metals. This occurs because some parts of the surface of metals act as anodes and corrode, whereas other parts act as cathodes and do not corrode.

Cathodic protection is provided if the whole surface is made cathodic to a separate anode and sufficient voltage is available between the two electrodes. This type of protection is used to inhibit corrosion of boilers, condensers, underground pipelines, ships, and water tanks. Sacrificial anodes of zinc, magnesium, or aluminum alloy may provide the potential, or inert anodes such as graphite, stainless steel, or platinum-plated titanium may be used with power supplied from a rectifier.

Anodic protection can be used to create a passive layer on the surface of some metals, such as steel and stainless steel in some environments. It is a practical method of controlling corrosion of tanks in the chemical industry, but is not feasible for copper or brass vessels. The tank is made anodic. An inert cathode, such as platinum-clad metal, is installed in the liquid in the tank. Current is applied so as to maintain a predetermined voltage between the anodic surface and a reference electrode, such as silver–silver chloride, in the liquid. Equipment to maintain precise potential control at high current output has made anodic protection practical. For example, a current of 0.0015 A/ft^2 (0.016 A/m^2) at +0.900 V to a silver–silver chloride electrode will maintain passivity of a carbon-steel tank holding 93% sulfuric acid at 27°C (80°F).

Alkali-chlorine processes. Electrolysis of alkali halides is the basis of the alkali-chlorine and chlorate industries. Chlorine, Cl_2, and caustic soda, NaOH (or caustic potash, KOH), are made by electrolysis of brine, a solution of sodium chloride, NaCl, in water. This is represented by reaction (1).

$$2NaCl + 2H_2O \xrightarrow{electricity} 2NaOH + Cl_2 + H_2 \quad (1)$$

Two processes are used to prevent the products from mixing: the diaphragm cell and the mercury cell. In the diaphragm cell process (**Fig. 3**) an asbestos diaphragm is interposed between a graphite anode and an iron screen cathode. Saturated purified brine fed around the anode passes through the diaphragm to the cathode. Chlorine is formed at the anode. Hydrogen is released at the cathode, leaving NaOH as a 10–15% solution and 10–15% residual NaCl in the cell liquor. By evaporation to 50% NaOH, the NaCl crystallizes out and is recycled. The decomposition voltage of brine to form chlorine and hydrogen is 2.3 V. At 0.75 A/in.^{2} (11.6 A/dm^2) the average voltage components of a diaphragm cell are as follows:

Anode potential	1.50 V
Cathode potential	1.25
Anolyte	0.47
Diaphragm	0.30
Conductors	0.18
Total	3.70 V

The current efficiency is 95.5–96.5%, due to some oxygen discharge at the anode and some chlorine being carried through the diaphragm. An installation is pictured in **Fig. 4**.

In the mercury cell process brine is electrolyzed between graphite anodes and a flowing mercury cathode, forming a dilute (0.2–0.4%) sodium amalgam which is decomposed in another

Table 3. Energy consumption of electrochemical products

Industry	kWh/lb	lb/kWh	Voltage/tank, cell, or furnace	Range, A/unit	Line voltage	Range, kVA/cell or furnace	kVA/line
Electrolytic refining							
Copper: multiple system,	0.09–0.2	5–13.3	0.2–0.3	6000–15,000	80–200	1.2–4.5	480–3000
series system	0.074	13.5	120–200				
Gold (troy lb)	0.15	6.6	0.5–2.8	150–500	3–10	0.1–1.4	0.5–5
Lead	0.07–0.09	11.1–14.3	0.35–0.5	5000–6000	100–185	1.75–3.0	500–1200
Nickel	1.1	0.9	2.4–2.6	5000–6000	220–230	12–16	1300–1500
Silver, Moebius (troy lb)	0.27–0.3	3.3–3.7	2.3–2.8	400–500	45–250	1–1.4	18–125
Silver, Thum (troy lb)	0.33–0.6	1.67–3.0	1.3–5.4	150–200	200–220	0.3–1.2	30–50
Solder	0.08	12.5	0.34				
Tin	0.085	11.8	0.3–0.35	5000	100–200	1.5–2.0	500–1000
Electrowinning							
Antimony			2.5–3	1500	300	4	450
Cadmium	0.65–0.97	1.03–1.54	2.5–3.1				
Chromium	5.0–8.4	0.12–0.2	4.2–4.3	10,000	250–300	40–45	2500–3000
Cobalt	1.2–1.56	0.64–0.83	3.7–4.5				
Copper	0.89–1.34	0.77–1.12	2.0–2.12	10,000–25,000	150–200	20–30	1500–5000
Manganese	4–4.5 dc	0.22–0.25	5–5.4	6000	600	30–40	3600
	5–5.3 dc	0.19–0.2					
Silver	0.58–1.0	1–1.73	1–1.5	300		0.3–0.5	0.3–0.5
Zinc	1.4–1.6	0.61–0.71	3–3.7	5000–10,000	600–800	15–40	3000–8000
Metal melting							
Copper	0.12–0.15	6.67–8.33	85–225	12,000–30,000		6000	
Copper alloys	0.15–0.3	3.33–6.67					
Steel: cold charge,	0.237–0.35	2.86–4.22	80–250	30,000–100,000		25,000–33,000	
hot charge	0.05–0.2	5–20	80–250	30,000–100,000		25,000–33,000	
Zinc	0.045	22.2					
Metal powders							
Copper	0.3–0.5	2–3.3	0.3–0.5	10,000–15,000			
Iron	1.20	0.83	2.5				
Nickel	1.13	0.83	1.4–1.5				
Zinc	1.37	0.73	3.4				

*1 kWh/lb = 7.937 MJ/kg; 1 lb/kWh = 0.1260 kg/MJ.

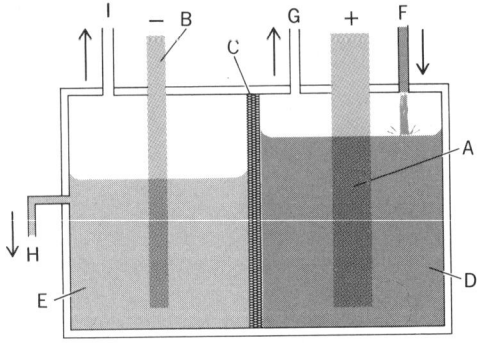

Fig. 3. Diagram of diaphragm cell for chlorine and caustic soda. A = graphite anode, B = iron screen cathode, C = asbestos diaphragm, D = anode compartment for brine and chlorine, E = cathode compartment for NaOH-NaCl cell liquor, F = brine inlet, G = chlorine outlet, H = cell liquor (caustic soda) outlet, I = hydrogen outlet.

compartment by water in contact with graphite surfaces to form H_2 and NaOH (**Figs. 5–7**). The products of the mercury cell are purer than those of the diaphragm cell. To offset the cost of mercury, a much higher current density is used in mercury cells. Typical components of voltage in a mercury cell at 5.12 A/in.^{2} (80 A/dm^2) are as follows:

Anode potential, reversible	1.34 V
Cathode potential, reversible	1.76
Decomposition voltage	3.10 V
Anode polarization	0.35 V
Cathode polarization	0.06
Electrolyte	0.60
Conductors and contacts	0.29
Total cell voltage	4.40 V

Economic factors dictate the use of higher current densities, equal to or exceeding 6.5 amperes/in.^{2} (100 A/dm^2). Cell voltage at these higher current densities can be calculated for good cell designs on the market from the equation: $V = 3.20 + 0.015C$, where C is the cathode current density in amperes per square decimeter. The current efficiency is approximately 95%. Inefficiency

Fig. 4. Photograph of an installation of diaphragm alkali-chlorine cells. (*Hooker Chemical Corp.*)

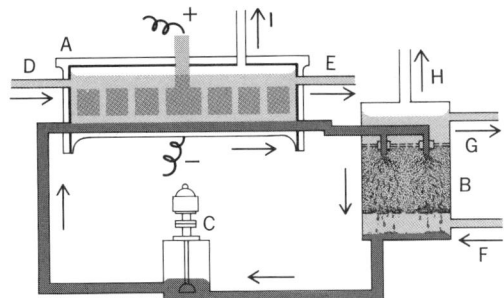

Fig. 5. Diagram of a mercury alkali-chlorine cell. A = electrolyzer, B = decomposer with graphite packing, C = mercury pump, D = feed brine, E = spent brine, F = water, G = 50% caustic soda, H = hydrogen, I = chlorine.

reactions are demonstrated in reactions (2) through (6). Adverse conditions can increase these inefficiencies.

$$2OH^- \longrightarrow 1/2\, O_2 + H_2O + 2e \quad \text{(at anode)} \tag{2}$$

$$H^+ + e \longrightarrow 1/2 H_2 \quad \text{(at cathode)} \tag{3}$$

$$Cl_2 + 2Na(Hg) \longrightarrow 2NaCl + (Hg) \quad \text{(at cathode)} \tag{4}$$

$$Cl_2 + H_2O \longrightarrow Cl^- + ClO^- + 2H^+ \tag{5}$$

$$3ClO^- \longrightarrow ClO_3^- + 2Cl^- \tag{6}$$

Two factors have been important in the development of chlorine technology: (1) The Nafion diaphragm of the synthetic resin type replacing the deposited asbestos diaphragm. (2) The dimensionally stable anode replacing the graphite anode. The dimensionally stable anode is a titanium substrate with a platinum-group coating of metals and oxides. Its use eliminates the continuously necessary voltage increase or anode-cathode spacing adjustment needed because of graphite-anode wear.

In addition, there was a swing away from mercury cells because of widespread publicity of so-called mercury poisoning by mercury discharges. These discharges have been reduced by better hydrogen cooling and recycle of metal dross, cutting mercury losses as well as permitting better "housekeeping."

Sodium hypochlorite is formed when the products of the electrolysis of brine are mixed. Electrolytic cells have been built for this purpose, but have limited or special use, such as for sterilization of swimming pools and algae control in power plant condensers. Sodium hypochlorite is usually made chemically.

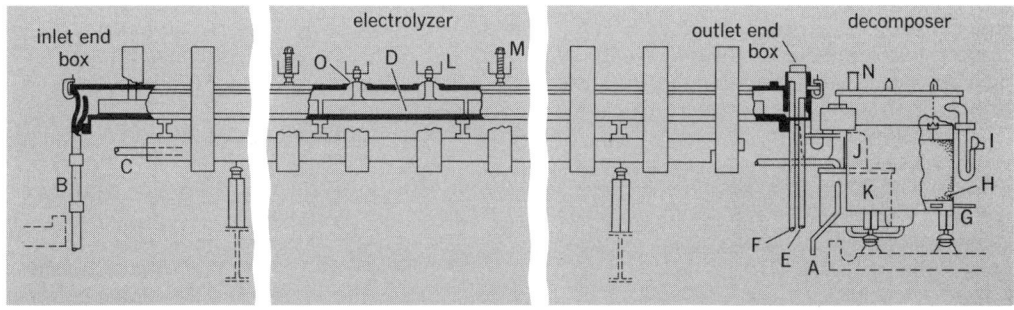

Fig. 6. Longitudinal section of Olin Mathieson E-11 mercury cell. A = dilute caustic outlet, B = brine inlet, C = mercury return, D = anode, E = brine-chlorine outlet, F = outlet end box vent, G = water inlet, H = graphite packing, I = caustic outlet, J = mercury pump, K = mercury pump sump, L = anode support bus, M = lifting screws, N = hydrogen outlet, O = anode seal. (*Olin Mathieson Chemical Corp.*)

Fig. 7. Chlor-alkali mercury cell; 300,000-A capacity. (*Olin Mathieson*)

Sodium chlorate is made in cells with graphite or lead peroxide anodes and steel cathodes. When mixing is encouraged, changes take place according to reactions (7) through (9). The overall reaction is labeled reaction (10). The temperature is kept below 40°C (104°F) in cells using graphite

$$6NaCl + 6H_2O \xrightarrow{electricity} 6NaOH + 3Cl_2 + 3H_2 \qquad (7)$$

$$6NaOH + 3Cl_2 \longrightarrow 3NaClO + 3NaCl + 3H_2O \qquad (8)$$

$$3NaClO \longrightarrow NaClO_3 + 2NaCl \qquad (9) \qquad NaCl + 3H_2O \xrightarrow{electricity} 3H_2 + NaClO_3 \qquad (10)$$

anodes to prevent excessive attack. The optimum efficiency is at pH 6.8; hydrochloric acid is added as required. Sodium dichromate prevents reduction of chlorate at the steel cathode. The conversion of hypochlorite to chlorate is a somewhat slow chemical reaction, occurring partly in the cells and partly in a rundown tank. Salt is added and electrolysis is continued until the sodium chloride is down to about 100 g/liter and the chlorate has reached the desired concentration. It is then recovered by crystallization. Cells operate at 3–3.5 V, 1–3 A/dm^2, and 80–85% current efficiency. Energy consumption is about 2.5 kWh/lb (20 megajoules/kg).

Hydrochloric acid electrolysis is of interest for recovery of chlorine from HCl resulting as a byproduct from organic chlorinations. A filter-press electrolyzer is used which has 30–50 unit cells with polyvinyl-cloth diaphragms and graphite electrodes. Hydrochloric acid of 30–33% concentration is fed to the anode compartment. Weak acid is withdrawn at about 20% and reconcentrated by absorbtion of HCl gas. The graphite anode is not attacked as long as the concentration is kept at 20% HCl or higher. The current efficiency is 92–96%, the loss being due to electrical leakage. Energy consumption is 1800 kWh/2000 lb (7.1 MJ/kg) chlorine (direct current). The voltage balance of a unit cell is as follows, for a total of 2.30 V:

Anode potential	1.02 V	Cathode polarization	0.5
Cathode potential	0.28	Electrolyte, diaphragm	0.3
Anode polarization	0.2		

Oxidations and reductions. These reactions occur in all cells, but in a narrower sense oxidation reactions are those in which oxygen or chlorine at the anode oxidizes some material to form a new compound; reduction reactions are those in which hydrogen, liberated at the cathode, reduces a material to a new product. There are no commercial applications of inorganic electrochemical reductions by this narrow definition.

Sodium perchlorate is made by oxidation of a solution containing $NaClO_3$ at pH 6.1–6.4 by use of a platinum anode and an iron cathode, with chromate in the electrolyte. A lead dioxide anode may be used with a stainless-steel or nickel cathode, with no chromate in the electrolyte. Energy consumption is 1.4–1.6 kWh/lb (11–13 MJ/kg) $NaClO_4$ (direct current). Other perchlorates are made by metathesis with $NaClO_4$.

Persulfuric acid, $H_2S_2O_8$, is made by oxidizing sulfuric acid as an intermediate in the production of hydrogen peroxide. The reactions for the process are shown in reactions (11) and (12).

$$2HSO_4^- \xrightarrow{\text{electricity}} H_2S_2O_8 + 2e \qquad (11)$$

$$H_2S_2O_8 + 2H_2O \longrightarrow 2H_2SO_4 + H_2O_2 \qquad (12)$$

Alkali persulfates can be made in the same way. The cell has smooth platinum anodes, a porous stoneware diaphragm, and a lead cathode cooled to 30°C (86°F). The reversible potential is 2.18 V and the operating voltage 5.0–5.5 V. The energy that is released as heat in the cell must be removed by cooling with stoneware or glass coils in the cell. Hydrogen peroxide is recovered by distillation.

Lead dioxide anodes are used in sodium chlorate, sodium perchlorate, sodium bromate, and periodate or periodic acid regeneration cells. The material is dense and is made from a lead nitrate solution. For some applications it is produced on a steel base, from which it is removed mechanically and chemically. This material is also applied to tantalum, platinum-clad tantalum, or graphite.

Periodic acid is used in producing dialdehyde starch, and the spent solution from the oxidation can be regenerated in a cell by use of lead dioxide anode, a porous ceramic diaphragm, and an iron cathode.

Electrolytic manganese dioxide for batteries is made by electrolyzing hot $MnSO_4$ solutions at pH 6.5–7.5 by use of graphite electrodes. The MnO_2 deposited on the anode is pulverized with the graphite and separated mechanically. Energy consumption is 1 kWh/lb (8 MJ/kg) MnO_2. The quality of the MnO_2 for battery use depends on cell temperature and anode current density.

Ion-permeable membrane cells. These utilize diaphragms made of ion-exchange resins. Cation-permeable membranes permit cations to pass through but not anions, whereas the reverse holds for anion-permeable membranes. A diagram of the movement of ion and water in an electric membrane stack is shown in **Fig. 8**. Purification of sea water is the most important application. Salt has been recovered from sea water which has been concentrated in this way. SEE ION-SELECTIVE MEMBRANES AND ELECTRODES.

Fused-salt electrolysis. Aluminum, barium, beryllium, cerium and misch metal, fluorine, lithium, magnesium, sodium, molybdenum, thorium, titanium, uranium, and zirconium are obtained by electrolysis of fused salts, because water interferes with the desired reaction. Raw materials must all be purified before addition to fused-salt electrolytes as in aqueous electrolytes. Metallizing is a process of depositing a metal as an alloy on a substrate from a fused complex metal salt.

Aluminum is produced in steel pots, lined with carbon, graphite, or silicon carbide, containing an electrolyte of alumina dissolved in fused cryolite, $AlF_3 \cdot 3NaF$, at 950–1000°C (1740–1830°F). The pool of aluminum in the bottom of the pot is the cathode. Contact is made by iron bars buried in the carbon lining or by titanium diboride contacts. Anodes are prebaked carbon blocks or the Söderberg type made from carbon paste baked in place. The arrangement of the apparatus for aluminum production is shown in **Fig. 9**. Aluminum is siphoned out periodically. Oxygen released at the carbon anode forms carbon monoxide.

Barium, used in the electronics industry and in alloys, is obtained by aluminum reduction at low pressures and high temperatures.

Beryllium is made by batch electrolysis of fused salt, starting with 25% $BeCl_2$ and 75% NaCl in a chrome-iron pot, which acts as cathode to a graphite anode. Beryllium is deposited on the

266 PHYSICAL CHEMISTRY SOURCE BOOK

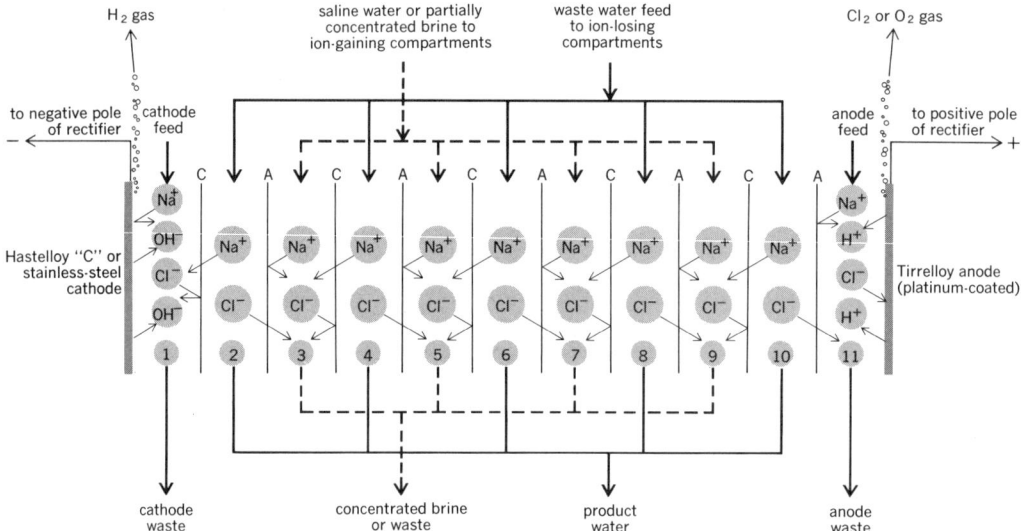

Fig. 8. Diagram of the basic ion and water flow in electric membrane stack. C = cation membrane; A = anion membrane; Na^+ = any cation, such as sodium; Cl^- = any anion, such as chloride. Numbers = compartments.

wall of the pot and is cleaned out and broken up when cold. Salt is washed out with water. The metal is in the form of bright crystalline flakes. A beryllium-copper eutectic can be made by using a copper cathode. Beryllium alloys containing copper and nickel are made from BeO in arc furnaces.

Calcium has been made by electrolysis of pure fused calcium chloride at about 800°C (1470°F). In this case the cathode is solid calcium, and is mechanically withdrawn from the cell

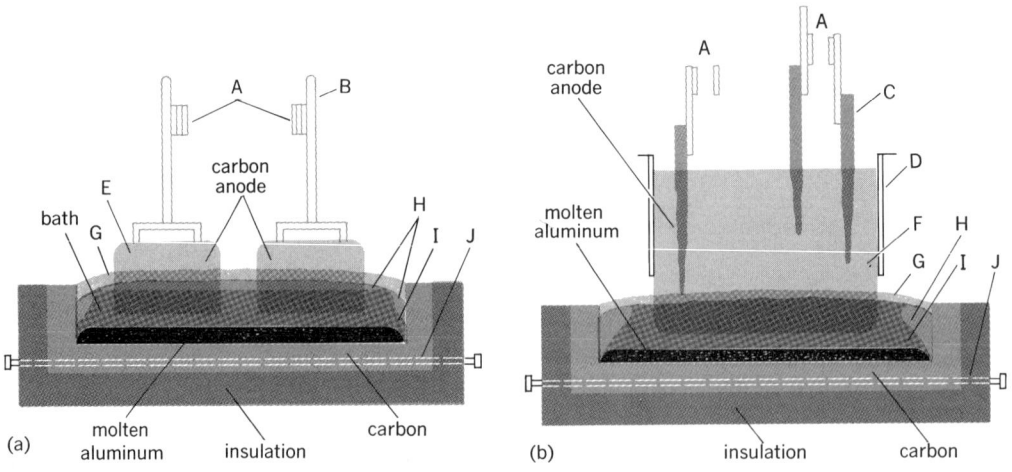

Fig. 9. Two types of aluminum cells, (a) utilizing prebaked carbon anode and (b) utilizing Söderberg carbon anode. A = anode bus, B = anode rod, C = anode stub, D = anode casing, E = prebaked carbon anode, F = Söderberg carbon anode baked in place, G = crust of frozen bath and alumina, H = frozen bath, I = bath (electrolyte), J = steel pin cathode collector.

as a "carrot." Calcium is now made by aluminum reduction at low pressures and high temperatures.

Cerium and misch metal are made from $CeCl_3$ or mixtures of chlorides of cerium, lanthanum, and neodymium in fused-salt electrolysis with NaCl. Misch metal is used for lighter flints.

Fluorine for separation of uranium isotopes is produced by electrolysis of 40% HF in KF between carbon anodes and steel cathodes at 88–100°C (190–212°F). A diaphragm of Monel screen keeps the products H_2 and F_2 separated. Dry HF gas is bubbled continuously into the electrolyte. At a current density of 1 A/in.^{2} (15 A/dm^2) the cell operates at 9–12 V and 96% current efficiency. Energy consumption is 3.0 kWh/lb (24 MJ/kg) fluorine. The theoretical decomposition potential is 2.85 V.

Lithium is made by electrolysis of fused 60% LiCl and 40% KCl at 450–500°C (840–930°F) in a cell similar to the Downs sodium cell.

Magnesium is produced by electrolysis of fused 25% $MgCl_2$ and 75% NaCl at around 700°C (1290°F). The Dow process for making magnesium from sea water uses material approximating $MgCl_2 \cdot 2H_2O$, which is fed around the graphite anodes, where dehydration occurs (**Fig. 10**). Gas from the anode compartment is wet chlorine, air, and hydrogen chloride; the latter is used to make fresh magnesium chloride from magnesium hydroxide. Magnesium metal is deposited on steel cathodes, which direct the metal to a collecting zone. The cell is a cast-steel pot in a furnace setting. Other cells use molten anhydrous magnesium chloride feed. They have brick-lined steel bodies with graphite anodes (**Fig. 11**). Magnesium chloride, with other chlorides, is separated in vacuum crystallizers from brines, dehydrated, and melted in electric resistance cells from which molten $MgCl_2$ is tapped periodically to feed the cells. Molten magnesium is ladled from the cells and cast into molds. The cells operate at 6–7 V and 80–88% current efficiency and use 8–8.5 kWh/lb (63–67 MJ/kg) metal.

Sodium was once made by electrolysis of fused NaOH, but since 1929 it has been made by electrolysis of NaCl in the Downs cell. The electrolyte is 40% NaCl and 60% $CaCl_2$ at 590°C (1090°F). The cell consists of a brick-lined steel vessel. Four graphite anodes project upward from the bottom. The cathode is made of steel cylinders concentric with the anodes and supported from iron arms extending through the sides of the cell, which also conduct current. A diaphragm of 26-mesh-per-inch (10 mesh-per-centimeter) iron screen directs the sodium into an inverted trough leading to a riser pipe, which cools the metal and conducts it to a collecting tank beside the cell. Chlorine is collected in a nickel cone inverted over the anode. Pure dry salt is fed to the

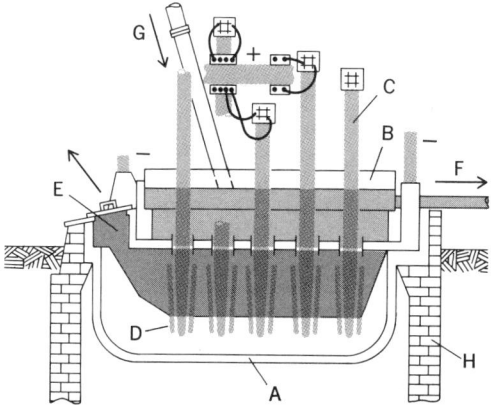

Fig. 10. Cross-sectional of the Dow magnesium cell. A = steel container, B = ceramic cover, C = graphite anodes, D = steel cathodes, E = magnesium collecting well, F = chlorine outlet, G = magnesium chloride feed, H = furnace setting.

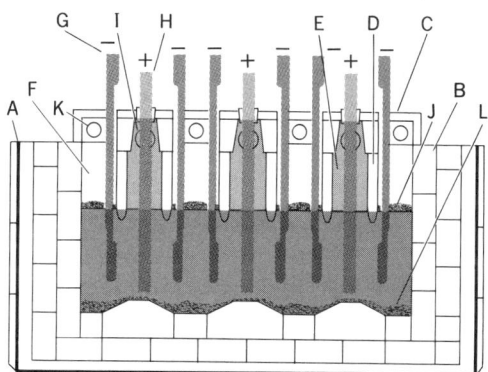

Fig. 11. Cross-section of anhydrous magnesium chloride cell. A = steel box, B = ceramic lining, C = ceramic cover, D = ceramic separators, E = anode (chlorine) compartment, F = cathode (metal) compartment, G = iron cathode, H = graphite anode, I = chlorine outlet, J = magnesium, K = cathode compartment vent, L = mud.

cell. The reactions in the cell are Eqs. (13) and (14). The metal at cell temperature is 5% Ca, but

$$2NaCl + 2e \longrightarrow 2Na + Cl_2 \quad (13) \qquad 2Na + CaCl_2 \rightleftharpoons Ca + 2NaCl \quad (14)$$

as it cools, the second equation is reversed so that the cool metal is about 1% Ca. It is filtered just above the melting point of the sodium, and the final product is under 0.04% Ca. Cells of 38,000 amp operate at 7 V and 83% current efficiency. Energy consumption is about 4 kWh/lb (32 MJ/kg) metal.

Molybdenum, thorium, titanium, uranium, and zirconium can all be made by electrolysis of their complex halides, K_3MoCl_6, $ThF_4 \cdot KF$, K_2TiF_6, KUF_5, and K_2ZrF_6, in molten NaCl. Inert atmospheres are required, and the cell is usually a graphite crucible acting as anode with a graphite or molybdenum cathode, on which the metal deposits as crystals or powder. The batch of metal on the cathode is cooled in an inert atmosphere, then broken off, pulverized, and leached with water. The metal powder is used as such or melted in a vacuum arc furnace.

In the case of tantalum, pure K_2TaF_7 is heated to 900°C (1650°F) in a graphite pot acting as anode with a removable metal cathode. When the cathode is loaded with deposited metal, it is removed and quickly replaced. The bath is replenished with K_2TaF_7. The cathode deposit is pulverized and washed with acid. The metal powder can be compacted by sintering.

Electrothermics. The manufacture of many products requires temperatures higher than can be obtained by combustion methods. Electric heat can usually be developed at, or close to, the point where it is required, so that it is relatively quick. It permits easy control of the atmosphere for oxidizing, reducing, or neutral conditions.

Products of the electric furnace include iron and steel; ferralloys; nonferrous metals and alloys; the exotic metals titanium, zirconium, hafnium, thorium, and uranium; and nonmetallic products such as calcium carbide, calcium cyanamide, sodium cyanide, silicon carbide, boron carbide, graphite, fused alumina, magnesium oxide, quartz, silica, thoria, zirconia, lime, spinel, kyanite, sodium aluminate, dolomite, boric acid and borides, carbides of zirconium and titanium and related metals, graphite, phosphorus and phosphoric acid, and chlorides of magnesium, boron, zirconium, and titanium. An electric smelting process converts ilmenite into iron and a titanium slag for pigment manufacture. Zinc metallurgy uses an electric furnace. Steam is generated in electric boilers where economically feasible.

Processes in gases. Electrical discharge through gases has industrial application in ozone production and nitrogen fixation. Ozonizers consist of two metal electrodes with an air gap and a dielectric, such as gas, between them. Very dry air passed through an air gap will then contain 10–12 mg ozone per liter. Fixation of nitrogen by passing air through an arc furnace, thus forming oxides of nitrogen, was once practiced when power was cheap in Norway, France, and Italy, but has been replaced by conventional processes.

Electromagnetic separation. Magnetic separation removes tramp iron from mixtures of granular solids, suspensions, and solutions. Magnetic separation is also used to separate solids of various magnetic susceptibilities, such as mineral fractions.

Electrodialysis. This is the separation of low-molecular-weight electrolytes from aqueous solutions by migration of the electrolyte through semi-permeable membranes in an electric field. It is used on an industrial scale for deashing starch hydrolyzates and whey, and in many municipalities for producing potable water from saline water. Its uses also include the concentration of liquid foods such as dairy products and citrus juices, the recovery of sulfite pulp waste and pickling acid, and the isolation of proteins. *See* COLLOID; DIALYSIS.

Electrophoretic deposition. This is the deposition of a nonconductive material in a finely divided state from a suspension in an inert medium. Electrophoresis is the migration of colloidal particles, which acquire positive or negative charges in an electric field. The process is useful in electropainting; for instance, electropainting of automobile bodies and other objects has now been adopted on a large scale. Rubber latex is an example of negatively charged colloid which can be plated on an anode. Electronic components can be coated with inorganic salts, oxides, and ceramics suspended in organic media. Bitumen can be electrodeposited out of an aqueous dispersion onto the anodic surface of steel pipe by use of an axially placed cathode. *See* ELECTROPHORESIS.

Electroendosmosis. This is the movement of a liquid with respect to an immobilized colloid in an electric field. The process is used in the dehydration of peat, dye pastes, and clay.

Dies in a clay extrusion press can be "lubricated" by making the die cathodic, which attracts a film of water to it from the wet clay. It is also used commercially for dewatering soils in mining, road building construction, and other civil engineering works.

ORGANIC PROCESSES

Organic electrochemistry was once regarded as a tantalizing area with many important laboratory achievements but few successes in commercial practice. This situation has changed, however, in that electroorganic processes are commercially advantageous if they can fulfill either of two conditions: (1) performance under conditions of voltage corresponding thermodynamically to the conversion of an organic group to a reduced or oxidized group, with the cell products relatively easy to isolate and purify; (2) performance of a highly selective, specific technique to make an addition at a double bond, or to split a particular bond (for example, between carbon atoms 17 and 18 of a complex molecule having 25 carbon atoms).

Selectivity and specificity are highly important in electroorganic processes for the manufacture of complicated molecules of vitamins and hormones—as well as for the medicinal products whose action on pathogenic organisms is a function of their spatial arrangement, steric forms, and resonance.

The electrolytic approach can also be competitive for some low-cost, tonnage products. Here continuous processing is important, and only a single phase should be present, that is, a solution rather than an emulsion, dispersion, or mechanical mixture. Only for fairly valuable products is it practical to find a conducting solvent and then to engineer around it.

The electrolytic oxidation and reduction of organic compounds differ from the corresponding and more familiar inorganic reactions only in that organic reactions tend to be complex and have low yields. The electrochemical principles are precisely those of inorganic reactions, while the procedures for handling the chemicals are precisely those of organic chemistry.

Most organic molecules are insoluble in the aqueous solutions that are the best electrical conductors. Unless solubility can be increased, the only other approach is to use organic solvents. These make relatively poor conductors; hence are encountered power loss, heat build-up, chemical inversion, and often, stepwise, complex reactions.

Oxidations. Commercial success in inorganic electrochemistry has come about by well-engineered combinations of organic and inorganic techniques in areas where strictly chemical methods are either impossible or inefficient, for example, in catalytic hydrogenation or oxidation.

The conventional oxidation reagents of the organic chemist are expensive. There is no market for the oxidant once it is reduced, and chemical regeneration is prohibitive in cost. Accordingly, these reagents are avoided. Electrolytic regeneration, however, can be relatively inexpensive if linked to a carefully controlled organic operation. Typical of this approach is manganese dioxide oxidation of anthraquinone with electrolytic regeneration of the oxidant. In the electrolytic oxidation of anthracene there is a cost-efficient process that utilizes a 20% sulfuric acid suspension with a small quantity of ceric sulfate as a catalyst. Other examples include chromic acid oxidation of oleic acid to perlargonic and azelaic acids, where the oxidant is regenerated electrolytically, and the electrolytic regeneration of periodic acid (a costly oxidizing reagent) in the dialdehyde starch process. These involve savings not only in the purchase price of the oxidants but also in the disposal cost of products that cannot be marketed.

In electrolytic reactions the cell surface is the only source of reductants or oxidants, and these must be produced at highest efficiencies. The mass action effects, concentrations, temperature of reaction, reaction velocities, diffusion, and equilibria between the initial and final products of the reaction all apply to electrochemical reduction and oxidation in the same manner as do the corresponding reactions carried on outside the electrolytic cell.

Materials that are reduced absorb hydrogen at the cathode, and may be considered cathodic depolarizers; those that are oxidized absorb oxygen at the anode, and are anodic depolarizers. Oxidation reactions may involve substances other than oxygen, such as chlorine.

Anodes are selected with a high oxygen or halogen overvoltage, and cathodes are selected with a high hydrogen overvoltage. (Because the accumulation of electrolysis products at anode and cathode causes polarization, the overvoltages are needed to move the products away and keep the process going.)

Reductions. Substances that are easy to reduce may be acted on, at the interface of cathodes, with low hydrogen overvoltage. Hard-to-reduce materials may require much higher overvoltages, which are reached through either the cathode composition or the current density.

The aromatic nitrogen-containing organic compounds have been extensively studied by many researchers. As early as 1900, F. Haber and K. Elbs showed that the nitro compounds of the type RNO_2, where R is an organic radical, could either be directly reduced to the amine RNH_2 (that is, aniline), or successively reduced to the nitroso product RNO, the beta aryl hydroxylamine RNHOH, the azoxy product RN—O—NR, and the material containing an azo group RN=NR, which in turn is reduced to the hydrazo form RNH=NHR or the idene type as H_2NR=RNH_2.

Although commercial processes based on these reactions were successful, they were eventually replaced by improved nonelectrolytic processes.

Bibliography. L. Baizer, *Organic Electrochemistry* 2d ed., 1983; D. R. Crow, *Principles and Applications of Electrochemistry*, 2d ed., 1979; H. Gerisher and C. W. Tobias (eds.), *Advances in Electrochemistry and Electrochemical Engineering*, vol. 13, 1984; A. Kuhn (ed.), *Industrial Electrochemical Processes*, 1971; U. Landau and E. Yeager (eds.), *Electrochemistry in Industry*, 1982; C. L. Mantell, *Batteries and Energy Systems*, 2d ed., 1983; C. L. Mantell, *Solid Wastes: Origin, Collection, Processing, and Disposal*, 1975; D. J. Pickett, *Electrochemical Reactor Design*, 2d rev. ed., 1979; I. Rousar et al., *Electrochemical Engineering I–II*, 1985; A. Varma, *Characterizing Electrodes and Electrochemical Processes*, 1985.

ELECTROLYSIS
PAUL DELAHAY

A method by which reactions are carried out in solutions of electrolytes or in molten salts by use of electricity. As shown in the **illustration**, the electrodes of an electrolytic cell are immersed in an electrolyte solution or in molten salts and are connected to a direct-current power supply. One or several reactions occur at each electrode when current flows through the cell. Reduction, a reaction in which electrons are consumed, occurs at the electrode called the cathode; oxidation occurs at the anode. For instance, sodium is produced at the cathode by reduction, and chlorine at the anode by oxidation in the electrolysis of molten sodium chloride.

Applications are important and varied: industrial production of chemicals, metallurgical extraction of metals, electroplating of metals, metal finishing, and production of electricity in batteries. Metallic corrosion often involves electrolytic processes. For application to analytical chemistry *SEE* ELECTRODEPOSITION ANALYSIS; POLAROGRAPHIC ANALYSIS.

Theory. The quantity of electrolysis products, their rate of production, and quite often their nature depend on electrolysis conditions. According to Faraday's law, the quantity of substance being consumed or produced by a single electrode reaction is proportional to the quantity of electricity consumed in electrolysis. This quantity of electricity is equal to the product of the current multiplied by the duration of electrolysis for a constant current or to the integral of the current over the duration of electrolysis for a variable current.

The nature and the relative abundance of electrolysis products at each electrode generally depend on the electrode potential. Direct control of potential is rarely used, and electrode potentials in industrial cells are controlled indirectly by adjustment of the current density (current per

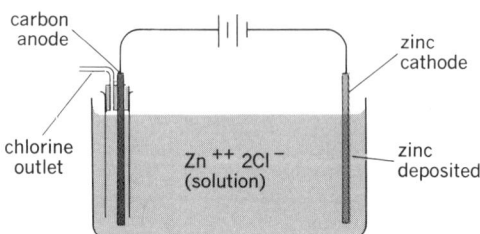

Electrolysis of zinc chloride solution.

unit area) at each electrode. Control is achieved because the current density depends on the electrode potential. Control of the electrolysis current has the advantage of allowing connection of several identical electrolytic cells in series in industrial installations.

Electrolytic cells are characterized by their current efficiency and their power consumption efficiency. The current efficiency is the ratio of the quantity of a substance being consumed or produced to the theoretical quantity of this substance as calculated from Faraday's law. The current efficiency of a single electrode reaction occurring without losses, such as side reaction, electrolysis of the solvent, and evaporation, is 100%.

The second efficiency characteristic of an electrolytic cell, the power consumption efficiency, is the ratio of the theoretical electrical power to the actual electrical power that is consumed in the production or consumption of a given quantity of substance. The power consumption efficiency is smaller than 100% because of overvoltage phenomena, losses of products by side reactions, and ohmic drop (voltage drop) in the cell. Power efficiencies as low as 50% or even lower are not uncommon.

Applications. Industrial applications for inorganic substances are many. Such important chemicals as hydrogen, oxygen, hydrogen peroxide, chlorine, and sodium hydroxide are produced by electrolysis. Water is enriched in deuterium oxide (heavy water) by electrolysis. (There is isotopic separation because the overvoltage for discharge of deuterium ions is larger than for hydrogen ions.) Certain metals such as aluminum, magnesium, and sodium are produced by electrolysis of molten salts. Deposition of these metals from aqueous solution is impossible because this reaction requires higher cathodic potentials than hydrogen evolution. Likewise, fluorine is produced by oxidation of fluoride ion in anhydrous hydrofluoric acid; electrolysis of aqueous solution of fluoride produces oxygen because this reaction occurs at lower anodic potentials than fluorine evolution.

Electroplating of thin layers of a corrosion-resistant metal on fabricated objects is an important technique. Chromium and nickel are most commonly used, but electroplating of other metals such as gold, silver, and copper also has applications. The composition of the electrolytic bath influences the structure and surface finish of the metallic coating; numerous formulas involving metallic complexes and organic additives have been developed empirically. Electroplating is also applied to the industrial refining of copper, silver, gold, and nickel. Certain metals (copper, zinc, and cadmium) are extracted from low-grade ores by electrolysis of a solution of their ores (electrowinning). The opposite reaction of electroplating—anodic oxidation—is applied to metal finishing in electropolishing.

Electrolysis of organic compounds has found only a few industrial applications, although numerous electrode reactions have been studied. Purely chemical preparative methods are more economical and often simpler than electrolysis. In some cases, however, electrolysis involves reactions which are more easily controlled than purely chemical methods. Important reactions include reduction of nitro compounds, aldehydes, ketones, carboxylic acids, unsaturated compounds, and halogenated substances; oxidation of fatty acids (Kolbe reaction), alcohols, aldehydes, ketones, and sugars; and halogenation by anodic oxidation. SEE ELECTROCHEMISTRY.

Bibliography. B. E. Conway, *Theory and Principles of Electrode Processes*, 1965; J. I. Duffy (ed.), *Electrodeposition Processes: Equipment and Compositions*, 1982; K. Grjotheim, *Aluminum Electrolysis: Chemistry of Hall-Heroult Process*, 2d ed., 1982; A. Kuhn (ed.), *Industrial Electrochemical Processes*, 1971; F. A. Lowenstein (ed.), *Modern Electroplating*, 3d ed., 1974; K. J. Vetter, *Electrochemical Kinetics: Theoretical Aspects*, 1967.

ELECTROKINETIC PHENOMENA
QUENTIN VAN WINKLE

Phenomena associated with the movement of charged particles through a continuous medium or with the movement of a continuous medium over a charged surface. The four principal electrokinetic phenomena are electrophoresis, electroosmosis, streaming potential, and sedimentation potential, or Dorn effect. These phenomena are related to one another through the zeta potential ζ of the electrical double layer which exists in the neighborhood of the charged surface.

Electrically charged layers. The distribution of electrolyte ions in the neighborhood of a negatively charged surface and the variation of potential ψ with distance from the surface are shown in **Fig. 1**. According to O. Stern, two different layers of ions are associated with the charged surface. The layer of ions immediately adjacent to the surface is called the Stern layer. The ions of this layer are held to the charged surface by a combination of electrostatic attraction and specific adsorption forces, such as short-range van der Waals interactions and chemical bonds. The thickness δ of this layer is assumed to be equal to the ionic radius of the adsorbed ion species. The second layer of ions is the Gouy layer. The boundary between the two layers is the limiting Gouy plane. The ions in the Gouy layer are acted upon only by electrostatic forces and thermal motions of the liquid environment (Brownian motion), and they form a diffuse atmosphere of opposite charge (positive charge in Fig. 1) to the net charge at the limiting Gouy plane. The net charge density of the diffuse ion atmosphere of the Gouy layer decreases exponentially with distance from the limiting Gouy plane. The Gouy layer forms the positive half of an electrical double layer, and the charged surface plus the Stern layer form the negative half. The effective distance of separation $1/\kappa$ between the two halves of the double layer is determined by the concentration of electrolyte (ionic strength). For an electrolyte of univalent ions in water at 25°C (77°F), the relationship for $1/\kappa$ from the Debye-Hückel theory is Eq. (1), where c is the concentration of electrolyte (moles/liter).

$$\frac{1}{\kappa} = \frac{3 \times 10^{-8}}{\sqrt{c}} \tag{1}$$

Variation of potential ψ with the distance x from the charged surface is shown by a solid curve in Fig. 1. Here ψ_0 represents the thermodynamic reversible electrode potential which is independent of the properties of the electrical double layer and dependent only on the activity of the ion which is in reversible electrochemical equilibrium with the substance of the charged surface. The potential ψ decreases linearly with increasing distance x in the region of the Stern layer. In the region of the Gouy layer, ψ decreases exponentially with increasing distance x.

Displacement of charged layers. In the four listed electrokinetic phenomena, a displacement occurs at some plane (plane of shear) between the charged surface and its atmosphere

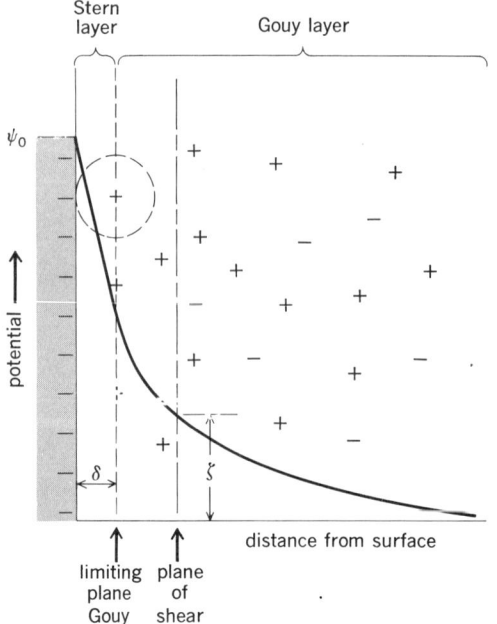

Fig. 1. Electrical double layer.

of ions. The position of the slipping plane in Fig. 1 is shown to be located in the Gouy layer. The potential of the plane of shear is the ζ-potential. From the theories of Gouy and Chapman, for spherical particles Eq. (2) holds. Here $1/\kappa$ is the effective thickness of the double layer, q the net

$$\zeta = \frac{q}{Da}\left(\frac{1}{1 + \kappa a}\right) \quad (2)$$

charge of the particle inside the plane of shear, D the dielectric constant of the liquid, and a the particle's radius at the plane of shear. For flat surfaces Eq. (3) holds where e is the charge per

$$\zeta = \frac{4\pi e}{D\kappa} \quad (3)$$

unit area of surface. Equations (2) and (3) show that ζ-potential is determined by the net charge at the plane of shear and $1/\kappa$, the effective thickness of the ion atmosphere. In turn, ζ-potential controls the rate of transport between the charged surface and the adjacent liquid. The relationship between rate of transport v_E and ζ-potential which is valid for all four electrokinetic phenomena is Eq. (4), where v_E is the velocity of the liquid at a large distance from the charged surface,

$$v_E = \frac{D\zeta E}{4\pi\eta} \quad (4)$$

E is the field strength (V/cm), and η is the viscosity of the liquid. The conditions for validity of Eq. (4) are that the double layer thickness $(1/\kappa)$ must be small compared to the radius of curvature of the surface; the substance of the surface must be nonconducting; and the surface conductance of the interface must be negligible. The equations which relate ζ-potential to electroosmotic flow rate and streaming potential may be obtained from Eq. (4) by use of Poiseuille's law for laminar flow through a capillary. For electrophoresis and sedimentation potential (Dorn effect), v_E is the velocity of the particles. E is the applied field strength for electrophoresis, whereas it is the gradient of potential developed by the sedimentation of charged particles in the Dorn effect.

Electrophoresis, electroosmosis, and streaming potential experiments have been shown to yield identical ζ-potentials for several different interfaces, particularly glass-water and protein-water systems. The sedimentation potential has not been significantly studied.

The effect of electrolytes on the ζ-potential of glass-water interfaces is shown in **Figs. 2** and **3**. As shown in Fig. 2, an increase in electrolyte concentration produces a decrease in ζ-

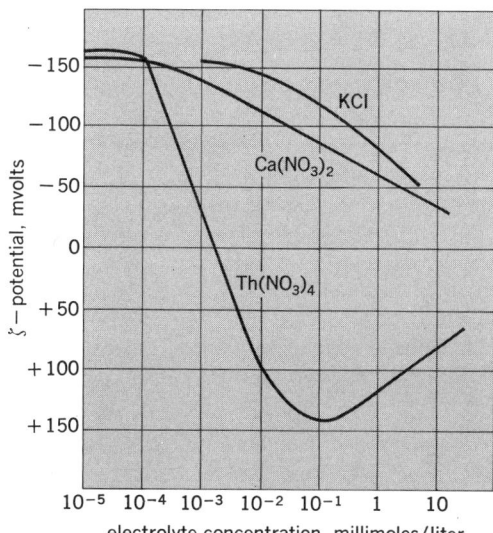

Fig. 2. Effect of electrolyte concentration on the ζ-potential of glass-water interfaces.

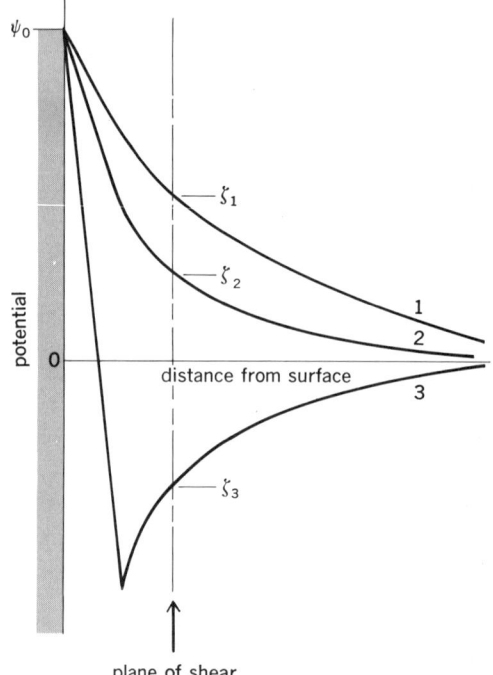

Fig. 3. Reversal of ζ-potential by ion adsorption.

potential, and ions of high charge of opposite sign to that of the surface can completely reverse the sign of the ζ-potential. The explanations for these two effects are given in Fig. 3, where the variation in ψ with distance from the surface is shown for low concentration of electrolyte in curve 1; moderate concentration of electrolyte in curve 2; and charge reversal by adsorption of ions (Th^{4+} on glass) in curve 3. Curves 1 and 2 show that an increase in electolyte concentration reduces ζ-potential by reducing $1/\kappa$, as indicated by Eqs. (1), (2), and (3). Curve 3 shows that reversal of charge by ion adsorption occurs in the Stern layer and that this gives rise to a ζ-potential of opposite sign to the original value. SEE COLLOID; ELECTROPHORESIS; STREAMING POTENTIAL.

ELECTROPHORESIS
B. R. WARE

The migration of electrically charged particles in solution or suspension in the presence of an applied electric field. Each particle moves toward the electrode of opposite electrical polarity. For a given set of solution conditions, the velocity with which a particle moves divided by the magnitude of the electric field is a characteristic number called the electrophoretic mobility. The electrophoretic mobility is directly proportional to the magnitude of the charge on the particle, and is inversely proportional to the size of the particle. An electrophoresis experiment may be either analytical, in which case the objective is to measure the magnitude of the electrophoretic mobility, or preparative, in which case the objective is to separate various species which differ in their electrophoretic mobilities under the experimental solution conditions.

Tiselius cell. The phenomenon of electrophoresis was first observed in 1807 by the Russian physicist F. F. Reuss, but electrophoresis was not employed as an experimental technique until the introduction of a new electrophoresis apparatus by Arne Tiselius in 1937. The apparatus of Tiselius detected electrophoretic motion by the moving-boundary method, in which a boundary is created between the solution of particles to be examined and a sample of pure solvent. As the particles migrate in an electric field, the boundary between solution and solvent can be observed to move, and if there are a number of species in the solution with different electrophoretic mobilities, a series of boundaries of various shapes and magnitudes can be detected. An electrophoresis experiment in a Tiselius cell is depicted in **Fig. 1**. Using his apparatus, Tiselius demonstrated the heterogeneity of human blood plasma, and showed for the first time that the globulin molecules could be separated into different classes, which were designated alpha, beta, and gamma globulin. The moving-boundary method was used for three decades to separate complex mixtures of charged macromolecules in solution and to study the physical characteristics of solutions of proteins and other macromolecules of biological and industrial importance.

Gel techniques. The resolving power of electrophoresis was greatly improved by the introduction of the use of gel supporting media. The gel matrix prevents thermal convection caused by the heat which results from the passage of electric current through the sample. The absence of convection reduces greatly the mixing of the various parts of the sample, and therefore allows for more stable separation. The dimensions of the cross-links of the gel may also provide a molecular sieving effect, which increases the resolving power of the electrophoretic separation of molecules of different size. In addition, the gel media may support a gradient of a separate reagent, which assists in the separation of macromolecules. Gradients of pH and of reagents of various types may be combined in two-dimensional arrays for even greater resolving power. A very successful derivative of the gel technique is the determination of the molecular weights of protein molecules by electrophoresis of the molecules in a gel medium which contains substantial amounts of detergent. The detergent denatures the protein molecules, changing them from globular, compact structures to long, flexible polymers which are coated with detergent molecules. These polymers move in the electric field through the gel medium with a velocity which is determined by the length of the polymer, and therefore by the molecular weight of the protein unit. This method is the most common technique for the determination of molecular weights of proteins in biochemical studies.

Isoelectric focusing. An important variation of the electrophoresis technique is isoelectric focusing. In this technique the medium supports a pH gradient which includes the isoelectric pH of the species being studied. Many charged macromolecules have both positive and negative charges on their surfaces, and the electrophoretic mobility is related to the net excess of charge of one type or the other. As the pH becomes more acidic, the number of positive charges increases, and as the pH becomes more basic, the number of negative charges increases. For each molecule of this type, there is one pH at which the net charge on the surface is zero, so that the molecule does not move when an electric field is applied and thus has an electrophoretic mobility of zero. This pH is called the isoelectric pH. If the molecule is introduced into a pH gradient which includes its isoelectric pH, it will migrate to the position of the isoelectric pH and then become stationary. In this way, all molecules of a given isoelectric pH will migrate to the same region—hence the term isoelectric focusing. The method of isoelectric focusing is particularly good for the analysis of microheterogeneity of protein species and other species which may differ slightly in their chemical content. SEE ISOELECTRIC POINT.

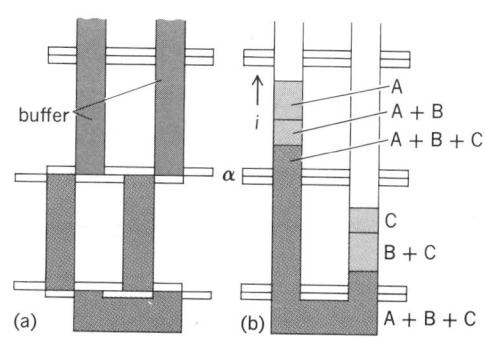

Fig. 1. Diagram of Tiselius electrophoresis cell, showing (a) formation and (b) motion of electrophoretic boundaries of a mixture of proteins. buffer + proteins A, B, and C

Isotachophoresis. One important variant of electrophoresis is the phenomenon of isotachophoresis, in which ionic species move with equal velocity in the presence of an electric field. The isotachophoretic condition is maintained when particles of different mobilities form boundaries within the solution so that the most mobile ions form the leading edge of the moving sample, followed by the less mobile ions in the order of their mobility. The reason that the ions with different mobility can move with the same velocity is that the electric field in each of the regions is inversely proportional to the mobility of the ionic species in order to maintain a constant current throughout the sample. Isotachophoresis has a number of important analytical and preparative applications which are featured by its advantages of high resolving power, sensitivity, speed, and the ability to concentrate rather than to dilute the components which are being analyzed.

Particle electrophoresis. The methodology of electrophoresis may be modified considerably when the particles undergoing analysis are of sufficient size to be viewed either with the naked eye or with the assistance of an optical microscope. This general area of particle electrophoresis has its most important applications in the analysis of the surface charge of living cells and in the study of various types of particles used in industrial coating processes. The most straightforward technique, called optical cytopherometry or microelectrophoresis, is that in which a human experimenter views the particles in an electric field under an optical microscope and determines manually the amount of time necessary to traverse a given distance. Although time-consuming and tedious, this technique has had many important applications. Attempts to modernize this method have included the introduction of high-speed photography, television technology, and the study of particle electrophoresis by the laser Doppler effect. Particle electrophoresis can be conducted under many of the conditions which were originally developed for smaller macromolecules. The use of gradients of density and pH and the methods of isoelectric focusing and isotachophoresis are commonly applied to particles which are even large enough to be viewed with the naked eye.

A common interference effect in the performance of particle electrophoresis is the sedimentation of the particles in the field of the Earth's gravity. This effect can be minimized in importance by performing the experiments in a medium of equal density, by performing the electrophoresis in a vertical direction and accounting for the effect of sedimentation, or by performing the experiment in such a short period of time that the degree of sedimentation in the Earth's gravity is not a significant interfering effect.

Laser applications. Application of the optical laser to electrophoretic detection resulted in the development of a technique which can be used for analytical electrophoresis experiments

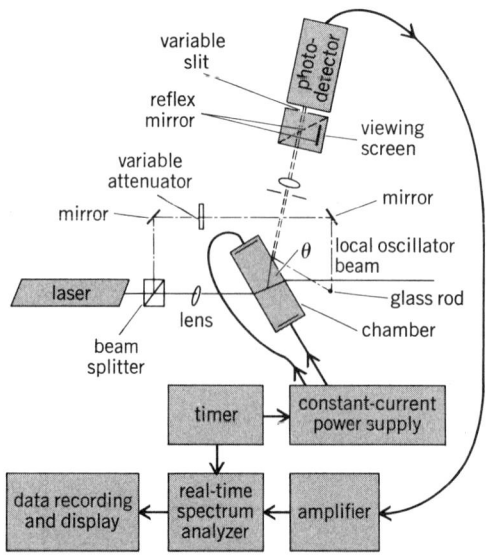

Fig. 2. Diagram of an electrophoretic light-scattering apparatus. Doppler-shifted light from the particles moving in the chamber is mixed with an unshifted reference beam (the local oscillator). The beat frequencies from the photodetector are spectrum-analyzed to produce the electrophorectic mobility histogram. (*After G. M. Hieftje, ed., New Applications of Lasers to Chemistry, ACS Symposium Ser. 85, 1978*)

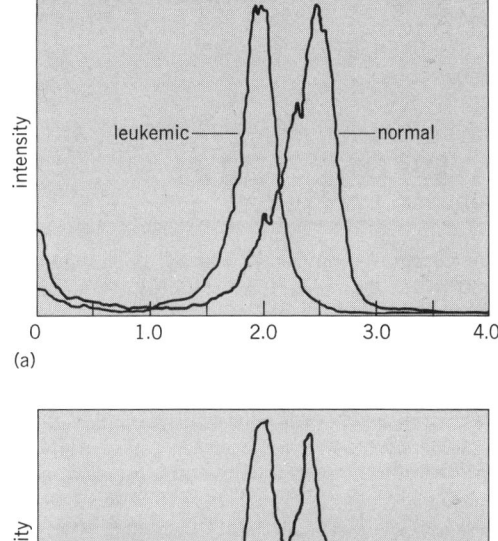

Fig. 3. Characterization of leukemic cells with electrophoretic light scattering. (a) Spectra of leukemic cells and normal cells. (b) Spectrum for mixture of the two cell types. (*After B. A. Smith, B. R. Ware, and R. S. Weiner, Electrophoretic distributions of human peripheral blood molecular white cells from normal subjects and from patients with acute lymphocytic leukemia, Proc. Nat. Acad. Sci. U.S.A., 73:2388, 1976*)

on particles of all sizes. The basic principle is that the highly monochromatic (single-frequency) laser light impinges upon the particles, and is scattered from the particles in all directions. When observing the laser light which has been scattered from a moving particle, one can detect that there is a slight shift in the frequency of the light as a result of the motion of the particle. This is the Doppler effect, which causes the change in the apparent tone of passing trains or cars and which is the operating principle of other familiar techniques such as radar. The application of the laser Doppler principle to electrophoresis experiments, often called electrophoretic light scattering (ELS) is an important method for the rapid determination of electrophoretic velocities. The complete electrophoretic mobility distribution of a sample of many particles can be determined in a time as short as 1 s with a precision heretofore unobtainable by standard technology.

An electrophoretic light-scattering apparatus with a sample Doppler spectrum is shown in **Fig. 2**. The very slight Doppler shifts caused by the electrophoretic motions are detected by the electronic "beats" which result when the scattered light is incident simultaneously with an unshifted reference beam, or "local oscillator," on a photodetector. An important application of electrophoretic light scattering is the characterization of leukemic cells (**Fig. 3**). The normal lymphocytes and leukemic cells isolated in the same way have distinctly different electrophoretic mobilities and mobility distributions, as shown by the two spectra in Fig. 3a. The spectrum in Fig. 3b shows the simultaneous detection of leukemic and normal cells in a mixture, based solely on their Doppler-detected electrophoretic mobilities. Electrophoretic light scattering has been used for the study of many types of living cells, cell organelles, viruses, proteins, nucleic acids, and synthetic polymers. SEE ELECTROLYTIC CONDUCTANCE.

Bibliography. Z. Deyl et al., *Electrophoresis*, pt. A, 1980; Z. Deyl and A. Crambach (eds), *Electrophoresis*, pt. B, Applications, 1983; P. J. Karol and M. H. Karol, Isotachophoresis, *J. Chem. Ed.*,

55:626–630, 1978; D. H. Leaback, Electrophoresis in protein analysis, *Chem. Brit.*, 10(10):376–383, 1974; B. R. Ware, The study of biological surfaces by laser electrophoretic light scattering, in G. M. Hieftje (ed.), *New Applications of Lasers to Chemistry*, Amer. Chem. Soc. Symp. Ser. 85, 1978.

STREAMING POTENTIAL
Quentin Van Winkle

The potential which is produced when a liquid is forced to flow through a capillary or a porous solid. G. H. Quincke (1859) found that the electromotive force produced by the streaming of pure water under a given pressure through a clay plate is independent of the size and thickness of the diaphragm and of the amount of water forced through the diaphragm; the electromotive force, is however, proportional to the pressure.

The streaming potential is one of four related electrokinetic phenomena which depend upon the presence of an electrical double layer at a solid-liquid interface. This electrical double layer is made up of ions of one charge type which are fixed to the surface of the solid and an equal number of mobile ions of the opposite charge which are distributed through the neighboring region of the liquid phase. In such a system the movement of liquid over the surface of the solid produces an electric current, because the flow of liquid causes a displacement of the mobile

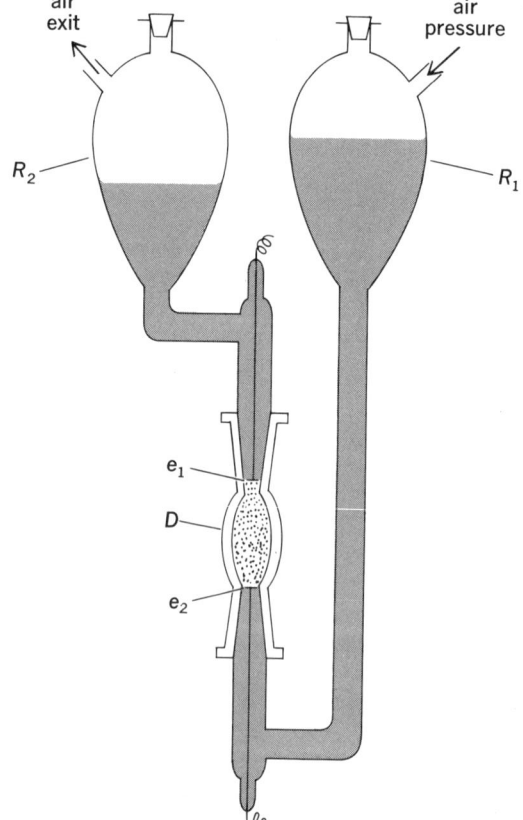

Fig. 1. Streaming potential apparatus.

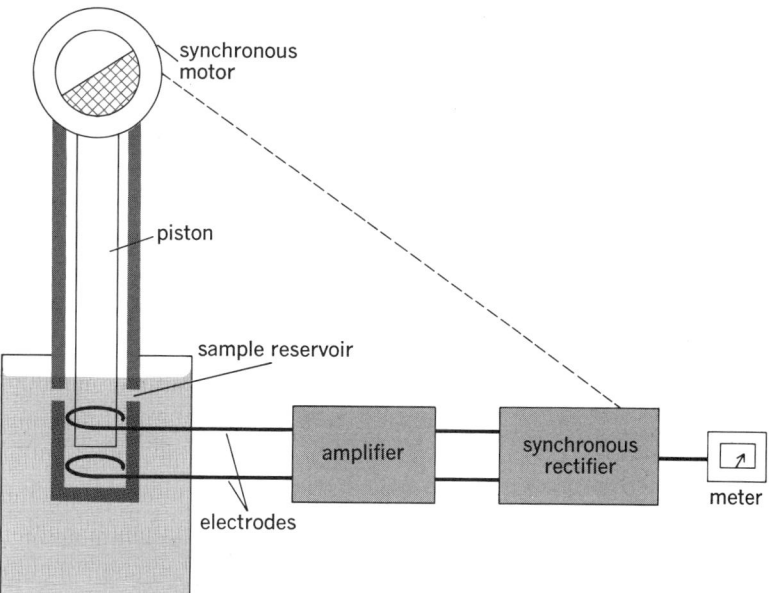

Fig. 2. Streaming current device.

counterions with respect to the fixed charges on the solid surface. The applied potential necessary to reduce the net flow of electricity to zero is the streaming potential.

The principal objective of streaming potential measurements is the evaluation of zeta potentials at solid-liquid interfaces. Equation (1) may be used for this purpose. Here ζ is the zeta

$$\zeta = \frac{4\pi\eta\kappa E}{PD} \qquad (1)$$

potential, E is the streaming potential, η is the viscosity of the liquid, κ is the conductance of the liquid as it exists in the capillary system, P is the applied pressure, and D is the dielectric constant of the liquid.

An apparatus used by R. A. Gortner for measurement of streaming potentials at cellulose-water and alumina-organic liquid interfaces is shown in **Fig. 1**. Perforated gold or platinum electrodes e_1 and e_2 are located on either side of a pad of compacted powder or fibers of a selected solid in diaphragm D. Liquid is forced by compressed air to flow from reservoir R_1 through the solid and into reservoir R_2. The potential between the electrodes e_1 and e_2 is measured with an electrometer-potentiometer system. This potential is the streaming potential E.

In systems containing concentrations of electrolyte above 10^{-3} N, streaming potentials are too low to be measured accurately. Then, the current produced by the streaming liquid may be used to evaluate the zeta potential. For capillaries of known dimensions, Eq. (2) applies for the

$$\zeta = \frac{4\pi\eta LI}{DAP} \qquad (2)$$

zeta potential. Here I is the streaming current, L is the length of the capillary, and A is the cross-sectional area of the capillary. For porous solids of unknown capillary dimensions, the ratio L/A in Eq. (2) may be evaluated by measuring the resistance R of the diaphragm impregnated with a liquid of known electrical conductance κ. The ratio is obtained by means of Eq. (3).

$$L/A = \kappa R \qquad (3)$$

The zeta potentials obtained from Eqs. (1) and (2) are valid only when the flow of the liquid through the diaphragm is laminar and when the radius of curvature of the pores is greater than the thickness of the double layer.

Zeta potentials are useful in predicting the stabilities of lyophobic sols such as aqueous colloidal suspentions of oil, clay, gold, metal oxides, and so on. For these colloidal systems, electrostatic repulsion operating between particles of like charge stabilizes the particles against collisions that lead to irreversible coagulation. It is sufficient to say that particle charge is directly proportional to zeta potential and that there is a certain minimum zeta potential (the critical potential) above which a particular type of colloidal sol is stable for an indefinite length of time, and below which coagulation occurs in a relatively short time.

In one versatile streaming current device instead of a capillary, the flow system utilizes a loosely fitted piston, driven by a synchronous motor and reciprocating cam assembly, moving up and down in a plastic block containing a cylindrical, dead-ended bore. This forces the liquid sample to flow back and forth through the annular space between the cylinder wall and the piston. A diagram of the system is shown in **Fig. 2**.

Since the walls possess an electrical charge, a four-cycle alternating current is generated and collected at the electrodes. The measuring circuit consists of an output transformer (amplifier), a synchronous rectifier, and a microammeter (meter). As given in Eq. (2), the streaming current I is proportional to zeta potential ζ of the cylinder and piston walls.

When a liquid suspension of colloidal particles is placed in the measuring head, the walls of the piston and cylinder take on the charge characteristics of the charged particles by attachment of the particles to the walls. As a consequence, the streaming current developed at the walls reflects the zeta potential of the suspended particles. The attachment of particles to the walls is a rapid and essentially reversible process which makes it possible to follow changes in zeta potential of suspended particles as reagents of various types are added to the suspension.

It is suggested that the device is useful in controlling a wide range of processes involving coagulation and stabilization of colloidal systems in water purification, in sewage treatment, and in the manufacture of paper, paint, adhesives, photographic film, pharmaceuticals, cosmetics, and textiles.

ELECTROANALYTICAL CHEMISTRY

Ion-selective membranes and electrodes	282
Coulometric analysis	286
Polarographic analysis	288
Electrodeposition analysis	292

ION-SELECTIVE MEMBRANES AND ELECTRODES
RICHARD P. BUCK

Membrane-based devices, involving permselective, ion-conducting materials, used for measurement of activities of species in liquids or partial pressures in the gas phase. Permselective means that ions of one sign may enter and pass through a membrane.
   Properties. Ion-selective electrodes are generally used in the potentiometric mode, and they superficially resemble the classical redox electrodes of types 0 (inert), 1 (Ag/Ag$^+$), 2 (Ag/AgCl/Cl$^-$), and 3 (Pb/PbC$_2$O$_4$/CaC$_2$O$_4$/Ca^{2+}). The last, while ion-selective, depends on a redox couple (electron exchange) rather than ion exchange as the principal origin of interfacial potential difference. Ion-selective electrodes have the typical shorthand form:

				Membrane
Cu;	Ag;	AgCl;	Electrolyte; M$^+$ Cl$^-$	permselective to M$^+$
Lead wire		Inner reference electrode	Inner filling solution	

or   Cu; Ag; Membrane permselective to M$^+$

The former is the ionic-contact "membrane" configuration, and the latter is the "all-solid-state" membrane configuration. In the former, both membrane interfaces are ion-exchange-active, and the potential response depends on M$^+$ activities in both the test solution and the inner filling solution. In the latter, the membrane must possess sufficient electronic conductivity to provide a reversible, stable electron-exchange potential difference at the inner interface, with ion exchange only at the test solution side.
   Potentiometric responses of ion-selective electrodes take the form in Eq. (1) when an ion-

$$V(\text{measured}) = V^0 + \frac{RT}{F} \ln \left[a_i^{1/z_i} + \sum_j (k_{ij} a_j)^{1/z_j} \right] \quad (1)$$

R = the universal gas constant
T = the absolute temperature
F = the Faraday constant (96,487 coulombs/equivalent)
V^0 = formal reference potential

selective electrode is used with an external reference electrode, typically a saturated-calomel reference electrode with a salt bridge, to form a complete electrochemical cell. Activities of the principal ion, a_i, and interfering ions, a_j, are in the external "test" solution and correspond to ions M^{z_i} and M^{z_j}, where z_i and z_j are charges with sign. The ion M^{z_i} is written first because it is the principal ion favored in the membrane, for example, high ion-exchange constant and high mobility. The k_{ij} values are "selectivity coefficients" which are experimentally determined, but can be related to extrathermodynamic quantities such as single-ion partition coefficients and single-ion activity coefficients and mobilities. When only one ion is present in a test solution, or the membrane is ideally permselective for only one ion, this equation simplifies as Eq. (2). V^0 can be written

$$V(\text{measured}) = V^0 + \frac{RT}{z_i F} [\ln a_i] \quad (2)$$

explicitly in terms of activities of species at the inner interface, or in terms of solid-state activities for the all-solid-state configuration. Equation (1), variously known as the Horovitz, Nicolsky, or Eisenman equation, resembles the Nernst equation, Eq. (2), but the former originates from different factors. Equation (1) cannot be derived from first principles for ions of general charge. However, when $z_i = z_j$, the equation can be derived by various means, including thermodynamic (Scatchard) equations and transport models (Nernst-Planck equations), mainly. The response slope 2.303$RT/z_i F$ is 59.14/z_i mV/decade of activity at 77°F (25°C). Measurements reproducible to ±0.6 mV are typically achieved, and activities can be reproducible to ±2% for monovalent ions. A normal slope is considered "nernstian," and can persist over a wide activity range, especially for solid electrodes, for example, 24 decades in the Ag^+, S^{2-} system, and 12 decades of H^+ using Li^+-based glass membrane electrodes. Less than nernstian slopes, and in the limit zero slope, can occur at low activities of sensed species, and can occasionally occur at very high activities. Ultimate low-level response (detection limit) is determined by the solubility of the membrane ion-exchanging material, although impurities may cause premature failure. Because of the logarithmic dependence of potential response on activities, activity measurements using ion-selective electrodes are particularly suited to samples with wide activity variations. Standardization against pure samples or samples of the type to be determined is required. Precise measurements of concentrations over narrow ranges are not favorable, but are possible by elaborate standardization schemes: bracketing, standard additions, and related methods involving sample pretreatment. SEE DONNAN EQUILIBRIUM.

Ion-selective electrodes are most often cylindrical, 6 in. (15 cm) long and 0.25 or 0.5 in. (6 or 13 mm) in diameter, with the lead wire existing at the top and the membrane sensor at the lower end. However, inverted electrodes for accommodating single-drop samples, and solid electrodes with a drilled hole or a special cap for channeling flowing samples past a supported liquid membrane, are possible configurations. The conventional format is intended for dip measurements with samples large enough to provide space for an external reference electrode. Single combination electrodes are useful for smaller samples, because the external electrode is built-in nearly concentrically about the ion-selective electrode. Drilled and channeled-cap electrodes are intended for use with flowing samples. Microelectrodes with membrane-tip diameters of a few tenths of a micrometer have been constructed for single-cell and other measurements in the living body (see **illus**. for construction details).

Ion-selective electrodes are intended to be used to monitor and measure activities of flowing, or stirred, solutions because electrodes detect and respond to activities only at their surfaces. The time responses of solid and liquid membrane electrodes to ideal step activity changes of the principal ion (already present in the membrane) can be very rapid: 200 milliseconds for glass to about 30 microseconds for AgBr. Generally this fast response cannot be observed or used, because sample mixing or diffusion of a fresh sample to an electrode surface is the limiting process. Also, almost any time two ions are simultaneously determining response, interior diffusion potential generation is involved in reaching a new steady-state potential. Similarly, formation of hydrated surface layers or layers of adsorbed matter layers introduces diffusion barriers. Response times from 2 to 20 s can be expected. About 20–30 samples per hour may be analyzed manually by the dip method, and about 60 per hour when samples are injected into a flowing stream of electrolyte. SEE ELECTRODE POTENTIAL; ELECTROLYTIC CONDUCTANCE.

Classification and responsive ions. Ion-selective electrodes are classified mainly according to the physical state of the ion-responsive membrane material, and not with respect to the ions sensed. It has also proved superfluous to distinguish between homogeneous membranes and those that are made from a homogeneous phase supported physically in voids of an inert polymer, or from two homogeneous phases intimately mixed, so-called heterogeneous membranes.

Glass membrane electrodes. These are used for hydrogen ion activity measurements. Glass electrodes are based on alkali ion silicate compositions. Superior pH-sensing glasses (pH 1 to 13 or 14) result from lithium silicates with addition of di-, tri-, and tetravalent heavy-metal oxides. The latter are not chain formers. Membranes responsive to Na^+, K^+, NH_4^+, and some other cations use additional Al_2O_3 or B_2O_3, or both. The pH glasses are highly selective for H^+ over other monovalent ions. The Na^+-sensing glasses are not intrinsically very selective for Na^+ over H^+, but useful pNa measurements can be made, even in excess K^+ at pH 7 or above. No glasses with high selectivity of K^+ over Na^+ have been found. Chalcogenide glasses containing low contents of Cu^{2+} or Fe^{3+}, while called glasses, are thought to be semiconductor electrodes with a high component of electron exchange, rather than ion exchange, for establishment of interfacial potential responses to Cu^{2+} and Fe^{3+}.

Electrodes based in water-insoluble inorganic salts. These electrodes include sensors for F^-, Cl^-, Br^-, I^-, CN^-, SCN^-, S^{2-}, Ag^+, Cu^{2+}, Cd^{2+}, and Pb^{2+}. The compounds used are silver salts, mercury salts, sulfides of Cu, Pb, and Cd, and rare-earth salts. All of these are so-called white metals whose aqueous cations (except La^{3+}) are labile. The salts themselves are Frenkel-defect solids which possess the necessary ionic conductivity. Ag_i^+ (interstitials) or Ag_i^- (vacancies) are the mobile species in the silver salts, while F^- interstitials are mobile in LaF_3.

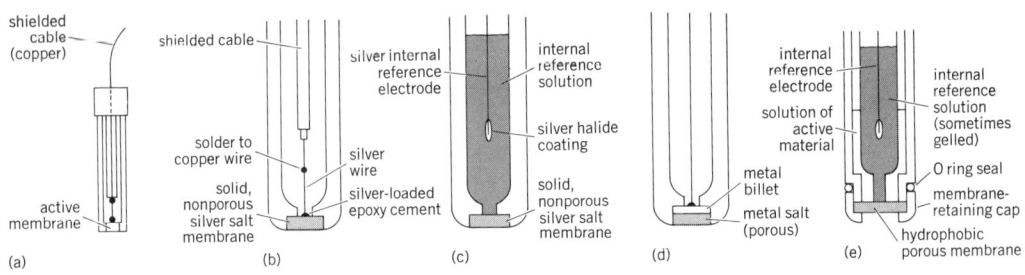

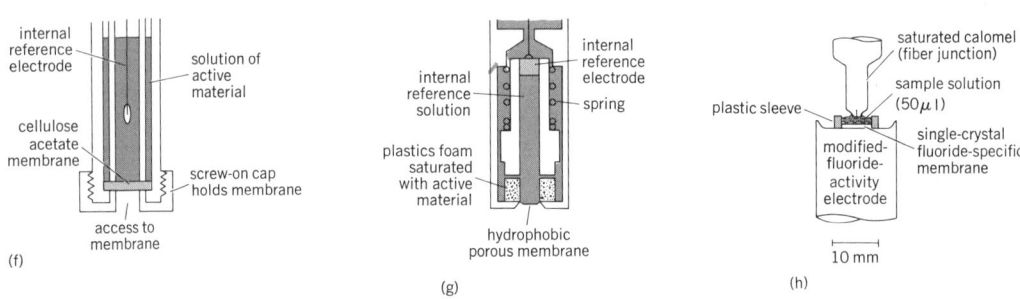

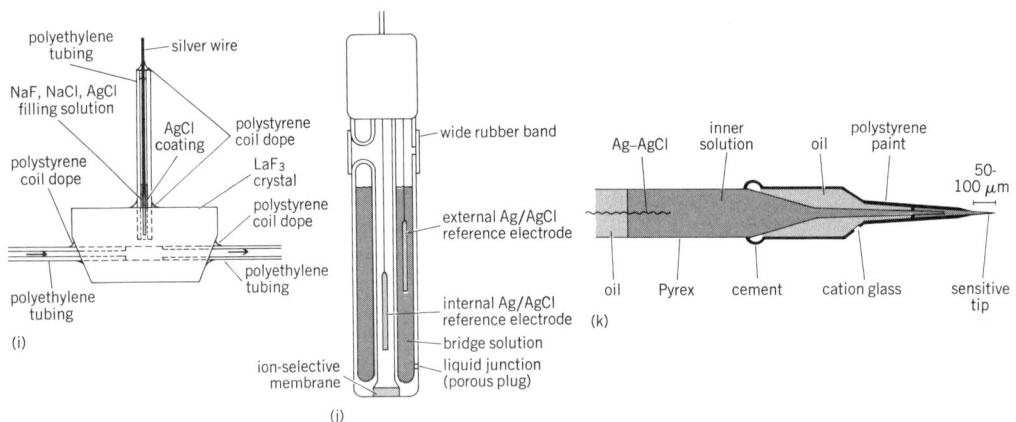

Diagrams of ion-selective electrodes. (a) Typical electrode configuration—in this example, an all-solid-state ion-selective electrode. (b) Enlarged view of the construction for metal-contacted-membrane ion-selective electrodes. (c) Enlarged view of the construction for internal electrolyte-contacted-membrane ion-selective electrodes. (d) Enlarged view of the construction for an electrode. (e–g) Enlarged views of constructions for liquid ion-exchanger-membrane ion-selective electrodes. (h) An inverted electrode microcell using a fluoride-sensing material; the reference electrode (external) is saturated calomel [SCE] (after R. A. Durst and J. K. Taylor, Anal. Chem., 39:1483, 1967). (i) Construction of a flow-through crystal electrode (after H. I. Thompson and G. A. Rechnitz, Anal. Chem., 44:300, 1972). (j) A combination electrode illustrating the usual active membrane surrounded by an attached Ag/AgCl external reference electrode. (k) An example of a cation-sensing microelectrode used in biological research (after R. N. Khuri, W. J. Flanagan, and D. E. Oken, J. Appl. Physiol., 21:1568, 1966).

These materials are ion exchangers, and show no diffusion potential. Single crystals, doped and undoped, may be used as membranes. Pressed pellets using inert binders such as polyethylene or an insoluble salt such as Ag_2S (for the silver halide electrodes) are popular.

In addition, powdered salts may be suspended in silicone rubber or polyvinyl chloride (about 50:50% by weight) to form heterogeneous flexible membranes. $CuS-Ag_2S$, $CdS-Ag_2S$ and $PbS-Ag_2S$ pressed pellets formed at about 480°F (250°C) are indirectly responsive to the divalent metal ion activities through control of Ag^+ activities at the electrode surface and in leached layers or surface pores by means of the common ion effect.

Electrodes using liquid-ion exchangers. These are electrodes supported in the voids of inert polymers such as cellulose acetate, or in transparent films of polyvinyl chloride, and provide extensive examples of devices for sensing. Fewer cation-sensing liquid-ion exchanger systems have been found. The principal example (and among the most important) is the Ca^{2+}-responsive electrode based on calcium salts of diesters of oil-soluble phosphonic acids. Anion-sensing electrodes typically use an oil-soluble cation Aliquat (methyltricaprylammonium) or a metal ion-uncharged organic chelating agent (Ni^{2+} or Fe^{2+} phenanthroline or substituted phenanthroline cations) in a support matrix. Sensitivity is virtually assured if the salt is soluble in a mediator solvent, typically a nitro aromatic or esters of difunctional carboxylic acids: adipic, sebacic, or phthalic. Selectivity poses a severe problem since these electrodes, based on hydrophobic materials, tend to respond favorably to many oil-soluble anions. Thus construction of electrodes for the simple inorganic anions F^-, OH^-, HCO_3^-, and HPO_4^{2-} is difficult. Yet many electrodes respond to SCN^-, I^-, Br^-, NO_4^-, ClO_4^-, and BF_4^- in accordance with the Hofmeister lyotropic series. Surfactant anion sensors use salts such as hexadecylpyridinium dodecylsulfate in o-dichlorobenzene; surfactant cation sensors use a picrate salt of the species to be measured. Acetylcholine may be measured in the presence of choline, Na^-, and K^+ using the tetra-p-chlorophenylborate salt in a phthalate ester in polyvinyl chloride, for example.

Neutral carrier-based sensors for monovalent and divalent cations are closely related to ion-exchanger-based electrodes. Both systems may involve ion-exchange sites, particularly negative mobile sites arising from mediators or negative fixed sites arising from hydrolysis of support materials. All of the available neutral carriers are hydrophobic complex formers with cations, and they may be either cyclic or open-chain species. These compounds permit selective extraction (leading to permselectivity) for ions such as K^+, Na^+, NH_4^+, and Ca^{2+} that would ordinarily exist as simple inorganic salts in the hydrocarbonlike membrane phase. Valinomycin is the best-known example, and its use in supported solvents such as diphenylether and $p-NO_2$ cymene provides an electrode with sensitivity of 10^5 for K^+/Na^+.

Electrodes with interposed chemical reactions. These electrodes, with chemical reactions between the sample and the sensor surface, permit a new degree of freedom in design of sensors for species which do not directly respond at an electrode surface. Two primary examples are the categories of gas sensors and of electrodes which use enzyme-catalyzed reactions. Gas sensors for CO_2, SO_2, NH_3, H_2S, HCl, and others can be made from electrodes responsive to H^+, S^{2-}, or Cl^-. By enclosing a pH glass membrane in a thin layer of dilute $NaHCO_3$, an electrode for partial pressure of CO_2 is formed, since H^+ increases in a known way with increasing dissolved CO_2. Similarly, immobilized enzymes convert a substrate such as urea or an amino acid to ammonia, which can be sensed and monitored by the underlying electrode. However, increased sensitivity is accompanied by an increased response time. Each diffusion and diffusion-reaction barrier slows the transport and increases the time constant of the overall sensor electrode. SEE ELECTRODE.

Applications. Electrodes for species identified above are, for the most part, commercially available. In addition, electrodes have been made and reported that are responsive to many other species. A few of these are: Cs^+, Tl^+, Sr^{2+}, Mg^{2+}, Zn^{2+}, Ni^{2+}, UO_2^{2+}, $Hg(II)$, HSO_3^-, SO_4^{2-}, IO_4^-, ReO_4^-, halide anion complexes of heavy metals (for example, $FeCl_4^-$), pyridinium, pyrocatechol violet, vitamins B_1 and B_6 and many cationic drugs, aromatic sulfonates, salicylate, trifluoroacetate, and many other organic anions. Applications may be batch or continuous. Important batch examples are potentiometric titrations with ion-selective electrode end-point detection, determination of stability constants of complexes and speciation identity, solubility and activity coefficient determinations, and monitoring of reaction kinetics, especially for oscillating reactions. Ion-selective electrodes serve as liquid chromatography detectors and as quality-control monitors in

drug manufacture. Applications occur in air and water quality (soil, clay, ore, natural-water, water-treatment, sea-water, and pesticide analyses); medical and clinical laboratories (serum, urine, sweat, gastric-juices, extracellular-fluid, dental-enamel, and milk analyses); and industrial laboratories (heavy-chemical, metallurgical, glass, beverage, and household-product analyses).
Bibliography. P. L. Bailey, *Analysis with Ion-Selective Electrodes*, 2d ed., 1980; P. W. Cheung et al., *Theory, Design and Applications of Solid State Chemical Sensors*, 1978; A. K. Covington, *Ion-Selective Electrodes*, 1981; H. Freiser (ed.), *Ion-Selective Electrodes in Analytical Chemistry*, 1978; J. Koryta and K. Stulik, *Ion-Selective Electrodes*, 2d ed., 1984; N. Lakshiminarayanaiah, *Membrane Electrodes*, 1976.

COULOMETRIC ANALYSIS
GEORGE W. O'DOM

An electroanalytical chemistry technique in which the amount of a substance is determined quantitatively by measuring the total amount of electricity required to deplete a solution of this substance. The total electricity or charge consumed is related to the concentration by Faraday's law. The faraday, F, is 96,487 coulombs per equivalent. Coulometric analyses are classified as primary or secondary. Primary coulometric analyses are those in which the substance to be determined is electrolyzed directly. Secondary coulometric analyses are those in which an electrolytically generated intermediate reacts stoichiometrically with the substance to be determined. When the solution is completely depleted, the total charge used is related directly to the amount of starting material. The basic equation for coulometry is stated in Eq. (1). Here Q is the total charge, i_t is

$$Q = \int_0^\infty i_t dt = nFVC_0 \tag{1}$$

the current at time t, n is the number of electrons exchanged per molecules or equivalents per mole, F is the faraday based on the carbon-12 scale, V is the solution volume, and C_0 is the initial concentration of the substance determined. Secondary coulometric processes are often referred to as coulometric titrations. However, any coulometric analysis is a titration with electrons and is therefore a coulometric titration.

A condition of 100% current efficiency must exist for a coulometric analysis; that is, all the current must be used either directly or indirectly to deplete the desired substance. For a coulometric analysis either the applied current or the applied potential can be controlled. The controlled-current method is seldom used in primary coulometric analysis, for, as the reaction proceeds, the electroactive substance is depleted and is unable to maintain the required current. Another reaction must then provide for part of the current, and there is no longer 100% current efficiecy. The controlled-potential method, however, is frequently used for primary coulometric analyses. *See* ELECTROCHEMICAL TECHNIQUES.

Controlled-potential analysis. In controlled-potential coulometry the potential of the working electrode is maintained constant with respect to a reference electrode. The potential can usually be selected from voltametric data, so that only the desired reaction occurs. For an electrode reaction which is diffusion limited and not complicated by chemical side reactions, the current i_t decays according to Eq. (2).

$$i = i_0^{-kt} \tag{2}$$

In this relation i_0 is the initial current; k is a constant which depends on the geometry of the cell and the electrode, the diffusion coefficient of the electroactive species, and the solution stirring rate; and t is the electrolysis time. The time required for complete depletion is independent of the initial concentration, but it is strongly dependent on the ratio of electrode area to solution volume (cell geometry) and the stirring rate. The end of the reaction is indicated when the current decays to the value of the background current. The total charge is determined with a coulometer. Controlled-potential coulometric analysis is both sensitive and selective. Its sensitivity is, however, limited by the background current. Background currents arise from the charging of the electrode-solution interface whenever there is a change in the potential of the electrode (double-layer charging), and from unwanted faradaic processes usually caused by electroactive impurities. Double-layer charging causes no great errors unless very dilute solutions are being analyzed. The contribution of the unwanted faradaic currents can often be determined by titrating a blank solution.

An increase in the ratio of electrode area to solution volume will result in an increase in the constant k in Eq. (2), and hence a decrease in the time required for complete electrolysis of the substance. Two coulometric techniques which take advantage of this are the flow electrolysis and thin-layer techniques.

Flow electrolysis. In the flow electrolysis technique the solution to be analyzed is passed at a constant rate through a column packed with electrode material whose potential is controlled. During passage through the column the electroactive substance is depleted from the solution. Under conditions of 100% current efficiency, the concentration of the electroactive substance C can be obtained by Eq. (3), where i is the current, G is the flow rate of solution, and n and F have their usual significance.

$$C = \frac{i}{nFG} \quad (3)$$

Thin-layer methods. The thin-layer technique is one in which a small volume of solution is entrapped in the vicinity of the working electrode. Electrolysis occurs at the working electrode. The solution thickness is quite small (usually less than 0.1 mm), and all the entrapped electroactive species is able to diffuse to the electrode surface rapidly and to be reacted upon.

Both controlled-current and controlled-potential coulometry have been used in the thin-layer technique. For controlled current, Eq. (4) holds. Here T is the electrolysis time, i is the

$$T = \frac{nFAlC_0}{i} - \frac{l^2}{3D} \quad (4)$$

applied current, A is the electrode area, C_0 is the initial concentration of the electroactive substance, D is the diffusion coefficient, and l is the average solution thickness. Conditions can usually be selected so that $l^2 < 3D$, and $l^2/3D$ in Eq. (4) can be neglected. The thin-layer equation then reduces to Eq. (1). The time required for coulometric analyses can be shortened by this technique.

Controlled-current analysis. For quantitative analysis controlled-current coulometry is used extensively. In this method a constant current is passed between two electrodes. It is useful in primary coulometric analyses only when the substance to be determined is deposited on the electrode surface, such as the reduction of an oxide layer. It has greater usefulness in secondary coulometric analyses. In these processes an intermediate is generated electrolytically and allowed to react stoichiometrically with the substance to be determined. The end point of the reaction can be detected by any method sensitive to the reaction. Electrochemical methods are frequently employed.

Acid-base, complexometric, and redox reactions can be analyzed by this technique. Among the intermediates which can be generated are the ions hydronium, hydroxide, silver(I), silver(II), copper(I), mercury(I), titanium(III), and uranium(V); and the molecules chlorine, bromine, and iodine. The advantages of this method over standard titration techniques are the elimination of the need for storage and standardization of standard solutions, and the ability to generate and use in solution unstable substances, such as silver(II).

Reaction mechanisms. Controlled-potential coulometry can also be used to help elucidate reaction mechanism. For a system in which the only reaction occurring is the electron transfer at the electrode surface, the number of electrons involved in the reaction can be determined directly from Eq. (1), providing the concentration and the solution volume are known. For systems in which chemical reactions complicate the reaction occurring at the electrode surface, coulometry can often be used in conjunction with other methods to determine the reaction mechanism. For example, when a slow chemical reaction occurs between two electron-transfer reactions, the n value measured by coulometry would correspond to both electron-transfer steps. Voltametry, on the other hand, would given an n value corresponding to the first electron-transfer step. The stirring rate and the concentration of electroactive substance can also affect the n value determined by coulometry.

Reaction kinetics. Controlled-current coulometry often provides a convenient means of measuring the rate at which chemical reactions occur. The reactant can be added as a coulometrically generated intermediate and allowed to react with a substrate in solution. The concentration of the intermediate can be monitored. The rate of generation can be made to equal the rate at which it is depleted by the chemical reaction, and the rate constant for the reaction thus calculated. This method provides for convenient addition of reactant at low concentrations, which allows rather rapid reactions to be measured. SEE ELECTRODEPOSITION ANALYSIS; ELECTROLYSIS.

POLAROGRAPHIC ANALYSIS
PETER ZUMAN

An electrochemical technique used in analytical chemistry. Polarography involves measurements of current-voltage curves, obtained when voltage is applied to electrodes (usually two) immersed in the solution being investigated. One of these electrodes is a reference electrode: its potential remains constant during the measurement. The second electrode is an indicator, and its potential varies in the course of measurement of the current-voltage curve, because of the change of the applied voltage. In the simplest version, so-called dc polarography, the indicator electrode is a dropping mercury electrode, consisting of a mercury drop hanging at the orifice of a fine-bore glass capillary (usually about 0.08 mm inner diameter). The capillary is connected to a mercury reservoir so that mercury flows through the capillary at the rate of a few milligrams per second. The outflowing mercury forms a drop at the orifice, which grows until it falls off. The lifetime of each drop is several seconds (usually 2 to 5). Each drop forms a new electrode; its surface is practically unaffected by processes taking place on the previous drop. Hence each drop represents a well-reproducible electrode with fresh, clean surface.

Apparatus. The dropping-mercury electrode is immersed in the solution to be investigated and placed in a cell containing the reference electrode (**Fig. 1**). Polarographic current-voltage curves can be recorded with a simple instrument consisting of a potentiometer or another source of voltage and a current-measuring device (such as a sensitive galvanometer). The voltage can be varied by manually changing the applied voltage in finite increments, measuring current at each, and plotting current as a function of the voltage. Alternatively, commercial instruments are available in which voltage is increased linearly with time (a voltage ramp), and current variations are recorded automatically (**Fig. 2**).

Polarographic curves. Oscillations of current result from growth of the individual mercury drops. Their mean value is usually measured. Three portions can be observed on a typical polarographic curve. At sufficiently positive potentials, only a small current flows. Then, in a region characteristic of the particular species and solution, an S-shaped current rise is observed. Finally, at sufficiently negative potentials, the current is again independent of applied voltage over a potential range. This current is called the limiting current and is usually a few microamperes or a fraction of a microampere. The limiting current (also called the height of the polarographic wave) increases usually with concentration of the investigated species in the solution. The wave height can be thus used for determination of concentration (that is, how much of the species is present in the solution). The potential at the point of the polarographic wave, where the current reaches half of the limiting value, is called the half-wave potential (Fig. 2). This potential is characteristic of the species studied (that is, it can be used to confirm which species are present).

Solutions. The solutions investigated polarographically must contain the species to be studied in 10^{-3} to 10^{-6} M concentration. In addition, the solution must contain a large excess (50-fold or greater) of a supporting electrolyte which does not react at the electrode in the potential region of interest and which can be a neutral salt, acid, base, or a mixture (such as a buffer). The function of the supporting electrolyte is to reduce the resistance of the solution, to ensure that species are transported to the surface of the electrode by diffusion rather than by migration in the electric field, to keep conditions at the electrode surface unchanged in the course of polarographic electrolysis, and to keep or convert the species into a form most suitable for electrolysis (for example, by complex formation or protonation).

Evaluations. The heights of polarographic waves are measured and compared with heights obtained with standard solutions of the species, the concentrations of which are known.

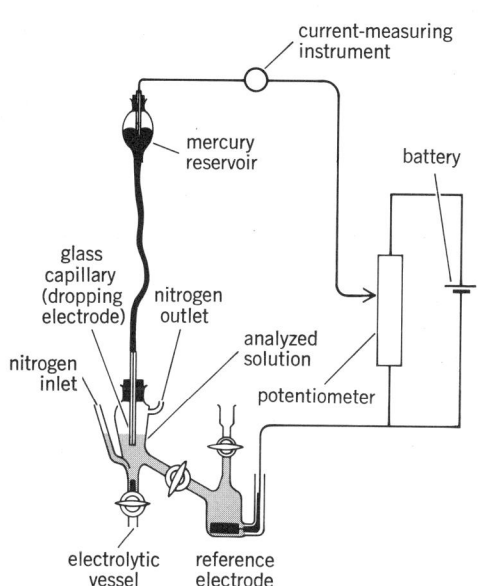

Fig. 1. Polarographic circuit.

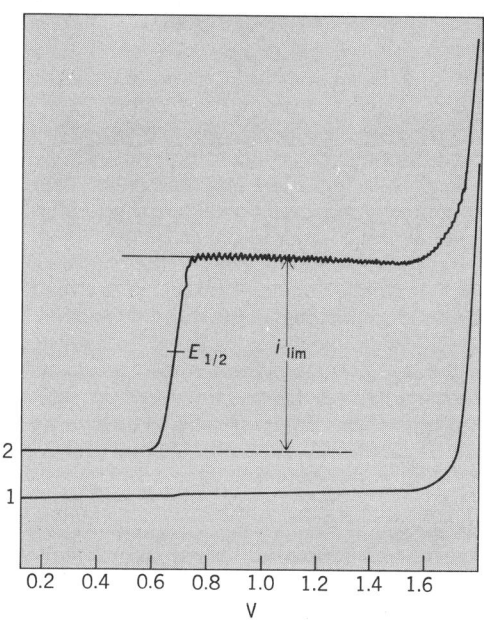

Fig. 2. Polarographic current-voltage curves: 1 = supporting electrolyte; 2 = curve in the presence of an electroactive (reducible) compound; $E_{1/2}$ = half-wave potential; i_{lim} = limiting current (wave height).

Accuracy varies from 2 to 5%, according to the potential range, shape of polarographic waves, and presence of other components. Under strictly controlled conditions, even 1% accuracy can be achieved.

Polarographically active species. To be studied polarographically, a species must undergo reduction or oxidation at the surface of the mercury drop, must form compounds with mercury, or must catalytically affect electrode processes. From changes in the shape of polarographic curves, it is also possible to determine various surfactants.

To be able to follow a given reducible species, its reduction must occur at more positive potentials than reduction of the component of the supporting electrolyte or solvent. Oxidations can be followed if they occur at potentials more negative than that of oxidation of the solvent, supporting electrolyte, or electrode material. Reduction processes are indicated by cathodic waves (above the zero current line), and oxidation processes by anodic waves (below the zero current line).

Polarography has been used for the determination of most metals (typically lead, copper, zinc, cadmium, iron, uranium, cobalt, nickel, manganese, potassium, sodium, and so forth), some inorganic acids and their anions (for example, iodic and periodic acids, nitrates, and nitrites), and some gases (for example, sulfur dioxide and oxygen). The wave of oxygen forms the basis of a convenient oxygen determination in technical gases, waters, or biological material. Alternatively the reducibility of oxygen means that oxygen must be removed (usually by purging with nitrogen) from solutions to be analyzed for other constituents. Reduction of inorganic species results in the lowering of the oxidation state, which may be accompanied by amalgam formation.

Electrode processes involving organic compounds at the dropping-mercury electrode result in a cleavage or formation of a chemical bond in the investigated organic molecule. Polarographic behavior depends primarily on the nature of the bond involved, even when the molecular environment (that is, the presence and kind of neighboring groups and other substituents, the type of the molecular frame to which the electroactive group is bound, and the spatial arrangement) also affects the electrode process. The reducibility of multiple bonds (for example, C=O, C=N, C=C, and N=N) or of strongly polarizable groups (such as NO_2, NO, and C-halogen), especially when conjugated, was recognized early. Recently the reducibility of some single bonds such as C—O, C—N, C—S, or C—P when the molecule contains activating groups was proved.

Because mercury ions react with various anions, anodic dissolution current of mercury can be used, for example, for the determination of halides, thiocyanate, cyanide, and sulfide among inorganic species; and of mercaptans, urea and thiourea derivatives, and dithiocarbamates among organic compounds.

Most important among catalytic waves are those which correspond to catalytic hydrogen evolution. Among compounds giving catalytic waves in buffered ammoniacal solutions of cobalt salts are some types of proteins.

The protein molecule must contain a suitably situated thiol or disulfide group and must be able to form a complex with cobalt and ammonia. It is assumed that the proton transferred by this complex undergoes reduction more easily than H_3O^+.

When blood serum is alkali-denatured and the proteins with largest molecular weight are separated by precipitation with sulfosalicyclic acid, the remaining filtrate contains proteins and their fragments, the contents of which differ in the serum of healthy and pathological individuals. Increase of the catalytic wave is observed for cancer and inflammatory diseases, and decrease of the catalytic wave for hepatitis and other liver diseases. In Europe the polarographic test is used in clinical analysis as a general screening test, in connection with nine other tests as a proof of malignancy, and in controlling cure of cancer and measuring the effectiveness of treatments such as surgery or irradiation.

Applications. Polarographic studies can be applied to investigation of electrochemical problems, to elucidation of some fundamental problems of inorganic and organic chemistry, and to solution of practical problems.

In electrochemistry polarography allows measurement of potentials, and yields information about the rate of the electrode process, adsorption-desorption phenomena, and fast chemical reactions accompanying the electron transfer. Since polarographic experiments are simple and not too time-consuming, polarography can be used in a preliminary test designed to find the most suitable model compound for detailed electrochemical investigations.

In fundamental applications, polarography allows one to distinguish the form and charge of the species (for example, inorganic complex or organic ion) in the solution. Polarography also permits the study of equilibria (complex formation, acid-base, tautomeric), rates, and mechanisms. For equilibria established in the bulk of the solution in more than 15 s, measurement of wave heights of individual components makes possible the evaluation of equilibrium constants. For equilibria that are very rapidly established at the electrode surface, equilibrium constants can be determined from shifts of half-wave potentials. Finally, for some equilibria between these two extremes, which are established in times comparable to the drop time (3 s), calculation of rate constants is possible (for example, for dehydration of hydrated aldehydes, or for protonation of anions derived from C-acids, such as $C_6H_5COCH_2COCH_3$). In this way, rate constants of very fast reactions of the order 10^5–10^{10} liters mole^{-1} s^{-1} can be determined.

For slower reactions, rate constants can be found from changes of wave heights with time. Moreover, as some reaction intermediates giving separate waves can be detected, identified, and followed polarographically (if their half-lives are longer than 15 s), polarography can prove useful in mechanistic studies. Elimination of Mannich bases, hydration of multiple bonds in unsaturated ketones, and aldolization are examples studied.

Finally, polarography can be used for investigation of the relationship between electrochemical data and structure. Reduction of most organic systems, particularly in aqueous media, involves steps with a high activation energy and is therefore irreversible. Half-wave potentials of such systems are a function of the rate constant of the electrode process. This heterogeneous rate constant is frequently influenced by structural effects, as are, in a similar manner, rate constants of homogeneous reactions. Therefore, effects of substituents on half-wave potentials in aromatic systems and in aliphatic systems can be treated by the Hammett relations and the Taft substituent constants, respectively. Among steric effects, for example, steric hindrance of coplanarity or effects of cistrans isomerisms affect polarographic curves. Polarography also makes it possible to distinguish between some epimers, for example, bearing axial or equatorial halogen.

Practical applications are predominantly analytical procedures. In inorganic analysis, polarography is used predominantly for trace-metal analysis (with increased sensitivity of differential pulse polarography and stripping analysis). In organic analysis, it is possible in principle to use polarography in elemental analysis, but such applications are deservedly infrequent. More frequent are applications in functional group analysis. Either the reaction product (for example, semicarbazone in the determination of carbonyl compounds, or N-nitrosoamine in determinations of secondary amines) is measured, or a decrease in concentration of a reagent (for example, chromic acid in determination of alcohols, or mercuric or silver ions in determination of thiols) is followed. Or, alternatively, the polarographic limiting current can be taken as a measure of the amount of a given electroactive group (for example, quinones, aliphatic nitro compounds, phenazines, or thiols).

More frequent is determination of individual compounds. If the species is electroactive, the analysis frequently consists of dissolving the material in a proper supporting electrolyte, recording the waves, and evaluating them. Electroinactive species can be determined indirectly, by converting them into electroactive by a suitable chemical reaction (for example, aromatics by nitration, secondary amines by nitrosation).

The most important fields of application of inorganic determinations are in metallurgy, environmental analysis (air, water, and sea-water contaminants), food analysis, toxicology, and clinical analysis. The possibility of being able to determine vitamins, alkaloids, hormones, terpenoid substances, and natural coloring substances has made polarography useful in analysis of biological systems, analysis of drugs and pharmaceutical preparations, determination of pesticide or herbicide residues in foods, and so forth. Polarography also makes possible determination of monomers, catalysts, and even some reactive groupings in polymers.

Other polarographic techniques. To eliminate unwanted charging current and to increase the sensitivity of polarography, the voltage is applied in regular pulses instead of gradually. When the current is measured only during the second half of the pulse, the technique is called pulse or differential pulse polarography. This technique is more sensitive by two orders of magnitude than dc polarography, and in inorganic trace analysis competes with atomic absorption and neutron activation analysis. With smaller dosage of more effective drugs, the sensitivity of differential pulse polarography has found application in drug analysis.

When an alternating voltage of small amplitude (a few millivolts) is superimposed on the dc voltage ramp, the technique is called ac polarography. It is particularly useful for obtaining information on adsorption-desorption processes at the surface of the dropping-mercury electrode. An increase in accuracy of polarographic methods can be achieved when the dropping-mercury electrode (or some other indicator electrode) is used in titrations. At constant applied voltage the current is measured as a function of volume of added titrant in amperometric titrations. The amperometric titration curve frequently has a shape of two linear sections. Their intersection corresponds to titration end point.

Related methods. Methods in which a potential sweep is used (instead of the practically constant potential of the electrode during the life of a single drop in polarography) are frequently called voltammetry. Either an electrode with nonrenewed surface—solid electrode, mercury pool, or mercury drop—is used, or the whole scan is carried out during the life of a single drop when the dropping-mercury electrode is used. The use of solid electrodes (for example, platinum, gold, or various forms of graphite) makes it possible to extend the voltage range to more positive potentials, so that numerous substances which cannot be oxidized when mercury is used can be oxidized. Solid electrodes can be used when they are stationary, rotating, or vibrating, but the change at the electrode surface in the course of electrolysis often decreases the reproducibility of the current-voltage curve compared with polarographic curves.

When extremely small concentrations of amalgam-forming metals (such as lead, cadmium, or zinc) are to be determined, a preconcentration can be carried out. The metal is deposited into a hanging mercury drop or into mercury plated on surfaces of solid electrodes. After a chosen time interval, a voltage sweep is applied, and the dissolution current of the metal ion from the analgam is measured. Such anodic stripping, particularly when combined with the differential pulse technique, allows determination up to 1 part in 10^{12} parts of solution.

When the applied voltage sweep is first increased and then decreased and the plot of dependence of voltage on time resembles a triangle, the technique is called cyclic voltammetry, and offers in particular information about products and intermediates of electrode processes and their reactions.

Finally, when current is applied and potential of an electrode measured, another group of techniques is developed. Most important among them is chronopotentiometry. In this technique, a constant current is applied to the indicator electrode in an unstirred solution, and its potential change is measured as a function of time. The time at which a sudden potential change is observed is called transition time, and its square root is proportional to the concentration of the reacting species. The method has found only limited application but has proven useful for the study of reactions of products of electrolysis. SEE ELECTROCHEMICAL TECHNIQUES.

Bibliography. R. N. Adams, *Electrochemistry at Solid Electrodes*, 1969; A. Bard (ed.), *Electroanalytical Chemistry*, vols. 1–12, 1966–1982; L. Meites, *Polarographic Techniques*, 2d ed., 1965; L. Meites et al., (eds.), *CRC Handbook Series in Inorganic Electrochemistry*, vols. 1–6, 1980–1986; J. A. Plambeck, *Electroanalytical Chemistry: Basic Principles and Applications*, 1982.

ELECTRODEPOSITION ANALYSIS
GEORGE W. O'DOM

An electroanalytical chemistry technique in which the product of the electrode reaction is insoluble and is either deposited upon or dissolved into the electrode. This method usually results in the complete depletion in solution of the deposited substance. Electrodeposition is applied most commonly to the separation of metals from solution and to their subsequent quantitative determination. However, it may be used simply as a separation technique. SEE ELECTROLYSIS.

The quantities of metals deposited may be determined by weighing a suitable electrode before and after deposition (electrogravimetry). Among the elements that may be determined by deposition are cadmium, nickel, zinc, copper, cobalt, antimony, tin, silver, and gold. At the anode the oxides of lead, PbO_2, and manganese, MnO_2, may be deposited. Halides may be deposited on a silver anode as the respective silver halides.

For a strictly reversible reaction, the potential of an electrode in equilibrium (no net electrodeposition occurring) with a solution of its ions of given concentration may be determined by using the Nernst equation, Eq. (1), which relates the formal potential of an electrode to the equi-

$$E = E^{\circ\prime} + \frac{RT}{nF} \ln[M] \tag{1}$$

librium potential in the presence of a solution of its ions. In Eq. (1) E is the equilibrium potential (versus a reference electrode) of the electrode, $E^{\circ\prime}$ is the formal potential, n is the number of electrons needed for the reduction of one metal ion, F is the faraday (96,487 coulombs/equivalent), R is the gas constant, T is the absolute temperature, and $[M]$ is the molar concentration of metal ions in equilibrium with the electrode. Variation of equilibrium electrode potentials with concentration of several ions is shown in **Fig. 1**. SEE ELECTRODE POTENTIAL.

When the electrode potential is made more cathodic than the equilibrium value, electrodeposition occurs until the metal ion concentration is lowered to that value which is in equilibrium with the electrode at the applied potential. A tenfold change in concentration at 30°C (86°F) may be affected by a $0.060/n$ volt change in potential. By making the electrode sufficiently cathodic, the metal ions remaining in the solution may be reduced to a negligible concentration. Two metallic species may be separated by adjusting the potential of the cathode so that it is less cathodic than the equilibrium potential of the metal to be left in solution, and more cathodic than the initial equilibrium potential of the metal to be removed. Practically all of the latter metal will be deposited as equilibrium is reestablished. In cases where the formal potentials of the simple metal ions are not sufficiently far apart to allow complete separations, it is often possible to lower the concentration of the interfering free metal ions in solution by adding a reagent to convert most of them to a complex ion species. After selectively complexing the metal to remain in solution, the potential required to deposit the low concentration of free ions of this metal is sufficiently more cathodic than that required for the other metal to make complete separation possible. **Figure 2** is a diagram of the electrodeposition apparatus.

Equation (1) cannot be used to select a deposition potential for a reaction which proceeds irreversibly at the electrode surface. For such a reaction the potential must be selected from current-potential curves.

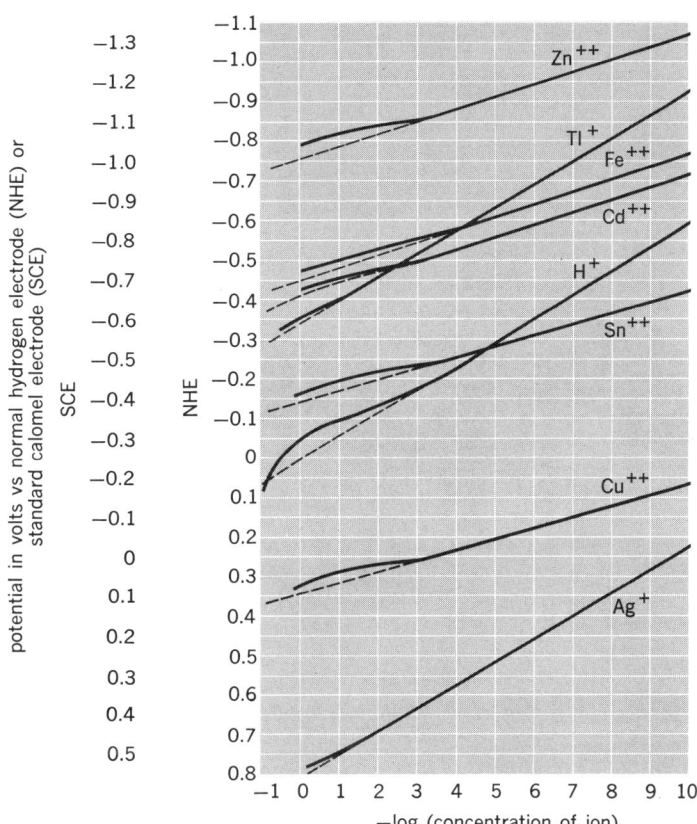

Fig. 1. Equilibrium electrode potentials. (*After G. W. Ewing, Instrumental Methods of Chemical Analysis*, 2d ed., McGraw-Hill, 1960)

The electrode material most commonly used for electroseparations is platinum. This metal is used as anode for most oxidations, and as cathode for depositions of the more readily reduced metals. Because of its low overvoltage on platinum, hydrogen is evolved and thereby interferes with the deposition of metals requiring very cathodic potentials for reduction.

Mercury is frequently used as a cathode material because its extremely high hydrogen overvoltage permits most metals to be deposited without interference from hydrogen evolution. As an anode, mercury finds very limited use because it is oxidized quite readily. Original procedures required removal of mercury from the deposited metals by distillation. However, coulometry is often effective for removing the metal from the mercury and obtaining a measurement of it.

Constant-potential electrolysis. The potential of the working electrode (the cathode when a metal is being deposited) is the most critical factor in obtaining a desired separation. It is not possible, however, to obtain constant potential at an electrode by applying a constant voltage to the cell. The applied voltage is expended in accordance with Eq. (2). In this expression, E_{ap} is

$$E_{ap} = E_{an} - E_c + \omega_{an} + \omega_c + IR \tag{2}$$

the voltage applied to the cell, E_{an} is the equilibrium potential of the anode, E_c is the equilibrium potential of the cathode, ω_{an} is the anodic overvoltage or increase in anode potential beyond the equilibrium value needed to pass the current of I amperes, ω_c is the corresponding cathodic overvoltage, and r is the ohmic resistance of the solution. The overvoltages for metal deposition and dissolution are usually less than 0.1 V, with notable exceptions for iron, cobalt, and nickel. Because they depend upon solution conditions and electrode form in a manner which is not fully understood, overvoltages cannot be predicted with certainty, thus limiting the accuracy with which the current in a given electrolysis may be calculated.

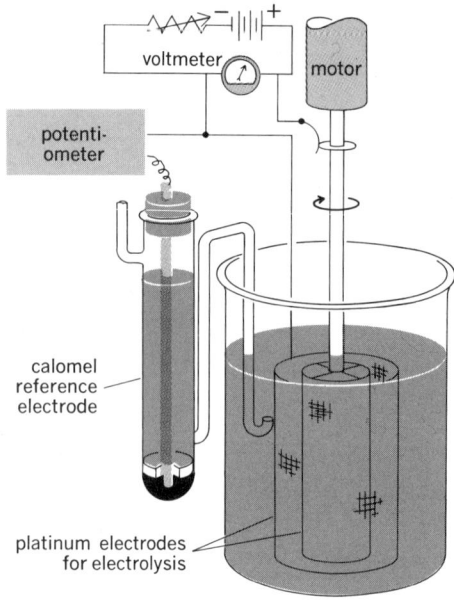

Fig. 2. Electrodeposition apparatus. (*After G. W. Ewing, Instrumental Methods of Chemical Analysis, 2d ed., McGraw-Hill, 1960*)

In electrolysis at constant applied voltage, the potential of a working cathode is less than the total applied voltage. When the applied voltage is adjusted so that the critical potential for a separation cannot be exceeded, the potential of the electrode is significantly less than this critical value throughout most of the deposition due to IR drop, and the electrolysis proceeds very slowly. If a stable reference electrode is placed in the solution and the voltage between this and the working electrode is maintained constant at the critical value by periodic adjustment of the voltage applied to the cell, much more rapid deposition results. Such manual control, however, is tedious. Instruments which automatically perform this task are known as potentiostats, and several types have been described in the literature. When the cathode potential is to be controlled, the desired voltage between the cathode and reference is maintained constant by an electronic or electromechanical servo system which changes the total voltage applied between the anode and cathode. When the anode potential is controlled, the voltage between the anode and the reference is kept constant by the same means. Anode and cathode potentials cannot both be controlled at the same time.

Internal electrolysis. Electrodeposition without externally applied voltage was an early but simple method of approximating a controlled cathode potential. An active metal, such as magnesium, which dissolves spontaneously is made the anode of a cell, and an inert electrode, such as platinum, is made the cathode. When the electrodes are shorted together, the potential of the cathode is equal to, and created by, the potential of the anode. By judicious choice of the anode metal and the concentration of the reagent in which it dissolves, the cathode potential may be made to assume predetermined values over most of the useful range. Deposition at the cathode occurs at the expense of dissolution of the anode. The current flow is limited by the magnitude of the spontaneous voltage of the cell, and by its internal resistance. Internal electrolysis has been used relatively little in recent years.

Constant-current electrolysis. This process precludes the possibility of electrode-potential control by electrical means. While the concentration of metal ions remains large, the potential of the cathode stays near the equilibrium potential. However, as the ions in the solution are depleted, the desired cathode reaction is not able to use all of the current forced through the cell, so the potential becomes more cathodic until an additional reaction, such as the depositon of another metal or the evolution of hydrogen, takes place to maintain the current. In most electroseparation devices a large voltage is applied to the cell. In order to effect a clean separation of two metals under these conditions, it is necessary to interpose a harmless reaction to limit the potential of the cathode before it exceeds the equilibrium potential of the metal to be left in solution. The reaction most often employed is hydrogen evolution, which may be made to occur at various potentials by proper adjustment of the pH. Through use of selective complexing agents to change the relative effective concentrations of the ions (and indirectly their equilibrium potentials), and also through control of pH to limit the cathode potential, many simple combinations of metals can be separated by this form of electrolysis.

Bibliography. A. J. Bard, *Electroanalytical Chemistry: A Series of Advances*, vols. 1–13, 1966–1983; A. J. Bard and L. R. Faulkner, *Electrochemical Methods: Fundamentals and Applications*, 1980; I. M. Kolthoff and P. J. Elving (eds.), *Treatise on Analytical Chemistry*, pt. 1, vol. 4, 1963; B. H. Vassos and G. W. Ewing, *Electroanalytical Chemistry*, 1983.

CELLS AND BATTERIES

Electromotive force	298
Battery	301
Primary battery	302
Reserve battery	304
Wet cell	307
Dry cell	310
Fuel cell	314
Mercury battery	317
Lithium primary cell	318
Storage battery	320

ELECTROMOTIVE FORCE
Walter J. Hamer

When two dissimilar electrodes are connected through an external conducting circuit, a difference in electric potential exists between them. Although this difference in potential is sometimes called the potential difference of the electrode couple, it is customary to say that the galvanic cell composed of the two dissimilar electrodes exhibits an electromotive (driving) force. This electromotive force (emf) is the resultant of the relative potential forces of the two dissimilar electrodes at which electrochemical reactions occur during cell operation. The two dissimilar electrodes need not be of unlike metals; for example, the metals may both be copper, but with the two coppers immersed in solutions of different concentration or composition. Likewise, both electrodes may be of the same gas, but with the pressure of the gas different at the two electrodes. *See Electrode potential.*

Cell types. Galvanic cells are of two general types, reversible and irreversible. If the chemical reactions at the electrodes can be exactly reversed by reversals in the direction of the current flow at the electrodes, the cell is said to be reversible; if the chemical reactions cannot be reversed, or if entirely different reactions occur on current reversal, the cell is then of the irreversible type. An example of a reversible cell is

$$(-)Pb(s) \mid PbSO_4(s) \mid H_2SO_4(aq) \mid PbO_2(s) \mid PbSO_4(s) \mid Pb(s)(+)$$

where $s =$ solid, $aq =$ aqueous solution, and the vertical lines represent the interfaces between different phases. This cell is the familiar lead-acid storage cell (or battery), widely used for a variety of purposes, including starting, ignition, and lighting for automobiles. When the cell is discharged, that is, when electric energy is being drawn from it, reactions (1) and (2) occur at the negative and positive electrodes, respectively. Here $l =$ liquid.

$$Pb(s) + SO_4^{2-}(aq) \rightarrow PbSO_4(s) + 2e^- \qquad (1)$$

$$PbO_2(s) + 4H^+(aq) + SO_4^{2-}(aq) + 2e^- \xrightarrow{Pb} PbSO_4(s) + 2H_2O(l) \qquad (2)$$

The overall cell reaction is (3), whereby lead sulfate and water are formed in the cell reaction. The

$$Pb(s) + PbO_2(s) + 2H_2SO_4(aq) \rightarrow 2PbSO_4(s) + 2H_2O(l) \qquad (3)$$

lead in the reaction at the positive electrode merely serves as an electronic conductor.

Now when the cell is charged, the reverse of the above reactions occurs, and lead sulfate is converted back to the initial state. The cell is, therefore, called a reversible one.

If, however, a cell is prepared by immersing zinc and platinum electrodes in perchloric acid, an irreversible cell results. This cell may be represented as

$$(-) \; Zn(s) \mid HClO_4(aq) \mid Pt(s) \; (+)$$

When the cell is discharged, reactions (4) and (5) occur at the negative and positive electrodes,

$$Zn(s) \rightarrow Zn^{2+}(aq) + 2e^- \qquad (4) \qquad\qquad 2H^+(aq) + 2e^- \rightarrow H_2(\text{gas, on platinum surface}) \qquad (5)$$

respectively. The overall cell reaction is (6). Here $g =$ gas.

$$Zn(s) + 2H^+(aq) \rightarrow Zn^{2+}(aq) + H_2(g) \qquad (6)$$

Now, if the cell is charged, that is, subjected to an electrolyzing current, instead of discharged, reactions (7) and (8) occur at the negative and positive electrodes, respectively. The overall cell reaction is then (9), or the simple electrolysis of water. Since the electrode

$$2H^+(aq) + 2e^- \rightarrow H_2(g, \text{on zinc surface}) \qquad (7)$$

$$2H_2O(l) \rightarrow O_2(g, \text{on platinum surface}) + 4H^+(aq) + 4e^- \qquad (8)$$

$$2H_2O(l) \rightarrow 2H_2(g) + O_2(g) \qquad (9)$$

reactions at each electrode for the charge differ from those obtained for the discharge, the cell is of the irreversible type.

In many cases the reversibility of a cell, or of the electrodes composing the cell, can be determined only by means of precise electrical measurements. Reversibility in the strict sense is determined by the response of a cell to very small (infinitesimal) discharging or charging current. In practice, measurements are made simultaneously of the current and the electromotive force of the cell for different values of the current. The slope of the electromotive force of the cell versus the current is ascertained for both the charging and the discharging currents. For reversible cells, the two slopes should be identical; furthermore, these slopes should be reproducible for repeated reversals in the direction of the flow of current through the cell. Also, for reversible cells, the internal resistance is small, and the magnitude of the slopes will be nearly zero. This means that the electromotive force of a reversible cell is insignificantly affected by the passage of very small currents through it in either direction. The reversibility of a galvanic cell is primarily a function of the reversibility of the two electrodes composing the cell, and the same criteria may be used to establish the reversibility of the electrodes.

Energy relations. Conditions may be chosen whereby practically no current flows through a cell in either direction. These conditions may be achieved by balancing the electromotive force of the cell against the electromotive force of another cell of known and steady value by using a highly sensitive galvanometer. When this state is achieved, the measured electromotive force of the cell E represents its reversible or maximum value. This value represents the maximum driving force of the cell. When it is multiplied by the total number of coulombs corresponding to the cell reaction, it gives the maximum electrical work, or free energy, which the cell is capable of producing. Thus, $E \times nF = -\Delta G$, where n is the number of electrons (or valence change) involved in the cell reaction, F is the faraday, and ΔG is the change in (Gibbs) free energy associated with the cell reaction. Furthermore, if the variation of the electromotive force of the cell with temperature is measured, the heat of reaction ΔH for the cell may be computed by the Gibbs-Helmholtz relation shown in Eq. (10), which may also be written as $\Delta H = \Delta G + T\Delta S$, where ΔS

$$\Delta H = -nFE + nFT \frac{dE}{dT} \qquad (10)$$

is the entropy change for the cell reaction since $nF(dE/dT) = \Delta S$.

If a galvanic cell has a negative emf-temperature coefficient, the heat of the reaction exceeds the free-energy change. Thus, the available electrical work is less than that corresponding to the heat of the reaction for the cell; the cell warms in operation, and heat is lost to the surroundings. Conversely, if the emf-temperature coefficient is positive, the free energy exceeds the heat of reaction, and the cell tends to cool when in operation; this cooling is overcome if the cell is maintained under isothermal conditions by absorption of heat from the surroundings. Thus, the total available electrical energy from such a cell is a resultant of the inherent changes in the heat content of the cell and the heat absorbed from the surroundings.

Electromotive force measurements. As stated above, in measuring the emf of a cell, a reference cell of known emf must be available to effect a comparison. This reference cell should have an emf that is known in terms of physical laws and units, and not one chosen arbitrarily. The emf of reference cells (and then, through comparisons, the emf of all cells) is known in terms of the mks (meter-kilogram-second) system of electromagnetic units through Ohm's law, $E = IR$, where E is emf in volts, I is current in amperes, and R is resistance in ohms. In practice, then, the emf of a cell is equal to IR, and it may be balanced against the IR drop across a resistor of known value through which a known current is flowing. If the values of the resistor and the current are known in mks electromagnetic units, then the emf of a cell is given in like units.

Standard cells. The standard cells for this purpose are the cadmium amalgam standard cells of the saturated type proposed by Edward Weston in 1892. This type of cell is the most reversible galvanic cell known and retains a constant emf to within a few microvolts for many decades. This cell consists of a cadmium amalgam anode (negative element), a mercury–mercurous sulfate cathode (positive element), and a saturated solution of cadmium sulfate containing crystals of $CdSO_4 \cdot 8/3 H_2O$. This cell may be represented by

$Cd-Hg(10\%)(2p) \mid CdSO_4 \cdot 8/3H_2O(c) \mid CdSO_4(sat.\ aq) \mid CdSO_4 \cdot 8/3H_2O(c) \mid Hg_2SO_4(s) \mid Hg(l)$

where $2p$ = two phase, c = crystals, $sat.\ aq$ = saturated aqueous solution, and the other symbols have the meaning given above. This cell has an emf, when freshly made, of 1.018636 volts at 20°C

(68°F) and for precise measurements must be maintained at a constant temperature. Its emf at other temperatures between 0 and 43.5°C (110°F) may be calculated from Eq. (11), where t is in

$$E_t = 1.018636 - 0.0000406 (t - 20) - 0.00000095 (t - 20)^2 + 0.00000001(t - 20)^3 \quad (11)$$

degrees Celsius. Saturated standard cells should not be used above 43.5°C, at this temperature, the crystals of $CdSO_4 \cdot 8/3H_2O$ are converted to $CdSO_4 \cdot H_2O$. Although standard cells prepared with the monohydrate are stable, they can be used with confidence only at temperatures above 43.5°C; when such cells are cooled, the monohydrate reverts to $CdSO_4 \cdot 8/3H_2O$, but the rate of conversion is slow and the emf of such cells is erratic for indefinite periods.

The cadmium standard cell is also made in the unsaturated type, that is, with no crystals of $CdSO_4 \cdot 8/3H_2O$, and is the type widely used in recording instruments, with potentiometers, and in pH meters. A solution of cadmium sulfate that is saturated at 4°C (39.2°F) is used in its preparation; the solution is then unsaturated at higher and normal temperatures. It is made portable by placing cork or plastic septa over the positive and negative elements. This cell has a very low emf-temperature coefficient (0.000005 volt/°C or 0.000009 volt/°F), about one-tenth that of the saturated type. On the average, this cell decreases in emf at the rate of 20 microvolts per year, and its ultimate life is about 10 years.

For the saturated type of cell, the overall reaction of the cell is (12), where x moles of Cd

$$xCd(yHg)(l + s) + Hg_2SO_4(s) + \frac{8/3}{M - 8/3} CdSO_4 \cdot mH_2O(l)$$
$$= (CdSO_4 \cdot 8/3H_2O) + 2Hg(l) + (x - 1)Cd(yHg)(l + s) \quad (12)$$

are associated with y moles of Hg in the amalgam, and m is the number of moles of water associated with 1 mole of $CdSO_4$ in the saturated solution. For the unsaturated cell the cell reaction is simply (13).

$$xCd(yHg)(l + s) + Hg_2SO_4(s) = CdSO_4(aq) + 2Hg(l) + (x - 1)Cd(yHg)(l + s) \quad (13)$$

A cross-sectional sketch of the saturated type of cell is shown in the **illustration**. The unsaturated type is made similarly except that no crystals are used and cork or plastic septa are placed above the amalgam and the mercurous sulfate paste. The H-form container is made of Kimball glass with platinum wires sealed in the bottom of each limb. Mercury purified by several

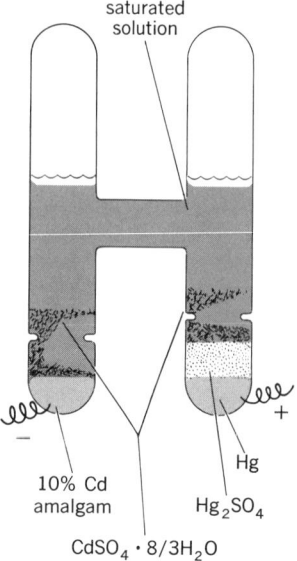

Cross-sectional diagram of Weston cell.

vacuum distillations is placed at the bottom of one limb and is used to prepare the amalgam used in the bottom of the other limb. The cadmium is purified by sublimation. A 10% (sometimes 12.5%) amalgam of cadmium is added while warm and in a single phase; on cooling, it becomes two-phased, with the solid phase being an isomorphous mixture of cadmium and mercury. Mercury(I) sulfate, prepared electrolytically, is then placed over the mercury, and crystals of $CdSO_4 \cdot \frac{8}{3}H_2O$ are added to both limbs. Then a saturated solution of cadmium sulfate is added to a level slightly above the crossarm, and the cell is sealed. The unsaturated cell is usually mounted in a nontransparent case, and the saturated cell is housed in oil baths or in thermoregulated air baths. After proper aging, these cells serve as reliable standards of emf with which the emf of all other cells are compared. *See* BATTERY; ELECTROCHEMISTRY.

Bibliography. M. Barak (ed.), *Electrochemical Power Sources: Primary and Secondary Batteries*, 1980; P. Delahay, *New Instrumental Methods in Electrochemistry*, 1954, reprint 1980; W. J. Hamer, *Standard Cells: Their Construction, Maintenance, and Characteristics*, 1965; D. G. Ives and G. J. Janz, *Reference Electrodes*, 1969; C. L. Mantell, *Batteries and Energy Systems*, 2d ed., 1983; R. White, *Electrochemical Cell Design*, 1984.

BATTERY
Harold C. Riggs

A device which transforms chemical energy into electric energy. The term is usually applied to a group of two or more electric cells connected together electrically. In common usage the term battery is also applied to a single cell, such as a flashlight battery.

Types. There are in general two types of batteries, primary batteries and secondary storage or accumulator batteries. Primary types, although sometimes consisting of the same plate-active materials as secondary types, are constructed so that only one continuous or intermittent discharge can be obtained. Secondary types are constructed so that they may be recharged, following a partial or complete discharge, by the flow of direct current through them in a direction opposite to the current flow or discharge. By recharging after discharge, a higher state of oxidation is created at the positive plate or electrode and a lower state at the negative plate, returning the plates to approximately their original charged condition.

Primary and secondary cells may be constructed from several materials. For the more important of these types *see* PRIMARY BATTERY; STORAGE BATTERY.

Applications. Primary cells or batteries are used as a source of dc power where the following requirements are important.

1. Electrical charging equipment or power is not readily available.
2. Convenience is of major importance, as in the case of the hand or pocket flashlight.
3. Stand-by power is desirable without cell deterioration during periods of nonuse for days or years. Reserve-electrolyte designs may be necessary, as in torpedo, guided-missile, and some emergency light and power batteries. *See* RESERVE BATTERY.
4. The cost of a discharge is not of primary importance.

Secondary cells or batteries are used as a source of dc power where the following requirements are important.

1. The battery is the primary source of power and numerous discharge-recharge cycles are required, as in industrial hand or rider trucks, electric street trucks, mine or switching locomotives, and submarines.
2. The battery is used to supply large, short-time (or relatively small, longer-time), repetitive power requirements, as in automotive and airplane batteries.
3. Stand-by power is required and the battery is continuously connected to a voltage-controlled dc circuit. With proper voltage the battery is said to "float" (drawing from the dc circuit only sufficient current to compensate automatically for the battery's own internal self-discharge). Telephone exchange, central-station circuit breaker, and emergency light and power batteries are in this category.
4. Long periods of low-current-rate discharge followed subsequently by recharge are required, as in buoy service.
5. The very large capacitance is beneficial to the circuit, as in telephone exchanges.

Size. Both the primary and secondary cells are manufactured in many sizes and designs, from the small electric wristwatch battery and the small penlight battery to the large submarine battery, where a single cell has weighed 1 ton (0.9 metric ton). In all applications the cell must be constructed for its particular service, so that the best performance may be obtained consistent with cost, weight, space, and operational requirements. Automotive and aircraft batteries generally use thin positive and negative plates with thin separation to conserve space and weight and to provide high rates of current discharge at low temperatures. Stand-by batteries use thick plates and thick separators to provide long life. Notable size and weight reductions were made through use of new plastic materials, active materials, and methods of construction.

Ratings. Since the power that can be obtained from a cell varies with its temperature and the rate of current discharge, the power-output rating is very important. Common secondary-battery practice is to rate cells in terms of ampere-hours (discharge rate in amperes times hours of discharge) and to specify the hourly rate of discharge. A popular automotive battery capable of giving 2.5 A for 20 h is rated at 50 ampere-hours at the 20-h rate. This same battery may provide an engine-cranking current of 150 A for only 8 min at 80°F (27°C) or for 4 min at 0°F (-18°C), giving a service of only 20 and 10 Ah, respectively. By multiplying ampere-hours by average voltage during discharge, watt-hour rating is obtained. Ratings must be made to a specified final voltage, which is either at the point of rapid voltage drop or at minimum usable voltage. The rating of primary batteries is generally stated as the number of hours of discharge which can be obtained when discharging through a specified fixed resistance to a specified final voltage.

Life. Life of cells varies from the single discharge obtainable from primary types to 10,000 or more discharge-charge cycles obtainable from some secondary cells operating at very high rates for very short times. Automotive batteries may generally be expected to give approximately 300 cycles, or to last 2 years. Industrial-truck sizes may be expected to give 1500 to 3000 cycles in 5–10 years. Stand-by sizes may be expected to float across the dc bus 8–30 years. Generally the most costly, largest, heaviest cells are the longest-lived.

To obtain life from batteries, certain precautions are necessary. The stated shelf life and temperature of wet primary cells must not be exceeded. For dry reserve-electrolyte primary cells and secondary cells of the dry construction with charged plates, the cell or battery container must be protected against moisture, and storage must be within prescribed temperature limits. Wet, charged secondary batteries require periodic charging and water addition, depending upon the kind of construction.

Reliability. Batteries are probably the most reliable source of power known. In fact, most critical electric circuits are protected in some manner by battery power. There are no moving parts and, with good quality control in component materials and construction, one can be assured of power, particularly since adequate checks to indicate the condition of the cells usually exist. To ensure reliability, manufacturer's stipulations on storage and maintenance must be followed.

PRIMARY BATTERY
Jack Davis

An electric battery designed to deliver only one continuous or intermittent discharge. It cannot be recharged efficiently. Primary batteries are designed to deliver limited amounts of electric energy, determined by the materials used and the size of the cell. When the available energy drops to zero, the battery is usually discarded. Primary batteries may be classified by the type of electrolyte used.

Aqueous-electrolyte batteries. These batteries use solutions of acids, bases, or salts in water as the electrolyte. These solutions have ionic onductivities of the order of 1 mho/cm (0.4 mho/in.) and practically no electronic conductivity. Practical cells, such as the common Leclanche dry cell and the alkaline-manganese-zinc cell use aqueous electrolytes. Disadvantages of such cells include corrosion of the electrode materials by the electrolyte, a relatively high evaporation rate of water vapor which can cause cell failure, and the difficulties of preventing leakage. For examples of cells with aqueous electrolytes SEE BATTERY; DRY CELL; MERCURY BATTERY; RESERVE BATTERY. For an example of a cell with a nonaqueous electrolyte SEE LITHIUM PRIMARY CELL.

Solid-electrolyte batteries. These use electrolytes of solid crystalline salts which have predominantly ionic conductivity. The conductivity is small compared with aqueous electrolytes, and the current output is of the order of 10^{-7} A/in.^{3} ($0.4 \cdot 10^{-7}$ A/cm^3).

Solid-electrolyte batteries may be classified in two broad categories: (1) cells with solid crystalline salt, such as silver iodide, as the electrolyte; (2) cells with ion-exchange membrane as the electrolyte. In either category, the conductivity must be nearly 100% ionic. Any electronic conductivity causes a continuous discharge of the cell and will limit the stand or shelf life.

A typical cell with solid crystalline salt electrolyte is the lead–lead chloride–silver chloride cell in **Fig. 1**. Here lead is the anode, lead chloride is the electrolyte, and silver chloride is the cathode. This cell has a potential of 0.49 volt. During discharge, lead is oxidized to lead ion and silver chloride is reduced to silver.

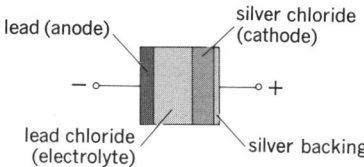

Fig. 1. Typical solid-electrolyte cell with solid crystalline salt electrolyte.

Cells with solid salt electrolyte have been developed into miniature batteries. One type delivers 90–100 V at 10^{-11} A, and has a capacity of 1 ampere-second. This is over 10^6 days at 10^{-11} A. The practical life of the cell is much less but may be as much as 10 years at room temperature. It can be stored at 160°F (71°C) for at least 30 days and will operate over the range −65 to +165°F (−54 to +74°C). The battery is ⅜ in. (9.5 mm) in diameter and 1 in. (25 mm) in length. With the increasing use of electronic devices and consequent miniaturization, solid-state batteries delivering low currents are finding newer applications. In addition, the use of electrolytes such as Ag$_3$SI or MAg$_4$I$_5$, (where M is K, Rb, NH$_4$, or Cs), which have better ionic conductivity than the lead or silver halides previously employed, gives cells with flash currents in the low milliampere range.

An example of a cell with ion-exchange membrane as electrolyte is the zinc–zinc ion exchange membrane, silver ion exchange membrane–silver cell shown in **Fig. 2**. Physically, the metal electrodes are in contact with the solid membrane which contains two regions. The region adjacent to the zinc is in the zinc ion state. The region adjacent to the silver is in the silver ion state. The discharge reaction increases the zinc ion quantity and decreases the silver ion quantity, in proportion to the amount of charge transferred. This cell has a potential of about 1.5 V.

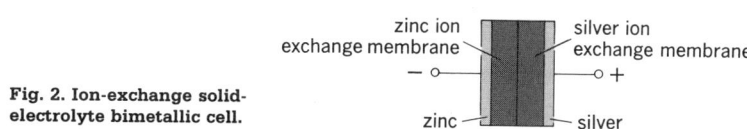

Fig. 2. Ion-exchange solid-electrolyte bimetallic cell.

The zinc-silver cell described has serious short-comings. The shelf life is poor, indicating internal self-discharge, and the capacity is limited by the available supply of silver ions. In strongly ionized types of ion-exchange material, the volume density of ionizing sites is about 1 equiv/liter (3.8 equiv/gal), or 0.4 ampere-hour/in.^{3} (0.0064 ampere-hour/cm^3). This is very low compared with metal-oxide cathodes. A cell with higher capacity can be made by replacing the silver ion exchange material and silver by manganese dioxide plated on an inert metal, such as tantalum. This gives a capacity of about 100 times as much, for equal volume.

Ion-exchange electrolytes are also used with hydrogen and oxygen gas electrodes (**Fig. 3**). The electrodes consist of platinized metal screens. The electrolyte is a hydrogen ion exchange material. The room temperature emf of this cell is 0.96 V. *See* I*on-selective membranes and electrodes.*

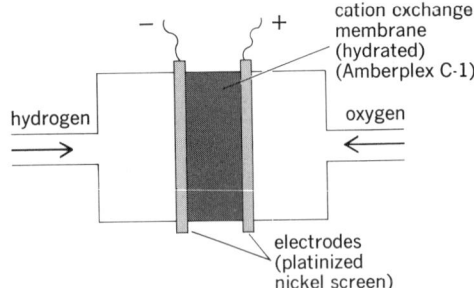

Fig. 3. Solid-electrolyte cell using an ion-exchange membrane as the electrolyte.

Waxy-electrolyte batteries. These use waxy materials, such as polyethylene glycol, in which a small amount of a salt is dissolved in the molten wax. At room temperatures these materials are solid. The conductivity is small and the current output is limited to about 10^{-6} A/in.^{2} (1.55×10^{-7} A/cm^2).

Figure 4 shows a battery stack of cells using a waxy electrolyte. The electrodes are sheet zinc and manganese dioxide. The electrolyte is made of polyethylene glycol in which is dissolved a small amount of zinc chloride. This electrolyte is melted and painted on a paper sheet to form the separator.

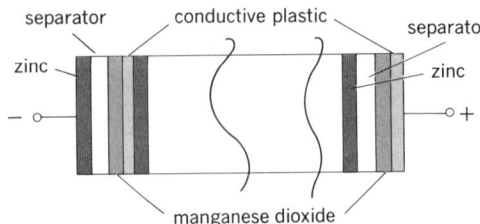

Fig. 4. Waxy-electrolyte battery stack.

A 25-cell stack, built as shown in Fig. 4 and measuring 0.34 in. (8.5 mm) in length and 0.25 in. (6.5 mm) in diameter, weighed 0.053 oz (1.5 g). A 0.50-in.-diameter (12.5-mm) stack weighed 0.20 oz (6.0 g). The initial open-circuit voltage was 37.5 V (1.5 V per cell).

The internal resistance of this cell is high, and it increases as temperature decreases. This high internal resistance limits the usefulness of the cell, but it may be suitable for long-life potential sources of miniature size.

Fused-electrolyte batteries. These use crystalline salts or bases which are solid at room temperature. In use, the cell is heated and maintained at a temperature above the melting point of the electrolyte.

Bibliography. M. Barak (ed.), *Electrochemical-Power Sources: Primary and Secondary Batteries*, 1980; T. R. Crompton, *Small Batteries: Primary Cells*, 1983.

RESERVE BATTERY
Jack Davis

A battery which is inert until an operation is performed which brings all the cell components into the proper state and location to become active.

High-energy primary batteries are often limited in their application by poor retention characteristics which are caused by self-discharge between the electrolyte and the active electrode materials. One method of overcoming this problem, particularly when the battery is required to operate at high current levels for relatively short periods of time (minutes or hours), is the use of a reserve battery.

Several types have been developed. In water-activated or electrolyte-activated batteries the water or electrolyte component is not present during storage. It is added just before the cell is put into use. In thermal batteries the electrolyte is a solid at room temperature and has very low conductivity. If the temperature is raised above the melting point, the conductivity of the electrolyte becomes excellent and the cell is capable of delivering significant power.

Water-activated batteries. Practical battery systems have been developed using magnesium anodes against silver chloride or cuprous chloride cathodes (**Fig. 1**). Cuprous chloride cathodes are less expensive than silver chloride cathodes, but they are also bulkier and less stable, particularly in a humid atmosphere. *Meta*-dinitrobenzene is also finding use as cathode material on account of its high ampere-hour capacity.

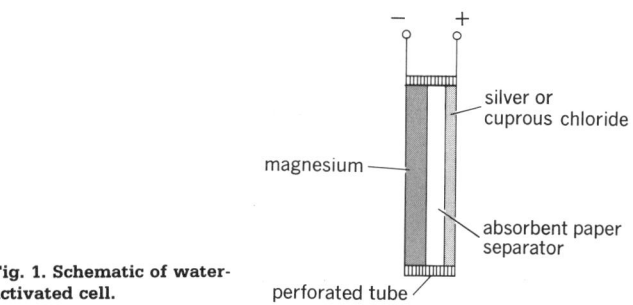

Fig. 1. Schematic of water-activated cell.

The batteries are assembled dry. The active elements may be separated by porous paper or another inert media. Water may be poured into a container holding the elements or may flow continuously through the element. Either fresh or salt water may be used.

The most important design factor for all reserve batteries is to ensure that the electrolyte is delivered as quickly as possible at the time of activation, at the same time avoiding chemical short-circuiting of the cells.

Cells with absorbent separators can be activated by immersion. Subsequent operation may either be in air, using only the water retained in the cells, or while immersed. Performance of a two-cell battery immersed in sea water showed an output of 18 watt-hours/lb (39.6 W-h/kg) when fully discharged in 6 min.

The dry elements stored in a sealed container are capable of indefinite storage life.

Electrolyte-activated batteries. Any cell can be made as a reserve-electrolyte cell. If the electrodes are in place, it is necessary only to add the electrolyte to make a complete cell. In practice, however, the separation of the electrolyte is done only when excessive deterioration would occur during wet storage prior to use. Great ingenuity has been shown in designing complete battery packages in which an aqueous electrolyte is stored in a separate chamber. The package contains a mechanism, which may be operated from a remote location, which drives the electrolyte out of the reservoir and into all the cells of the battery. In general, these packaged batteries have been used only in military applications. The following couples have been used in reserve cells containing electrolytes: Zn/Cu, Pb/PbO$_2$, Zn/AgO, Mg/*meta*-dinitrobenzene, Zn/PbO$_2$, Zn/MnO$_2$, Cd/PbO$_2$. However, not all these couples are amenable to heavy rate discharge.

Gas-activated batteries. The liquid-activated batteries previously described have disadvantages which are difficult to overcome; for example, automatic electrolyte-charging equipment may cause intercell shorting. Unless the design is quite complex, it is difficult to avoid flooding or uneven filling. An alternative approach is to introduce a gas which reacts with the spacer material to form a conducting electrolyte.

Boron trifluoride gas reacts with dry, hydrated barium hydroxide to form a highly acid

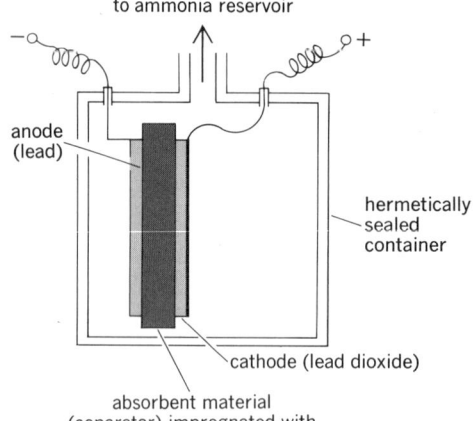

Fig. 2. Schematic of ammonia-vapor-activated reserve-type primary cell.

solution containing barium salts, borates, and fluoborates. Ammonia gas reacts with ammonium salts to form a solution having good conductivity (**Fig. 2**). Suitable electrode couples are Zn/MnO_2 and Pb/PbO_2. These gas-activated batteries operate well over a wide temperature range.

Thermal batteries. These are also known as heat-activated or fused-electrolyte batteries. Some compounds, such as sodium chloride and potassium hydroxide, show very low conductivity in the solid state at room temperature but very good conductivity in the molten state—a mixture of sodium hydroxide and potassium hydroxide becomes an excellent ionic conductor when heated above 338°F (170°C). If a zinc anode and a silver oxide cathode are combined with solid pads of the eutectic mixture, all the elements of a cell are present. The electrolyte has an appreciable amount of entrained moisture, which plays a role in the discharge, but the cell will also work with carefully dried materials. Such cells are capable of high power output for a few minutes, when heated to 392°F (200°C) or higher. At 1 A/in.^{2} (0.15 A/cm^2) of positive plate, the cell voltage is 1.16 at 392°F (200°C), 1.23 at 482°F (250°C), 1.30 at 572°F (300°C).

Thermal batteries are capable of operation at very low ambient temperatures, provided that a suitable heat source is available to melt the electrolyte. For ordinary temperatures, they are not advantageous as compared with reserve aqueous-electrolyte types. Because the electrolyte in thermal batteries is insert and nonconductive at normal temperatures, they can be stored indefinitely, acting as primary batteries when the temperature of the electrolyte is raised to that at which they become ionically conductive.

A magnesium and manganese dioxide cell with sodium hydroxide electrolyte can operate for longer discharge times than the zinc–silver oxide cell mentioned previously because of the greater stability of the reactants at high tempratures (**Fig. 3**).

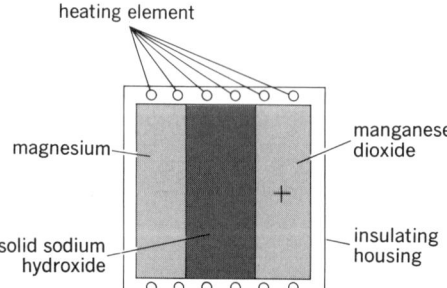

Fig. 3. Schematic of thermal cell.

In addition to the couples already mentioned, successful systems have been employed using Mg, Ca, or Li alloys as anodes and K_2CrO_4, WO_3, MoO_3, or $PbCrO_4$ as cathodes.

Thermal batteries are essentially of low energy efficiency because the heat absorbed in melting the electrolyte is not available in the electrical output. Consequently, although they are restricted to small cell sizes, they are extremely useful because of their high energy output, long storage retention over wide temperature ranges, ruggedness, and ability to be used in any position. *See* BATTERY.

Bibliography. D. G. Fink and H. W. Beatty, *Standard Handbook for Electrical Engineers*, 12th ed., 1987; G. W. Heise and N. C. Cahoon, *The Primary Battery*, vol. 1, 1971, vol. 2, 1976; R. Jasinski, *High Energy Batteries*, 1967; C. L. Mantell, *Batteries and Energy Systems*, 2d ed., 1983.

WET CELL
Kenneth Franzese, L. Rozeanu, and Jack Davis

A primary cell in which there is a substantial amount of free electrolyte in liquid form. Important examples of wet cells are the Lalande or caustic soda cell, the air-depolarized alkaline cell, the Weston standard cell, and the organic electrolyte cell.

Lalande cell. This cell (**Fig. 1**) uses a zinc anode, a cupric oxide cathode, and an electrolyte of sodium hydroxide in aqueous solution (caustic soda).

The amalgamated zinc electrodes are cast as flat plates or as hollow cylinders with thin sections, which corrode through as the copper oxide electrode approaches exhaustion. The bright copper color of the cathode can be seen through these openings to warn the user of approaching exhaustion of the cell.

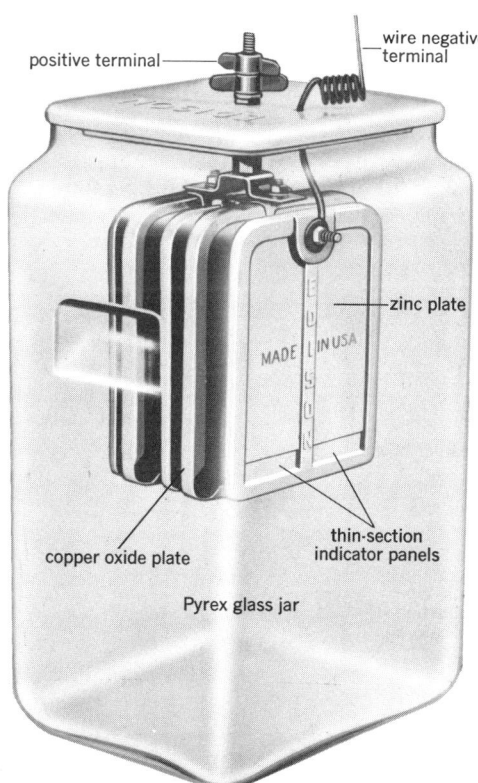

Fig. 1. Lalande-Edison copper oxide primary battery.

The cathode is made by molding cupric oxide into flat plates or hollow cylinders. The oxide is mixed with a binder, pressed, and roasted. As it is used, the cupric oxide is partially reduced to metallic copper, which greatly increases the conductivity of the cathode.

The electrolyte is a solution of sodium hydroxide in water of good purity. The specific gravity is about 1.21. Normally, the surface of the electrolyte is covered with a layer of oil which retards evaporation of water and absorption of carbon dioxide from the atmosphere.

The anode reaction in the Lalande cell is the oxidation of zinc to form zinc oxide, which dissolves in the electrolyte to form sodium zincate. With sufficient electrolyte, no solid phase forms until the cell is nearly exhausted. Then precipitation occurs. Because of this reaction, it is necessary to provide about 8 ml (0.27 fl oz) of 1.21 sp gr solution per rated ampere-hour (Ah). The zinc must be mounted near the top of the electrolyte to prevent premature cutting off of the discharge by anodic polarization.

The cathodic reaction is the reduction of cupric oxide to metallic copper. For this reason, the plate always has the appearance of metallic copper.

The cell potential is 0.95–1.0 V, but the closed-circuit voltage on normal discharge starts at about 0.65 and decreases slowly to a cutoff voltage of about 0.50. Commercial cells are intended for relatively long continuous service. A 500-Ah light-duty cell has an output rating of 1.75 A at 70°F (21°C). A heavy-duty cell of the same ampere-hour rating is capable of continuous output at 6.5–12.0 A. Unit energy output is about 1.1 kWh/ft^3 (0.039 Wh/cm^3) of cell.

At temperatures below 70°F (21°C), full output (ampere-hours) can be obtained at a reduced current. Current is reduced by 40% at 40°F (4°C), by 67% at 20°F (-6°C), and by 83% at 0°F (-17°C).

Air-depolarized alkaline cell. This cell (**Fig. 2**) uses a zinc anode, a porous carbon cathode exposed to air on one face, and an alkaline electrolyte. The carbon cathode utilizes atmospheric oxygen. Its zinc anode and alkaline electrolyte are like those of the Lalande cell, but it has twice the operating voltage and twice the watt-hour output of an equal-size Lalande cell.

The cathodic reactions have been shown to be reactions (1)–(3). The oxygen reacts to form

$$2e + O_2 + H_2O \rightarrow O_2H^- + OH^- \quad (1) \qquad O_2H^- + OH^- + 2H^+ \rightarrow H_2O_2 + H_2O \quad (2)$$

$$H_2O_2 \rightarrow H_2O + 1/2 O_2 \quad (3)$$

hydrogen peroxide, which decomposes readily to water and oxygen. The net amount of oxygen, then, is 0.3 g/Ah. At standard temperature and pressure, 210 ml (7.1 fl oz) of oxygen is consumed per ampere-hour.

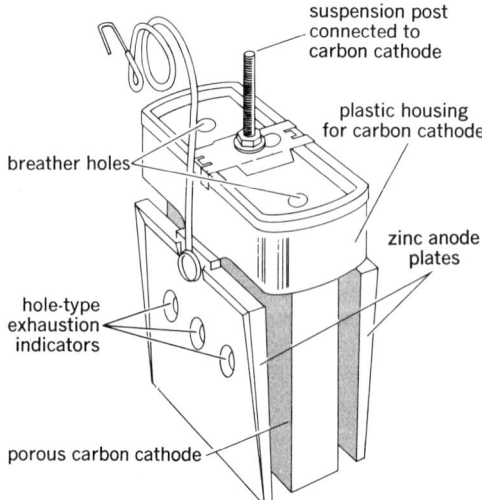

Fig. 2. Construction of air-depolarized alkaline cell.

The porous carbon has been reported to have an apparent density of only 0.65 with a porosity of 60%. To function properly, the inner surfaces should be dry. To resist penetration of electrolyte, the pore size must be small and the surfaces partially waterproofed by impregnation with paraffin. The cathode acts as a pump, drawing oxygen from the air. If too great a current is drawn, the pressure in the pore may drop sufficiently to allow electrolyte penetration. This reduces the activity of the carbon and may cause cell failure. Hence it is important that the cell should not be overloaded. For a railway cell, rated at 500 Ah, the recommended continuous drain is 2.0 A at temperatures above 45°F (7°C). This is a current density of 3.1 A/ft^2 (3.3 mA/cm^2). Much higher current densities can be obtained with porous carbon electrodes by special design.

The cell open-circuit voltage is 1.46. The railway cell at room temperature (75°F or 23°C), rated at 500 Ah, will deliver 2 A at an average of 1.13 V to a cutoff of 1.05 V. On 3-A intermittent signal test, the cell delivers rated capacity at 1.09 V, average at 75°F (23°C), 0.98 V average at 32°F (0°C).

Air-depolarized cells are now available in sizes up to 2500 Ah. The 2500-A-h cell delivers 5.64 kWh/ft^3 (0.20 Wh/cm^3) and 0.074 kWh/lb (0.16 kWh/kg) when discharged at low rates.

Weston standard cell. The Weston cell of 1893 has become the accepted standard of electromotive force (emf). The Weston normal or saturated cadmium cell has an emf of 1.01864 absolute volts at 68°F (20°C). When purified materials are used, cells having the same emf to within a few microvolts may be made. These cells maintain their emf very well. Reference standards, in daily use for many years, are remarkably constant. These standard cells are made with materials of spectroscopic purity and are maintained under diffuse light in a thermostatically controlled oil bath in a room maintained at 77°F (25°C) and 60% relative humidity.

The cell uses a two-phase amalgam of cadmium as the anode. For a 10% amalgam, one part of cadmium by weight and nine parts of mercury are required. These materials may be combined either by heating them together or by electrolytic deposition of cadmium into mercury. At ordinary temperatures a liquid phase is in equilibrium with a solid phase. This gives a very stable potential which depends only on the temperature.

The cathode is mercurous sulfate, Hg_2SO_4, in contact with mercury.

The electrolyte is a saturated solution of cadmium sulfate in equilibrium with the solid phase $CdSO_4 \cdot 8/3 H_2O$. In some cells, sulfuric acid is added to prevent hydrolysis of mercurous sulfate.

The cell is usually made of glass in the form of an H, as in **Fig. 3**. Platinum-wire leads are sealed in the base of each arm. Mercury, carefully purified, is placed at the bottom of one arm and a 10% cadmium amalgam is placed, while warm and in a single phase, in the other arm. When the amalgam has cooled and separated into two phases, crystals of mercurous sulfate are placed above the mercury and crystals of cadmium sulfate are placed above the amalgam. A saturated solution of cadmium sulfate is then added to about 0.1 in. (2–3 mm) above the crossbar, and the cell is hermetically sealed.

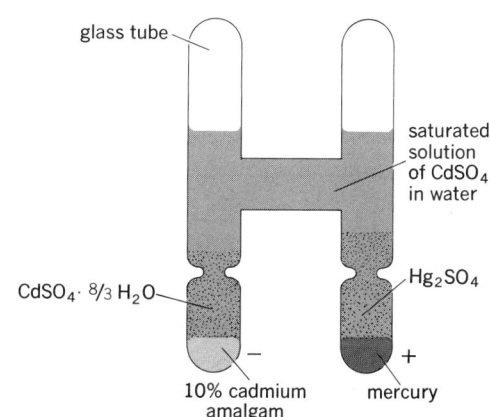

Fig. 3. Schematic of Weston saturated standard cell.

The best-saturated cells of this type may be measured to the ten-millionth part of a volt at specified temperatures which must be known to within 0.01°C (0.018°F).

The saturated Weston cell has a relatively large temperature coefficient of emf. For portable use, it is general practice to use a cell with an unsaturated electrolyte. This has a temperature coefficient which is only one-fourth as great as that of the saturated cell.

Standard cells are not intended as power sources. They should be used only for comparison of voltages. Ordinary voltmeters put too heavy a drain on the cell for any reliable voltage measurements.

Organic electrolyte cell. A different class of cells is that based on the use of particularly reactive metals (Li, Ca, Mg) in conjunction with organic electrolytes. The best-known type in this class is the lithium-cupric fluoride cell theoretically capable of delivering over 700 Wh/lb (1.54 kWh/kg), more than three times the capacity of the highly ranked Zn-AgO cell.

The lithium anode is usually made in sheet form, but variants using lithium powder trapped in appropriate grids are also known.

The cathode is a mixture of CuF_2 and various conductive materials, most often graphite and carbon black either together or separately, in order to' obtain the electronic conductivity that CuF_2 does not possess.

The electrolyte considered most compatible with these electrodes is lithium perchlorate (1 M) in propylene carbonate or butyrolactone.

The cell works with 80% cathodic current efficiency, delivering 25% of its energy to the 2-V end point, the initial voltage being 3.2 V. This represents 80 Wh/lb (176 Wh/kg), that is, 320 Wh/lb (704 Wh/kg) available energy. On the other hand the current density is only 0.5–3 mA/cm^2, too little except for special applications. Nonetheless, owing to their considerable thermodynamic energy density, the organic electrolyte cells remain the great hope for the future.

For the present, the most favored anode material is lithium which, besides providing fully reversible anodic reaction, possesses the exclusive properties of a low specific gravity (0.534) and a high electrode potential (3.045 V). In fact after beryllium, lithium has the highest energy availability per atom and is by far the lightest metal known.

The cathode problem is still under investigation, and at least two other salts, cobalt (111) fluoride (CoF_3) and nickel chloride ($NiCl_2$), are claimed to be superior, working with cathodic current efficiencies close to 100%.

The critical problem remains the electrolyte because of the difficulty of combining high specific conductivity, low viscosity, hydrophobicity, chemical inertness toward electrodes, and adequate electrolytic properties.

Bibliography. M. Barak (ed.), *Electrochemical Power Sources*, 1980; D. G. Fink and H. W. Beaty (eds.), *Standard Handbook for Electrical Engineers*, 12th ed., 1987; G. W. Heise and N. C. Cahoon (eds.), *The Primary Battery*, vol. 1, 1971, vol. 2, 1976.

DRY CELL
Kenneth Franzese and Jack Davis

A primary cell in which the electrolyte is absorbed in a porous medium, or is otherwise restrained from flowing. Common practice limits the term dry cell to the Leclanché cell, which is the major commercial type. Other dry cells, not discussed in this article, include the mercury cell, the alkaline-zinc-manganese dioxide cell, and the air-depolarized cell. These cells all use aqueous electrolytes immobilized in absorbent materials or gels. By order of the Federal Trade Commission, "leakproof" must not be printed on any dry cell. *See Primary battery.*

Construction. The Leclanché cell is made in a variety of sizes in either round or flat shapes. The energy-volume ratio of flat-cell batteries is about twice that of batteries made with round cells because of the absence of an expansion chamber and carbon rod and because of their rectangular form, which eliminates waste space in assembled batteries and voids between cells.

The negative electrode is usually solid zinc, either in cup form (cylindrical cell) or in sheet form (flat cell). The carbon positive electrode is embedded in a black mixture of manganese dioxide and carbon black. The carbon is either a rod located in the center of a cylindrical cell or a coating on the back of the flat zinc electrode in the flat cell. The separator between the black mix and the zinc electrode consists of a paper barrier coated with cereal or methyl cellulose. In the older construction the separator was composed of a gelatinous paste which also held the electrolyte. Not only does the paper barrier give about 30% more volume for the black mix, but it also decreases the internal resistance of the cell and reduces the number of manufacturing operations.

The electrolyte is a solution of ammonium chloride and zinc chloride in water. The cell enclosure consists of a top seal in the cylindrical cell or thin plastic wrappings for the flat cell.

Zinc electrode. Zinc cans (cup form of electrode) are made by drawing or impact extrusion. A typical alloy for extruded cans contains 1.0% lead, 0.05% cadmium, and remainder zinc. The zinc used is of 99.99% purity. The alloying elements improve the mechanical properties and decrease wasteful corrosion. The cell capacity is less than that theoretically possible from the weight of zinc. This is because zinc is used as the container, the limiting factor being the manganese dioxide. For example, the no. 6 can weighs 4 oz (110 g), giving a theoretical capacity of 90 ampere-hours. The actual capacity is in the range of 40–50 Ah, depending on the black mix composition and rate of discharge.

Black mix. The black mix is composed of manganese dioxide mixed with carbon black. The manganese dioxide is usually obtained from natural ore, but may be a synthetic product prepared by chemical precipitation or by electrolytic methods. Mixtures of the natural and synthetic oxides are also used. The carbon black is usually acetylene black made by the thermal decomposition of acetylene. Graphite is used to a lesser extent. Manganese dioxide has a theoretical capacity of about 8.6 Ah/oz (0.3 Ah/g). The practical capacity is somewhat less. The carbon black is used in varying proportions, depending on design factors. It serves the double purpose of increasing the conductivity of the manganese dioxide and absorbing the electrolyte. For cells that require a high capacity at low current drains (for example, transistor use), the ratio of manganese dioxide to carbon may be 10:1; and, at the other extreme, for cells requiring a very high flash current (for example, photoflash), the ratio may be as low as 1:1. In addition to these components, the black mix also contains electrolyte amounting to about 25% of the total weight.

Carbon rod. The carbon rod used in a cylindrical cell serves as the conductor of electricity for the positive electrode; it also serves as a vent to allow gas to escape. Carbon rods are usually made from petroleum coke which is calcined, ground, and mixed with pitch. The "green" rods are baked to form a hard carbon, having low electrical resistance. They may be partially waterproofed by impregnation with oil or paraffin wax to prevent capillary creepage of electrolyte out of the cell.

Flat cells are usually made with duplex electrodes. The zinc is coated on one side with a carbonaceous coating, which serves to conduct electricity between the zinc and the black mix of the adjacent cell.

Separator. The modern method of separating the two electrodes is to use kraft paper which has been coated on the side adjacent to the zinc with a film of cereal or methyl cellulose containing mercurous (or mercuric) chloride. The latter corresponds to a mercury concentration of 3.45×10^{-5} oz/in.^{2} (0.155 mg/cm^2) of paper area.

In the older method a paste was made from a mixture of electrolyte with corn starch and wheat flour. The paste was added to the cell in liquid form and gelatinized by heating the cell in a water bath. By increasing the zinc chloride concentration in the electrolyte, the gelatinization of the paste could take place at ambient temperatures. Mercuric chloride was added to the paste, but it was not possible to control the intensity or uniformity of amalgamation to the same degree as with the paper-lined cells. In a typical assembly line the cereal-coated paper is formed into a cylinder on a mandrel and inserted into the zinc can with a bottom washer. The calculated amount of wet black mix is injected into the lined can, followed by insertion of the carbon rod. Simultaneously with insertion of the carbon rod, the black mix is compressed. This gives a solid mass and forces sufficient electrolyte out of the mix to completely wet the cereal-coated paper barrier. A cardboard washer is placed on the black mix and an air space (expansion chamber) is enclosed over this washer by a further washer over which is poured a layer of bitumen. This seal is necessary to minimize moisture loss. Flat cells are made by placing treated paper containing the paste between the black-mix cake and the zinc of each cell.

Electrolyte. The electrolyte is made by dissolving ammonium chloride and zinc chloride in water. For paste cells a small amount of mercuric chloride is usually added. This component, however, converts to zinc chloride as soon as the zinc and electrolyte come into contact. Mercury then plates out on the zinc. The composition of the electrolyte depends on the cell designs.

During discharge the composition of the electrolyte changes. In one test in which a D-size cell was discharged through a 4-ohm resistance, the pH of the paste layer next to the zinc changed from 5.7 to 3.8 (more acid) while the pH of the innermost portion of the mix went from 5.8 to 10.1 (more alkaline).

Ordinary dry-cell electrolyte has a resistivity of 2.42 ohm-cm at 68°F (20°C). For low-temperature operation special electrolytes have been developed. An electrolyte of 12% zinc chloride, 15% lithium chloride, 8% ammonium chloride, and 65% water is fluid at −40°F (−40°C). Other electrolytes for low-temperature operation use a mixture of calcium chloride, zinc chloride, and ammonium chloride solutions.

Cell enclosure. Whereas the cylindrical cell was originally wrapped in a paper jacket, modern methods are more sophisticated in order to resist leakage which could damage valuable equipment, for example, cameras, tape recorders, and record players. One method uses an absorbent board wrap and an outer jacket of sheet steel. The cells are finally sealed in these containers with tinplate top and bottom closures. The other method (see **illus**.) uses a jacket consisting of a laminate of absorbent paper, polyethylene, and kraft paper covered with a cellulose acetate–coated label. As with the steel-jacketed cells, tinplate covers are placed top and bottom to make contact with the carbon rod and zinc can, respectively.

Flat cells use thin plastic wrappings around the edges of each cell. This confines the electrolyte to individual cells and avoids internal discharge in a stack. The wrappings are sufficiently gas-permeable to prevent the building up of pressure in the cell. After the requisite number of cells are stacked, the stack is bound together by tapes, and dipped in molten wax for further moistureproofing.

Cell chemistry. At the anode (zinc) the zinc oxidizes to zinc ion and simultaneously liberates electrons to the external circuit, at a rate proportional to the current. For each ampere which flows, 0.04 oz. (1.2 g) of zinc per hour is converted to zinc ion.

At the cathode (manganese dioxide), the electrons from the external circuit reduce the manganese dioxide to three different substances, depending on circumstances which have not yet been thoroughly explained. Studies have shown, however, that the total ampere-hour output of the cell can be accounted for by analyzing the cathode mix for the following substances: soluble manganese (Mn^{2+}), each gram of which accounts for nearly 1 Ah of discharge; insoluble mangan-

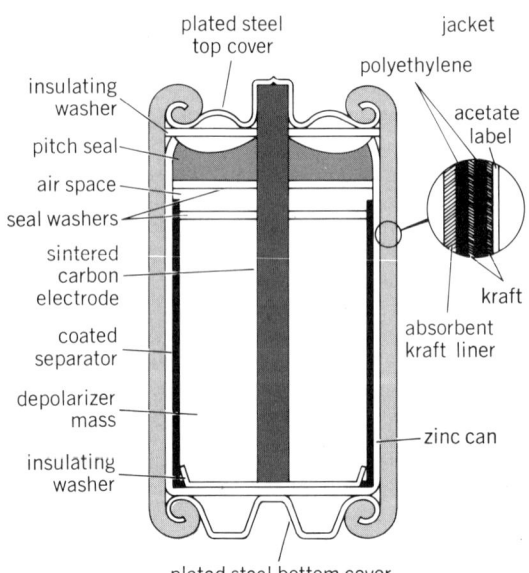

Modern Leclanche dry cell.
(*Bright Star Industries Inc.*)

ite (MnOOH), each gram of which accounts for about 0.3 Ah of discharge; insoluble hetaerolite ($ZnO \cdot Mn_2O_3$), each gram of which accounts for about 0.22 Ah of discharge.

The electrochemical reduction of the manganese dioxide (MnO_2) has been reported to occur as the reaction to form soluble manganese, shown as reaction (1). This occurs only when the cell delivers current.

$$MnO_2 + 4H^+ + 2e^- \rightarrow Mn^{2+} + 2H_2O \tag{1}$$

Two secondary reactions, (2) and (3), can then occur.

$$MnO_2 + Mn^{2+} + 2OH^- \rightarrow 2MnOOH \tag{2}$$

$$MnO_2 + Mn^{2+} + 4OH^- + Zn^{2+} \rightarrow ZnO \cdot Mn_2O_3 + 2H_2O \tag{3}$$

Reaction (3) can occur only if zinc is in solution in the cathode mix.

Operating characteristics. The service capacity of dry cells is not a fixed number of ampere-hours, but varies with current drain, operating schedule, cutoff voltage, operating temperature, and storage conditions prior to use. Most cells are tailor-made for their rated end use. For example, D-size cells can be rated as general-purpose, industrial flashlight, transistor, electronic flash, and photoflash. No. 6 cells are specially formulated for bell ringing, telephone, protective alarm, ignition, and general purpose. Minimum outputs of D cells as specified by the General Service Administration are as follows:

(1) General-purpose flashlight—2¼ ohms for 5-min periods at 24-h intervals until the closed-circuit voltage drops to 0.65 V; 400 min. (2) Light industrial flashlight—4 ohms for 4-min periods beginning at hourly intervals for 8 consecutive hours each day until the closed circuit voltage drops to 0.9 V; 950 min. (3) Heavy industrial flashlight—4 ohms for 4-min periods beginning at 15 min intervals for 8 consecutive hours each day until the closed-circuit voltage drops to 0.9 V; 800 min. (4) Photoflash bulb test—0.15 ohm for 1 s each minute for 1 h at 24-h intervals for 5 consecutive days each week until the closed-circuit voltage falls below 0.5 V; 800 s. (5) Electronic photoflash—1.0 ohm for 15 s each minute for 1 h at 24-h intervals for 5 consecutive days each week until the closed-circuit voltage falls below 0.75 V; 275 15-s discharges. (6) Transistor test—83⅓ ohms during a continuous period of 4 h daily until the closed-circuit voltage falls below 0.9 V; 200 h.

Temperature effect. The higher the temperature during discharge, the greater is the energy ouput. Conversely, the lower the temperature, the lower is the output. At $-9°F$ ($-23°C$) the battery is virtually inoperative. However, shelf life is influenced in the reverse direction by environmental temperatures.

Better low-temperature output can be obtained with special electrolytes and cell structures giving a high ratio of electrode area to mix thickness, and special types of manganese dioxide.

Shelf life. This is the period of time that a battery can be stored before it drops to 90% of its capacity when tested fresh at 70°F (21°C) and 50% relative humidity.

Deterioration in a dry cell occurs in a number of ways: (1) Zinc can oxidize by reaction with the electrolyte; this reaction produces hydrogen. (2) Manganese dioxide can be reduced by carbon and by the organic materials used in the cells; this can produce carbon dioxide. (3) Water can be evaporated from the electrolyte; this increases the cell resistance and alters the composition of the electrolyte unfavorably.

In general, shelf life decreases as the cell size becomes smaller: with well-constructed cells a shelf life of 3 years with a no. 6 telephone cell and 10 months with a penlight cell. Other sizes can be prorated. Flat cells have a shorter shelf life than cylindrical cells: 6–9 months for all sizes. It has been shown that 30 h is the typical minimum initial output for a 9-V transistor radio battery (six flat cells), and 28 h after 6 months' delay. High temperatures reduce shelf life, and at 90°F (32°C) the shelf life of a battery is about one-third that of one stored at 70°F (21°C). Low-temperature storage increases the shelf life of batteries considerably. Batteries stored at 39°F (4°C) [sealed polyethylene bags should be used to prevent condensation with subsequent corrosion of the terminals and metal jacket or degradation of the paper jacket] have their shelf life increased two or three times. Tests have been conducted by military agencies in many countries showing that, when batteries are frozen, they suffer no deterioration for about 10 years. They must, however, be allowed to reach room temperature before use.

Recharging. It is possible to recharge dry-cell batteries for five or six cycles provided the following precautions are taken: The battery should be subject only to shallow discharges between cycles, used immediately after charging, charged over a 10- to 15-h period at constant current, and not overcharged.

Zinc chloride cell. The zinc chloride cell is a variation of the standard Leclanché carbon zinc cell, differing mainly in the quantity of zinc chloride and ammonium chloride in the electrolyte. In fact, whereas the standard carbon zinc cell uses mostly ammonium chloride with a small percentage of zinc chloride, the zinc chloride cell uses primarily zinc chloride with little or no ammonium chloride.

Most physical and electrical characteristics of the zinc chloride cell are essentially the same as those of the standard carbon zinc cell, for example, watthours per kilogram, watthours per cubic centimeter, emf (open circuit voltage), voltage under load, and flash current. However, the zinc chloride cell does have significant advantages in the following areas: (1) better low-temperature performance; for example, at $-0.4°F$ ($-18°C$) typically 45% of the 70°F (21°C) capacity is available; (2) better continuous and high drain capacity due to more efficient depolarization; and (3) better leakage resistance because water is consumed along with the active materials, making the cell practically dry at the end of discharge. The major disadvantage of the zinc chloride cell is the need for an improved seal because of its high sensitivity to moisture loss.

Modifications. In addition to those systems discussed, the following combinations have been developed: (1) The magnesium + magnesium perchlorate or bromide + manganese dioxide modification is more expensive than the Leclanché type. It exhibits the delayed voltage at the commencement of discharge characteristic of magnesium cells and has excellent storage properties. It has a higher initial voltage than the standard dry cell. (2) Another example of cell modification is magnesium + magnesium perchlorate or bromide + metadinitrobenzene. The organic cathode used can be produced at a cost that gives the same number of watt-hours per dollar as electrolytic manganese dioxide. This factor becomes important as supplies of battery-grade natural ore diminish. (3) The cell modification of aluminum + aluminum and chromic chlorides and ammonium chromate + manganese dioxide is attractive on account of the lower density and electrochemical equivalent of aluminum compared with zinc.

Bibliography. T. R. Crompton, *Small Batteries: Primary Cells*, vol. 2, 1983; A. Fleischer (ed.), *Proceedings of the Power Sources Conference*, annually; C. L. Mantell, *Batteries and Energy Systems*, 2d ed., 1983.

FUEL CELL
Jack Davis, L. Rozeanu, and Kenneth Franzese

An electric cell that converts the chemical energy of a fuel directly into electric energy in a continuous process. The efficiency of this conversion can be made much greater than that obtainable by thermal-power conversion. In the latter the chemical reaction is made to produce heat by combustion. The heat is then transformed partially into mechanical energy by a heat engine, which drives a generator to produce electric energy. Further loss is involved if the direct current generated is converted into alternating current.

Although, in principle, the nature of the reactants is not limited, the fuel-cell reaction almost always involves the combination of hydrogen with oxygen, reaction (1). At 25°C (77°F) and

$$H_2(g) + 1/2 O_2(g) \rightarrow H_2O(l) \tag{1}$$

1 atm (10^5 pascals) pressure, that is, standard temperature and pressure (STP), the reaction takes place with a free energy change (ΔG) of $\Delta G = -56.69$ kcal/mole, that is 237,000 joules/mole water.

If the reaction is harnessed in a galvanic cell working at 100% efficiency, a cell voltage of 1.23 volts results. In actual service, such cells have shown steady-state potentials in the range 0.9–1.1 volts, with reported coulombic efficiencies of the order 73–90%.

The most popular and successful type is the classical H_2-O_2 fuel cell of the direct or indirect type. In the direct type, hydrogen and oxygen are used as such, the fuel being produced in independent installations. The indirect type employs a hydrogen-generating unit which can use as raw material a wide variety of fuel. The reaction taking place at the anode is as in reaction (2), and at the cathode as in reaction (3).

$$2H_2 + 4OH^- \rightarrow 4H_2O + 4e^- \quad (2) \qquad O_2 + 2H_2O + 4e^- \rightarrow 4OH^- \quad (3)$$

Because of the low solubility of H_2 and O_2 in electrolytes, the reactions take place at the interface electrode-electrolyte, requiring a large area of contact. This is obtained with porous materials called upon to fulfill the following main duties: The materials must provide contact between electrolyte and gas over a large area, catalyze the reaction, maintain the electrolyte in a very thin layer on the surface of the electrode, and act as leads for the transmission of electrons. The catalytic effect is obtained with metals, mainly platinum, silver, nickel, cobalt, and palladium, and certain special acids.

The thickness of the electrolyte layer, on which depends the internal resistance of the cell, is controlled by pore size, wetting properties, and pressure of the fuel gas. When pressure is used, care must be taken not to increase it to the extent that gas is allowed to bubble through the electrolyte because of the danger of forming an explosive hydrogen-oxygen mixture. The fuel cells may work with acid or alkaline electrolytes.

The acid electrolytes require costly corrosion-resistant construction materials but are not sensitive to CO and CO_2 in the fuel, which may lead to the buildup of carbonates. Some models using phosphoric acid have been quite successful. The alkaline electrolytes are more practical, and they are found in most fuel cells produced industrially.

Some fuel cells are designed to work with molten carbonates as electrolyte, at temperatures as high as 800°C (1470°F). These cells are attractive because they can use reformed hydrocarbon fuels, require a small investment, and can be made as large units. They are insensitive to carbon oxides but have important shortcomings, such as excessive size, rapid corrosion of metallic parts, and long periods of heating required before useful service.

The cells may use any alkali metal carbonate or eutectic mixtures of the same. The reactions taking place are those known for the systems involving hydrogen and oxygen or carbon monoxide and oxygen. In the case of carbon monoxide, the reaction step at the anode is as in reaction (4) and at the cathode as in reaction (5).

$$CO + CO_3^{2-} \rightarrow 2CO_2 + 2e^- \quad (4) \qquad CO_2 + 1/2 O_2 + 2e^- \rightarrow CO_3^{2-} \quad (5)$$

The total cell reaction is given by reaction (6), with a free energy change of $\Delta G = (257$

$$2CO + O_2 \rightarrow 2CO_2 \quad (6)$$

kJ/mole) of CO. At 100% efficiency, this gives a theoretical cell voltage of 1.34 V. at standard temperature and pressure.

The principal overall reactions which have been employed in fuel-cell work are summarized in **Tables 1** and **2**.

The direct anodic use of carbon has been practically abandoned in modern fuel-cell work. Carbon potentials seem entirely due either to carbon monoxide, CO, or to hydrogen, H_2, formed at high temperature by direct reaction between the carbon and the electrolyte. For example, in

Table 1. Theoretical cell potentials at various temperatures

	Cell potential, V					
Reaction	25°C (77°F)	100°C (212°F)	250°C (480°F)	500°C (930°F)	750°C (1400°F)	1000°C (1800°F)
$C + O_2 \rightarrow CO_2$	1.02	1.02	1.02	1.02	1.02	1.01
$2C + O_2 \rightarrow 2CO$	0.71	0.75	0.82	0.93	1.04	1.15
$2CO + O_2 \rightarrow 2CO_2$	1.33	1.30	1.23	1.11	1.00	0.88
$2H_2 + O_2 \rightarrow 2H_2O$	1.23	1.18	1.12	1.05	0.97	0.90

Table 2. Theoretical material consumption

Reaction	Temperature, °C (°F)	Consumption, g/kWh (oz/kWh)		
		Anode	Cathode	Total
$C + O_2 \rightarrow CO_2$	750 (1400)	109.7 (3.9)	292.7 (10.3)	402.4 (14.2)
$2C + O_2 \rightarrow 2CO$	750 (1400)	215.3 (7.6)	287.0 (10.1)	502.3 (17.7)
$2CO + O_2 \rightarrow 2CO_2$	750 (1400)	522.4 (18.4)	298.5 (10.5)	820.9 (29.0)
$2H_2 + O_2 \rightarrow 2H_2O$	100 (212)	31.6 (1.2)	253.0 (8.9)	284.6 (10.1)
$2H_2 + O_2 \rightarrow 2H_2O$	750 (1400)	38.5 (1.4)	307.7 (10.9)	346.2 (12.2)

the Jacques cell, which consists of carbon electrodes and iron (air) electrodes in molten sodium hydroxide, H_2 is liberated at the carbon by reaction with the electrolyte. Also, in methane fuel cells, methane decomposes at high temperatures, as shown by reaction (7). It is this H_2 which is

$$2CH_4 \rightarrow 2C + 4H_2 \quad (7)$$

responsible for the observed potential.

For technical reasons, it is simpler to use the carbon or hydrocarbon fuel in a chemical reactor to produce the active gases, H_2 and CO, than to attempt to operate a cell under the conditions best suited for the chemical reaction. Typical chemical production of the active gases might be as in reaction (8). A mixture of H_2 and CO can also be produced by reacting carbon with steam, as shown by reaction (9).

$$2C + O_2 \rightarrow 2CO \quad (8) \qquad\qquad C + H_2O \rightarrow CO + H_2 \quad (9)$$

The complete engineering design of the chemical reactor in conjunction with the fuel cell has been extensively studied. The main difficulties in the past were due to the large concentration of CO in the reformed fuel, which poisons the Pt catalyst often used in the fuel cell proper. Various new catalysts have been developed, such as Pt-Rh, Pt-Ir, and Pt-Ru, well capable of processing H_2-CO fuel mixtures.

Work with H_2 has established that it can operate efficiently at moderate temperatures, with polarization decreasing as the temperature is increased. This permits the use of aqueous solutions. One has been reported to have, at 25°C (77°F), the following characteristics:

Current density, mA/cm^2	0	1.1	11	54
Cell voltage	1.12	1.01	0.95	0.70

Some general characteristics are listed in **Table 3**.

The hydrazine-air-fuel cell is based on the reaction shown in (10). The unit cell voltage is

$$N_2H_4 + O_2 \rightarrow N_2 + 2H_2O \quad (10)$$

0.6–0.7 V, and its greatest advantage is that it uses a condensed fuel, convenient in some applications.

Table 3. Characteristics of three fuel-cell systems

Type	Principle	Temperature, °C (°F)	Power, W	Application
General Electric	Ion-exchange membrane	25–35 (77–95)	100–1000	Gemini
Allis-Chalmers	Porous Ni electrodes and porous electrolyte vehicle	90–100 (194–212)	2000	Space flight
Bacon (Pratt and Whitney)	Porous Ni electrodes	200–220 (390–430)	500–1500	Apollo

A different concept is used in alkaline metal–oxygen fuel cells developed by the M. W. Kellogg Co. In an actual unit the electrochemical process involves oxidation of Na with oxygen from air, with NaOH as electrolyte. The reaction taking place at the anode is as in reaction (11) and at the cathode as in reaction (12), with the cell reaction given as (13).

$$4Na \rightarrow 4Na^+ + 4e^- \quad (11) \quad O_2 + 4e^- + 2H_2O \rightarrow 4OH^- \quad (12) \quad 4Na + O_2 + 2H_2O \rightarrow 4NaOH \quad (13)$$

Because Na as such is too reactive, the cell uses a sodium amalgam which is quite stable in concentrated NaOH. The sodium amalgam–oxygen fuel cell possesses some remarkable features. It provides almost 1.5 V at steady stage and very high current densities of the order of 200 mA/cm^2, is insensitive to water quality, and requires a small gas consumption. It is expected to play an important role in some applications. *See* D$_{RY}$ $_{CELL}$; E$_{LECTRODE}$ $_{POTENTIAL}$.

Bibliography. E. Findl and M. Kein, Electrolytic regenerative hydrogen-oxygen fuel-cell battery, *Proceedings of the 20th Annual Power Sources Conferences,* 1966; D. Linden (ed.), *Handbook of Batteries and Fuel Cells,* 1984; R. Noyes, *Fuel Cells for Public Utility and Industrial Power,* 1978; L. Oniciv, *Fuel Cells,* transl. by J. Hammel, 1976.

MERCURY BATTERY
J$_{ACK}$ D$_{AVIS}$

A primary dry-cell battery consisting of a zinc anode, a cathode of mercuric oxide (HgO) mixed with graphite, and an electrolyte of potassium hydroxide (KOH) saturated with zinc oxide (ZnO). With carefully purified materials and balanced amounts of ZnO and HgO, the cell has very low self-discharge and makes efficient use of the active materials.

In some cells which require long-term continuous drains, for example, in hearing aid use, MnO$_2$ is added to the HgO. In these cells, the open circuit voltage is slightly above 1.4 V, compared with 1.35 V for cells using 100% HgO as depolarizer.

Within the steel can, the active materials are separated by a porous material which prevents migration of conducting particles from the mercuric oxide pellet. Dense dialysis paper and porous polyvinyl chloride have been used for this purpose. The electrolyte is completely absorbed in the active materials, separator, and absorbent materials. The steel can serves as the contact to the HgO. A metal top with a concentric neoprene grommet closes off the top of the can and serves as the contact to the zinc.

The electrochemical system may be written as below. This does not involve the electrolyte. The cell potential, therefore, does not change appreciably with different concentrations of alkali.

$$\text{Anode: } Zn + 2OH^- \rightarrow ZnO + H_2O + 2e^-$$
$$\text{Cathode: } HgO + H_2O + 2e^- \rightarrow Hg + 2OH^-$$
$$\text{Overall: } Zn + HgO \rightarrow ZnO + Hg$$

The cutaway view shown in **Fig. 1** is of the flat pellet structure. Two other types are manufactured: the cylindrical structure, which also uses pressed amalgamated zinc powder but is made in sizes corresponding to the N, AA, and D Leclanche cells; and the wound anode flat structure, with large diameter and high surface area, giving superior performance at low temperatures.

When current flows, the ZnO formed in the cell reaction quickly saturates the small amount of electrolyte and then precipitates out. This maintains a constant composition of the electrolyte. Offsetting this is the transport of water away from the anode by the solvated potassium ions. The equilibrium under steady current flow results from complex exchanges through the separator, making cell-voltage characteristics under load dependent on initial electrolyte composition.

The mercury cell has a theoretical output of 7.015 ampere-hour/oz (0.247 A-h/g) of HgO. In practice, the cathode pellet contains about 95% HgO and 4% graphite, having a theoretical output of 6.646 A-h/oz (0.234 A-h/g). The anode is 90% zinc and 10% mercury. This has a theoretical output of 20.959 A-h/oz (0.738 A-h/g).

As built, the cells have slight excess of cathodic capacity. A discharged cell will then have no zinc left to react with the electrolyte and evolve hydrogen. Thus a cell with 0.44 oz (12.5 g) of

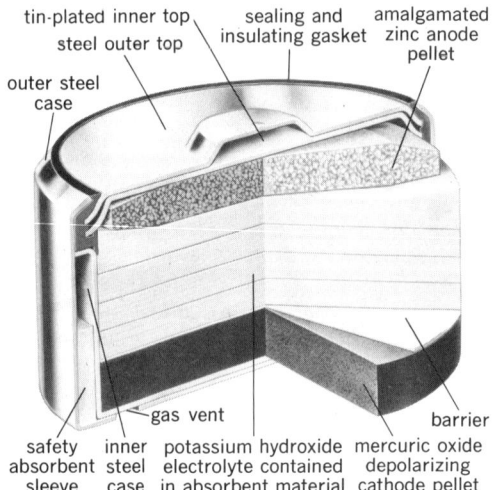

Fig. 1. Cutaway view of mercury cell.

cathodic material has 0.13 oz (3.6 g) of zinc amalgam, compared with 0.14 oz (3.98 g). needed for exact balance.

The electrolyte used is about 0.06 in.^{3}/A-h (1 ml/A-h). One composition is 3.5 oz (100 g) KOH, 6.1 in.^{3} (100 ml) H$_2$O, 0.56 oz (16 g) ZnO. The actual cell capacity is only slightly less than the theoretical. The overall cell output is approximately 0.36 A-h/lb (0.79 A-h/kg) and 5 A-h/in.^{3} (0.3 A-h/cm^3).

The ampere-hour capacity of mercury cells is relatively unchanged with variation of discharge schedule and to some extent with variation of discharge current. The cells have a relatively flat discharge characteristic, as shown in **Fig. 2**. The outstanding features of the mercury cell

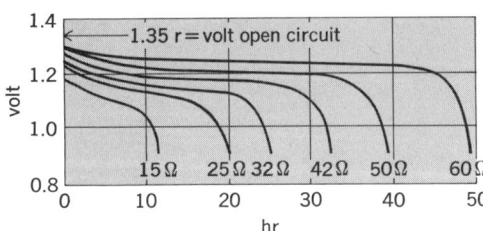

Fig. 2. Voltage-discharge characteristics of mercury cells under continuous load conditions at 70°F (21°C). At 1.25 V, equivalent current drains for resistances are: 15 ohms, 83 mA; 25 ohms, 50 mA; 32 ohms, 40 mA; 42 ohms, 30 mA; 50 ohms, 25 mA; 60 ohms, 20 mA.

include flat discharge curve, small variation in capacity with intermittent or continuous discharge, shelf life of several years, and good high-temperature characteristics. SEE DRY CELL; PRIMARY BATTERY; RESERVE BATTERY.

Bibliography. M. Barak (ed.), *Electrochemical Power Sources*, 1980; N. C. Cahoon and G. W. Heise, *The Primary Battery*, 2 vols., 1971, 1976.

LITHIUM PRIMARY CELL
JACK DAVIS AND KENNETH FRANZESE

A primary cell whose anode is composed of lithium. The lithium cell is a development which has a number of advantages over other primary cell systems. Lithium is an attractive anode because of its reactivity, light weight, and high voltage (between 1.6 and 3.6 V, depending on the other electrode).

Lithium primary cell systems

Type	Nominal voltage	Nominal energy density, Wh/cm^3†
Li-SO_2	2.9	.5
Li-CuO	1.6	.6
Li-$(CF)x$*	2.8	.6
Li-CuS	2.0	.5
Li-Ag_2CrO_4	3.3	.6
Li-I_2	2.8	.25–.8
Li-PbI_2PbS	1.9	.5
Li-$SOCl_2$	3.5	.9
Li-V_2O_5	3.4	.7
Li-MoO_3	2.9	.6
Li-MnO_2	2.8–3.4	.6

*Variable chemical composition.
†$1 Wh/cm^3 = 16.4 Wh/in.^3$

The advantages include high energy density, flat discharge characteristics, excellent service over a wide temperature range (as low as $-40°F$ or $-40°C$), and good shelf life (up to 5 years without refrigeration).

Nonaqueous solvents are used as the electrolyte because of the solubility of lithium in aqueous solutions. Organic solvents, such as acetonitrile and propylene carbonate, and inorganic solvents, such as thionyl chloride, are typical. A compatible solute is added to provide the necessary electrolyte conductivity. A number of different materials—sulfur dioxide, carbon monofluoride, vanadium pentoxide, manganese dioxide, copper sulfide, and so forth—are used as the active cathode materials.

The **table** lists the important cell systems. The three systems thought to be the most important are the types that use sulfur dioxide (SO_2), thionyl chloride ($SOCl_2$), and iodine (I_2).

Li-SO_2 cell. In the Li-SO_2 system the SO_2 is used for the cathode; acetonitrile (CH_3CN) and lithium bromide (LiBr) are used for the electrolyte; lithium foil is used for the anode; and polypropylene is used as the separator. The cell reactions are given by reaction (1).

$$\text{Anode: } 2Li \rightarrow 2Li + 2e^- \tag{1a}$$

$$\text{Cathode: } 2SO_2 + 2e^- \rightarrow S_2O_4^{2-} \tag{1b}$$

$$\text{Overall: } 2Li + 2SO_2 \rightarrow Li_2S_2O_4 \tag{1c}$$

Lithium dithonite

The good shelf life of this cell is attributed to the protective film formed by the initial reaction of lithium and sulfur dioxide, which prevents further reaction or loss of capacity during storage. Claimed energy densities have been reported to be as high as 8.2 $Wh/in.^3$ (0.5 Wh/cm^3) and 148.5 Wh/lb (330 Wh/kg).

This cell is initially pressurized from 2 to 4 atm (200 to 400 kilopascals) and is capable of very high currents (50 amperes for D size). The combination of these two facts leads to a serious safety problem because excessive heat generated by continuous-high-rate or short-circuit discharge can cause extremely high internal pressures (over 30 atm or 3 megapascals). Therefore, the use of a vent to limit the pressure buildup to preset levels is required. In addition, when multiple-cell arrangements are needed, the use of diodes and fuses is recommended to prevent accidental cell charging, short circuits, and damage from cell reversals.

Li-$SOCl_2$ cell. In the Li-$SOCl_2$ system the $SOCl_2$ is used as both the cathode and the electrolyte; lithium foil is used as the anode; and lithium aluminum chloride ($LiAlCl_4$) is used as the solvent. The cell reaction is given by reaction (2). The cathodic reaction forms sulfur monoxide

$$4Li + 2SOCl_2 \rightarrow 4LiCl + SO_2 + S \tag{2}$$

(SO), an unstable biradical which dimerizes and decomposes, undergoing exothermic reaction (3).

$$SOCl_2 + 2e \rightarrow SO + 2Cl \quad (3a) \qquad 2SO \rightarrow (SO)_2 \quad (3b) \qquad (SO)_2 \rightarrow S + SO_2 \quad (3c)$$

Claimed energy densities have been reported to be as high as 14.76 Wh/in.^{3} (0.9 Wh/cm^3) and 189 Wh/lb (420 Wh/kg).

This cell is not initially pressurized, but has a protective film similar to the Li-SO$_2$ cell which assures excellent shelf life. Some variations in construction have significantly reduced the danger of explosion, but nevertheless this cell (like the Li-SO$_2$) should never be deliberately charged, forced open, or disposed of in fire.

One other disadvantage in lithium cells is a delay in voltage brought about by the same film which aids shelf life. The delay can be only seconds (Li-SO$_2$) or up to 10 or more minutes (Li-SOCl$_2$).

Li-I$_2$ cell. In the Li-I$_2$ system the I$_2$ is used as the cathode, LiI as the solid-state electrolyte, lithium for the anode, and poly-2-vinylpyridine (P2VP) as the separator. The P2VP and LiI are actually bound together in a charge transfer complex and as such simultaneously serve the functions of cathode, depolarizer, and separator. The cell reactions are given by reaction (4). Claimed

$$\text{Anode: } 2Li \rightarrow 2Li^+ + 2e^- \quad (4a)$$

$$\text{Cathode: } 2Li^+ + 2e^- + P2VP \cdot nI_2 \rightarrow P2VP \cdot (n-1)I_2 + 2LiI \quad (4b)$$

$$\text{Overall: } 2Li + P2VP \cdot nI_2 \rightarrow P2VP \cdot (n-1)I_2 + 2LiI \quad (4c)$$

energy densities have been reported to be as high as 13.12 Wh/in.^{3} (0.8 Wh/cm^3) and 103.5 Wh/lb (238 Wh/kg).

This cell is designed primarily for cardiac pacemakers and works with a typical current drain of 30 microamperes, with a self discharge said to be approximately 10% in 10 years. SEE BATTERY; PRIMARY BATTERY.

Bibliography. P. Bro, Heat generation in Li/SOCl$_2$ cells, *Power Sources 7: Proceedings of the 11th International Power Sources Symposium*, pp. 571–582, 1979; J. B. Gabano (ed.), *Lithium Batteries*, 1983; R. W. Graham (ed.), *Primary Electrochemical Cell Technology: Advances Since 1977*, Chem. Technol. Rev. 191, 1981; Energy Technol. Rev. 25, pp. 126–187, 1978; C. C. Liang and C. F. Holmes, The lithium/iodine pacemaker battery, *Progress in Batteries and Solar Cells*, vol. 2, pp. 50–53, 1979; R. T. Mead, C. F. Holmes, and W. Greatbatch, Design evolution of the lithium iodine pacemaker battery, *Proceedings of the Symposium on Battery Design and Optimization: Electrochemical Society Proceedings*, vol. 79–1, pp. 327–333, 1979.

STORAGE BATTERY
W. W. SMITH

An assembly of identical voltaic cells in which the electrochemical action is reversible so that the battery may be recharged by passing a current through the cells in the opposite direction to that of discharge. While many nonstorage batteries have a reversible process, only those that are economically rechargeable are classified as storage batteries. SEE BATTERY; PRIMARY BATTERY.

Storage batteries, sometimes known as electric accumulators or secondary batteries, have two general classifications: lead-acid and alkaline. Active materials and electrolytes for both classes of batteries will be explained later. The **table** gives an approximate comparison of the several principal types of storage battery couples in terms of output per unit weight and unit volume.

Some of the important uses of storage batteries are to start gasoline and diesel engines; to operate communications circuits; switch tripping and closing in power-generating and -handling systems; emergency lighting; emergency power both with and without conversion to alternating current; railway car lighting and air conditioning; rapid transit car controls; marine power systems; power for underwater exploratory vehicles and submarines; to activate photographic and portable sound systems as well as portable TV and radio; and various military applications.

Comparison of the principal types of storage battery

Battery type	Volts per cell	Energy, Wh/lb*	Density, Wh/in.³†
Lead-acid	2.0	10–15	0.6–1.3
Nickel-iron	1.2	10–14	0.6–1.0
Nickel-cadmium	1.2	8–11	0.4–0.8
Nickel-cadmium sintered	1.2	10–13	1.0
Silver-zinc	1.5	20–100	3
Silver-cadmium	1.1	15–50	2.5

*1 Wh/lb = 7.94 kJ/kg.
†1 Wh/in.³ = 0.22 kJ/cm³.

LEAD-ACID STORAGE BATTERY

The lead-acid type of storage battery is so classified because the electrolyte is an acid and the plates are largely lead. The positive active material is lead peroxide and the negative active material is lead sponge. The active materials are supported by grids made of lead alloys.

The lead-acid battery maintains a preeminent place among all commercial types of storage batteries in volume of manufacture.

Principles of operation. A great many types of lead-acid cells are produced, but all have certain features in common. One is the open-circuit cell electromotive force (emf), which exists between a positive lead peroxide (PbO_2) electrode and a negative sponge lead (Pb) electrode when the two are immersed in sulfuric acid electrolyte ($H_2SO_4 + H_2O$). This value is independent of the quantities of lead peroxide, lead, or electrolyte present but does vary with temperature and sulfuric acid (H_2SO_4) concentration. At 25°C (77°F) the emf varies from 2.050 V with acid at 1.200 sp gr to 2.148 V with acid at 1.300 sp gr. The relatively small variation with temperature is given in millivolts/°C over a range 0–40°C (32–104°F) as 0.30 for 1.200 sp gr electrolyte, 0.22 for 1.250 sp gr, 0.19 for 1.280 sp gr, and 0.18 for 1.300 sp gr (in millivolts/°F, respectively 0.16, 0.12, 0.11, and 0.10).

Reaction (1) represents the cell reactions insofar as beginning and end materials are concerned. It is known as the double-sulfate theory, since lead sulfate ($PbSO_4$) is formed at both electrodes.

$$PbO_2 + Pb + 2H_2SO_4 \underset{\text{charge}}{\overset{\text{discharge}}{\rightleftharpoons}} 2PbSO_4 + 2H_2O \tag{1}$$

Reaction (1) can be split into reactions (2) and (3), indicating the reactions at the two electrodes.

$$PbO_2 + 2H^+ + H_2SO + 2e^- \underset{\text{charge}}{\overset{\text{discharge}}{\rightleftharpoons}} PbSO_4 + 2H_2O \text{ (at positive)} \tag{2}$$

$$Pb + SO_4^- \underset{\text{charge}}{\overset{\text{discharge}}{\rightleftharpoons}} PbSO_4 + 2e^- \text{ (at negative)} \tag{3}$$

On discharge the overall effect is a reduction of PbO_2 at the positive electrode and an oxidation of Pb at the negative electrode, accompanied by sulfation in both cases. In charging, a counter voltage is imposed on the cell terminals, and current is forced through the cell in a direction opposite to that in which the cell discharges. This reverses the ionic movements in relation to the electrodes and, in effect, reverses the cell reactions. On discharge the electrolyte specific gravity decreases, and on charge it increases. Specific gravity serves as a measure of the sulfuric acid concentration and thus as an index of state of charge.

Reactants. For a given quantity of electricity, such as ampere-hours, the three reactants, PbO_2, Pb, and H_2SO_4 take part in the reaction in amounts governed by Faraday's law. Thus, for a 1 ampere-hour discharge, 3.866 g (0.136 oz) of sponge lead are converted to $PbSO_4$, 4.463 g (0.158 oz) PbO_2 are converted to $PbSO_4$, and 3.660 g (0.129 oz) of H_2SO_4 are consumed.

A cell constructed to contain exactly the amounts of reactants given above, however, would not yield 1 Ah of capacity even under optimum practical conditions. Action at each electrode is slowed drastically when the concentration of H_2SO_4 in the electrolyte approaches a low figure because it is required in the electrode reaction. But even if ample H_2SO_4 were present, 1 Ah of capacity still would not be attained, since there will always remain an appreciable amount of PbO_2 or Pb, or both, in the solid electrodes, which cannot be reached by the electrolyte. The capacity attained in practice divided by what should be obtained in principle from the amount of reactants present is known as the utilization coefficient. This coefficient varies with types of cells, rate of discharge, and temperature. Unfortunately, it is a low value even under the best of conditions.

This problem is aggravated by the coating of nonconducting sulfate that forms on the active materials. Another is the diminishing conductivity of the electrolyte as the H_2SO_4 content decreases.

The utilization coefficient is decreased by the use of high-current rates. At higher current densities, the electrode reaction is concentrated at the surface of the plates. As a result, the pore openings at the plate surface become blocked with sulfate, restricting conduction and diffusion to the interiors of plates. Plates destined for high-discharge current densities are therefore made relatively thin. By substituting many thin plates for a few thick plates containing the same amounts of active materials, the utilization coefficient at high rates, and hence the capacity attainable, will be increased.

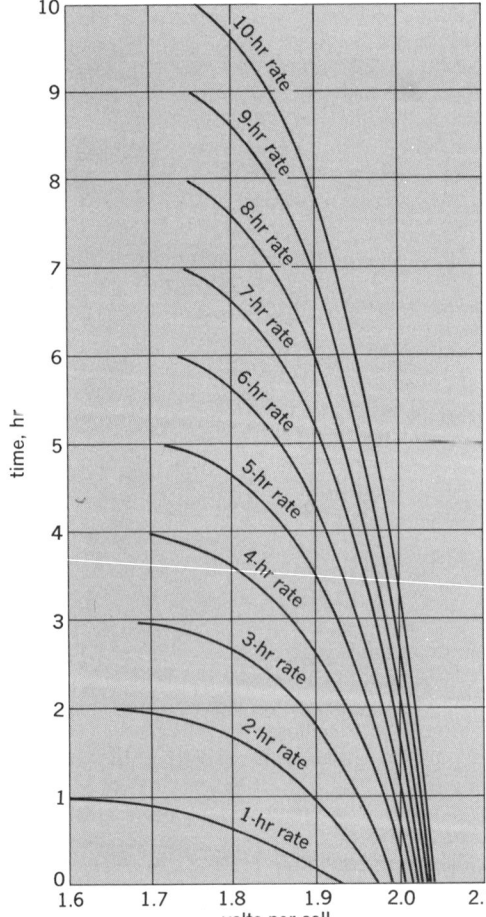

Fig. 1. Typical volt-time curves for a lead-acid cell for various discharge rates. (*ESB Inc.*)

Figure 1 shows a typical set of volt-time curves for different discharge rates of a lead-acid cell, illustrating the variations from a 1-h rate at high discharge current to a 10-h rate at low discharge current. **Figure 2** illustrates the decrease in ampere-hour capacity with increase in discharge current. If a short time is allowed for diffusion after a high-rate discharge, more of the unused possible capacity of the cell becomes available.

Cell temperature has an appreciable effect on capacity, largely because the viscosity of the electrolyte changes. Thus the diffusion of H_2SO_4 is retarded at low temperatures and the capacity is lowered.

Also, the capacity is decreased if the acid concentration becomes too low. On the other hand experience has shown that negative plates do not function well if the full-charge specific gravity is over 1.300, although positive plates operate more efficiently in high specific gravity. The usual range of full-charge specific gravity is 1.200–1.280, the choice depending on the application of the cell, the ambient operating temperature, and susceptibility of the cell to self-discharge. Specific gravity is usually determined by a hydrometer. It can also be measured by chemical methods.

Cell construction. Aside from cost, first consideration must be given to the kind of service for which a cell is destined, and second consideration to design features that reduce operational troubles. A compromise, for example, between life and weight or between life and cost, is usually required.

Pasted plates. In the most familiar type of plate construction, the basic structural number is a diecast grid, such as the ladder type shown in **Fig. 3**. This grid is made of a lead alloy

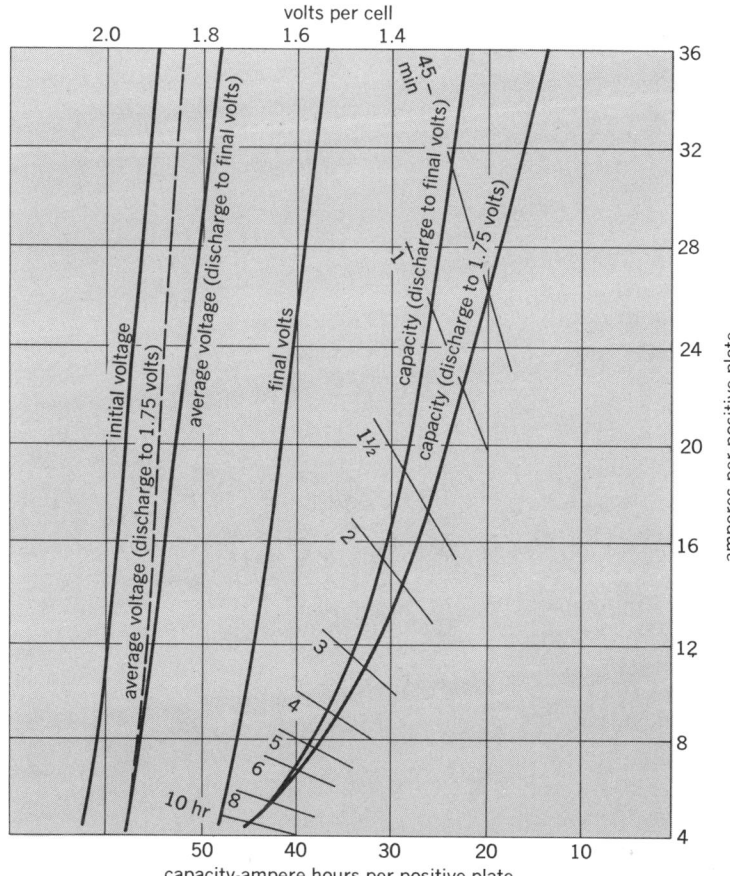

Fig. 2. Rated-discharge curves for a lead-acid cell. (*ESB Inc.*)

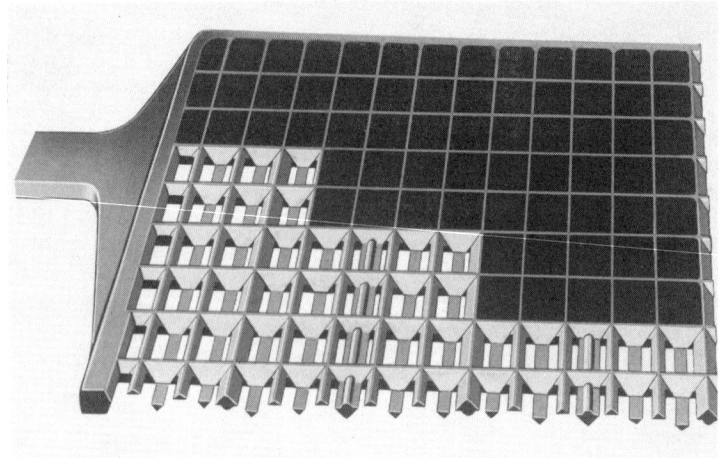

Fig. 3. Typical ladder-type grid showing a portion of it pasted. (*ESB Inc.*)

containing, for example, (3–11%) antimony or (0.01–0.1%) calcium. The grid is pasted with a slurry of lead oxide, sulfuric acid, and water. This is followed by a processing which finally converts the active material to PbO_2 for positive plates (chocolate-brown color) and to sponge lead for negative plates (gray color).

The negative plates are always sponge lead whether used with pasted positive plates or with positive plates of other types.

The life of a lead-acid battery plate is closely related to the thickness of the metal bars used in the positive plate. In applications such as engine cranking where light weight and high-rate performance are of more importance than life, the battery plates are usually made as thin as possible. Thicker plates are used where long life and reliability are more important than first cost,

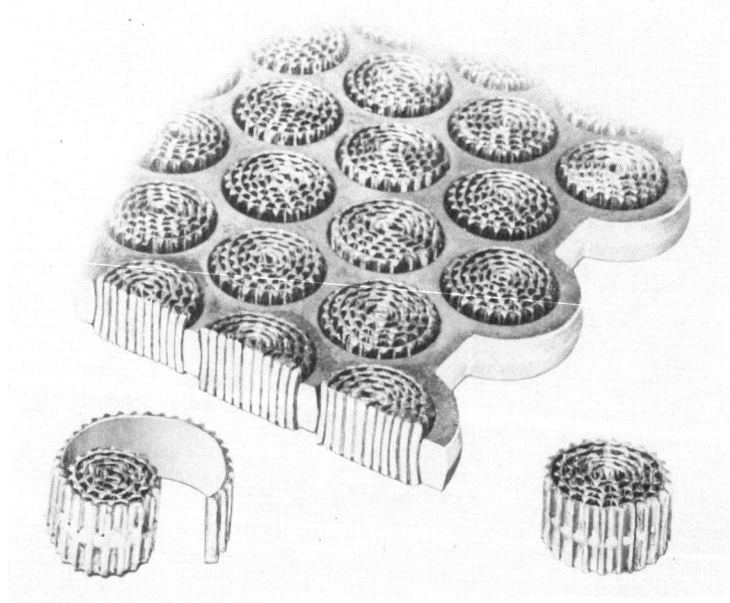

Fig. 4. Section of Manchester plate with detail of lead button. (*ESB Inc.*)

space, and weight. The thinnest plates in use are about 1.3 mm (0.05 in.) thick; the thickest plates range up to 19 mm (0.75 in.).

Manchester plate. These plates consist of a heavy alloy grid with circular openings into which pure lead "buttons" are pressed. These buttons are made from lead tape by crimping and rolling to develop a large surface area (**Fig. 4**). A forming agent in dilute sulfuric acid electrochemically forms a layer of PbO_2 on the surface of the button. Manchester plates are usually mounted in a cell with pasted negatives and in a relatively large quantity of low-gravity acid.

The cells are heavy and bulky. They are used in stationary installations, as for telephones, switch operation, and large emergency lighting, where they are "floated" on a line of constant voltage or trickle-charged with a constant current and are only occasionally discharged. Under such conditions of service, Manchester plates give exceptionally long life.

Gould spun plates. This type of positive plate, shown in **Fig. 5**, is manufactured from heavy sheet lead by passing the plate between disks which cause the lead to flow in between them to form leaves and spaces. After PbO_2 is formed on this developed surface, the plates are assembled with pasted negative plates and used in substantially the same types of service as Manchester plates. An advantage of this plate is elimination of antimony, hence local action, from the cell construction. This advantage is usually gained at some sacrifice of life.

Tubular-type plate. In this positive plate the active material is held in a porous-walled tube with a central alloy spine as conductor. The tube is made of felted or woven chemically inert fibers. This plate has many applications but is particularly successful where the service calls for repeated or routine deep-discharge cycles, such as in industrial trucks and mine locomotives.

Freezing of electrolyte. The freezing points of the usual range of sulfuric acid electrolytes as full-charge specific gravities [from $-52°C$ ($-62°F$) for 1.250 at 15°C (59°F) to $-70°C$ ($-94°F$) for 1.300 at 15°C (59°F)] are well below most arctic temperatures, but the end-of-discharge specific gravities can result in freezing points above arctic temperatures unless precautions are taken. With the proper choice of separator, a high-gravity acid can be used in severe arctic conditions without detriment to the negative electrodes and yet can have a relatively high end-of-discharge gravity.

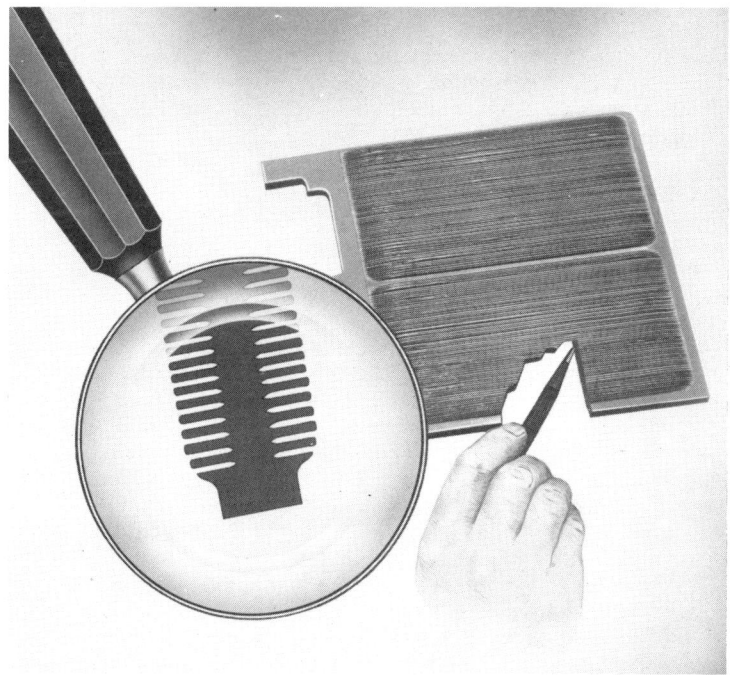

Fig. 5. Gould spun Planti positive plate. (*Gould Inc., Industrial Battery Division*)

Charging. For fast, yet efficient and noninjurious, charging, the modified constant-potential method is recommended. A high current rate is used until a voltage, such as 2.38, is obtained. This voltage is then maintained with a decreasing current until the finishing rate recommended by the manufacturer is reached. The finishing rate is continued to the end of the charge. Several methods are used for setting the end of the charge, the best known being arrival at a constant potential, arrival at a constant specific gravity, or charging a certain percent of ampere-hours in excess of the ampere-hours that have been taken out. A lengthy but efficient charge can be made using the finishing rate from the start. A two-step charge can be made using first a high current and then the finishing rate, the change being made automatically by a voltage relay or ampere hour meter.

Cell containers are currently made of hard rubber or plastic. For automobile, railway, and motive power batteries the cell containers are made of highly shock-resistant materials such as semihard rubber, modified polystyrene, or polypropylene. Stationary batteries use jars made of clear modified polystyrene. For extreme shock resistance such as submarine service, rubber-lined polyester fiber-glass jars are used.

ALKALINE-TYPE STORAGE BATTERY

The alkaline-type storage battery is so classified because the electric energy is obtained from chemical action of an alkaline solution. One type of battery has positive plates of some nickel compound and negative plates of iron. Another type uses a nickel compound and cadmium. A third uses silver oxide and zinc.

Nickel-iron alkaline cell. This battery is composed of cells having a hydrated nickel oxide and iron in an alkaline solution. It was invented by Thomas Edison early in the twentieth century. The positive active material in this cell is a higher oxide or hydroxide of nickel. The negative material is fine iron powder. The electrolyte is 1.200 sp gr (at 15°C or 59°F) potassium hydroxide, to which a little lithium hydroxide is sometimes added.

The chemical behavior of the nickel-iron cell is shown in reaction (4).

$$2NiOOH \cdot H_2O + Fe \underset{\text{charge}}{\overset{\text{discharge}}{\rightleftharpoons}} 2Ni(OH)_2 + Fe(OH)_2 \qquad (4)$$

The nickel hydrate formed by charging the battery is not an exact chemical compound. Directly after charging, it contains some excess dissolved oxygen. The dissolved oxygen is not tightly held and is released in the 10–24-h period following the charge. It is electrically active, and a battery discharged immediately after charge will have a greater output than if it stands until the oxygen is lost.

The KOH electrolyte supplies ions for conductivity, but unlike the lead-acid battery, the concentration of electrolyte in alkaline cells (nickel-iron, nickel-cadmium, silver-zinc, and silver-cadmium) does not undergo any net change in the chemical action of the cell. As a consequence, the specific gravity of the electrolyte does not change and cannot be used to indicate the state of charge of an alkaline battery, as in the case of the lead acid battery. However, it also means that the gravity stays up at all times, and the battery is much less susceptible to accidental damage from freezing than lead acid.

In the tubular nickel-iron cell a perforated steel tube is tightly packed with alternate layers of nickel hydrate and thin nickel flake to provide electrical conductivity. The layers are thin and there are about 300 layers in a 10-cm (4-in.) tube.

The negative material is packed into long pockets of perforated sheet steel. The pockets are laced together and pressed to form a single structural member. The top rail and bottom rail are welded in place to complete the plate. Positive and negative plates are shown in **Fig. 6**. Typical discharge curves of a 100-Ah nickel-iron cell are shown in **Fig. 7**.

It has been found desirable to charge the tubular nickel-iron cell at comparatively high rates (10–20 A for a 100-Ah battery) in order to reduce the time required for a full charge. However, great numbers of tubular-iron cells are floated at rates of 0.002–0.004 mA per ampere-hour of capacity with excellent results.

The open-circuit potential of the negative iron electrode is very close to the hydrogen potential. This makes the electrode susceptible to rather high local action or self-discharge. The

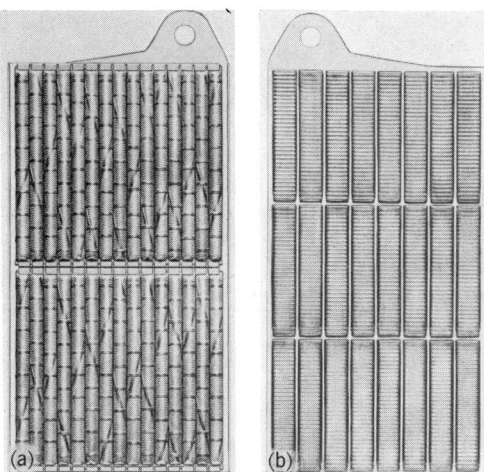

Fig. 6. Plate construction of nickel-iron cell. (*a*) Positive plate. (*b*) Negative plate. (*ESB Inc.*)

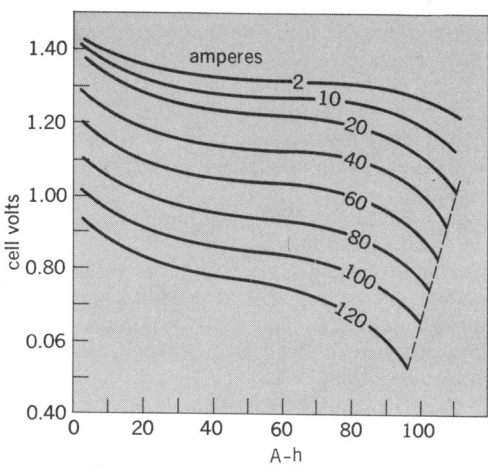

Fig. 7. Typical volt-time curves of nickel-iron alkaline cells for various discharge rates. (*After D. G. Fink and J. M. Carroll, eds., Standard Handbook for Electrical Engineers, 10th ed., McGraw-Hill, 1968*)

battery when on open circuit or on charge continually gives off hydrogen. Therefore nickel-iron cells must be well ventilated. They require higher float currents and more frequent watering than most other types of cell.

Manufacturers warn against operating nickel-iron cells about 46°C (115°F). Also, depending somewhat on discharge rate, capacities drop rapidly below a critical temperature of about 2.2°C (36°F).

Nickel-cadmium alkaline cells. Reaction (5), for the nickel-cadmium cell, is exactly the

$$2\text{NiOOH} \cdot \text{H}_2\text{O} + \text{Cd} \underset{\text{charge}}{\overset{\text{discharge}}{\rightleftarrows}} 2\text{Ni(OH)}_2 + \text{Cd(OH)}_2 \tag{5}$$

same as the one for the previous cell type except for the use of cadmium instead of iron. The earlier remarks about positive electrode and electrolyte apply equally to this cell. The cadmium negative electrode differs from the iron negative in that its potential is below the hydrogen potential. Therefore, the cadmium electrode is completely inert to the electrolyte. It requires almost no float current to keep charged, and consequently the water consumption and float charge currents are extremely low.

The original nickel-cadmium cell, now known as the "pocket" type, was invented by Waldmar Jungner at about the same time that Edison invented the tubular cell. In this cell the positive and negative plates are of the same construction as that described for the iron electrode used in the tubular-iron cell. The positive electrode uses graphite as conductor instead of the nickel flake used in the tubular cell.

The pocket nickel-cadmium battery is widely used for emergency power use. The cadmium electrode has extremely low stand loss, and it can be kept at a state of full charge with very little maintenance. It most forms it is not well suited for cycle service. It complements the tubular-iron cell. Nickel-cadmium cells may be floated at voltages of 1.40–1.45 per cell.

After an emergency discharge it is desirable to fully charge the battery before it is shifted to the float circuit.

Sintered plate cells. During World War II the Germans developed a sintered-plate type of nickel-cadmium cell. Extremely fine nickel powder, obtained from decomposition of nickel carbonyl, is sintered in a mold around a nickel or nickel-plated screen. For positive plates these plaques are impregnated with a nickel salt (usually nitrate) and processed to produce nickel hydrate in the pores. Plaques for the negative electrodes are impregnated with a cadmium salt (nitrate or chloride) and processed in a manner like that for the positive. The electrolyte is a solution of KOH made with specific gravities in the range 1.240–1.300.

Sintered-plate cells are displacing the original types, being superior in several respects. They have less internal resistance and a higher utilization coefficient, and they perform better at both higher and lower temperatures.

They are especially suited to extremely high rate discharges, low-temperature operation, and other severe applications. They are used for aircraft and diesel starting and for many military services.

Sealed cells. It has been found that the smaller sizes of nickel-cadmium cells can be operated in the fully sealed state. Sealed cells are made from very small hearing-aid sizes up to the larger flashlight sizes. In order to work in the sealed state, the cell must have a very limited amount of electrolyte, and the ratio and relative states of charge of positive and negative plates must be carefully controlled. Containers are made of nickel-plated steel or plastic. In the flashlight types the plates and separators are often rolled up in a spiral coil. They have many of the features of sintered-plate cells.

Charging is not critical. It can be done rapidly and efficiently by constant-current, constant-potential, and modified constant-potential methods; gassing begins around 1.47 V, and when using normal charge rate (5 h), the end voltage will be 1.75.

Typical discharge and charge curves are shown in **Figs. 8** and **9**.

Silver oxide–zinc alkaline cell. Silver oxide positive plates and sponge-zinc negative plates came into use during the late 1940s. They have high ampere-hour and watt-hour capacities per unit of volume or weight. A high-specific-gravity KOH solution, up to 1.450 has been found advantageous in minimizing local action. The cell reaction can be expressed as reaction (6).

$$AgO + Zn + H_2O \underset{charge}{\overset{discharge}{\rightleftharpoons}} Ag + Zn(OH)_2 \qquad (6)$$

Charging can be accomplished by a constant-current or modified constant-potential charge, as long as the cell voltage does not exceed 2.1 V at any time. Typical discharge and charge curves are shown in **Figs. 10** and **11**.

Silver oxide–zinc cells are used both as primary and secondary cells for military use and for non-military applications where battery power with minimum weight is an essential consideration.

Freezing of alkaline electrolyte. The use of high-gravity KOH electrolyte for nickel-cadmium and silver oxide-zinc cells eliminates freezing under severe arctic conditions. High-specific-gravity electrolyte cannot be used with nickel-iron cells.

Venting of storage cells. Venting must be provided for all storage cells to permit escape of local-action gas or gas generated in the charging process. The only exceptions are the special sealed cells, in which gassing is held to a minimum and any hydrogen or oxygen generated is recombined through catalysis.

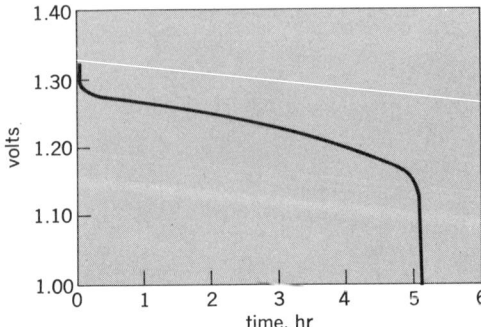

Fig. 8. Typical discharge curve for sintered-plate nickel-cadmium cell. (*ESB Inc.*)

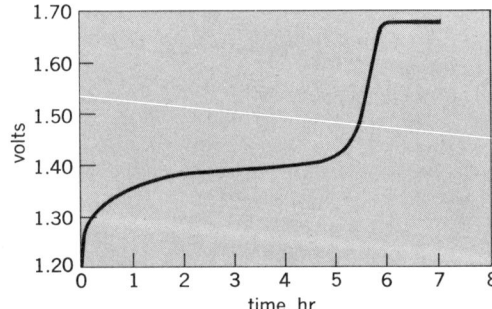

Fig. 9. Typical charge curve for sintered-plate nickel-cadmium cell. (*ESB Inc.*)

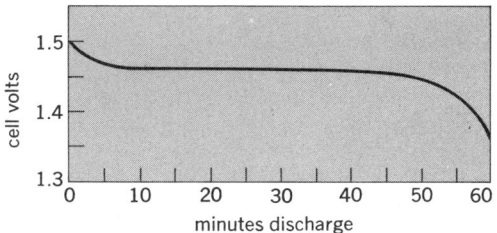

Fig. 10. Typical discharge curve exhibited by a silver oxide-zinc cell. (*ESB Inc.*)

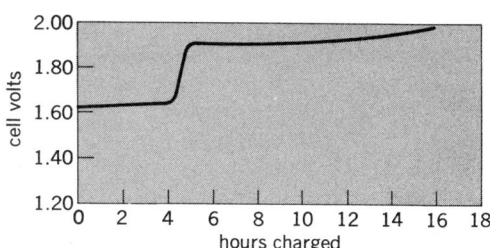

Fig. 11. Typical charge curve exhibited by a silver oxide-zinc cell. (*ESB Inc.*)

The provision for escape of gas has led to numerous devices to prevent spillage of electrolyte from cells in aircraft and other applications.

Bibliography. H. Bode, *Lead Acid Batteries*, 1977; S. U. Falk and A. J. Salkind, *Alkaline Storage Batteries*, 1969; D. G. Fink and H. W. Beaty (eds.), *Standard Handbook for Electrical Engineers*, 12th ed., 1987; C. L. Mantell, *Batteries and Energy Systems*, 2d ed., 1983; G. W. Vinal, *Storage Batteries*, 4th ed., 1955.

OPTICAL PHENOMENA

Optical activity	332
Quasielastic light scattering	335
Tyndall effect	339
Chemiluminescence	339
Opalescence	340
Photochemistry	342
Photolysis	347
Inorganic photochemistry	349
Laser photochemistry	353

OPTICAL ACTIVITY
VINCENT MADISON

The effect of asymmetric compounds on polarized light. To exhibit this effect, a molecule must be nonsuperimposable on its mirror image, that is, must be related to its mirror image as the right hand is to the left hand. An optically active compound and its mirror image are called enantiomers or optical isomers. Enantiomers differ only in their geometric arrangements; they have identical chemical and physical properties. The right-handed and left-handed forms of a molecule can be distinguished only by their optical activity or by their interactions with other asymmetric molecules. Optical activity can be used to probe other aspects of molecular geometry, as well as to identify which enantiomer is present and its purity.

As an example of optical isomers, consider tartaric acid (**Fig. 1**), which was one of the first synthetic molecules to be separated into its enantiomers. In this case the asymmetry of each isomer is magnified when trillions of molecules form a crystal; two types of asymmetric crystals are formed.

The physical basis of optical activity is the differential interaction of asymmetric substances with left versus right circularly polarized light. If solids and substances in strong magnetic fields are excluded, optical activity is an intrinsic property of the molecular structure and is one of the best methods of obtaining structural information from a sample in which the molecules are randomly oriented. The relationship between optical activity and molecular structure results from the interaction of polarized light with electrons in the molecule. Thus the molecular groups that contribute most directly to optical activity are those that have mobile electrons which can interact with light. Such groups are called chromophores, since their absorption of light is responsible for the color of objects. For example, the chlorophyll chromophore makes plants green.

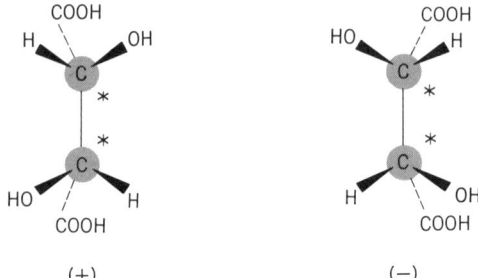

Fig. 1. Enantiomers of tartaric acid.

Methods of measurement. Optical activity is measured by two methods, optical rotation and circular dichroism.

Optical rotation. This method depends on the different velocities of left and right circularly polarized light beams in the sample. The velocities are not measured directly, but both beams are passed through the sample simultaneously. This is equivalent to using plane-polarized light. The differing velocities of the left and right circularly polarized components yield a rotation of the plane of polarization. A polarimeter for observing optical rotation consists of a light source, a fixed polarizer, a sample compartment, and a rotatable polarizer. A cell containing solvent is placed between the polarizers, and one of them is adjusted to be perpendicular to the other, excluding the passage of light. The solvent in the cell is then replaced by a solution of the sample, and the polarizer is rotated to again exclude passage of light. The optical rotation a is the number of degrees the polarizer was rotated. A positive or negative sign indicates the direction of rotation. Enantiomers have rotations of equal magnitude, but opposite signs. The optical rotation depends on the substance, solvent, concentration, cell path length, wavelength of the light, and temperature. Standardized specific rotations $[\alpha]$ are reported as defined in Eq. (1), where T is the temper-

$$[\alpha]_\lambda^T = \frac{a}{cl} \tag{1}$$

ature (°C), λ the wavelength (often the orange sodium D line), l the cell path length in decimeters, and c the concentration in grams per milliliter. Alternatively, M_ϕ is defined by normalizing to the rotation for a 1-molar solution, Eq. (2), where M_ϕ is the molar rotation and MW the molecular

$$M_\phi = [\alpha]_\lambda^T \text{MW}/100 \tag{2}$$

weight. For polymers, the mean residue rotation, m_ϕ, may be defined by the right side of Eq. (2) by using the mean residue (monomer unit) weight for MW. The variation of optical rotation with wavelength is known as optical rotatory dispersion (ORD).

Circular dichroism. Circular dichroism (CD) is the difference in absorption of left and right circularly polarized light. Since this difference is about a millionth of the absorption of either polarization, special techniques are needed to determine it accurately. Circular dichroism spectrometers consist of a light source, a monochromator to select a single wavelength, a modulator to produce circularly polarized light, a sample compartment, a phototube to detect transmitted light, and associated electronic components. The modulator rapidly switches (typically 50,000 times per second) between left and right circular polarization of the light beam. The absorption of an optically inactive sample is independent of polarization, so that the light intensity at the phototube is constant; thus a constant direct current is generated. The absorption of an optically active sample depends on the polarization, so that the light intensity at the phototube varies at the frequency of the modulator; thus an alternating current is generated. The circular dichroism is proportional to the amplitude of the alternating current. The proportionality constant is determined through calibration by using a compound of known circular dichroism.

Circular dichroism is reported as a difference in absorption, Eq. (3), or as an ellipticity (a measure of the elliptical polarization of the emergent beam), Eq. (4), for a 1-molar solution, where

$$\Delta\epsilon = \epsilon_L - \epsilon_R = (A_L - A_R)/(c'l') \tag{3} \qquad M_\theta = 3300\Delta\epsilon \tag{4}$$

ϵ is the extinction coefficient, A is the absorbance [log (I_0/I)], subscripts L and R indicate left or right circular polarization, c' is the concentration in moles per liter, l' is the path length in centimeters, I_0 and I are the light intensities in the absence and presence of the sample, respectively, and M_θ is the molar ellipticity. Either $\Delta\epsilon$ or ellipticity, m_θ, may be expressed per residue by making c' the concentration of residues (monomer units). As in the case of optical rotation, enantiomers have circular dichroism spectra of equal magnitude but opposite signs.

Variation with wavelength. Optical rotation and circular dichroism are two manifestations of the same interactions between polarized light and molecules. They are related by a mathematical transformation. An important difference between the two measurements is the way in which they vary with wavelength. Optical rotation extends to wavelengths far from any absorption of light. Thus colorless substances still have significant optical rotation at the sodium line. However, all groups which absorb light (chromophores) contribute at all wavelengths, and it can be difficult to extract the contribution of a single group. On the other hand, circular dichroism is confined to the narrow absorption band of each chromophore. Thus it is easier to determine the contribution of individual chromophores, information vital to structural analysis.

Correlation with molecular structure. In synthesizing enantiomers, chemists focus on an asymmetric center, that is, a locus which imparts asymmetry to the whole molecule. A common asymmetric center is a tetrahedral carbon atom with four different groups attached, such as the carbons marked with asterisks in tartaric acid (Fig. 1). However, in correlating optical activity with molecular structure, the focus is on the three-dimensional arrangement of the chromophores which interact most strongly with light.

As examples, consider the nucleoside adenosine and its dimer (**Fig. 2**). The most mobile electrons are in the aromatic ring system, the chromophore (Fig. 2a). The electrons in the sugar ribose are more tightly bound and interact less strongly with visible and ultraviolet light. However, all the asymmetric centers are in the ribose part of the molecule. For adenosine, light interacting with the aromatic chromophore is only weakly influenced by the asymmetric centers in ribose, so that small circular dichroism bands are observed (Fig. 2c).

In the covalently linked dimer of adenosine, the observed circular dichroism bands are

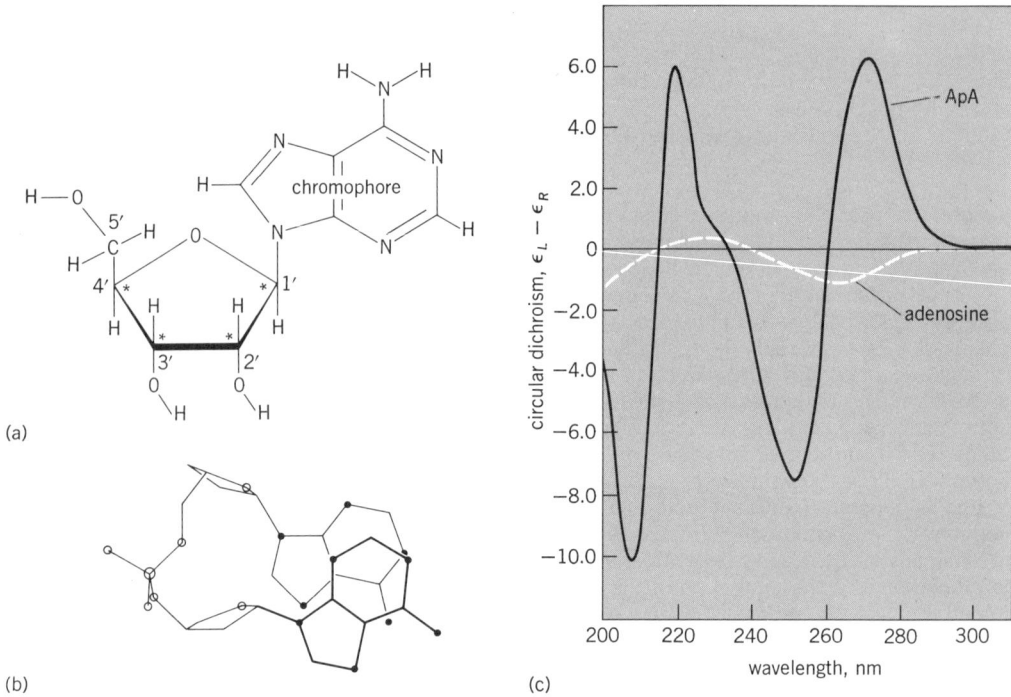

Fig. 2. Adenosine and its dimer. (a) Structure of adenosine. Asymmetric centers are marked by an asterisk. (b) Stacked arrangement of adenosine dimer (ApA). The 3' carbon of one adenosine is linked to the 5' carbon of the other by a phosphate group. (c) Circular dichroism spectra of ApA and adenosine at neutral pH in aqueous solution at room temperature.

about 10 times larger than those of the monomer. In the 240- to 300-nm region of the spectra (Fig. 2c), two bands are observed for the dimer, but only one for the monomer. This indicates strong interaction of the two aromatic chromophores, and hence their close proximity in the dimer. Analysis of the circular dichroism spectra expected for various arrangements of the two chromophores, as well as other types of experimental data, indicates that the aromatic rings are stacked (Fig. 2b). The asymmetric centers in ribose cause the formation of the stacked arrangement shown rather than its mirror image.

Stacking of aromatic rings, as exemplified by the adenosine dimer, is a common feature of nucleic acid polymers (deoxyribonucleic acid and ribonucleic acid) isolated from biological sources. Slight differences in the stacking geometry gives each of these polymers a characteristic circular dichroism spectrum. Alterations in the stacking arrangement caused by some pharmacologically active agents can be detected through alterations in the circular dichroism spectra. These structural changes may in turn be related to the pharmacological action.

A derivative of the amino acid proline (**Fig. 3**) can be used to illustrate another way in which optical activity depends on molecular structure. In this molecule only the OCN group (amide chromophore) which is in the horizontal plane of the drawing and the hydrogen which is marked $H^{\neq}$ need be considered. By forming the N-H bond, $H^{\neq}$ acquires a charge of about $+\frac{1}{3}$ electron. It has been predicted that such a positive charge will perturb the motion of the electrons in the amide chromophore in a manner which will produce a negative circular dichroism band when the charge is above the plane of the amide group and to the right of the oxygen. Only for the arrangement shown is the magnitude of the circular dichroism band expected to be as large as observed. Furthermore, it has been shown that there will be no circular dichroism if $H^{\neq}$ is in either of the two planes shown, and that for $H^{\neq}$ in adjacent quadrants the sign of the circular

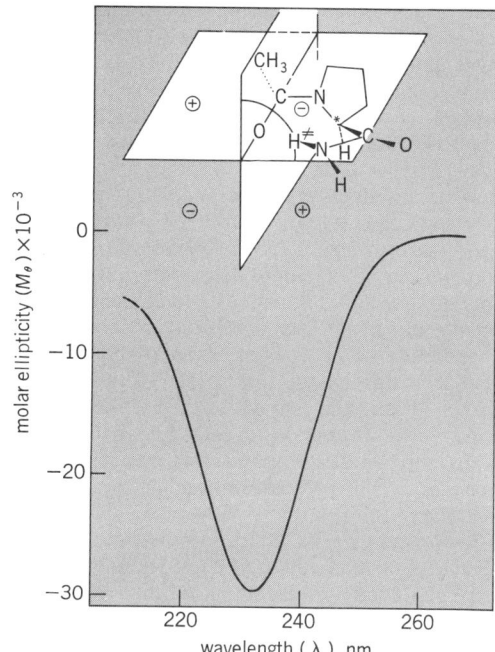

Fig. 3. Folded arrangement of L-proline derivative (*N*-acetyl-*L*-proline-amide) and its circular dichroism spectrum in *p*-dioxane solution at room temperature.

dichroism band alternates (Fig. 3). For this compound, reflection through the horizontal plane will generate the enantiomer. This would place H$^{\neq}$ in the lower right quadrant and generate a positive circular dichroism band with magnitude equal to that of Fig. 3. SEE STEREOCHEMISTRY.

QUASIELASTIC LIGHT SCATTERING
ROBERT PECORA

Small frequency shifts or broadening from the frequency of the incident radiation in the light scattered from a liquid, gas, or solid. The term quasielastic arises since the frequency changes are usually so small that, without instrumentation specifically designed for their detection, they would not be observed and the scattering process would appear to occur with no frequency changes at all, that is, elastically. The technique is used by chemists, biologists, and physicists to study the dynamics of molecules in fluids, mainly liquids and liquid solutions.

Several distinct experimental techniques are grouped under the heading of quasielastic light scattering (QLS). Intensity fluctuation spectroscopy (IFS) is the technique most often used to study such systems as macromolecules in solution and critical phenomena where the molecular motions to be studied are rather slow. This technique, also called photon correlation spectroscopy and, less frequently, optical mixing spectroscopy, is used to measure the dynamical constants of processes with relaxation time scales slower than about 10^{-6} s. For faster processes, dynamical constants are obtained by utilizing techniques known as filter methods, which obtain direct measurements of the frequency changes of the scattered light by utilizing a monochromator or filter much as in Raman spectroscopy.

Static light scattering. If light is scattered by a collection of scatterers, the scattered intensity at a point far from the scattering volume is the result of interference between the wavelets scattered from each of the scatterers and, consequently, will depend on the relative positions and orientations of the scatterers, the scattering angle θ, and the wavelength λ of the light used. The structure of scatterers in solution whose size is comparable to $(4\pi\lambda) \sin \theta/2$ ($\equiv q$) where q is the length of the scattering vector, may be studied by this technique, variously called static light scattering, integrated intensity light scattering, or in the older literature simply light scattering. It was, in fact, developed in the 1940s and 1950s to measure equilibrium properties of polymers both in solution and in bulk. Molecular weights, radii of gyration, solution virial coefficients, molecular optical anisotropies, and sizes and structure of heterogeneities in bulk polymers are routinely obtained from this type of experiment. Static light scattering is a relatively mature field, although continued improvements in instrumentation (mainly the use of lasers and associated techniques) are steadily increasing its reliability and range of application.

Both static and quasielastic light scattering experiments may be performed with the use of polarizers to select the polarizations of both the incident and the scattered beams. The plane containing the incident and scattered beams is called the scattering plane. If an experiment is performed with polarizers selecting both the incident and final polarizations perpendicular to the scattering plane, the scattering is called polarized scattering. If the incident polarization is perpendicular to the scattering plane and the scattered polarization lies in that plane, the scattering is called depolarized scattering. Usually the intensity associated with the polarized scattering is much larger than that associated with the depolarized scattering. The depolarized scattering from relatively small objects is zero unless the scatterer is optically nonspherical.

Intensity fluctuation spectroscopy. The average intensity of light scattered from a system at a given scattering angle depends, as stated above, on the relative positions and orientations of the scatterers. However, molecules are constantly in motion due to thermal forces, and are constantly translating, rotating and, for some molecules, undergoing internal rearrangements. Because of these thermal fluctuations, the scattered light intensity will also fluctuate. The intensity will fluctuate on the same time scale as the molecular motion since they are proportional to each other.

Figure 1 shows a schematic diagram of a typical intensity fluctuation apparatus. Light from a laser source traverses a polarizer to ensure a given polarization. It is then focused on a small volume of the sample cell. Light from the scattering volume at scattering angle θ is passed through an analyzer to select the polarization of the scattered light, and then through pinholes and lenses to the photomultiplier (PM) tube. The output of the photomultiplier is amplified, discriminated, sent to a photon counter, and then to a hard-wired computer called an autocorrelator, which computes the time autocorrelation function of the photocounts. The autocorrelator output is then sent to a computer for further data analysis.

The scattered light intensity as a function of time will resemble a noise signal. In order to facilitate interpretation of experimental data in terms of molecular motions, the time correlation function of the scattered intensity is usually computed by the autocorrelator. The autocorrelation function obtained in one of these experiments is often a single exponential decay, $C(t) = \exp(-t/\tau_r)$, where τ_r is the relaxation time.

The upper limit on decay times τ_r that can be measured by intensity fluctuation spectroscopy is about a microsecond, although with special variations of the technique somewhat faster

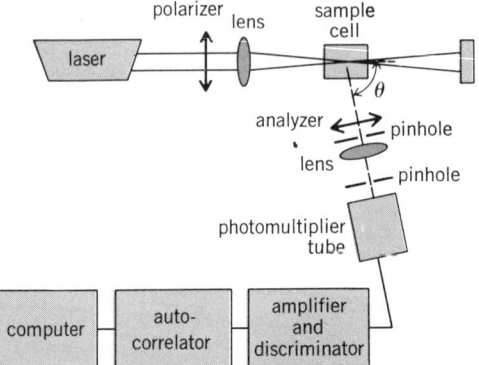

Fig. 1. Schematic diagram of an intensity fluctuation spectroscopy apparatus.

decay times may be measured. For times faster than this, filter experiments are usually performed by using a Fabry-Perot interferometer.

Fabry-Perot interferometry. Light scattered from scatterers which are moving exhibits Doppler shifts or broadening due to the motion. Thus, an initially monochromatic beam of light from a laser will be frequency-broadened by scattering from a liquid, gas, or solid, and the broadening will be a measure of the speed of the motion. For a dilute gas the spectrum will usually be a gaussian. For a liquid, however, the most common experiment of this type yields a single lorentzian line with its maximum at the laser frequency $I(\omega) = A/\pi[(1/\tau_r)/(\omega^2 + 1/\tau_r^2)]$. **Figure 2** shows a schematic of a typical Fabry-Perot interferometry apparatus. The Fabry-Perot interferometer acts as the monochromator and is placed between the scattering sample and the photomultiplier. Fabry-Perot interferometry measures the (average) scattered intensity as a function of frequency change from the laser frequency. This intensity is the frequency Fourier transform of the time correlation function of the scattered electric field. Intensity fluctuation spectroscopy experiments utilizing an autocorrelator measure the time correlation function of the intensity (which equals the square of the scattered electric field). For scattered fields with gaussian amplitude distributions the results of these two types of experiment are easily related. Sometime intensity fluctuation spectroscopy experiments are performed in what is sometimes called a heterodyne mode. In this case, some unscattered laser light is mixed with the scattered light on the surface of the photodetector. Intensity fluctuation spectroscopy experiments in the heterodyne mode measure the frequency Fourier transform of the time correlation function of the scattered electric field.

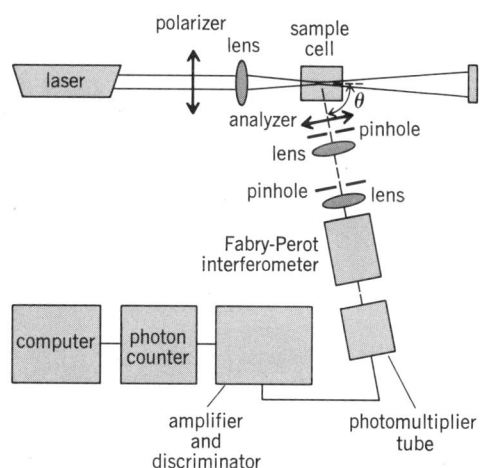

Fig. 2. Typical filter apparatus.

Translational diffusion coefficients. The most widespread application of quasielastic light scattering is the measurement of translational diffusion coefficients of macromolecules and particles in solution. For particles in solution whose characteristic dimension R is small compared to $q^{-1} = (4\pi/\lambda \sin \theta/2)^{-1}$, that is, $qR < 1$, it may be shown that the time correlation function measured in a polarized intensity fluctuation spectroscopy experiment is a single exponential with relaxation time $1/\tau_r = 2q^2D$, where D is the particle translational diffusion coefficient. For rigid, spherical particles of any size an intensity fluctuation spectroscopy experiment also provides a measure of the translational diffusion coefficient.

Translational diffusion coefficients of spherical particles in dilute solution may be used to obtain the particle radius R through use of the Stokes-Einstein relation [Eq. (1)], where k_B is

$$D = \frac{k_B T}{6\pi\eta R} \quad (1)$$

Boltzmann's constant, T the absolute temperature, and η the solvent viscosity. If the particles are shaped like ellipsoids of revolution or long rods, relations known, respectively, as the Perrin and Broersma equations may be used to relate the translational diffusion coefficient to particle dimensions. For flexible macromolecules in solution and also for irregularly shaped rigid particles, the Stokes-Einstein relation is often used to define a hydrodynamic radius (R_H).

This technique is routinely used to study such systems as flexible coil macromolecules, proteins, micelles, vesicles, viruses, and latexes. Size changes such as occur, for instance, in protein denaturation may be followed by intensity fluctuation spectroscopy studies of translational diffusion. In addition, the concentration and, in some cases, the ionic strength dependence of D are monitored to yield information on particle interactions and solution structure.

Intensity fluctuation spectroscopy experiments are also used to obtain mutual diffusion coefficients of mixtures of small molecules (for example, benzene–carbon disulfide mixtures) and are also used to measure the behavior of the mutual diffusion coefficient near the critical (consolute) point of a binary liquid mixture. Experiments of this type have proved to be very important in formulating theories of phase transitions.

Rotational diffusion coefficients. Rotational diffusion coefficients are most easily measured by depolarized quasi-elastic light scattering. The instantaneous depolarized intensity for a nonspherical scatterer depends upon the orientation of the scatterer. Rotation of the scatterer will then modulate the depolarized intensity. In a similar way, the frequency distribution of the depolarized scattered light will be broadened by the rotational motion of the molecules. Thus, for example, for dilute solutions of diffusing cylindrically symmetric scatterers, a depolarized intensity fluctuation spectroscopy experiment will give an exponential intensity time correlation function with the decay constant containing a term dependent on the scatterer rotational diffusion coefficient D_R [Eq. (2)]. A depolarized filter experiment on a similar system will give a single lorentzian

$$1/\tau_r = 2(q^2 D + 6D_R) \tag{2}$$

with $1/\tau_r$ equal to one-half that given in Eq. (2). For small molecules (for example, benzene) and relatively small macromolecules (for example, proteins with molecular weight less than 30,000) in solution, filter experiments are used to determine rotational diffusion coefficients. In these cases, the contribution of the translational diffusion to τ_r is negligible. For larger, more slowly rotating macromolecules, depolarized intensity fluctuation spectroscopy experiments are used to determine D_R.

Quasielastic light scattering is the major method of studying the rotation of small molecules in solution. Studies of the concentration dependence, viscosity dependence, and anisotropy of the molecular rotational diffusion times have been performed on a wide variety of molecules in liquids as well as liquid crystals.

Rotational diffusion coefficients of very large (≥ 100 nm) nonspherical particles may also be measured from polarized intensity fluctuation spectroscopy experiments at high values of q. SEE DIFFUSION IN GASES AND LIQUIDS.

Other applications. There are many variations on quasi-elastic light scattering experiments. For instance, polarized filter experiments on liquids also give a doublet symmetrically placed about the laser frequency. Known as the Brillouin doublet, it is separated from the incident laser frequency by $\pm C_s q$, where C_s is the hypersonic sound velocity in the scattering medium. Measurement of the doublet spacing then yields sound velocities. This technique is being extensively utilized in the study of bulk polymer systems as well as of simple liquids.

In a variation of the intensity fluctuation spectroscopy technique, a static electric field is imposed upon the sample. If the sample contains charged particles, the molecule will acquire a drift velocity proportional to the electric field strength $v = \mu E$, where μ is known as the electrophoretic mobility. Light scattered from this system will experience a Doppler shift proportional to v. Thus, in addition to particle diffusion coefficients, quasielectric light scattering can be used to measure electrophoretic mobilites. SEE ELECTROPHORESIS.

Quasielectric light scattering may also be used to study fluid flow and motile systems. Intensity fluctuation spectroscopy, for instance, is a widely used technique to study the motility of microorganisms (such as sperm cells); it is also used to study blood flow.

Bibliography. L. P. Bayvel and A. R. Jones, *Electromagnetic Scattering and its Applications*, 1981; B. J. Berne and R. Pecora, *Dynamic Light Scattering*, 1976; B. Chu, *Laser Light Scattering*, 1974.

TYNDALL EFFECT
QUENTIN VAN WINKLE

Visible scattering of light along the path of a beam of light as it passes through a system containing discontinuities. The luminous path of the beam of light is called a Tyndall cone. An example is shown in the **illustration**. In colloidal systems the brilliance of the Tyndall cone is directly dependent on the magnitude of the difference in refractive index between the particle and the medium. In aqueous gold sols, where the difference in refractive index is high, strong Tyndall cones are observed.

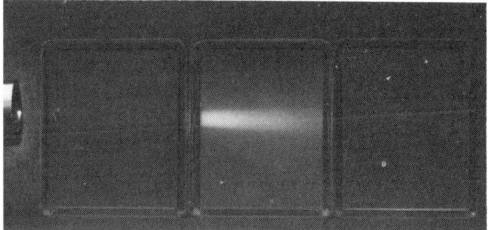

Luminous light path known as the Tyndall cone or Tyndall effect. (*Courtesy of H. Steeves and R. G. Babcock*)

For systems of particles with diameters less than one-twentieth the wavelength of light, the light scattered from a polychromatic beam is predominantly blue in color and is polarized to a degree which depends on the angle between the observer and the incident beam. The blue color of tobacco smoke is an example of Tyndall blue. As particles are increased in size, the blue color of scattered light disappears and the scattered radiation appears white. If this scattered light is received through a nicol prism which is oriented to extinguish the vertically polarized scattered light, the blue color appears again in increased brilliance. This is called residual blue, and its intensity varies as the inverse eighth power of the wavelength. SEE COLLOID.

CHEMILUMINESCENCE
THERESE WILSON

The type of luminescence wherein a chemical reaction supplies the energy responsible for the emission of light (ultraviolet, visible, or infrared) in excess of that of a blackbody (thermal radiation) at the same temperature and within the same spectral range. Below 900°F (500°C), the emission of any light during a chemical reaction is a chemiluminescence. The blue inner cone of a bunsen burner or the Coleman gas lamp are examples.

Many chemical reactions generate energy. Usually this exothermicity appears as heat, that is, translational, rotational, and vibrational energy of the product molecules; whereas, for a visible chemiluminescence to occur, one of the reaction products must be generated in an excited electronic state (designated by an asterisk) from which it can undergo deactivation by emission of a photon. Hence a chemiluminescent reaction, as shown in reactions (1) and (2), can be regarded

$$A + B \rightarrow C^* + D \quad (1) \qquad C^* \rightarrow C + h\nu \quad (2)$$

as the reverse of a photochemical reaction.

The energy of the light quantum $h\nu$ (where h is Planck's constant, and ν is the light frequency) depends on the separation between the ground and the first excited electronic state of C; and the spectrum of the chemiluminescence usually matches the fluorescence spectrum of the emitter. Occasionally, the reaction involves an additional step, the transfer of electronic energy from C* to another molecule, not necessarily otherwise involved in the reaction. Sometimes no discrete excited state can be specified, in which case the chemiluminescence spectrum is a structureless continuum associated with the formation of a molecule, as in the so-called air afterglow: $NO + O \rightarrow NO_2 + h\nu$ (green light). SEE PHOTOCHEMISTRY.

The efficiency of a chemiluminescence is expressed as its quantum yield ϕ, that is, the number of photons emitted per reacted molecule. Many reactions have quantum yields much lower (10^{-8} $h\nu$ per molecule) than the maximum of unity, Einsteins of visible light (1 einstein = $Nh\nu$, where N is Avogadro's number), with wavelengths from 400 to 700 nanometers, correspond to energies of about 70 to 40 kcal per mole (300 to 170 kilojoules per mole). Thus only very exothermic, or "exergonic," chemical processes can be expected to be chemiluminescent. Partly for this reason, most familiar examples of chemiluminescence involve oxygen and oxidation processes; the most efficient examples of these are the enzyme-mediated bioluminescences. The glow of phosphorus in air is a historically important case, although the mechanism of this complex reaction is not fully understood. The oxidation of many organic substances, such as aldehydes or alcohols, by oxygen, hydrogen peroxide, ozone, and so on, is chemiluminescent. The reaction of heated ether vapor with air results in a bluish "cold" flame, for example. The efficiency of some chemiluminescences in solution, such as the oxidation of luminol (I) and, especially, the reaction of some oxalate esters (II) with hydrogen peroxide, can be very high ($\phi = 30\%$).

It is believed that the requirements for chemiluminescence are not only sufficient exothermicity and the presence of a suitable emitter, but also that the chemical process be very fast and involve few geometrical changes, in order to minimize energy dissipation through vibrations. For example, the transfer of one electron from a powerful oxidant to a reductant (often two radical ions of opposite charge generated electrochemically) is a type of process which can result, in some cases, in very effective generation of electronic excitation. An example, with 9,10-diphenylanthracene (DPA), is shown in reaction (3). The same is true of the decomposition of four-

$$DPA^{-\cdot} + DPA^{+\cdot} \rightarrow DPA^* + DPA \qquad (3)$$

membered cyclic peroxides (III) into carbonyl products, shown in reaction (4), which may be the prototype of many chemiluminescences.

Bibliography. M. Deluca and W. McElroy (eds.), *Bioluminescence and Chemiluminescence*, 1981.

OPALESCENCE
BENJAMIN CHU

The milky iridescent appearance of a dense transparent medium when the system (or medium) is illuminated by polychromatic radiation in the visible range, such as sunlight. Slight changes in the rainbowlike color of the system can occur, depending on the scattering angle, that is, the angle between the directions of incident radiation and of observation. All dense transparent mediums have local density fluctuations due to the thermal motions of molecules, or concentration fluctuations due to the presence of a second component, such as colloidal suspensions or macromolecules in solution. Local fluctuations in density (or concentration) are accompanied by local fluctuations in the refractive index. Since the system is optically inhomogeneous, some of the light is scattered to the side. Normally, the amount of light scattered is very small, perhaps of the order of magnitude of 10^{-4} or less of its incident radiation. Whenever the amplitude of fluctuation becomes large, a significant portion of the incident light may be scattered. The transmitted light is then visibly weakened, and the medium look turbid.

Opalescence is a general term which applies to the optical phenomenon of intense scattering in the visible range of the electromagnetic radiation by a system with strong local optical inhomogeneities. The iridescence, or rainbowlike display of interference of colors, arises because the intensity of scattered light is approximately proportional to the reciprocal fourth power of the wavelength of incident light (Rayleigh's law).

Critical opalescence. A classical view of the critical point in gas-liquid phase transitions is that it is the state at which the densities of the coexisting gas-liquid phases are equal, and also that it is represented by a characteristic critical temperature above which gas cannot be liquified, no matter how great the applied pressure is. The corresponding pressure required to liquefy the gas at the critical temperature is the critical pressure. For a one-component system the compressibility becomes very large in the neighborhood of the critical point and infinite at the critical point itself. Thus the energy required in the compression of a gas to a given amplitude of fluctuations becomes smaller the closer one approaches the critical point. There the thermal motions of molecules can produce strong density fluctuations, resulting in a very impressive scattering, the so-called critical opalescence. Apart from the critical points of gas-liquid transitions, several other types of second-order phase transitions at which the second derivatives of the free energy are discontinuous, such as critical mixing (consolute) points of binary liquid mixtures, exhibit critical opalescent behavior.

One can study the nature of phase transitions by observing the size and shape of local fluctuations and their time-dependent changes. By using statistical models, one can then relate fluctuations to thermodynamic and transport properties of the system, such as isothermal compressibility and thermal conductivity. In the critical region any such properties become very difficult to measure by conventional means. The changes at the critical region in thermodynamic and transport properties are very dramatic; for example, both isothermal compressibility and heat capacity diverge at the critical point. While the scattered intensity is related to the amplitude of local fluctuations in the refractive index, the angular dependence of scattered light reveals the extent of such fluctuations. As one approaches the critical point, the scattered intensity increases because the isothermal compressibility becomes larger; the system consequently looks very turbid. Further analysis shows that the scattered light is concentrated more and more in the forward direction. This indicates that the extensions of fluctuations approach the wavelength of the incident radiation, which is a few thousand angstroms for light in the visible region. These large fluctuations result from long-range molecular interactions in the system. A suitable approach is to consider the incident electromagnetic radiation as a measuring scale. Thus, for smaller fluctuations or fluctuations in metal alloys near the consolute point, one can use x-rays with a wavelength of 0.15 nanometer or less instead of visible light with a wavelength of several thousand angstroms, to investigate fluctuation sizes ranging from tens to hundreds of angstroms, even when the system is barely opalescent to the naked eye or does not transmit visible light.

Time dependency. The density (or concentration) fluctuations that are produced by thermal motions of molecules are time-dependent, so that light is quasi-elastically or inelastically scattered. The spectral distribution of the scattered light characterizes the time dependence of such fluctuations and can be resolved by means of interferometric and optical beat frequency techniques, using laser as a light source. The divergence in heat capacity and probably also in thermal conductivity indicates a slowdown in the relaxation times of thermal fluctuations and an increasing difficulty in reaching thermal equilibrium. Experiments and theory are both difficult. Care should therefore be exercised when studying the phenomenon of critical opalescence.

Bibliography. B. Chu, *Molecular Forces: Based on the Baker Lectures of Peter J. W. Debye*, 1967; M. S. Green and J. V. Sengers (eds.), *Critical Phenomena*, Nat. Bur. Stand. Misc. Publ. 273, 1966.

PHOTOCHEMISTRY
Brian Stevens

The branch of chemistry concerned with reactions of excited molecules produced by the absorption of light. Chemical reactions involve the breaking (and formation) of chemical bonds which requires energies in the range of 200–600 kilojoules/mole (48–144 kilocalories/mole); this corresponds to the energy of light quanta in the ultraviolet (100–400 nanometers), visible (400–700 nm) and near-infrared (700–1000 nm) regions of the electromagnetic spectrum. Light of shorter wavelengths (x-rays, gamma rays) has sufficient energy to ionize and to dissociate molecules. The effects of this light constitute the field of radiation chemistry, while light of longer wavelengths (infrared) is not sufficiently energetic to produce electronic excitation in single quanta excitations.
SEE LASER PHOTOCHEMISTRY; RADIATION CHEMISTRY.

Electronic states of molecules possessing an even number of electrons with spins paired are termed singlet states (superscript 1) or triplet states (superscript 3) if two electrons have parallel spins. Excited singlet or triplet states are often designated $\pi\pi^*$ if they are produced by a transition in which a π-electron in an unsaturated molecule (containing one or more double bonds) is promoted to an antibonding π^*-orbital, or $n\pi^*$ if an electron in a nonbonding (lone-pair) n-orbital is promoted to a π^*-orbital as in the case of a carbonyl compound $>$C=O. The excited states of an unsaturated molecule may therefore be $^1\pi\pi^*$; $^3\pi\pi^*$, $^1n\pi^*$ or $^3n\pi^*$ in which the double bond is reduced to a single bond. Saturated molecules containing only single (σ) bonds undergo electronic excitation to $\sigma\sigma^*$ states in which a bond is broken to produce atoms or free radicals.

Electronic excitation reduces the ionization potential and increases the electron affinity of a molecule by an increment equal to the excitation energy, and therefore promotes electron transfer reactions. The change in electronic configuration produced by light absorption leads to dramatic changes in chemical properties; thus the $^1\pi\pi^*$ state of 2-naphthol is more acidic than the ground state by a factor of 10^6. Since the Gibbs free energy is also increased by light absorption, electronically excited molecules may undergo spontaneous reactions to products which are thermodynamically inaccessible from the ground state. This is exemplified by the photosynthetic reaction in which the conversion of carbon dioxide (CO_2) and water (H_2O) to carbohydrate $(CH_2O)_n$ and oxygen (O_2) is mediated by excited chlorophyll molecules. The reverse process of respiration leads to a decrease in Gibbs free energy and takes place spontaneously in the absence of light.

The study of photochemical reactions provides a firm molecular basis for the interpretation of those processes arising from the interaction of sunlight with the biosphere which are responsible not only for creating the conditions under which life on the Earth could develop, but also the provision of food and energy.

Quantum yield. The rate of a photochemical reaction is proportional to the rate at which electronically excited molecules M* are produced; since each absorbed photon (energy $h\nu$) of light produces one excited molecule, this excitation rate is equal to the rate of light absorption or the intensity of light absorbed. I_a. The reaction rate is therefore proportional to I_a, or rate = γI_a, where the proportionality constant, γ = rate/I_a, known as the quantum yield (or efficiency), is characteristic of the reaction under the conditions of examination. The quantum yield of the primary process (that involving the excited molecule) cannot exceed unity and is usually much lower than this since electronically excited molecules revert to the ground state (with or without the emission of luminescence within a period of $\sim 10^{-3}$ s (triplet state) to 10^{-8} s (singlet states). These so-called photophysical processes, summarized in **Fig. 1**, compete with photochemical reactions, effectively reducing the quantum yield. Additionally, the excited molecule may be quenched by other molecules Q in a process of electron transfer [reaction (1)] or energy transfer [reaction (2)], which leave the molecule M chemically unchanged.

$$M* + Q \rightarrow (M^+Q^-) \text{ or } (M^-Q^+) \rightarrow M + Q \quad (1) \qquad M* + Q \rightarrow M + Q* \quad (2)$$

On the other hand, if the primary photochemical products are atoms or free radicals, these may undergo secondary nonphotochemical chain reactions which amplify the primary process and lead to overall photochemical quantum yields much greater than unity. An example is reaction (3a), in which the primary process [reaction (3b)] is followed by the chain propagating steps

$$H_2 + Cl_2 \xrightarrow{light} 2HCl \quad (3a) \qquad Cl_2 + light \rightarrow 2Cl \quad (3b)$$

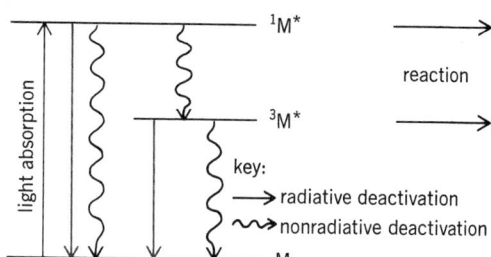

Fig. 1. Radiative and nonradiative deactivation of electronically excited molecule compete with reaction of singlet state $^1M^*$ and of lower-energy triplet state $^3M^*$.

[reactions (3c) and (3d)] to produce $\sim 10^6$ molecules of HCl for each light photon absorbed by Cl_2,

$$Cl + H_2 \rightarrow H + HCl \qquad (3c) \qquad H + Cl_2 \rightarrow Cl + HCl \qquad (3d)$$

that is, an overall quantum yield of 10^6. SEE CHAIN REACTION.

Mechanisms. Photochemical reaction mechanisms are deduced from measurements of the dependence of quantum yields on such different reaction variables as reactant concentration and the concentration of added substances. Primarily it is necessary to identify the nature of the electronically excited state responsible for the reaction, which is more often singlet ($^1M^*$) or triplet ($^3M^*$) following which a theoretical examination of the orbital transformations along the reaction coordinate may be attempted.

If the quantum yield is reduced by a selective triplet-state quencher (Q) which introduces the competing process [reaction (4)], then the triplet state $^3M^*$ may be regarded as the reactive

$$^3M^* + Q \rightarrow M + {}^3Q^* \qquad (4)$$

state; otherwise the singlet state $^1M^*$ is assigned the role of reactive intermediate. The rate constant of the primary process may then be obtained by monitoring the decay of the intermediate $^{1,3}M^*$ in absorption or emission following flash or laser pulse excitation. SEE FREE RADICAL; QUANTUM CHEMISTRY.

Photochemical reactions. These are classified as unimolecular, in which the excited molecule itself undergoes chemical change, or bimolecular if the excited molecule reacts with another molecule present.

Unimolecular processes. Such processes which lead to a single product are known as photoisomerization reactions. Examples are the formation of Dewar benzene (I) from benzene (II), shown in reaction (5), which involves a higher singlet state of benzene, and the photochromic

$$\text{(II)} \xrightarrow{h\nu} \text{(I)} \qquad (5)$$

reaction of crystalline dinitrobenzylpyridine [reaction (6)]. Ring closure is exemplified by reaction

(Colorless) $\xrightarrow{h\nu}$ (Colored) $\qquad (6)$

(7), which bears promise as a solar-energy storage process, whereas ring opening is involved in

Norbornadiene $\xrightarrow{h\nu}$ Quadricyclene $\qquad (7)$

the photochemical synthesis of vitamin D_2 [reaction (8)]. These reactions proceed via the $^1\pi\pi^*$ state.

$$\text{Ergosterol} \xrightarrow{h\nu} \text{Previtamin D}_2 \quad (8)$$

In a $\pi\pi^*$ state the reduction of a double bond to a single bond permits free rotation of the end groups resulting in cis-trans isomerization as in the case of stilbene [reaction (9)]. The photochemical transformation of 11-*cis*-retinal to all-*trans*-retinal is believed to trigger visual response, and the phototherapy of neonatal jaundice probably involves the cis-trans isomerization of bilirubin, a yellow pigment product of heme degradation, to a water-soluble isomer which can be excreted.

$$\text{cis} \xrightleftharpoons{h\nu} {}^{1,3}\pi\pi^* \longrightarrow \text{trans} \quad (9)$$

Unimolecular reactions producing two product radicals or atoms as the result of bond rupture are exemplified by the photolysis of ozone O_3 [reaction (10)] and of oxygen O_2 [reaction (11)]

$$O_3 \xrightarrow{h\nu} O_2 + O \quad (10) \qquad O_2 \xrightarrow{h\nu} O + O \quad (11)$$

in the upper atmosphere by absorbing solar radiation of wavelengths shorter than 300 nm. Ozone is reformed by the combination of O_2 molecules and O atoms and exists in photochemical equilibrium in the so-called ozone layer. Concern has been expressed that very stable aerosol spray propellants or Freons such as CF_2Cl_2 may find their way into the upper atmosphere where the products of photodissociation [reaction (12)] may deplete the ozone concentration by the secondary reaction (13) with the result that part of the solar radiation below 300 nm may reach the Earth's

$$CF_2Cl_2 \xrightarrow{h\nu} CF_2Cl + Cl \quad (12) \qquad Cl + O_3 \rightarrow ClO + O_2 \quad (13)$$

surface. Certain aromatic hydroxy compounds dissolved in water dissociate into ions when excited, for example, 2-naphthol [reaction (14)].

$$\text{2-naphthol-OH} \longrightarrow \text{2-naphtholate-O}^- + H_3O^+ \quad (14)$$

Bimolecular processes. When these involve electronically excited unsaturated molecules which form a single product the reaction is referred to as photoaddition [reaction (15)]. If the adduct A is the unexcited molecule M itself, the reaction is known as photodimerization [reaction (16)]. In either case the reaction is believed to proceed via the excited adduct MA*, known as an

$$M^* + A \rightarrow MA \quad (15) \qquad M^* + M \rightarrow M_2 \quad (16)$$

exciplex, or M_2^*, an excimer, both of which in certain cases emit a characteristic fluorescence spectrum. The simplest photodimerization reaction is that of ethylene to produce cyclobutane [reaction (17)], which is believed to involve the $^1\pi\pi^*$ state of ethylene. This type of process can

$$\begin{array}{c} H_2C=CH_2 \\ H_2C=CH_2 \end{array} \xrightarrow{h\nu} \begin{array}{c} H_2C-CH_2 \\ | \quad | \\ H_2C-CH_2 \end{array} \quad (17)$$

take place in the solid state if neighboring molecules are suitably oriented in the crystal lattice. For example, the α modification of transcinnamic acid, in which nearest neighbors have a head-to-tail configuration (IIIa), photodimerizes to α-truxillic acid (IV), as shown in reaction (18),

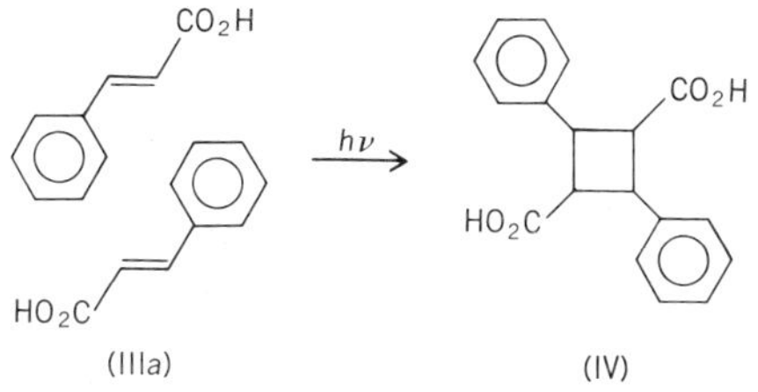

$$\text{(IIIa)} \xrightarrow{h\nu} \text{(IV)} \tag{18}$$

whereas the β-crystalline modification based on a head-to-head configuration of adjacent molecules (IIIb) produces β-truxinic acid (V), as shown in reaction (19). In the γ modification of crys-

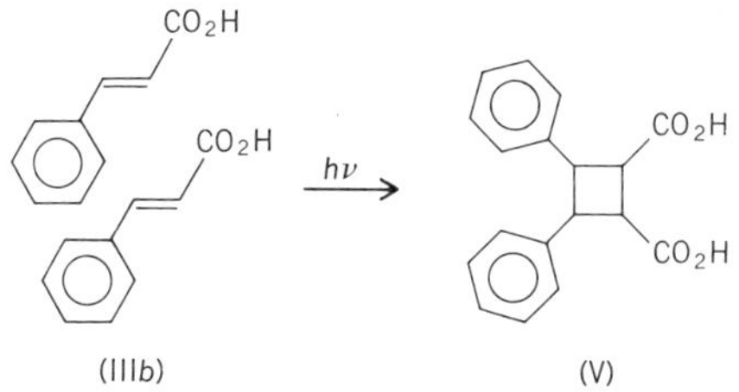

$$\text{(IIIb)} \xrightarrow{h\nu} \text{(V)} \tag{19}$$

talline transcinnamic acid, the distance between neighboring molecules is too great for dimerization to take place; these examples illustrate topochemical control of product formation by nearest-neighbor orientation. The photodimerization of adjacent thymine residues (VI) on the same strand is responsible for one type of photochemical lesion in DNA [reaction (20)]. The twin strands of

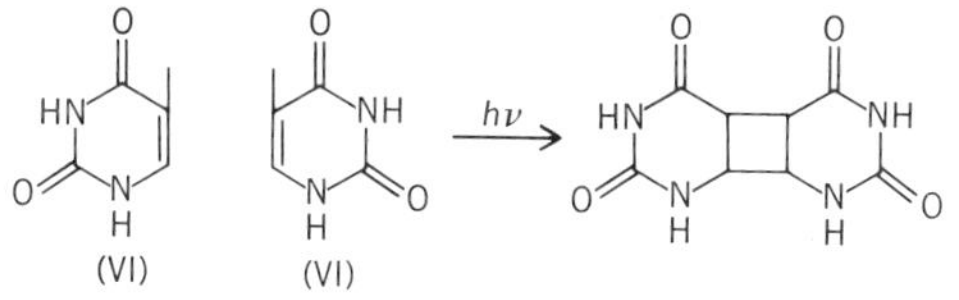

$$\text{(VI)} + \text{(VI)} \xrightarrow{h\nu} \tag{20}$$

DNA may be photochemically cross-linked by psoralen, which can intercalate with two base pairs and form cyclobutane linkages (**Fig. 2**) which inhibit DNA synthesis and cell division; this is

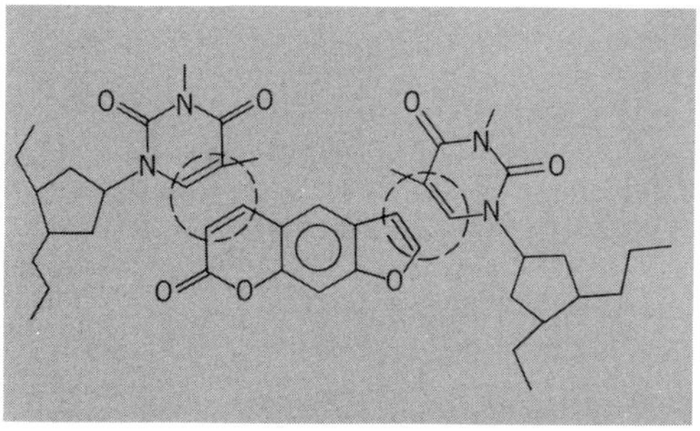

Fig. 2. Psoralen intercalating with two base pairs, forming cyclobutane linkages between strands of DNA.

believed to be the molecular basis for the phototherapy of psoriasis following oral ingestion of psoralen derivatives by the patient.

In the presence of light, a light-absorbing sensitizer (S), and molecular oxygen, many unsaturated molecules (A) are converted to peroxides (AO_2), as shown in reaction (21a). The primary photochemical process involves the addition of A to singlet molecular oxygen $^1O_2^*$ produced from its triplet ground state $^3O_2$ by energy transfer from the sensitizer triplet state $^3S^*$ [reaction (21b)].

$$A \xrightarrow[\text{light}]{S/O_2} AO_2 \qquad (21a) \qquad\qquad ^3S^* + {}^3O_2 \to S + {}^1O_2^* \qquad (21b)$$

The conversion of α-terpinene to ascaridole in this way was probably the first commercially exploited photosynthetic reaction (22). Cellular damage under the same conditions is termed photo-

$$\text{[α-terpinene]} + {}^1O_2^* \longrightarrow \text{[ascaridole]} \qquad (22)$$

dynamic action, in which the role of $^1O_2^*$ is strongly implicated; β-carotene which quenches $^1O_2^*$ in the energy transfer process [reaction (23)] is believed to protect plants from self-destruction,

$$^1O_2^* + \beta\text{-carotene} \to {}^3O_2 + {}^3\beta\text{-carotene}^* \qquad (23)$$

and has been successfully used in the treatment of erythropoietic protoporphyria, a hereditary inability to metabolize porphyrins which accumulate in the skin and act as sensitizers in sunlight.

An example of a bimolecular photochemical process which produces two products is afforded by the $^3n\pi^*$ states of certain ketones which abstract H atoms from an adjacent substrate RH [reaction (24) where · denotes free radical]. The free radicals thus produced subsequently

$$>\!C = O(^3n\pi*) + RH \to \; >\!\dot{C} - OH + R\cdot \qquad (24)$$

dimerize to yield the overall reaction products; however, if RH is a polymer chain, secondary reactions of R in the presence of oxygen result in chain breaking, a process which has been exploited in the production of photodegradable packaging and picnic items.

Theoretical aspects. The principle that orbital symmetry is conserved in concerted reactions has afforded an interpretation of many types of photochemical (and thermal) reactions at the molecular level. Ring closure of a π-electron system exemplified by the butadiene derivative may produce either of the two isomers shown [reaction (25)]. The Woodward-Hoffman rules for

$$\text{(butadiene derivative)} \longrightarrow \text{conrotatory / disrotatory products} \qquad (25)$$

this electrocyclic process for a system containing k π-electrons are summarized by

k	Thermal reaction	Photochemical reaction
$4q$	conrotatory	disrotatory
$4q + 2$	disrotatory	conrotatory

where $q = 1, 2, 3 \ldots$. The same rules apply to ring opening.

The concerted (bimolecular) cycloaddition reaction of two systems containing m and n π-electrons, an example of which is shown in reaction (26), is photochemically allowed if

$$\text{[diene]} + \text{[ene]} \longrightarrow \text{[cyclohexene]} \qquad (26)$$

$$m = 4 \quad n = 2$$

$m + n = 4q$, and photochemically forbidden if $m + n = 4q + 2$, where again q is an integer.

Rules have also been developed for sigmatropic reactions (atom migration in a π-electron system) and cheletropic (molecular elimination) processes.

The conservation of spin angular momentum limits singlet and triplet energy-transfer processes to those shown in reactions (27) and (28), which can be used to identify the photochemi-

$$^1M^* + {}^1Q \rightarrow {}^1M + {}^1Q^* \quad (27) \qquad\qquad {}^3M^* + {}^1Q \rightarrow {}^1M + {}^3Q^* \quad (28)$$

cally reactive state of M, and often affords an indication of preferred reaction pathways at the molecular level. Since the projected electron spin quantum number has possible values $\pm \frac{1}{2}$, the resultant spin angular momentum [$S = \frac{1}{2} - \frac{1}{2} = 0$ for two spin-paired electrons (singlet state) or $S = \frac{1}{2} + \frac{1}{2} = 1$ for two electrons with parallel spin (triplet state)] has the result that the concerted photochemical addition reactions involve either the excited singlet state [reaction (29)] or two excited triplet states [reaction (30)], which is less probable. The photochemical addition of

$$^1M^*(S = 0) + {}^1A(S = 0) \rightarrow {}^1MA(S = 0) \quad (29)$$

$$^3M^*(S = 1) + {}^3A(S = 1) \rightarrow {}^1MA(S = 1 - 1 = 0) \quad (30)$$

triplet and ground states produces the adduct triplet state [reaction (31)] unless it involves an intermediate diradical [reaction (32)] in a nonconcerted process. The formation of organic perox-

$$^3M^*(S = 1) + {}^1A(S = 0) \rightarrow {}^3MA^*(S = 1 + 0 = 1) \quad (31)$$

$$^3M^*(S = 1) + {}^1A(S = 0) \rightarrow MA(S = {}^1/_2 \pm {}^1/_2) \rightarrow {}^1MA(S = 0) \quad (32)$$

ides by the addition of unsaturated molecules to singlet oxygen $^1O_2^*$ (but not the $^2O_2$ ground state) is indicative of a concerted process in which orbital symmetry is conserved. Evidence presented to the effect that thymine dimerization and psoralen binding to DNA base pairs results from triplet excitation, indicates that these are nonconcerted reactions involving a diradical intermediate. SEE CHEMICAL DYNAMICS; MOLECULAR ORBITAL THEORY.

Bibliography. J. D. Coyle, *Introduction to Organic Photochemistry*, 1986; W. A. Noyes et al. (eds.), *Advances in Photochemistry*, vols. 1–12, 1963–1979; N. J. Turro, *Modern Molecular Photochemistry*, 1981; R. B. Woodward and R. Hoffmann, *The Conservation of Orbital Symmetry*, 1970; M. Wrighton (ed.), *Inorganic and Organometallic Photochemistry*, 1978; A. Zewail (ed.), *Photochemistry and Photobiology*, vols. 1–2, 1984.

PHOTOLYSIS
FREDERICK C. BROWN

Chemical decomposition by the action of radiant electromagnetic energy, especially light. Photolysis occurs in certain crystals, notably the silver halides, when they are exposed to radiation. When this occurs, the effect of the radiation is to produce a definite chemical change resulting in the separation of photolytic silver. Photolysis of the silver halides is discussed in this article because of the extensive investigations that have been carried out on these materials and because of their importance in the photographic process. Actually, photolysis occurs in many other materials, such as the lead and thallium halides, zinc oxide, the metallic azides, and in organic compounds such as the oxalates, styphnates, and fulminates.

A photographic emulsion consists of microcrystalline grains of silver bromide, AgBr, or silver chloride, AgCl, embedded in gelatin. Upon prolonged exposure to light, so-called print-out specks of silver form within and on the surface of the grains. Much shorter exposures produce a latent image which can be made visible by the process of development. Experiments have shown that the latent image consists of only a very few atoms of silver in each grain. The high sensitivity of the photographic system comes about because of the enormous gain (10^9) which can be achieved by reduction of each exposed grain that occurs during development.

Gurney-Mott theory. This theory of the photographic process proposes a two-stage mechanism as shown in **Fig. 1**. In the first stage a light quantum is absorbed at a point within

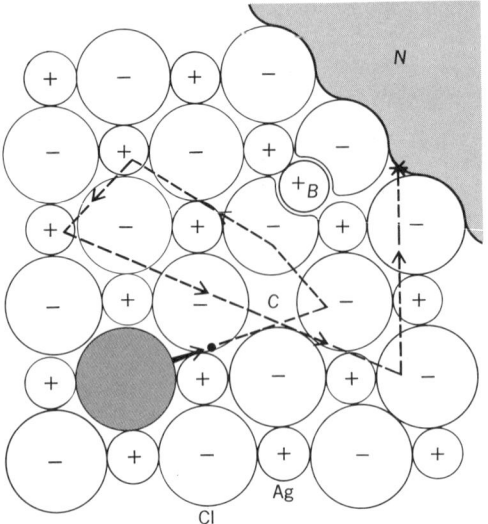

Fig. 1. Schematic representation of Gurney-Mott theory. Broken line shows path of photoelectron which is eventually trapped in vicinity of a silver speck N. The interstitial silver ion B has come up to neutralize the charge of the electron and thus to add to the silver speck.

the silver halide grain, releasing a mobile electron and a positive hole. These mobile defects diffuse to trapping sites (sensitivity centers) within the volume or on the surface of the grain. In the second stage, the trapped (negatively charged) electron is neutralized by an interstitial (positively charged) silver ion, which combines with the electron to form a silver atom. The silver atom at the sensitivity center is capable of trapping a second electron, and then the process repeats, causing the silver speck to grow. The positive holes are assumed to diffuse to the surface without recombining with electrons, where they escape or react with the gelatin.

The early Gurney-Mott theory has been criticized by J. W. Mitchell and others, especially because of the assumed lack of electron-hole recombination. The essential idea of an electronic process linking the initial absorption of light quanta with the formation of the image speck is nevertheless well founded. **Figure 2** shows an electron micrograph of an emulsion grain after a print-out level exposure to synchronized light and voltage pulses. The print-out silver specks occur on one side because of the action of the electric field, and there is experimental evidence for positive-hole migration in the opposite direction.

Practical emulsions contain a distribution of silver halide grain sizes and shapes. The slower fine-grain emulsions may have an average grain size of 0.05 micrometer or smaller, whereas the grain sizes in a high-speed emulsion may be of the order of several micrometers. The silver halide in a negative emulsion is usually AgBr that contains a small amount of silver iodide. Various substances, for example, gold and silver sulfide, act as sensitizers, and the spectral response can be extended by the addition of dyes. Formation of the latent image in a fast emulsion containing sensitizers is highly complicated and is thought to involve the role of structural imperfections such as dislocations, jogs on dislocations, and twin planes. There is increasing evidence that the surface plays a crucial role.

Large crystals. The photolysis of large crystals of the silver halides and of other compounds such as the lead halides has been studied in considerable detail. When interpreting the results on single crystals, it becomes necessary to distinguish between darkening produced by light in the volume of the crystal and darkening produced near the surface. A pure crystal of AgCl does not darken appreciably upon exposure to light absorbed below the surface. On the other hand, crystals which contain small traces of impurity, particularly copper in the monovalent state, darken with high efficiency up to a saturation level which depends upon the amount of impurity present. **Figure 3** shows the small amount of absorption for darkening of this type, which occurs in the early stages of exposure. Here the extinction of light is caused mainly by absorption. Prolonged exposure, however, produces darkening near the surface, which shows considerable light

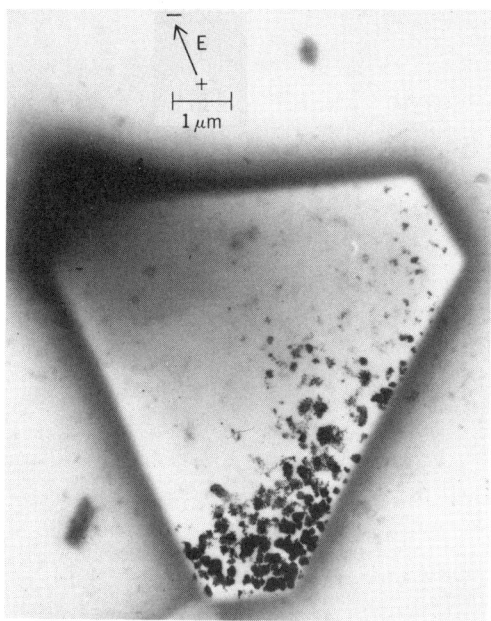

Fig. 2. Electron micrograph of a silver halide emulsion grain exposed to repetitive pulses of light and to an electric field E. Cloud in the gelatin is thought to be due to escape of bromine from the grain, as if positive holes were displaced upward and to the left, whereas electrons are displaced downward and to the right by the applied electric field.

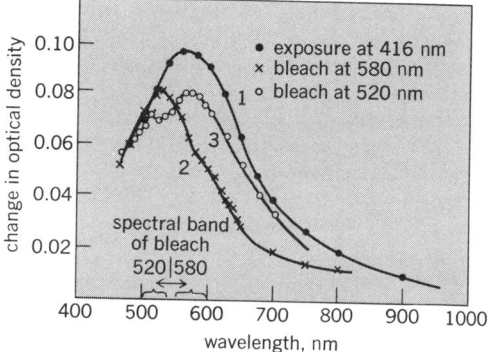

Fig. 3. Curve 1 shows the volume darkening produced at room temperature in a 3.7-mm-thick AgCl crystal by absorption of 4.2×10^{16} photons/cm^2 in the characteristic absorption edge at 416 nm. Upon continued exposure, this darkening proceeds with high efficiency to a level which depends upon impurity content. Illumination within the colloid band itself produces bleaching (curves 2 and 3). Optical density is equal to $\log_{10} I_0/I$, where I_0 is intensity of incident light and I intensity of transmitted light.

scattering. The centers responsible for the surface darkening are larger and may be similar to the colloidal metal particles formed in alkali halides by coagulation of F-centers during heat treatment. SEE PHOTOCHEMISTRY.

INORGANIC PHOTOCHEMISTRY
JOHN F. ENDICOTT

Principally the study of the light-induced behavior of various metal compounds. The physical and chemical properties of substances are generally altered by the absorption of light. Typical metal compounds have a characteristic number (coordination number) of molecules or ions (ligands) directly bonded to the metal center. This article will refer to six-coordinate compounds (for example, ML_6^{n+}). Many of these compounds are colored, and much interest has been aroused by speculation that some metal compounds could mediate the transformation of solar radiation into useful chemical or electrical energy.

The photochemistry of metal compounds has grown in concert with modern theories of the electronic structure of molecules and of chemical bonding in molecules. Photochemical studies are often designed to probe and test these theories. The range of pertinent studies spans most of the subdisciplines of chemistry and includes or bears on such topics as photophysics, the development of laser materials, catalysis, photosynthesis, oxidation-reduction chemistry, acid-base chemistry, organometallic chemistry, metalloenzyme chemistry, solid-state chemistry, and surface chemistry.

Excited states. The absorption of light results in a rearrangement of electrons within a molecule. In many molecules, the new electronic configuration can persist for a significant period of time. It is useful to regard these excited states of molecules as new chemical species with chemical properties distinctly different from those of the ground state. Chemical properties depend on electronic configurations, and the ground-state and excited-state electronic configurations differ. In general, the bond lengths (angles, and so on) of the excited state of a molecule will be different from those of the ground-state (or thermally equilibrated) molecule. As a consequence, some of the energy used to generate an excited state is degraded to heat as the excited molecule relaxes to a bonding arrangement compatible with the new electronic configuration. In addition, the initial excited state (*X) may rapidly convert (or cross) to a lower-energy excited state (*Y) with yet another electronic configuration (see **illus.**). After light absorption in metal compounds, the time required for generation of the lowest-energy excited state is usually very short (less than a nanosecond). As a consequence, any chemistry due to the higher-energy excited states (for example, *X in the illustration) must occur very quickly, and either is intramolecular or involves the nearest neighbors to the excited molecule in condensed phases. The lowest-energy excited states of some metal compounds (*Y in the illustration) can exist for nearly a millisecond. This is long enough for many collisions in condensed phases, and such excited-state species can often react with other molecules present in the medium (bimolecular reactions). The lifetimes of excited molecules are limited by: the probability of chemical reaction; the probability of return to the ground state with the emission of light; and the probability of return to the ground state without light emission, but with the generation of heat energy.

Unimolecular reactions. The initial steps of the simplest excited-state chemical reactions involve only the excited molecule or solvent molecules, and may be classified as: excitation [reaction (1)]; decrease in coordination number [reaction (2)]; increase in coordination number, for S, a solvent species [reaction (3)]; and homolysis, as in oxidation-reduction, in bonding electrons shared equally in the products, with ·X as a free radical [reaction (4)].

$$\begin{bmatrix} L & L & L \\ & | & \\ & M & \\ / & | & \backslash \\ L & L & X^- \end{bmatrix}^{(n-1)+} + h\nu \longrightarrow *[ML_5X]^{(n-1)+} \tag{1}$$

$$*[ML_5X]^{(n-1)+} \longrightarrow ML_5^{n+} + X^- \text{(or L)} \tag{2}$$

$$*[ML_5X]^{(n-1)+} + S \longrightarrow \begin{bmatrix} L & L & L \\ & | & \\ & M-S & \\ / & | & \backslash \\ L & L & X^- \end{bmatrix}^{(n-1)+} \tag{3}$$

$$*[ML_5X]^{(n-1)+} \longrightarrow \begin{bmatrix} L & L & L \\ & | & \\ & M & \\ / & | & \\ L & L & \end{bmatrix}^{(n-1)+} + \cdot X \tag{4}$$

Each of these chemical processes, (2), (3), or (4), gives rise to an unstable, therefore reactive, chemical intermediate. Such intermediates are often useful in the synthesis of new compounds, as in reaction (5). When these intermediates are generated on a surface or in an inert

$$ML_5^{n+} + Y^- \rightarrow [ML_5(Y^-)]^{(n-1)+} \tag{5}$$

matrix, their lifetimes may be enhanced, and they may function as reactive catalytic sites. *See* FREE RADICAL.

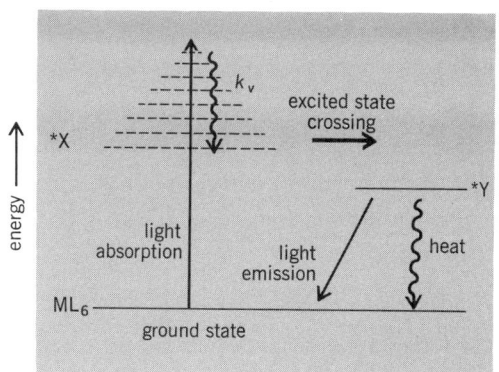

Qualitative electronic energy level scheme for a six-coordinate metal compound (ML_6). Absorption of light by ML_6 results in an excited state *(ML_6) in which electrons are rearranged (either *X or *Y). Any bond length difference between ML_6 and *(ML_6) results in the initial excited state being vibrationally excited (bonds stretched or compressed from the equilibrium position). Vibrational relaxation k_v is rapid in condensed phases.

The changes in the number of ligands coordinated to the central metal [reaction (2) or (3)] can be theoretically related to the relationship between the electronic configuration of the molecular species and the nature of the coordinate covalent bond. The net chemical changes observed depend on the chemistry of the intermediate species [$(ML_5)^{n+}$ or $(ML_5SX)^{(n-1)+}$] as well as the chemistry of the reactive excited state [*$(ML_5X)^{(n-1)+}$].

While excited-state reactions which result in changes in coordination number produce a single reactive species, the intermediate metallofragment, homolytic cleavage of a metal ligand bond results in formation of a pair of very reactive species: a reduced metallofragment and a one-electron oxidized ligand species. These very reactive substances tend to recombine to regenerate the original metal compound, in addition to reacting with themselves or other solution species. The recombination reaction may be very rapid (in less than a nanosecond) when these species are formed in close proximity, or it may be somewhat slower (of the order of 10^{-6} s) if the species manage to separate a few molecular diameters. In order for a simple homolytic process to be observed, the minimum energy difference between the reactive excited state and the ground state must exceed the M^{n+}—(X^-) bond energy. However, there are some systems in which the solvent can assist the homolytic process by means of a concerted displacement of the departing radical fragment. The net quantum yields for photohomolytic processes are reasonably large (that is, greater than 0.1) only when the recombination reactions are slow compared to other reactions of the reactive fragments. Homolytic reactions tend to dominate the chemistry of the excited states of cobalt(III) complexes.

Homolytic reactions turn out to be very efficient for methylcobalamin and other organocobalt complexes related to vitamin B_{12}. These materials are very highly colored (deep red), and about 40% of the light photons absorbed result in homolytic cleavage of the cobalt-carbon bonds. Various recombination reactions decrease the observed product yields to less than 30%. This photosensitivity results in degradation of these natural complexes, and photoinduced reactions are not important to their enzymatic function. However, the small photonic energy required for cobalt-carbon homolysis (80–200 kilojoules mol^{-1}) is a manifestation of the weakness of this chemical bond. Thermal cleavage of the cobalt-carbon bond has been postulated as a key step in the enzymic functioning of coenzyme-B_{12}.

Inevitably, many of the photoinduced unimolecular processes found for metal compounds cannot be neatly placed in the above categories. Among the most intriguing are processes which involve bond-breaking or bond-making processes on a ligand. For example, irradiations of some rhodium(III) or iridium(III) azide complexes (M^{III}—N_3^-) result in cleavage of the N^-—N_2 bond forming a coordinated nitrene, M^{III}—N^-. Another remarkable example is cleavage of the O_2C—(CO_2^{2-}) bond in some oxalate complexes

$$M\begin{matrix} O-C=O \\ | \\ O-C=O \end{matrix}$$

to form a metal-carbon bonded formate complex

$$M-C\overset{\displaystyle O}{\underset{\displaystyle OH}{\Big\|}}$$

There are a number of photoinduced isomerizations, for example,

$$M-N\overset{\displaystyle O}{\underset{\displaystyle O}{\diagdown}} + h\nu \rightarrow M-O-N-O$$

which might appear to fall outside the categories listed above, but in which the primary photoprocess appears to involve metal-ligand homolysis, followed by very rapid recombination of the fragments before they can be separated by one or more solvent molecules.

Biomolecular excited-state reactions. Reactions between excited states and molecules other than solvent species are very important when the electronically excited molecules are long-lived. Metal compounds are frequently facile oxidation-reduction reagents, and metal-compound excited states can be employed to displace oxidation-reduction equilibria. The important categories of bimolecular excited-state reactions are electron transfer, shown in reaction (6a) or (6b), and quenching by electronic energy transfer, as in reaction (7).

$$*[ML_6^{n+}] + A \rightarrow ML_6^{(n-1)+} + A^+ \quad (6a) \qquad *[ML_6^{n+}] + B \rightarrow ML_6^{(n+1)+} + B^- \quad (6b)$$

$$[ML_6^{n+}]* + C \rightarrow ML_6^{n+} + C* \quad (7)$$

The transfer of an electron from one species to another constitutes a net redistribution of electrical charge. As a consequence, there is necessarily some difference in the solvation of reactants and products; in addition, there will often be differences in the bond lengths of reactants and products. The larger these differences in bond length of solvation, the slower will be the rate of the electron transfer process. Excited-state lifetimes are necessarily short, so that electron transfer reactions must occur rapidly if they are to be of any consequence. Very rapid excited-state electron transfer reactions occur: most often between large molecules (thus minimizing solvation energies); most often between molecules in which the electron transfer does not result in large bond-length changes; and in part because the excitation energy stored in the excited molecule helps make the reactants appreciably less stable than the products. Typical of the metal compounds employed in the study of such reactions are *tris*-bipyridyl complexes, $M(LL)_3^{n+}$, where

$$(LL) = \text{2,2'-bipyridine structure}$$

and $M^{n+} = Ru^{2+}$ or Cr^{3+}. Enough energy is released in the electron transfer process to overcome some of the intrinsic factors limiting the rate. Indeed, typical metal compound excited states can store enough energy that even the recombination reactions $ML_6^{(n-1)+} + A^+ \rightarrow ML_6^{n+} + A$ or $ML_6^{(n+1)+} + B^- \rightarrow ML_6^{n+} + B$, could ideally produce 1–2 V of electrical energy in a battery. Such excited-state electron transfer reactions afford a convenient means of generating very reactive intermediate species (A^+, B^-).

Quenching by electronic energy transfer is a theoretically complex process since it requires simultaneous relaxation of a donor electron and excitation of an electron in the acceptor molecule. Although such reactions tend to be degradative in most systems, they can be employed to generate chemically interesting acceptor molecule excited states which cannot be easily populated by direct light absorption.

Attempts to utilize metal compounds to mediate the transformation of light energy from the Sun into a useful chemical fuel have focused largely on the cleavage of water, shown in reaction (8). Several metal-compound excited states store enough energy to promote this reaction

$$H_2O + ML_6^{n+} + h\nu \rightarrow H_2 + 1/2 O_2 + ML_6^{n+} \quad (8)$$

as it is written. However, detailed consideration of reaction (8) reveals that it must involve at least two water molecules and four electrons per product molecule formed, and the energy requirements for this reaction cannot be simply equated to the energy available from single-electron transfer processes of molecular excited states, as in reactions (6a) and (6b). On the other hand, reactive intermediates generated in electron transfer processes might be capable of transforming more than one equivalent of electrons (for example, by reduction of H^+ to a coordinated hydride, H^-). Appropriate intermediates might be homogeneous (for example, low-valent complexes in solution) or heterogeneous (for example, colloidal metals or metal oxides). By mounting the absorbing metal complex on the surface of an electrode, it might be possible to effect a catalytic cycle for photochemical fuel generation (the electrode could be used to replace any electrons transferred to or from the electrolytic medium). Related applications may involve photocurrents induced at semiconductor electrodes. SEE CHEMICAL DYNAMICS; LASER PHOTOCHEMISTRY; PHOTOCHEMISTRY.

Bibliography. A. W. Adamson and P. D. Fleischauer (eds.), *Concepts of Inorganic Photochemistry*, 1975, reprint 1984; M. S. Wrighton (ed.), *Inorganic and Organometallic Photochemistry*, 1978.

LASER PHOTOCHEMISTRY
DAVID J. NESBITT

The branch of physical chemistry in which chemical reactions are induced, altered, or monitored by laser light. Lasers have had an immense impact on the field of photochemistry by providing scientists with an extremely intense, nearly monochromatic source of light. There are lasers that extend from wavelengths of less than 110 nanometers (vacuum ultraviolet) on out to more than 100,000 nm (far-infrared); for comparison, the entire visible spectrum extends from only 400 nm (violet) to 700 nm (red). **Figure 1** displays some of the more common laser systems and their operating wavelengths.

Advantages of lasers. The main advantages of a laser over a conventional light source (such as discharge or arc lamps) for study of photochemistry are threefold. First, laser light can be collimated into a beam with a very small divergence angle, routinely less than 1/100 of a degree; over the length of a football field the beam diameter grows by less than 0.8 in. (2 cm). The high degree of collimation of laser light permits very efficient illumination far from the source.

Second, laser light is exceptionally pure in color. The spectral purity of light is measured in reciprocal centimeters or wave numbers, cm^{-1}, defined as the frequency width divided by the speed of light, $\Delta v/c$. While the full visible spectrum is 10,000 cm^{-1} wide, a typical laser may be

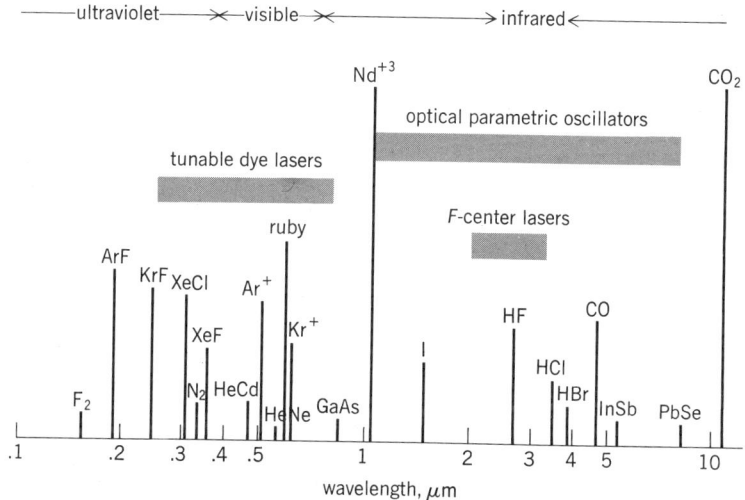

Fig. 1. Laser systems and their operating wavelengths.

as narrow as a few cm^{-1}, and with care, frequency widths as small as several trillionths of a wave number have been obtained. This is of central importance in photochemistry, since molecules may absorb over only an extremely small range of frequencies (for example, of the order of 10^{-2} cm^{-1} in the near infrared); only a miniscule fraction of the photons from a spectrally broad light source is actually utilized. Additionally, the spectral purity of a laser permits specific excitation of one out of thousands of closely spaced absorption features, a property which makes state-selected chemistry and isotope separation possible. This near monochromaticity can be disadvantageous as well, if the available laser frequencies do not overlap with absorption features in the molecule of interest. This difficulty has been largely solved for the visible, where organic dye lasers permit continuous tuning over a significant portion of spectrum. Nonlinear, pulsed laser techniques, such as frequency doubling, metal-vapor four-wave mixing, stimulated Raman scattering, and optical parametric oscillation, have extended the accessible frequencies to include the vacuum ultraviolet and near-infrared region of the spectrum. F-center, difference-frequency mixing, and diode laser techniques can generate continuous-wave radiation over a large portion of the near infrared.

Third, lasers are generally more powerful than conventional light sources. A continuous-wave argon ion laser will produce 10 W at 514.5 nm. Pulsed lasers, which compress the light energy into very short time periods (10^{-6} to 10^{-12} s), can generate correspondingly higher peak powers, typically from 10^3 to 10^9 W.

All in all, the laser provides the photochemist with a source of light between 10 and 20 orders of magnitude more spectrally bright than previously available.

Laser-initiated chemistry. One area of laser photochemistry is the study of single-collision chemical reaction dynamics. Ideally, the scientist would like to generate reagent molecules in a specific state and monitor the progress of the reaction as the product states are formed. This state-to-state picture of reaction dynamics requires extremely sensitive and selective species generation and detection, a task that typically demands the intensity and spectral purity of a laser light source. Additionally, the reagent state preparation must occur on a time scale significantly shorter than that of the reaction studies; fast pulsed lasers are therefore very convenient. Selective reagent generation may proceed by direct photolysis to form highly reactive radicals as in the laser-initiated chain reactions (1).

$$Cl_2 \xrightarrow[300 \text{ nm}]{\text{laser}} 2Cl^\cdot \quad (1a) \qquad Cl^\cdot + H_2 \longrightarrow HCl + H^\cdot \quad (1b) \qquad H^\cdot + Cl_2 \longrightarrow HCl + Cl^\cdot \quad (1c)$$

Since the chain reaction continues to propagate, a single laser photon in such a system can trigger the generation of thousands of product molecules.

Alternatively, the photons need not dissociate the molecule, but rather provide it with specific internal excitation, as in reactions (2), where the asterisk represents vibrational excitation

$$HBr \xrightarrow[3.46\mu m]{\text{laser}} HBr^* \quad (2a) \qquad HBr^* + I \longrightarrow HI + Br \quad (2b)$$

of the molecule. Without this vibrational excitation, the reaction is endothermic and proceeds extremely slowly at room temperature. The extra vibrational energy has been observed to accelerate the reaction by factors of 10^9. In this fashion, lasers can selectively pump energy into a chemical system; monitoring the subsequent reaction provides highly detailed information on the dynamics of the reactive collisions. This is in sharp contrast with early temperature-dependence studies of reactions, which do not isolate the specific effects of vibrational versus translational excitation of reagents, for example.

Provided that the reaction is exothermic, analyses of the final states can often be obtained from the light emitted by the excited products, or chemiluminescence. For typical reaction exothermicities, however, the product chemiluminescence is in the infrared, where photon detection is not very sensitive. An often-used alternative technique for molecules that absorb in the visible or near ultraviolet is laser-induced fluorescence. Here a separate laser illuminates the reaction zone and is tuned over the wavelength region where the products absorb. Molecules in different states absorb light at slightly different frequencies and can then emit via a process known as fluorescence. The fluorescence is typically in the visible, where very sensitive photon detectors can be used. Its intensity provides a measure of the concentrations of product molecules in a particular final state. This technique is limited to molecules that exhibit discrete absorption in the range of a tunable laser. SEE CHEMILUMINESCENCE.

A natural and practical application of this research is the use of lasers to drive specific chemical reactions. Successful examples of laser-induced chemistry are readily found and represent potential directions for industrial use. Carbon dioxide laser irradiation of diborane (B_2H_6) produces icosaborane ($B_{20}H_{16}$) in high yield, whereas conventional pyrolysis generates a mixture of products in which $B_{20}H_{16}$ is not found. Lasers have been used to generate catalysts in a gas-phase mixture, which in turn accelerate a selected chemical reaction. For example, photolytic ejection of CO can serve to catalyze the hydrogenation process [reactions (3)]. In some chemical

$$[Cr(CO)_6 + h\nu \rightarrow Cr(CO)_5 + CO]$$

$$\text{(cyclododecatriene)} + 2H_2 \longrightarrow (CH_2)_{10} \begin{array}{c} CH \\ \| \\ CH \end{array} \tag{3}$$

syntheses, the laser can be used to convert unwanted products into a desired product, as in the photoisomerization of cis-1–4-hexadiene [reaction (4)]; or the laser can be used to purify a final

$$\text{cis-1,4-hexadiene} \xrightarrow[10.6 \ \mu m]{laser} \text{trans isomer} \tag{4}$$

product by selective photodegradation of trace impurities. An interesting pharmaceutical application of lasers is in the key, light-induced reaction in vitamin D_3 synthesis, where the KrF laser may prove to be an economical alternative to the mercury-arc light source.

A particularly promising direction for laser-induced chemistry utilizes intense vibrational excitation by powerful infrared lasers to switch on a reaction pathway that normally would not proceed at all. For example, boron trichloride ordinarily reacts with benzene to form $C_6H_5BCl_2$ only at temperatures in excess of 600°C (1100°F) even in the presence of a palladium catalyst. On the other hand, a brief exposure of the same reagents to pulsed CO_2 laser light at 10.6 micrometers yields this product at room temperature without a catalyst. Bromination of pentafluorobenzene has also been induced by CO_2 laser light [reaction (5)]. On exposure to a 50-W continuous-wave CO_2

$$C_6HF_5 + Br_2 \xrightarrow[10.6 \ \mu m]{laser} C_6BrF_5 + HBr \tag{5}$$

laser, the reaction is carried out to 50% completion in several minutes, yielding no side products. The mechanism in both of these examples has been shown to involve highly vibrationally excited intermediates.

The potential large-scale industrial market for laser-initiated chemistry is limited by the significant cost of generating laser photons. A mole (6.023×10^{23}) of photons from a CO_2 laser costs considerably less than one from a visible Ar^+ laser. Consequently, only very efficient, and preferably nonstoichiometric (that is, catalytic), laser-assisted photochemical systems are considered to be economically feasible.

Picosecond photochemistry. Picosecond studies involve the use of state-of-the-art, ultrashort (10^{-12} s) laser pulses to probe the reaction dynamics of very fast processes. One picosecond is of the same order as the time between collisions in a liquid or the time required for a full rotation of a small molecule. These picosecond pulses are formed by mixing together several pulses of very slightly different frequencies, a method known as mode locking. For the majority of the pulse duration, the individual frequencies are out of phase, and destructively interfere with each other. But for a very short time, all the frequencies are in phase, resulting in a brief but intense burst of light. This pulse is then split into a pump and a probe pulse. The intense pump pulse passes directly through the sample, while the weaker probe pulse is first expanded into many subpulses and then delayed optically by a series of glass plates before it is passed through the sample (**Fig. 2**). Since light travels 50% more slowly in glass than in air, these pulses emerge

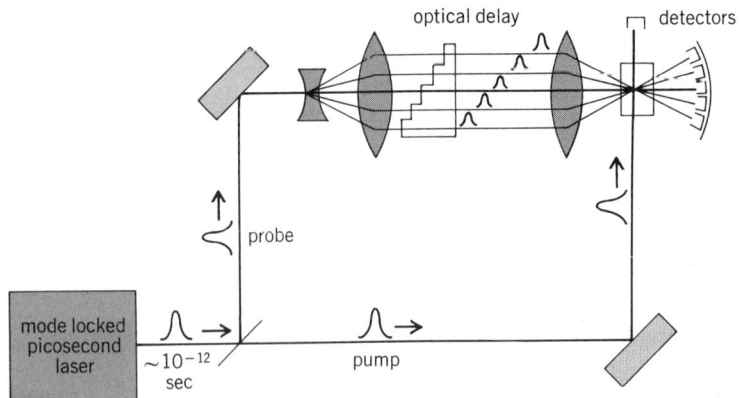

Fig. 2. Laser system for picosecond photochemistry.

staggered in time. As a result, the sample is illuminated by the initial pump pulse followed by a staccato burst of probe pulses. By measurement of the absorption of the delayed probe pulses, the fast chemical processes initiated by the pump pulse can be monitored on an unprecedented time scale. Some early applications of this technique investigated the recombination dynamics of photo-dissociated iodine in solution. Scientists have also been able to study the rapid light-induced structural changes in the large biological molecules responsible for vision.

Photon dissociation. A long-standing goal of the laser photochemist has been to manipulate the pathway of a photochemical process. One vigorously pursued approach to this problem involves using extremely intense lasers to induce dissociation of a specific bond in a molecule—in effect, molecular photosurgery. The electrons of a molecule in the ground state arrange themselves to shield the positively charged nuclei from one another, which tends to hold the molecular fragments together in a chemical bond. In many electronically excited states, the electrons cannot shield the nuclei effectively; a molecule in one of these states rapidly dissociates.

Single photon. This type of dissociation is simply the excitation of the electrons in a bond from their stable ground state to a repulsive upper state. In principle, therefore, selective bond rupture requires determining the ground- and upper-state energy separation and illuminating the molecule with a photon of the appropriate wavelength (typically in the ultraviolet; **Fig. 3**). In practice, however, the dissociation occurs so rapidly (less than a picosecond) that the molecule is somewhat insensitive to the precise photon energy. As a result, several different excited states can result from the absorption of the same photon, which in turn can cause the molecule to dissociate into many different undesired products.

Multiphoton. An alternative approach is to supply the requisite energy via a large number of less energetic photons. In the ground electronic state there are a number of individual energy levels corresponding to different amounts of excitation of a particular vibrational mode of the molecule. The energy spacing of these levels is approximately constant, typically between 1 and 10 kcal/mole (4 and 40 kilojoules/mole). An appropriate-wavelength infrared laser can therefore excite molecules sequentially up this vibrational manifold by using the individual states as one would use rungs on a ladder. Eventually the molecule becomes so vibrationally excited that it dissociates. The advantage of this multiphoton approach is that the individual vibrational states are relatively long-lived, and thus absorb over a very narrow range of frequencies. This permits very selective excitation of the molecules and thus greater control of the products formed.

The practicality of this technique is limited by the difficulty of obtaining a laser that generates the correct infrared frequency. Additionally, a typical bond strength of 50–100 kcal/mole (200–400 kJ/mole) requires consecutive absorption of over 20 infrared photons, which demands extremely high laser powers. This latter difficulty has been overcome with pulsed CO_2 lasers, which can generate up to 3×10^{11} W in a 10^{-9}s pulse. (Over the duration of these short pulses, such a laser generates more power than Niagara Falls.) The former difficulty is still largely un-

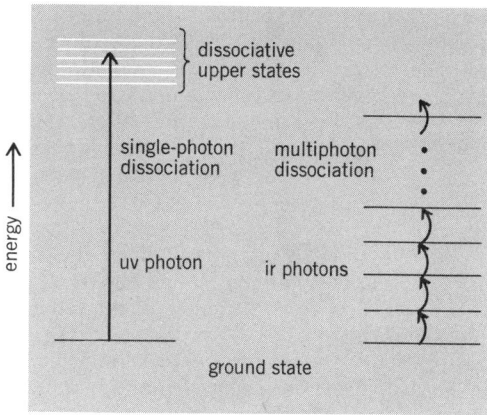

Fig. 3. Energy levels of excited molecules in single-photon and multiphoton dissociation.

solved, and thus experimentation has been limited mainly to molecular systems with infrared absorptions that happen to coincide with frequencies of the CO_2 laser.

In principle, infrared multiphoton dissociation could be an extremely selective, photosurgical tool, provided that the vibrational energy remains in the desired bond on the time scale of the excitation. In actuality, the initially localized vibrational energy probably extends throughout the molecule on a picosecond time scale. The multiphoton results most likely do not represent selective dissociation, but rather a collision-free molecular heating that ultimately leads to fragmentation.

Laser isotope separation. An early hope for large-scale practical application of laser photochemistry was to achieve the separation of isotopes. In typical chemical reactions, which involve only the electrons surrounding the nucleus, the isotopes of a given element behave nearly identically. (Fundamentally, this is why they are difficult to separate using standard chemical techniques.) The nuclear chemistry of the isotopes, however, can be vastly different. For example, ^{235}U is a fissionable isotope used in atomic reactors, whereas ^{238}U, with three extra neutrons, not only does not sustain the fission process but actually inhibits it. The elements occur naturally as a mixture of isotopes (the $^{235}U/^{238}U$ natural abundance ratio is less than 1%), and the difficulty and expense of any purification scheme is considerable. The conventional UF_6 gas diffusion separation method, which relies on fractions of a percent difference in diffusion rates of the two isotopic species, consumes 10^5 kWh per mole of ^{235}U recovered. (Compare this, however, with the 5×10^6 kWh/mole generated by the product in a nuclear reactor.) Consequently, an immense effort has been invested in developing cost-efficient techniques for laser-induced separation of isotopes.

In principle, this is a straightforward process. Because of the small differences in mass, compounds of different isotopes of the same element exhibit absorption spectra that are slightly shifted with respect to one another. If a laser can pump energy preferentially into one isotope and thereby induce a permanent change, than an efficient separation is feasible. One separation technique involves crossing a beam of molecules with an intense laser tuned to an absorption feature in one of the desired isotopes. Photons possess a small but finite momentum; molecules that absorb light from the laser will be deflected away from the rest of the beam. The magnitude of the deflection is very small unless multiple absorptions occur, significantly limiting this technique. Another method uses lasers to accelerate a chemical reaction by selective vibrational excitation of a particular isotope. This desired isotope is predominant among the reaction products, and can be readily separated from the original natural-abundance mixture. By far the most promising isotope separation scheme uses intense pulsed lasers for selective multiphoton dissociation. With this technique, the laser preferentially dissociates molecules of a particular isotope; the isotopically enriched radical fragments then react with added scavengers, and can be removed. Application of this technique has achieved spectacular isotopic separations for a number of elements, most notably sulfur, boron, nitrogen, and hydrogen.

For example, photodissociation of SF_6 by pulsed CO_2 laser excitation of the v_3 mode has produced dissociation yields in the beam near unity with isotopic enrichments of the $^{34}S/^{32}S$ ratio by a factor of 10. A far more dramatic demonstration is observed in the hydrogen/deuterium isotope separation in trifluoromethane, CF_3H (or CF_3D). In this system, a pulsed CO_2 laser selectively excites the v_5 mode of the deuterium-bearing molecules up to dissociation [reactions (6) and (7)]. The DF produce is then extracted as the source of enriched deuterium. The isotopic selectiv-

$$CF_3D \xrightarrow[10.2\ \mu m]{laser} :CF_2 + DF \quad (6) \qquad CF_3H \xrightarrow[10.2\ \mu m]{laser} \text{little or no photodissociation} \quad (7)$$

ity (that is, the D/H ration of products divided by the natural abundance ratio) has been shown to be 11,000 under optimum conditions, with essentially 100% dissociation of the sample per laser pulse, and a photon efficiency on the order of 10%. Estimated energy consumption per mole of deuterium generated is less then 28 kWh, which compares rather favorably with a cost of 30 kWh/mole by the conventional gas diffusion process.

Unfortunately, separation of the uranium isotopes has proven substantially more elusive. A major difficulty has been finding an intense laser source that emits at a wavelength absorbed by the few gaseous compounds of uranium. Much of the research thrust has been with uranium hexafluoride, UF_6, which absorbs predominantly in the 16-μm region, but has much weaker overtone and combination bonds (sums of the fundamental absorption frequencies) which can be accessed by a CO_2 laser. By stimulated rotational Raman scattering of CO_2 laser frequencies in a high-pressure H_2 cell, a powerful 16-μm laser source was developed for use in UF_6 isotope separation. The practical prospects of isotope-selective, multiphoton ionization of uranium metal vapors appear to be extremely promising and have been the focus of intense research and engineering efforts. SEE CHEMICAL DYNAMICS; PHOTOCHEMISTRY.

Bibliography. *Laser Photochemistry and Diagnostics: Recent Advances and Future Prospects*, NSF/DOE Seminar Report, June 4–5, 1979; B. A. Lengyel, *Lasers*, 1971; C. B. Moore (ed.), *Chemical and Biochemical Applications of Lasers*, vols. 1–5, 1974–1980.

10 SPECIALIZED FIELDS OF STUDY

Quantum chemistry	360
Computational chemistry	365
High-pressure chemistry	368
High-temperature chemistry	372
Magnetochemistry	375
Radiation chemistry	377
Solid-state chemistry	381

QUANTUM CHEMISTRY
HENRY F. SCHAEFER III

A branch of chemistry concerned with the application of quantum mechanics to chemical problems. More specifically, it is concerned with the electronic structure of molecules. Since 1960 the ease with which the quantum chemist may obtain reliable approximate solutions to the nonrelativistic Schrödinger equation has improved by at least six orders of magnitude. This article presents a brief review of developments in ab initio molecular electronic structure theory since 1960. The term ab initio implies that no approximations have been made in the one- and two-electron integrals, shown in Eqs. (1) and (2), arising from the ordinary nonrelativistic hamiltonian, Eq. (3).

$$I(i|j) = \int \phi_i^*(1) \left\{ \frac{-\nabla_1^2}{2} - \sum_A \frac{Z_A}{r_{1A}} \right\} \phi_j(1) \, dv(1) \quad (1)$$

$$(ij|kl) = \int \phi_i^*(1)\phi_j^*(2) \frac{1}{r_{12}} \phi_k(1)\phi_l(2) \, dv(1) \, dv(2) \quad (2)$$

$$H = \sum_i \left\{ \frac{-\nabla_i^2}{2} - \sum_A \frac{Z_A}{r_{iA}} \right\} + \sum_i \sum_{j>i} \frac{1}{r_{ij}} \quad (3)$$

In contrast, semiempirical methods resort to various approximate schemes, especially in evaluating the two-electron integrals $(ij|kl)$. The present discussion is restricted to the method which dominates the field of quantum chemistry, namely, the Hartree-Fock or self-consistent-field approximation.

Definitions. For closed-shell molecules, the form of the Hartree-Fock wave function is given by Eq. (4), in which $A(n)$, the antisymmetrizer for n electrons, has the effect of making a

$$\psi_{HF} = A(n)\phi_1(1)\phi_2(2) \ldots \phi_n(n) \quad (4)$$

Slater determinant out of the orbital product on which it operates. The ϕ's are spin orbitals, products of a spatial orbital χ and a one-electron spin function α or β. For any given molecular system, there are an infinite number of wave functions of form (4), but the Hartree-Fock wave function is the one for which the orbitals ϕ have been varied to yield the lowest possible energy [Eq. (5)].

$$E = \int \psi_{HF}^* H \psi_{HF} d\tau \quad (5)$$

The resulting Hartree-Fock equations are relatively tractable due to the simple form of the energy E for single determinant wave functions [Eq. (6)].

$$E_{HF} = \sum_i I(i|i) + \sum_i \sum_{j>i} [(ij|ij) - (ij|ji)] \quad (6)$$

To make this discussion more concrete, it should be noted that for singlet methylene (the CH_2 molecule), the Hartree-Fock wave function is of the form given in Eq. (7).

$$\psi_{HF} = A(8) \, 1a_1\alpha(1) \, 1a_1\beta(2) \, 2a_1\alpha(3) \, 2a_1\beta(4) \, 1b_2\alpha(5) \, 1b_2\beta(6) \, 3a_1\alpha(7) \, 3a_1\beta(8) \quad (7)$$

The same energy expression, Eq. (6), is also applicable to any open-shell system for which the open-shell electrons all have parallel spins. This follows from the fact that such Hartree-Fock wave functions can always be expressed as a single Slater determinant. A simple example is triplet methylene, shown in Eq. (8), for which the outer two $3a_1$ and $1b_1$ orbitals have parallel spins. For clarity it is often helpful to abbreviate Eq. (8) as Eq. (9). Although solution of the

$$\psi_{HF} = A(8) \, 1a_1\alpha(1) \, 1a_1\beta(2) \, 2a_1\alpha(3) \, 2a_1\beta(4) \, 1b_2\alpha(5) \, 1b_2\beta(6) \, 3a_1\alpha(7) \, 1b_1\alpha(8) \quad (8)$$

$$\psi_{HF} = 1a_1^2 2a_1^2 1b_2^2 3a_1\alpha 1b_1\alpha \quad (9)$$

Hartree-Fock equations for an open-shell system such as triplet methylene is more difficult than for the analogous closed-shell system, Eq. (7), the procedures are well established.

In fact, methods are available for the solution of the Hartree-Fock equations for any system for which the energy expression involves only coulomb and exchange integrals, Eqs. (10) and (11).

$$J_{ij} = (ij|ij) \quad (10) \qquad K_{ij} = (ij|ji) \quad (11)$$

Open-shell singlets are a class of systems that can be treated in this way, and one such example is the first excited singlet state (of $^1B_1$ symmetry) of methylene, Eq. (12). In addition, these same

$$\psi_{HF} = \frac{1}{\sqrt{2}} \; 1a_1^2 \; 2a_1^2 \; 1b_2^2 \; 3a_1\alpha \; 1b_1\beta - \frac{1}{\sqrt{2}} \; 1a_1^2 \; 2a_1^2 \; 1b_2^2 \; 3a_1\beta \; 1b_1\alpha \quad (12)$$

generalized Hartree-Fock procedures can be used for certain classes of multiconfiguration Hartree-Fock wave functions.

Basis sets. To solve the Hartree-Fock equations exactly, either the orbitals ϕ must be expanded in a complete set of analytic basis functions or strictly numerical (that is, tabulated) orbitals must be obtained. The former approach is impossible from a practical point of view for systems with more than two electrons, and the latter has been accomplished only for atoms and for a few diatomic molecules. Therefore the exact solution of the Hartree-Fock equations is abandoned for polyatomic molecules. Instead an incomplete (but reasonable) set of analytic basis functions is adopted and solved for the best variational [that is, lowest energy given by Eq. (5)] wave function of form (4). Such a wave function is referred to as of self-consistent-field (SCF) quality. For very large basis sets, then, it is reasonable to refer to the resulting SCF wave function as near-Hartree-Fock.

For large chemical systems, only minimum basis sets (MBS) can be used in ab initio theoretical studies. The term "large" includes molecular systems with 100 or more electrons. A large molecule treated by MBS-SCF methods is the carbazole-trinitrofluorenone complex, $C_{25}N_4O_7H_{14}$, with 232 electrons. A minimum basis set includes one function for each orbital occupied in the ground state of each atom included in the molecule. For the first row atoms B, C, N, O, and F, this means that a minimum basis set includes $1s$, $2s$, $2p_x$, $2p_y$, and $2p_z$ functions.

Traditionally, minimum basis sets have been composed of Slater functions, such as those seen in **Table 1** for the carbon atom. However, experience has shown that the evaluation of the molecular integrals, Eq. (2), arising when Slater functions are employed is extremely time-consuming. Therefore each Slater function in a minimum basis set is typically replaced by a linear combination of three or four gaussian functions. The resulting chemical predictions obtained with such STO-3G (Slater-type orbital–three gaussian functions) or STO-4G basis sets are usually indistinguishable from the corresponding Slater function results.

Minimum basis sets are inadequate for certain types of chemical predictions. Therefore a basis twice as large, and appropriately designated double zeta (DZ), is often used in theoretical studies. Here, however, it is not as fruitful to expand each Slater function as a linear combination of gaussians. Instead gaussian functions $x^p y^q z^r e^{-\alpha r^2}$ are used directly in atomic self-consistent-field calculations and then contracted according to the atomic results. Perhaps the most widely used contracted gaussian double-zeta basis sets are those of T. H. Dunning. His basis has $9s$ and $5p$ original (or primitive) gaussian functions, and is contracted to $4s$ and $2p$. Thus the basis may be designated C(9s5p/4s2p).

Table 1. Minimum basis set of Slater functions for the carbon atom

Label	Analytic form	Exponent ζ*
$1s$	$(\zeta^3/\pi)^{1/2} \exp(-\zeta r)$	5.673
$2s$	$(\zeta^5/3\pi)^{1/2} r \exp(-\zeta r)$	1.608
$2p_x$	$(\zeta^5/\pi)^{1/2} x \exp(-\zeta r)$	1.568
$2p_y$	$(\zeta^5/\pi)^{1/2} y \exp(-\zeta r)$	1.568
$2p_z$	$(\zeta^5/\pi)^{1/2} z \exp(-\zeta r)$	1.568

*The orbital exponents ζ are optimum for the ^3P ground state of the carbon atom.

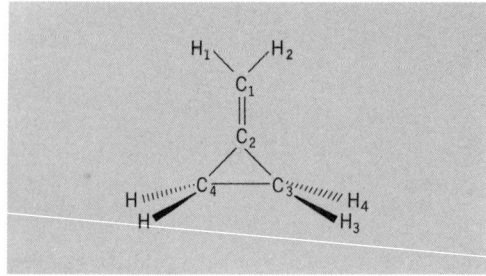

Fig. 1. Methylenecyclopropane structure.

Just as the double-zeta basis logically follows the minimum set, the logical extension of the double-zeta set involves the addition of polarization functions. Polarization functions are of higher orbital angular momentum than the functions occupied in the atomic self-consistent-field wave function. That is, for carbon, d, f, g, . . . functions will be polarization functions. Fortunately, d functions are far more important than f, f functions are far more important than g, and so on. For most chemical applications a double-zeta plus polarization (DZ + P) basis including a single set of five d functions ($d_{x^2-y^2}$, d_{z^2}, d_{xy}, d_{xz}, d_{yz}) will be quite adequate for first-row atoms.

Structural predictions. Ab initio theoretical methods have had the greatest impact on chemistry in the area of structural predictions. A good illustration of this is the methylene radical for which S. F. Boys reported the first ab initio study in 1960. Boys predicted the structure of triplet methylene to be r_e (C—H) = 0.112 nm, Θ_e(HCH) = 129°. Unfortunately, however, the work of Boys was largely ignored due to the spectroscopic conclusion of G. Herzberg that the lowest triplet state was linear. Herzberg's conclusion was greatly strengthened by a very influential semiempirical study of H. C. Longuet-Higgens, who concurred that the ground state of CH_2 was linear.

It was not until 1970 that a definitive theoretical prediction of the nonlinearity of $^3B_1$ CH_2 appeared. The prediction was swiftly verified by independent electron spin resonance experiments. For many chemists, the structure of triplet methylene was the first genuine example of the usefulness of ab initio theoretical chemistry.

Turning from the specific to the more general, the most encouraging aspect of ab initio geometry predictions is their perhaps surprising reliability. Essentially all molecular structures appear to be reliably predicted at the Hartree-Fock level of theory. Even more encouraging, many structures are accurately reproduced by using only minimum-basis-set self-consistent-field methods. This is especially true for hydrocarbons. A fairly typical example is methylenecyclopropane (**Fig. 1**), with its minimum-basis-set self-consistent-field structure compared with experiment in **Table 2**. Carbon-carbon bond distances differ typically by 0.002 nanometer from experiment, and angles are rarely in error by more than a few degrees. Even for severely strained molecules such

Table 2. Minimum-basis-set self-consistent-field geometry prediction compared with experiment for methylenecyclopropane

Parameter*	Theory	Experiment
$r(C_1=C_2)$	0.1298 nm	0.1332 nm
$r(C_2—C_3)$	0.1474 nm	0.1457 nm
$r(C_3—C_4)$	0.1522 nm	0.1542 nm
$r(C_1—H_1)$	0.1083 nm	0.1088 nm
$r(C_3—H_3)$	0.1083 nm	0.109 nm
$\theta(H_1C_1H_2)$	116.0°	114.3°
$\theta(H_3C_3H_4)$	113.6°	113.5°
$\theta(H_{34}C_3C_4)$	149.4°	150.8°

*Here r represents the carbon-carbon bond distance; θ represents the bond angle in degrees of H-C-H bonds; the numbers on C and H correspond to the numbered atoms in Fig. 1.

as bicyclo[1.1.0]-butane, very reasonable agreement with the experimental structure is obtained. It is noteworthy that experimental geometries are available for only half of the C_4 hydrocarbons studied. Thus, for many purposes, theory may be considered complementary to experiment in the area of structure prediction.

For molecules including atoms in addition to C and H, minimum-basis-set self-consistent-field results are sometimes less reliable. For example, the F_2N_2 molecule has minimum-basis-set self-consistent-field bond distances r_e (N—F) = 0.1384 nm, r_e (N≡N) = 0.1214 nm, which are respectively 0.0107 nm longer and 0.0169 nm shorter than the experimental values. Fortunately, vastly improved agreement with experiment results is obtained when a larger basis set is adopted for F_2N_2.

In general, double-zeta self-consistent-field structure predictions are considerably more reliable than those based on minimum basis sets. A noteworthy exception is the water molecule, for which minimum-basis-set self-consistent-field yields a bond angle of 100.0° and double-zeta self-consistent-field predicts 112.6°, compared to the well-known experimental value of 104.5°. More typical are the HF and F_2 molecules, for which the minimum-basis-set, double-zeta, and experimental bond distances are 0.0956, 0.0922, and 0.0917 nm (HF); and 0.1315, 0.1400, and 0.1417 nm (F_2). In fact it can be argued that if the calculations will not be made beyond the Hartree-Fock (single-configuration) approximation, double zeta self-consistent-field is often a reasonable stopping point.

Transition states are typically more sensitive to basis set than equilibrium geometries. This is true because potential energy surfaces are often rather flat in the vicinity of a saddle point (transition state). An example is the carboxime–cyanic acid rearrangement given in reaction (13).

$$\text{HONC} \rightarrow \text{HOCN} \qquad (13)$$

Minimum-basis-set self-consistent-field and double-zeta self-consistent-field transition state geometries are compared in **Fig. 2**; there it is seen that the minimum-basis-set and double-zeta prediction is presumably the more reliable. It should be noted that for several other transition states, better agreement is found between the two methods. More typical structural variations are ~0.005 nm in internuclear separations and 5° in angles.

As the larger basis sets within the Hartree-Fock formalism are considered, better agreement with experiment is frequently obtained. As implied above, the water molecule bond angle is much improved at the double-zeta plus polarization level, to 106.1°. However, it is often the case that adding polarization functions has only a marginal effect on predicted geometries. A reasonably typical comparison is given by the NH_2F and PH_2F molecules. In this example, the only polarization functions added were sets of d functions on the central N or P atom. The only pronounced improvement with respect to experiment is for the P—F separation in PH_2F, and this is improved by 0.0034 nm when d functions are added to phosphorus. The good agreement with experiment for NH_2F and PH_2F suggests a high degree of reliability for the comparable NHF_2 and PHF_2 predictions, where no experimental structures have been determined.

An interesting comparison of the three most frequently used basis sets is given in **Table 3** for the linear HCNO molecule. The most sensitive geometrical parameter is the N—O bond distance, for which the minimum basis set is 0.0095 nm too long, the double zeta still 0.0056 nm

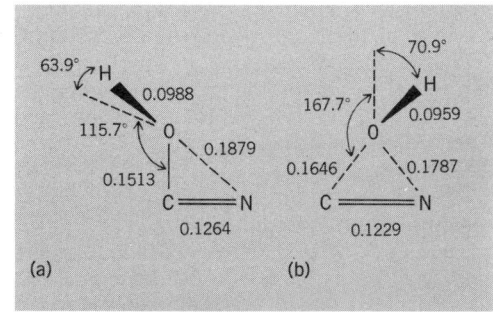

Fig. 2. Comparison of (a) minimum-basis-set and (b) double-zeta self-consistent-field transition-state geometries. Internuclear separations are given in nanometers.

Table 3. Equilibrium geometry of formonitrile oxide, HCNO, from self-consistent-field theory and experiment*

Bond distance	Basis set			Experiment
	MBS	DZ	DZ + P	
r_e (H—C)	0.1065	0.1049	0.1059	0.1027
r_e (C—N)	0.1155	0.1133	0.1129	0.1168
r_e (N—O)	0.1294	0.1255	0.1201	0.1199

*Values are in nanometers.

too long, but the double-zeta plus polarization result is in nearly perfect agreement with experiment. For the CN distance, the minimum-basis-set treatment actually gives the best agreement with experiment. The experimental microwave spectrum is difficult to unravel for a quasilinear molecule such as HCNO, and it has been suggested that the double-zeta plus polarization prediction for the C—H distance may be more reliable than experiment.

Energetic predictions. Among the most chemically important energetic quantities are conformational energy changes, exothermicities or heats of reaction, dissociation energies, and activation energies or barrier heights. In general, only the first of these, and sometimes the second, is reliably predicted at the Hartree-Fock level of theory. In other words, energetic quantities are often sensitive to the effects of electron correlation.

Conformational energy changes are, almost without exception, properly reproduced within the Hartree-Fock formalism. In fact, certain types of barriers, typified by the ethane rotational barrier, are quite satisfactorily predicted at the minimum-basis-set self-consistent-field level of theory. More sensitive problems, such as the ammonia inversion barrier and the rotational barrier of hydrogen peroxide, demand the inclusion of polarization basis functions.

Although Hartree-Fock exothermicities are often unreliable, there is at least one fairly large class of reactions for which consistently good agreement with experiment has been found. Generally speaking, heats of reaction for systems having closed-shell reactants and products are often predicted successfully. More specifically, even better agreement with experiment is found for isodesmic reactions, where the number of bonds of each type is conserved. In fact, reasonable predictions are often made at the minimum-basis-set self-consistent-field level for isodesmic reactions. Further, such information can sometimes be used indirectly (or in conjunction with other thermo-chemical information) to predict quantities that might be very difficult to evaluate by more straightforward ab initio methods.

The dissociation energies of covalent molecules are generally predicted poorly by single-configuration self-consistent-field methods. Certainly the best-known example is the F_2 molecule, for which the molecular Hartree-Fock energy lies about 1 eV above the Hartree-Fock energy of two fluorine atoms. This problem is often mistakenly attributed to the "perverse" nature of the fluorine atom. In fact, the near-Hartree-Fock dissociation energy of N_2 is 5.27 eV, only about half of the experimental value, 9.91 eV. In addition, the near-Hartree-Fock dissociation energy of O_2 is 1.43 eV, only one-third of the experimental value, 5.21 eV. Thus the Hartree-Fock dissociation energies of covalent molecules are consistently much less than experiment.

Another frequent failing of the Hartree-Fock method is in the prediction of the barrier heights or activation energies of chemical reactions. However, it must be noted that there are many classes of reactions for which Hartree-Fock theory does yield meaningful barrier heights. Two well-studied examples are the isomerizations shown in reactions (14) and (15). For the HNC

$$HNC \rightarrow HCN \quad (14) \quad\quad\quad CH_3NC \rightarrow CH_3CN \quad (15)$$

rearrangement, comparison between self-consistent field (40.0 kilocalories or 167.4 kilojoules) and configuration interaction (CI; 36.3 kcal or 151.9 kJ) barriers reveals good qualitative agreement. However, the inclusion of d functions in the C and N basis sets appears to be very important. For example, for the methyl isocyanide rearrangement, the self-consistent-field barrier decreases from

60 to 45 kcal (251 to 188 kJ) when polarization functions are added to a double-zeta basis. The remaining discrepancy with the experimental activation energy of 38 kcal (159 kJ) is about equally due to correlation effects and the fact that the zero-point vibrational energy at the transition state is ~3 kcal (13 kJ) less than for CH_3NC. Hydrogen isocyanide is one of three interstellar molecules (HNC, HCO^+, and HN_2^+) to be identified by ab initio theory prior to their laboratory detection. Generally speaking, unimolecular reactions seem to be treated more reliably by Hartree-Fock methods than bimolecular systems. A second example is the geometrical isomerization of cyclopropane. This system has been studied in considerable detail by L. Salem and coworkers, who totally resolved the structure of the transition state within the full 21-dimensional hypersurface. Although their work involved only a minimum basis set, it goes slightly beyond the self-consistent-field model in that 3×3 configuration interaction was included. The predicted barrier height is 53 kcal (222 kJ), in reasonable agreement with the experimental value, 64 kcal (268 kJ).

For many attractive potential energy surfaces, that is, those having no barrier or activation energy at all, Hartree-Fock methods are frequently reliable. An example that has been carefully documented is reaction (16). One of the most important features of the $H + Li_2$ surface is the fact

$$H + Li_2 \rightarrow LiH + Li \tag{16}$$

that the C_{2v} HLi_2 structure is a chemically bound entity. Self-consistent-field theory suggests that the dissociation energy relative to $LiH + Li$ is 20.2 kcal (84.5 kJ), in excellent agreement with the large-scale configuration interaction result of 22.4 kcal (93.7 kJ).

The Hartree-Fock formalism has very powerful predictive capabilities. However, large classes of chemical problems cannot be reasonably described, and a good deal of discretion is required on the part of the theoretician. Although state-of-the-art quantum chemistry has gone well beyond the Hartree-Fock model, the model is certain to remain the cornerstone of electronic structure theory for many years to come. *See* CHEMICAL BONDING; MOLECULAR ORBITAL THEORY; MOLECULAR STRUCTURE AND SPECTRA.

Bibliography. R. Dandel et al., *Quantum Chemistry*, 1984; I. N. Levine, *Quantum Chemistry*, 3d ed., 1983; P.-O. Lowdin (ed.), *Advances in Quantum Chemistry*, vols. 1–18, 1964–1987; H. F. Schaefer, *Quantum Chemistry*, 1984.

COMPUTATIONAL CHEMISTRY
ZELDA R. WASSERMAN

A branch of theoretical chemistry that uses a digital computer to model systems of chemical interest. In this discipline, the computer itself is the primary instrument of research. The use of computers for analysis of experimental data, and for the storage and display of results obtained with other tools, is distinct from computational chemistry. The latter permits calculation of quantities which can be measured experimentally, such as molecular geometries of ground and excited states, heats of formation, and ionization potentials. Alternatively, quantities not readily accessible by existing experimental techniques, such as geometries of transition states and detailed structure of liquids, may be evaluated.

Because of the increasing power and availability of computers, and the simultaneous development of well-tested and reliable theoretical methods, the use of computational chemistry as an adjunct to experimental research has increased rapidly. Calculations ranging from a few seconds to many hours of computer time can serve as a guide to exclude less favorable reactions or unstable products, or to select several more fruitful procedures from the many possible ones. In addition, modeling of chemical systems with a computer enables the researcher to examine them on a scale of space or time as yet unmeasurable by experimental techniques. This can give insight into a chemical system beyond that provided by experiment. Examples are examination of the dynamics of a chemical reaction or of detailed changes in conformation of a polymer in solution. Examination of the molecular orbitals occupied by the electrons of the molecule can provide insight into chemical bonding and the electronic interactions which determine specific geometric configurations. Thus computational chemistry can yield information which may not be experimentally available. *See* CHEMICAL DYNAMICS.

Computational chemistry may be the application of existing theory and numerical methods to new molecules, or it may be the development of new computational methods. The latter may include incorporating more physics into the mathematical model in order to provide a better theoretical description of the system being studied, for example, inclusion of interactions between individual electrons in molecular orbital calculations. These more complete studies, for "large" molecules, usually need to be performed on a supercomputer, or on a mid-size computer with an array processor. Another approach is to devise simpler methods which will approximate the accuracy of more complex calculations. These include semiempirical methods in which values of hard-to-calculate terms are derived from experiment. Such an approach allows fruitful work with a mid-size or even a desk-top computer, avoiding the need for expensive computer resources.

Before a computational study is begun, the suitable theoretical method must be determined. Some methods are known to yield more accurate results for systems of a certain type, for example, simple organic molecules. The preferred method must be balanced against the computing facilities available and the amount of computer time to be allocated.

Molecular orbital theory. The most active area of computational chemistry involves molecular orbital calculations which yield a description of the electronic distribution in molecules. In this method, a solution is sought for Eq. (1), the Schrödinger equation, where H is the hamil-

$$H\psi = E\psi \qquad (1)$$

tonian operator, ψ is the electronic wave function, and E is the energy of the wave function. The equation is solved self-consistently until there is no change in ψ or E. A molecular orbital is composed of linear combinations of atomic orbitals. These are expressed as Slater functions, Eq. (2), where η is the quantum number, or more often gaussian functions or linear combinations of

$$\eta = r^{n-1}e^{-\alpha r} \qquad (2)$$

gaussians, Eq. (3), and p, q, and s are integers. In Eqs. (2) and (3), $r = \sqrt{x^2 + y^2 + z^2}$ is the

$$\eta = \sum_i c_i x^p y^q z^s e^{-\alpha_i r^2} \qquad (3)$$

distance from the atom and the c's and α's are coefficients and exponents which define the basis function. Gaussian functions are used most often, as the necessary integrals may be evaluated analytically.

Molecular orbital computational methods vary in complexity from extended Hückel theory (which deals only with the valence electrons) to ab initio methods which calculate everything from first principles. In between are a number of semiempirical methods [complete neglect of differential overlap (CNDO), intermediate neglect of differential (INDO), and modified neglect of differential overlap)]. Many of the programs that have been developed have provision for geometry optimization using the gradient of the energy to determine a minimum-energy geometry.

Molecular orbital calculations can take from a few seconds of computer time for extended Hückel calculations to many hours on the largest supercomputers for detailed ab initio calculations. The time for any method usually increases as N^4, where N is the number of atomic orbitals.

An example of molecular orbital techniques applied to novel, possibly speculative, systems can be seen in the results of ab initio computations completed by P. v. R. Schleyer and coworkers. They concluded that the carbon-lithium compounds CLi_5 and CLi_6 may be stable and thus violate the octet rule. The computed C-Li (**Fig. 1**) distances are similar for these compounds and CLi_4. The pattern of electron density indicates that the extra electrons are involved in Li-Li bonding. This group cites experimental results which support the possible existence of these hypervalent species. SEE MOLECULAR ORBITAL THEORY; QUANTUM CHEMISTRY.

Molecular dynamics. Molecular dynamics calculations simulate the behavior of physical systems by integration of Newton's equations of motion. This method has been applied to studies of liquid water, deposition of molecules on surfaces, cluster formation, and conformational changes of alkanes and polymer chains, among others. As the time step of the integration must be small relative to the vibrations of the particles in order for the numerical integration to be stable, the method requires vast amounts of computer time. Because of this, it is usually not practical to include more that several hundred particles. However, the system may be considered infinite because of the imposition of periodic boundary conditions. This means that the box containing

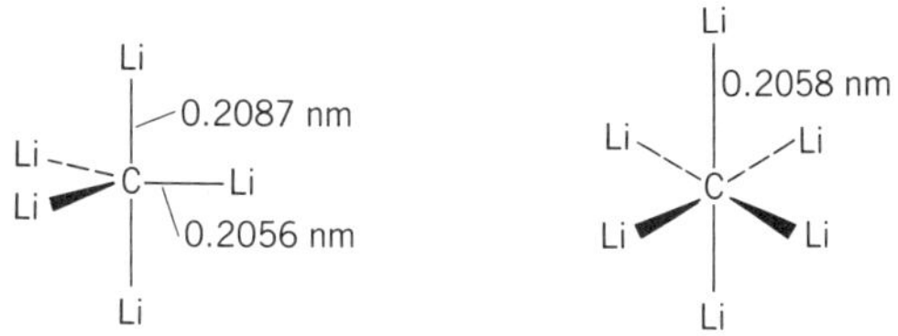

Fig. 1. Computed geometries of CLi_5 and CLi_6. (*After P. v. R. Schleyer et al., CLi_5, CLi_6, and the related effectively hypervalent first-row molecules, $CLi_{5-n}H_n$ and $CLi_{6-n}H_n$, J. Amer. Chem. Soc., 105:5930–5932, 1983*)

the particles is considered to be replicated in all dimensions, in the manner that strips of wallpaper continue a pattern around a room.

To decrease the amount of computer time needed for these simulations, a family of related stochastic (using random forces) methods has been developed. These range from Monte Carlo methods, where each particle movement is determined randomly, to Langevin dynamics methods, which treat forces between nearby particles exactly and approximate those due to remote particles stochastically.

This area of computational chemistry has grown rapidly. Research activity includes application of the method to more complex physical systems, such as freezing, melting, and phase separations, as well as comparisons of the accuracy and costs of the various related methods.

Polymer relaxation processes have been studied by a number of experimental techniques such as nuclear magnetic resonance, dielectric relaxation, ultrasonic attenuation, dynamical light scattering, fluorescence depolarization, and excimer fluorescence. While these studies detect transitions between trans and gauche conformations, they provide little information about the detailed process by which a bond undergoes a transition.

Most of the experiments detect an activation energy of a single barrier height (between trans and gauche states), indicating that transitions occur independently. However, a single transition in a polymer chain would necessitate a wide swing by the rest of the chain, which is unlikely to occur. To investigate the detailed mechanisms of these transitions, E. Helfand and coworkers performed brownian dynamics simulations (solution of the Langevin equations) of polymer conformational transitions. These studies confirmed an activation energy of a single barrier height. But in addition, they indicated a highly enhanced rate of transition of a bond whose second-nearest neighbor has recently undergone a transition. These favorable pairs are separated by a trans central bond. A clockwise transition of a bond on one side would be associated with a counterclockwise rotation on the other (**Fig. 2**). This pair of changes leads to lateral motions of the tails, rather than large swings. Evidence for these pairs of transitions was subsequently detected experimentally.

Molecular mechanics. This method has been in use since the early 1950s, particularly for small organic molecules. The molecular mechanics model expresses the energy of a system as potential functions of the bond lengths, bond angles, dihedral angles, and van der Waals interactions of the atoms in the system. Using the gradient of the energy, the programs adjust the coordinates of the atoms until a configuration of minimum energy is determined. The potential functions, or force fields, contain parameters which are optimized to produce results which agree with a set of experimental measurements. The assumption is that these parameter values will be equally valid for other molecules of similar structure which have not been studied experimentally.

While the existing programs contain parameters which apply to a wide range of molecules, difficulties arise when attempts are made to use them for calculations on novel systems with unusual interactions or geometric structures. Such parameter values may not be valid, or they may not be known. Research has continued to widen the class of molecules for which the method may be expected to yield reliable results.

When the relative stability of chemical structures is under consideration, molecular mechanics is an appealing technique. Calculations take several orders of magnitude less computer time than molecular orbital calculations, and the method provides the only feasible approach for large biological molecules. *See* BOND ANGLE AND DISTANCE; CHEMICAL BOND THEORY; CHEMICAL BONDING.

Interactive computer graphics. This field owes its existence to the rapid development of the capabilities of interactive computer graphic devices. The chemist is essentially working

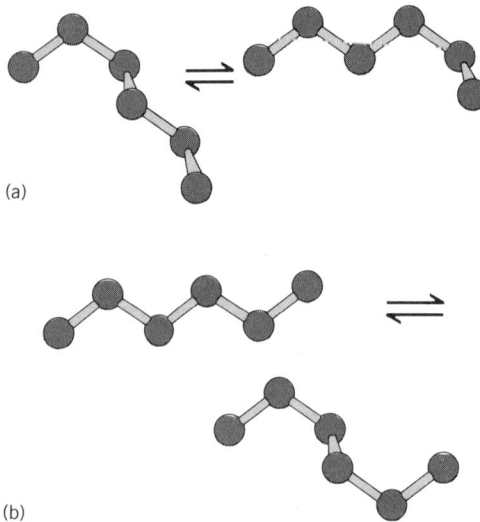

Fig. 2. Two pairs of transitions which result in translation of the polymer tails. (a) Gauche migration gtt ⇌ ttg, transitions on second-nearest neighbors which result in a shift of the gauche bond from one end to the other. (b) Pair gauche production ttt ⇌ g$^+$tg$^-$, transitions on second-nearest neighbors which transform an all trans chain to one with a gauche at either end. (*After E. Helfand, Theory of the kinetics of conformational transitions in polymers, J. Chem. Phys., 54:4651–4661, 1971*)

with a sophisticated set of molecular models which are stored in the computer memory and displayed on a screen. Moving a joystick or turning a dial moves atoms about on the screen as easily as tangible models can be moved physically. Stereoscopic views are produced either by placing left- and right-eye images next to each other on the screen and using a viewer to merge the images, or by alternately blinking left- and right-eye images and looking through a viewer which presents the correct image to each eye.

This technique is most useful when studying interactions between large molecules, and it is often used to fit potential drug molecules into the active sites of biomolecules. The intuition of the chemist takes the place of theoretical calculations in determining geometric configurations, and the most common application is in establishing starting geometries for subsequent molecular mechanics refinement. When faster computers become available, it may be combined with energetic calculations performed in real time.

P. A. Kollman and coworkers modeled the stereoselective hydrolysis of peptides catalyzed by enzymes. They used molecular mechanics calculations to simulate interactions of the enzyme α-chymotrypsin with both the substrate L-*N*-acetyltryptophanamide and the inhibiting D form. Considering both an initial nonbonded complex and the covalently bonded transition state, the calculations confirm the high stereoselectivity of the enzyme. The selectivity is not associated with the initial state but only with the bonded complex, and relaxation of the enzyme is essential for stereoselectivity. *See* STEREOCHEMISTRY.

The Kollman group used interactive graphic techniques to position the molecules for subsequent energy refinement. The efficiency of the graphics procedure allowed the sampling of large areas of the potential energy surfaces needed to locate the energy minima characteristic of the complexes.

Bibliography. U. Burkert and N. Allinger, *Molecular Mechanics*, American Chemical Society, 1982; W. H. Miller et al. (eds.), *Modern Theoretical Chemistry Series*, vols. 1–8, 1976–1977.

HIGH-PRESSURE CHEMISTRY
ROBERT H. WENTORF, JR.

Chemistry at very high pressures, arbitrarily chosen to be above 10,000 bars (10^5 pascals), and mainly concerned with solid and liquid states. A bar is 10^6 dynes/cm^2, or 1.0197 kg/cm^2, or 0.9869 atm, or 10^5 Pa. Multiples of the bar are the kilobar (1 kbar = 10^3 bars = 10^8 Pa) and the megabar

(1 Mbar = 10^6 bars = 10^{11} Pa). At 25°C (77°F) and 10 kbar (10^6 Pa), nearly all ordinary gases are liquid or solid, and only a few liquids are not frozen; thus most high-pressure chemistry involves either higher temperatures, at which chemical reactions can occur at appreciable rates, or studies of internal arrangements in solids.

Figure 1 illustrates the range of high pressures which exist in nature as well as those which can be attained in the laboratory. Three broad ranges of pressures are also shown. In the lowest range, from 1 bar (10^5 Pa) to about 10^5 bars (10^{10} Pa), normal low-pressure chemical behavior prevails, and only minor departures from the usual valence and coordination rules are found. However, as discussed later, many interesting changes in materials can be effected in this pressure range as atoms are forced into new bonding arrangements. In the second range, from 10^5 to 10^9 bars (10^{10} to 10^{14} Pa), the energy added by compression becomes comparable with chemical bond energies, so that outer-shell electronic orbits are distorted and atoms and molecules change in character. A general tendency toward more metallic behavior is observed as the electrons become less strongly fixed to particular atoms, and chemical bonds may be broken. In the third region, upward of about 10^9 bars (10^{14} Pa), the delocalization of electrons is extensive, and the material consists of a mixture of ions and electrons, so that chemical bonds are of little importance. The boundaries on these three pressure ranges are, of course, only approximate, and show some variation according to the temperature and the atoms involved.

Equipment. High-pressure chemical phenomena can be studied in a wide variety of types of equipment, depending on the pressure and temperature range and the object of the study. The highest laboratory pressures, several megabars, are achieved, albeit but for a few microseconds, by the accelerative forces generated by high explosives or high-velocity impacts in the so-called shock-wave techniques. Static pressures which may be exerted for minutes or hours have reached a maximum of about 400 kbar (4×10^{10} Pa) between diamond faces on specially supported anvils of the type shown in **Fig. 2**a, but the specimens are quite thin, with typical diameters of 1 mm, and the temperature range is limited. Larger specimens, 1 cm^3 (0.6 in^3.) or more in size, may be studied up to about 100 kbar (10^{10} Pa) and 2000°C (3600°F) or higher in cylindrical, tetrahedral, or cubical apparatus of the types that are indicated in Fig. 2a, c, and d.

There is no theoretical upper limit to the static pressures which may be achieved, but practical limits are imposed by the magnitude of the forces required, the materials of construction available, and the stress gradients in them. The changes occurring in the compressed material may be monitored in place, for example, by optical, x-ray, or electrical techniques, for in many cases the material reverts to its original state upon release of pressure. Pressure apparatus is conveniently calibrated by observing definite changes in certain substances, for example, resistance changes in bismuth at about 25 kbar (2.5×10^9 Pa) or in lead at 130 kbar (1.3×10^{10} Pa).

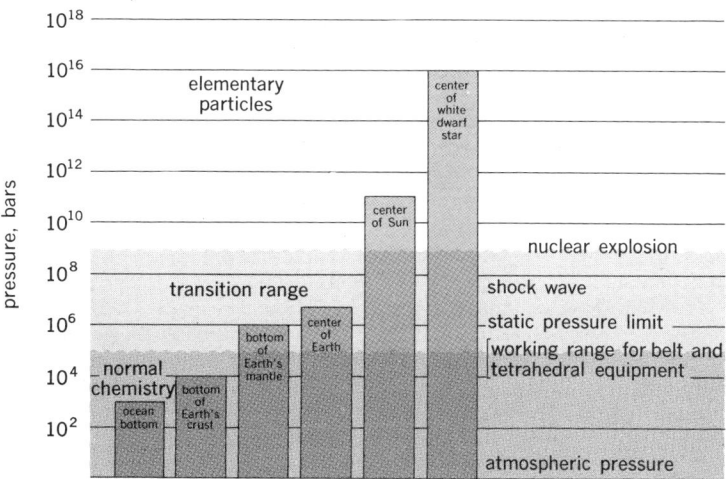

Fig. 1. Range of existing natural high pressures. 1 bar = 10^2 kPa.

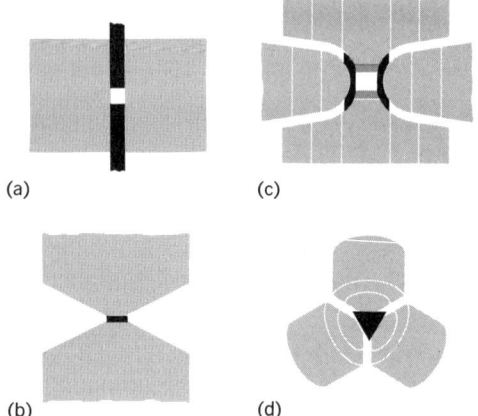

Fig. 2. Basic types of static high-pressure equipment. (a) Bridgman anvil. (b) Piston and cylinder. (c) Belt design. (d) Tetrahedral design.

High-pressure effects. The simplest effect of high pressure is the closer compression of atoms. The noble gases and alkali metals are quite compressible (potassium shrinks to half its original volume under a pressure of 100 kbar or 10^{10} Pa), whereas most oxides and the stronger metals are considerably stiffer. However, at a pressure exceeding about 100 kbar (10^{10} Pa), most of the easily compressed electronic clouds are tightened up, and the compressibilities of most substances approach each other. Usually, only minor amounts of energy compared with chemical bond energies can be added by compression to 100 kbar. Nevertheless, the atoms of the substance can thereby be forced much closer together than they would be by cooling to low temperatures. An immediate consequence is an increase in melting temperature with pressure for most substances, since nearly all solids, with the exception of unusual ones like ice and bismuth, expand when they melt, and the process of melting absorbs heat as the entropy of the substance increases. The thermodynamic relationship shown in Eq. (1) expresses the change in melting tem-

$$\frac{dT}{dP} = \frac{T\Delta V}{\Delta H} \qquad (1)$$

perature T with pressure P as a function of the volume increase ΔV and the heat absorbed in melting ΔH for a given quantity of the substance. Thus, pressures of 50–100 kbar (5–10 × 10^9 Pa) increase melting points about 100°C (212°F) for substances such as iron, whose ΔV of melting is small, to several hundred degrees for substances such as NaCl, whose ΔV of melting is large. At 1 bar (10^{10} Pa), NaCl melts in an iron crucible, but at 100 kbar (10^{10} Pa) iron can be melted in an NaCl crucible.

Substances which consist of large molecules are easily stiffened or frozen by high pressures. The mobility of the molecules is sharply decreased by a sort of interlocking and tangling effect; thus for the substance to be sheared, chemical bonds must be broken, a process which requires considerable energy. For example, ordinary oils become so stiff at a few dozen kilobars that they are useless for transmitting pressure, and a droplet of pressure-frozen oil is capable of denting a steel plate. This stiffening phenomenon limits the study of most reactions of organic molecules to low pressures because they are rather large and "freeze" easily, but yet are usually not stable enough to withstand the temperatures necessary for liquefaction or intermolecular reactions.

The thermodynamic relationship shown in Eq. (1) applies not only to melting but also to phase changes or internal structure changes in solids. In general, the higher-density forms are favored at high pressures, but since ΔH can be positive or negative, the effects of pressure and temperature may oppose or reinforce each other in determining the stability of a particular high-pressure phase. Phase changes between solids rarely run freely to follow the theoretical equilibrium line, but instead tend to be sluggish or exhibit a region of indifference. The stronger or more refractory the solid, or the greater the atomic displacements involved in the change, the broader the region of indifference. An outstanding example is diamond, which at room conditions persists at nearly 20 kbar (2×10^9 Pa) out of its stability field. The existence of a region of indifference can hamper the study of some high-pressure phenomena, but on the other hand, it may permit the recovery of high-pressure phases for more detailed analysis at room conditions. The region of indifference may shrink drastically in the presence of solvents or catalysts which can promote the phase change. One of the more effective solvents, especially for systems containing oxides, is hot water. For chemically reducing systems such as carbon, the diamond-graphite transformations are assisted by the group VIII metals, iron, nickel, platinum, and so on. Other carbon solvents, such as AgCl or CdO, exist but do not permit diamond to form, apparently because they do not favor carbon cations. For transformations in the BN system, nitrides, either molten or in metallic solution, are effective catalysts. The rule "like dissolves like" is a useful, though not infallible, guide in these matters.

Some interesting chemical effects of pressure are related to shifts in chemical equilibria. The free-energy change ΔG determines chemical equilibria at temperature T through an expression of the form of Eq. (2), where R is the gas constant, and a_m, a_n, etc., are terms related to the

$$\Delta G = RT \ln (a_m a_n \cdots / a_x a_y \cdots) \tag{2}$$

concentrations of reactants and products in a reaction of the type $m + n + \cdots = x + y + \cdots$. For most reactions, ΔG ranges about 50 kcal (200 kilojoules) each side of zero. At 43 kbar (43×10^8 Pa), a change in volume of 1 cm^3 (0.06 in.^{3}) corresponds to a change in energy of 10 kcal (40 kJ), and a change of 1 cm^3 (0.06 in^3.) per gram mole of reacting substance is not large, so that it is easy to find large shifts in chemical equilibria produced by pressure. Phase changes can also be regarded as special kinds of intramolecular reactions whose equilibria are shifted by pressure when interatomic bonds are broken and reformed.

Some general rules exist regarding phase changes or electronic bonding changes in solids produced by pressure.

The first rule recognizes that open structures stabilized by relatively weak ionic or Van der Waals forces can easily collapse under pressure to denser structures. For example, eight crystalline forms of water ice have been identified. KCl collapses from the NaCl structure to the more closely packed CsCl structure at about 25 kbar (2.5×10^9 Pa). White phosphorus, made up of separate rings of four P atoms, is permanently pushed into the denser, more stable, semiconducting black form at pressures of a few tens of kilobars.

The second rule results from the compression of the outermost electronic shells and the delocalization of the electrons so that they are not as firmly fixed to particular atoms, and so that greater numbers of atoms can cluster around a given atom as the strongly directed valence forces are weakened. This is the same behavior observed as atomic number increases: Most of the lighter elements, with a few valence bonds per atom, are nonmetallic, whereas most of the heavier elements are metallic with higher valence numbers. Thus, pressure favors not only more metallic behavior but also behavior more like that of elements of higher atomic number. For example, upon compression to 100–130 kbar ($1-1.3 \times 10^{10}$ Pa), silicon and germanium, normally semiconductors with the relatively open diamond structure, collapse to the white tin structure and become good metallic electrical conductors, equivalent to aluminum in that respect. As another example, silica, with four oxygen atoms about each silicon atom, changes at pressures of about 100 kbar (10^{10} Pa) and moderate temperature in the presence of water to stishovite, which has the rutile structure of TiO_2, with six oxygen atoms about each silicon. Many other examples of similar behavior in silicate systems are known wherein pressures of 100 kbar or so force common crustal minerals into crystalline forms embodying higher coordination numbers, similar to those found in oxide compounds of heavier atoms. These effects are important in considerations of the nature of the deeper layers of the Earth's crust.

The approach to the metallic state with increasing pressure may take different forms. Substances such as phosphorus, iodine, and selenium become substantially metallic at 100–150 kbar ($1-1.5 \times 10^{10}$ Pa). On the other hand, when normally insulating organic compounds such as pentacene or hexacene, which consist of five or six aromatic rings fused together, are compressed to 150–200 kbar ($1.5-2 \times 10^{10}$ Pa), they become semiconductors, since electrons are able to travel between the large molecules as the electronic clouds overlap. Certain complex compounds containing linear chains of metal atoms, such as Magnus's green salt, $Pt_2Cl_4(NH_4)_4$, increase in electrical conductivity up to 160 kbar (1.6×10^{10} Pa) and then show a decrease; evidently, conductivity is favored only in particular ranges of interatomic spacing. Mössbauer studies have shown that ferric ion is reversibly reduced to ferrous ion in many compounds at pressures in the 100–200 kbar (1–2×10^{10} Pa) range. Theoretical calculations indicate that even hydrogen could become metallic at pressures variously estimated to be 2–18 Mbar (2–18×10^{11} Pa), but such pressures would be extremely difficult to generate and use, even with shock-wave techniques.

By changing the environment of atoms, high pressures can have strong effects on cooperative phenomena such as magnetism and superconductivity. For example, the high-pressure form of iron found above about 110 kbar (1.1×10^{10} Pa) at 25°C (77°F) is not ferromagnetic. Superconducting transition temperatures are generally lowered by the application of high pressures.

Many chemical reactions proceed through an intermediate state whose volume may be larger or smaller than that of the reactants. When the volume is smaller, the rate of reaction may be increased by pressure. Usually the intermediate state is more voluminous, especially in solids and the rate of reaction is reduced by pressure. Here, as with chemical equilibria, the energy change due to pressure in the 30–100 kbar (3–10×10^9 Pa) range can be comparable with ordinary chemical bond or reaction-activation energies, and large pressure effects are possible. Certain reaction pathways may be effectively blocked so that new paths are followed. For example, at 130 kbar (1.3×10^{10} Pa), where diamond is stable at temperatures up to 3000°C (5400°F) or more, the pyrolysis of some organic compounds, especially those consisting mainly of aromatic rings, produces graphite as the initial product whereas paraffin or polyethylene loses hydrogen to form waxy, dense solids of increasing microcrystalline diamond content as the pyrolysis temperature increases. A hexagonal form of diamond can be prepared by subjecting highly crystalline graphite to pressure of 130 kbar (1.3×10^{10} Pa) and 1500°C (2700°F). The reaction proceeds to some extent at 25°C (77°F) but proceeds much further on heating. (Poorly crystalline or partly amorphous graphite yields ordinary cubic diamond.) The hexagonal diamond can be recovered at room pressure and temperature if it has been heated at high pressure. Hexagonal graphitic BN can be converted to a wurtzite form by the application of 130 kbar (1.3×10^{10} Pa) at 30°C (86°F), but at high temperatures, or in the presence of a molten catalyst-solvent, the cubic form of BN, which is slightly more stable than the wurtzite form, is obtained.

Bibliography. R. S. Bradley (ed.), *Advances in High Pressure Research*, vols. 1–4, 1966–1974, H. G. Drickamer and C. W. Frank, *Electronic Transitions and the High Pressure Chemistry and Physics of Solids*, 1973; N. Isaacs, *Liquid Phase High Pressure Chemistry*, 1981.

HIGH-TEMPERATURE CHEMISTRY
JOHN L. MARGRAVE

The study of chemical phenomena occurring above 500 K (440°F). High temperatures represent one of the important variables available to scientists for increasing the variety of possible chemical reactions over that expected for classical ground-state atoms and molecules. One can enhance the relative population of excited rotational, vibrational, and electronic states by increasing the temperature and thus effectively create new species and new mechanisms for reaction. The potentialities of this approach are well illustrated by the three laws of high-temperature chemistry: (1) At high temperatures everything reacts with everything. (2) The higher the temperature, the faster the reaction. (3) The products may be anything. With an infinity of species available at high temperatures, the "golden age" of chemical synthesis is probably still in the future.

High temperatures also provide a common tie among the various options for energy production, conversion, or storage. For maximum thermodynamic efficiency, an energy production cycle should operate with a working fluid at as high a temperature as possible, and exhaust the spent fluid at as low a temperature as possible. Thus, in the combustion of coal to produce electric power or in the combustion of gasoline or diesel fuel to propel a car or an airplane, there is a need for materials of construction which allow operation of such devices at high temperatures. In the evaluation of new fuels or propellants, higher flame temperatures are among the desirable properties often sought.

It is convenient to discuss temperatures in terms of energy and to note that 11,500 K (20,200°F) corresponds to 1 electronvolt. In this sense, the particles emitted by radioactive nuclei or accelerated in cyclotrons and synchrotrons, which have energies in the keV, MeV, and BeV ranges, are effectively at temperatures of $\sim 10^7$ K, $\sim 10^{10}$ K, and $\sim 10^{13}$ K, respectively, and "high-energy physics" is synonymous with "ultrahigh temperature chemistry."

Traditional high-temperature chemistry has been mainly concerned with phenomena in the range of 500–3000 K (440–4900°F), although exotic flames can produce temperatures up to $\sim$6000 K (10,300°F), shock waves can generate temperatures up to $\sim$25,000 K (45,000°F), electric arcs can be operated in constricted modes to produce temperatures of $\sim$50,000 K (90,000°F), and nuclear processes begin to occur at temperatures in the millions of degrees range. Laser-excitation of selected energy states can produce species with effective temperatures in the range of 10^8 K. The major goals of high-temperature scientists include (1) the characterization of all important gaseous molecules, ions, and condensed phases–molecular formulas and structures, energy levels, thermodynamic properties, and the details of chemical bonding; (2) the establishment of reaction rate parameters and correlations with molecular properties; (3) the development of unique approaches to chemical syntheses and the preparation of new materials; and (4) the development of new techniques for generating or containing or utilizing high-temperature fluids in connection with energy production, conversion, and storage.

Gaseous species. Studies of thermal decomposition, vaporization, and/or sublimation of inorganic compounds have shown that complex gaseous species are much more common than usually realized. For example, condensed alkali metal sulfates vaporize appreciably as M_2SO_4 (gas); alkali metal carbonates vaporize as M_2CO_3 (gas); and even NH_4X(solid) yields NH_4X(gas). At high temperatures the relative fraction of complex species in the vapor actually increases with the increasing temperature for many systems. Vapors at high pressures and high temperatures can thus be complex. Rapid mass spectrometric sampling techniques have demonstrated unequivocally the existence of polymers such as $(Ar)_m$, $(CO_2)_n$, and $(H_2O)_q$ where m, n, and q are 2,3,4, . . . , 8,9,10, ... As a further complication, various oxidation states can exist in high-temperature systems and thereby a still greater variety of possible molecular species. Typical systems which have been characterized by a combination of mass spectrometric, optical, infrared, nuclear magnetic resonance, and electron spin resonance spectroscopy and chemical studies are the difluorides of group IV (CF_2, SiF_2, GeF_2, SnF_2, PbF_2) and their polymers.

In particular, the technique of generating high-temperature gaseous species and then reacting them with low-temperature molecules on a surface which is maintained at low temperatures (~ -190°C or -310°F) has been very productive. Practical approaches to a great variety of organometallic syntheses have been developed through the use of molecules such as SiF_2, C_2, C_3, BF, SiO, and SiS, and various atoms—such as C, Si, Li, Mg, Fe, Cu, Ni, Pd, and Pt. Matrix-isolation spectroscopy at 4–100 K (-452 to -279°F) provides the opportunity for gaining detailed information about atomic/molecular parameters—energy levels, frequencies, bond angles—which are necessary for calculation of thermodynamic functions and the prediction of chemical reactions.

Condensed systems. Condensed systems at high temperatures provide additional versatility in the chemical world since one fails to observe almost uniquely in high-temperature systems that atoms do combine in the ratio of small whole numbers. Such observations on gases led to the formulation of the law of definite proportions and other basic rules of early chemistry, but these rules no longer hold exactly for condensed systems at temperatures in the 1000–3000 K (1300–4900°F) range. Experiments have established melting points of refractory solids (binary carbides, borides, silicides, and so forth) at 1 atm (10^2 kPa) pressure in the 2000–4000 K (3100–6700°F) range, and unless one goes to higher pressures than 1 atm there really is no sold-state chemistry at temperatures greater than $\sim$4000 K (6700°F). Obviously, there is a chemistry of liquid systems which might be of some technological importance, and ingenious techniques for generating and maintaining reactive high-temperature liquids have been described.

Phase diagrams have demonstrated conclusively that one can prepare crystalline solids in an almost infinite number of compositions. For example, there are "pure" compounds fitting classical valences, and then one can substitute, as in $MgAl_2O_4$, where Al^{3+} ions can occupy some of the Ms^{2+} sites or Mg^{2+} ions can occupy some of the Al^{3+} sites so that there is an almost continual variation in properties. One can prepare Ta-O-C phases in which the gross stoichiometry is

$Ta_1O_xC_{1-x}$ and, further, find it extremely difficult to detect the fact that oxygen is even present since the sizes of the oxygen and carbon atoms are so nearly the same in this lattice. Discovery of oxynitride phases and the establishment of essentially continuous solid solutions over wide ranges of compositions for many oxide, sulfide, carbide, boride, and other systems have been reported. The variations in compositions of alloys are similarly extremely complex; the variety possible is only beginning to be appreciated. Techniques for presenting phase diagrams have been described in which an attempt is made to present a single diagram which summarizes the behavior of whole classes of alloy systems rather than that of individual binary or ternary combinations.

Properties of high-temperature solids are, of course, not reliably interpreted until the exact stoichiometric and structural data have been obtained. The status of high-temperature thermodynamic properties is acceptably characterized by saying that no material is described thermodynamically to better than $\pm 0.5\%$ up to 1500 K (2200°F), or better than to ± 1–2% from 1500–2000 K (2200–3100°F). There are practically no reliable data at temperatures greater than 2000 K (3100°F) except for a handful of basic materials—tungsten, molybdenum, tantalum, and aluminum oxide.

There is a great need for experimental techniques by which thermodynamic measurements can be reliably extended into this high-temperature region. There is already experimental evidence which is not explainable by current theoretical viewpoints on solids and liquids at high temperatures. Vaporization, sublimation, and other heterogeneous equilibrium studies have been made for many systems, and these are, of course, sensitive to the exact nature of the surfaces, to variations in stoichiometry as gases are evolved from a condensed phase, and to other factors which may accompany deviations from equilibrium.

Electrochemical techniques. One of the unique ways in which high-temperature thermodynamic data have been obtained involves the use of electrochemical techniques of the same sort as those which were widely applied during the 1920–1940 period for the establishment of reliable thermodynamic reference data for aqueous systems and for simple solids at temperatures near 25°C (77°F). High-precision free energies of formation for nonstoichiometric phases can be derived over fairly wide ranges of temperature. However, eventually at 1500 K (2200°F) and higher, one begins to have difficulty with vaporization, melting, and high diffusion rates.

Calorimetry. Calorimetry at temperatures up to 3000 K (4900°F) became routine through the use of electron bombardment heating and levitation heating. In electron bombardment heating, one boils electrons out of a thermionic emitter, such as thoriated tungsten, and accelerates them across a potential drop of a few thousand volts to strike the conducting sample and raise its temperature by means of the kinetic energy transferred and also by the heating caused by the electric resistance of the material. Levitation heating of conducting samples is accomplished with standard radio-frequency induction heaters and a pair of oppositely wound (left-handed and right-handed) coils. Samples weighing up to 1000 g (35 oz) can be levitated, melted, and cast in a containerless, controlled atmosphere. Copper, gold, platinum, tantalum, graphite, and many other materials have been levitated and, in all cases except graphite, raised to temperatures above their melting points (3000 K or 4900°F and higher). Levitation calorimetry has provided heats of fusion and heat capacities for many metals, alloys, and conducting compounds in both solid and liquid states.

Chemical kinetics. From the viewpoint of heterogeneous kinetics, the high-temperature area has been studied extensively, with the practical concerns involving (1) the rates of interaction between various corrosive gases (oxygen, nitrogen, sulfides, halogens, and so on) and surfaces of various pure metals, alloys, and ceramic structural materials, and (2) the catalytic effects of solids of various stoichiometries on various gas reactions as in hydrocarbon refining, in the preparation of SO_3 or NH_3, or in the catalytic conversion of $CO + H_2$ to methanol, of CO to CO_2, or of NO_x to N_2 and O_2.

Homogeneous gas kinetics in the region above 1000 K (1300°F) has been mainly concerned with reactions of neutral atomic and molecular species which were stable at high temperatures, as in typical flames. The roles of H, O, OH, and of intermediate combustion products like HO_2, MO, and CHO have been established by direct mass-spectrometer probing of flames. The importance of electrons and atomic or molecular ions in flame kinetics has also been recognized. This knowledge is of interest in identifying the most efficient combustion systems for energy generation and in elucidating the chemical parameters which are crucial to the development of flame-retardant materials.

Another fertile area for high-temperature kinetics research has been the study of the reaction rates of atoms or molecules which have been produced by either photochemical, laser-pulse, or electric-arc excitation or by pulsed thermal dissociation processes and then allowed to react either in systems at relatively higher pressures by random collisions or in systems at low pressures by molecular-beam techniques. Thus, a hydrogen atom created from the H_2 molecule in an arc or by thermal dissociation, requires 2.24 eV/atom, and therefore is a chemical species of the sort that might be expected in a very high-temperature system. The thermodynamic potential of such atoms for reaction is much greater than that for most molecular species, and reaction rates are usually measurable although relatively fast. Selective excitation to specific energy levels is possible with tunable dye lasers.

Several types of monitoring techniques, rapidscan infrared spectroscopy, electron spin resonance spectroscopy, and mass spectrometry, are typically used in these studies. Gas reaction rates in mass spectrometers also have been widely explored by scientists interested in ion-molecule reactions. Of course, since ions are created endothermally by the impartation of several electron volts of energy, they are high-temperature species. Many new types of high-temperature reactants can be created. It is possible that a synthetic chemistry making use of ion-molecule reactions may someday be of economic significance. Ion sputtering is a widely used technique for the preparation of electronic circuitry. *See* CHEMICAL DYNAMICS.

Bibliography. L. Eyring (ed.), *Advances in High Temperature Chemistry*, vols. 1–4, 1967–1972; J. W. Hastie, *High Temperature Vapors*, 1975; J. L. Margrave (ed.), *Modern High Temperature Science*, 1984; E. T. Turkdogan, *Physical Chemistry of High Temperature Technology*, 1980.

MAGNETOCHEMISTRY
DIETER M. GRUEN

The branch of chemistry which studies the interrelationship between a magnetic field and atomic and molecular structures.

When a substance is placed in a magnetic field of strength H, the magnetic induction B is given by Eq. (1). The quantity I is the intensity of magnetization, and $I/H = \kappa$ is the magnetic

$$B = H + 4\pi I \tag{1}$$

susceptibility per unit volume. The magnetic susceptibility per unit mass is $\kappa/D = \chi$, where D is the density.

A substance in a magnetic field acquires an intensity of magnetization which may be either smaller or larger than that induced in a vacuum by the same field. In the first case, the substance is said to be diamagnetic. In the second case, the substance may be paramagnetic, ferromagnetic, or antiferromagnetic.

Diamagnetism, a universal property of matter, is usually of the order of magnitude 10^{-6} to 10^{-5}. Temperature-dependent paramagnetism, on the other hand, arises only when an atom, ion, or molecule possesses a permanent magnetic moment either in the ground state or in an excited state. A permanent magnetic moment is the result of the presence of one or more unpaired electrons. Paramagnetic susceptibilities are of the order of magnitude 10^{-4} to 10^{-3}.

A substance composed of atoms with permanent magnetic moments which are very near to one another (for example, iron metal) may display ferromagnetism. This phenomenon occurs when large numbers of the atoms with permanent magnetic moments interact so that their individual moments align in a parallel fashion, giving rise to a large resultant moment.

On the other hand, a similar substance (for example, manganese metal) may display antiferromagnetism. Here, the magnetic moments align in an antiparallel fashion, thus largely canceling the individual magnetic moments of the atoms. Parallel versus antiparallel alignment depends, among other factors, upon interatomic distances. Magnetic theories are not yet refined enough to enable one to predict whether a given substance will be ferromagnetic or antiferromagnetic.

In general, the susceptibility of diamagnetic substances is independent of temperature and of field strength. The susceptibility of paramagnetic substances is often inversely proportional to the absolute temperature but independent of field strength. The susceptibility of ferromagnetic and antiferromagnetic substances is dependent on both temperature and field strength in a rather complicated way.

There are many methods available for the measurement of magnetic susceptibility. Most of these methods involve measuring the force exerted on a sample by a magnetic field.

Atomic diamagnetism. The only important application of diamagnetic ionic susceptibilities is their use as correction factors for measured susceptibilities. All substances, even though paramagnetic, have an underlying diamagnetism. A precise determination of the paramagnetic susceptibility of sodium neptunyl acetate, for example, should include subtraction from the measured molar susceptibility of the diamagnetic ionic susceptibilities of sodium, acetate, and neptunyl ions.

The diamagnetism of atoms and ions can be calculated theoretically by considering electron density distributions summed for each electronic shell. In addition, a large number of diamagnetic susceptibilities have been determined empirically. There is, in general, agreement between the measured values and those calculated theoretically.

Molecular diamagnetism. Estimates of the diamagnetism of organic compounds are based primarily on empirical methods. From measurements on a large number of compounds, B. Pascal concluded that diamagnetic susceptibilities could be represented by Eq. (2), where n_A is

$$\chi_M = \sum n_A \chi_A + \lambda \qquad (2)$$

the number of atoms of susceptibility χ_A in the molecule, and λ is a constitutive correction depending on the nature of the bonds between the atoms. In this expression, χ_A is not the theoretical atomic susceptibility referred to in the previous section, but is a purely empirical constant derived from the measured susceptibilities. This procedure, when applied to organic compounds, often gives results that are within 1% of the experimentally determined values.

To illustrate the Pascal method, a simple example will suffice. According to this method, the molar susceptibility of ethyl bromide (C_2H_5Br) is given by $2\chi_C + 5\chi_H + \chi_{Br} + \lambda$. In this case, λ is the constitutive correction for the C-Br band. The magnitude of these quantities is shown in computation (3). The experimentally observed molar susceptibility is -53.3×10^{-6}.

$$\{-[(2 \times 6.00) + (5 \times 2.93) + 30.6] + 4.1\} \times 10^{-6} = -53.1 \times 10^{-6} \qquad (3)$$

The magnetic susceptibility of a noncubic substance varies along different crystal axes. Susceptibilities measured along different crystal axes are called principal susceptibilities, and the difference between principal susceptibilities is called the magnetic anisotropy of the substance. A large amount of structural information can be obtained from measurements of principal susceptibilites. For example, the principal susceptibility of graphite perpendicular to the hexagonal axis of a single graphite crystal is about -0.5×10^{-6}, or nearly the same as the powder susceptibility of diamond. However, along the hexagonal axis, the susceptibility of graphite is -21.5×10^{-6} at room temperature. It is thought that the large diamagnetic susceptibility along the hexagonal axis is a result of the diamagnetism of conduction electrons, which are present in graphite but not in diamond.

Atomic paramagnetism. As stated previously, an atom with unpaired electrons has a permanent magnetic moment and is therefore paramagnetic. The present discussion will restrict itself to the paramagnetism exhibited by transition element compounds of the iron group and the lanthanide and actinide series. The magnetic properties of the palladium and platinum group compounds and the magnetochemistry of coordination compounds are not discussed here.

The modern quantum-mechanical theory of magnetism was, to a great extent, developed by J. H. Van Vleck. One of the triumphs of this theory is the remarkable agreement between the theoretically calculated magnetic moments of the lanthanide ions and those experimentally determined. The paramagnetism of the lanthanides arises from unpaired electrons in the 4f shell which are unique because they are but little affected by the electric fields of the surrounding anions.

It is, therefore, a good first approximation near room temperature to calculate the magnetic moments of the lanthanide group ions on the assumption that their behavior is that of free gaseous ions.

A somewhat similar situation obtains with respect to the actinide series of compounds. Here, the paramagnetism arises from unpaired electrons in the incomplete $5f$ shell. These electrons are less well shielded from the crystalline electric fields than are the $4f$ electrons. Therefore, in calculating the magnetic moments of actinide ions one must take account of the splitting of the energy levels by the crystalline field. When this is done, very satisfactory agreement between calculated and observed moments is obtained in most cases.

The effect of the crystalline electric fields on the magnetic properties of ions is even more striking in the case of the iron group compounds with their unpaired $3d$ electrons. In general, the magnetic moment of an unpaired electron is proportional to the vector sum of the orbital and the spin angular momentum vectors. The electric fields in a crystal or a solution containing an iron group ion interact so strongly with the orbital part of the moment that they quench its contribution to the magnetic moment almost entirely. Therefore, the observed moments for these compounds are often in very close agreement with moments calculated using the so-called spin-only formula.

Molecular paramagnetism. Oxygen has two unpaired electrons in its normal state and its therefore paramagnetic. The molar susceptibility of oxygen over a wide range of pressures and temperatures is given by the simple equation $\chi_M = 0.993/T$ where T is the absolute temperature. Other paramagnetic gases are NO, NO_2, ClO_2, and ClO_3.

There are a large number of organic compounds which possess one or two unpaired electrons. These compounds are known as free radicals. One of the most famous examples is hexaphenylethane which dissociates in benzene solution to give the free radical, triphenylmethyl. Other compounds, such as α,α-diphenyl-β-picrylhydrazyl are stable free radicals even in the solid state.

Many organic compounds, of which fluorescein and naphthalene are examples, are normally diamagnetic. These materials, however, become paramagnetic when exposed to ultraviolet light. The reason for this is that light excites the molecules to triplet or phosphorescent states which are characterized by two unpaired electrons. Magnetic susceptibility measurements on such materials during irradiation have yielded important information the mechanism of phosphorescence. Paramagnetic resonance measurements on naphthalene have proved earlier static susceptibility measurements.

Ferro- and antiferromagnetism. Ferromagnetic substances are distinguished chiefly by their large susceptibilities at low magnetic fields and by the fact that their specific magnetization is a function of field, up to the field at which the substance is saturated. Above a certain temperature, the Curie point, all ferromagnetic substances lose their ferromagnetism and become paramagnetic.

Antiferromagnetic substances also undergo a transition at a temperature which is characteristic for each material. Above this temperature, known as the Néel point, the substance is paramagnetic. The susceptibility of the material in the antiferromagnetic state is field dependent and is smaller than the susceptibility above the Néel point.

Some examples of ferromagnetic materials are iron, cobalt, nickel, gadolinium, uranium hydride, and nickel disulfide. Examples of antiferromagnetic substances are manganese, titanium trichloride, uranium trichloride, and neptunium dioxide. SEE MOLECULAR STRUCTURE AND SPECTRA.

Bibiliography. R. L. Carlin, *Magnetochemistry*, 1985.

RADIATION CHEMISTRY
SHEFFIELD GORDON

The study of the chemical effects of the absorption of high-energy radiation in matter. High-energy radiation includes the emanation associated with radioactive decay and fission (that is, alpha particles, electrons, gamma rays, and neutrons), together with their related atom and fission recoils; and the artificial analogs of such emanations produced by accelerating, electrons, protons, deuterons, and helium nuclei, as well as charged nuclei of higher atomic number [currently as high as 18 (argon)] and x-rays.

Sources of high-energy radiations in the laboratory and industry include radioactive nuclides (for example, ^{60}Co, ^{90}Sr, and ^{3}H) and instruments such as x-ray tubes, Van de Graaff generators, the betatron, the cyclotron, and the synchrotron. An electron accelerator known as the linac (linear electron accelerator) has proved particularly valuable for the study of transient species which have lifetimes as short as 16 pico-seconds, and another electron accelerator, known as the Febetron, has been used for the study of the effects of single pulses of electrons with widths of several nanoseconds at very high currents.

An experimental apparatus using a single 25-ps pulse of electrons from a linear accelerator is shown in **Fig. 1**. The short pulse of 13-MeV electrons emerges from a thin window and enters an optical cell containing the chemical system being studied. Before penetrating the cell, part of the electron beam is intercepted by a cell containing xenon gas. A continuous spectrum of light due to the production of Cerenkov radiation emerges from this cell and, after suitable optical delays with respect to the electron pulse, impinges on the sample cell, allowing the absorption spectra and lifetime of the transient species produced by the electron pulse in the cell to be studied. In addition to the optical absorption method, a variety of fast reaction techniques are used to study the reactions of short-lived species produced by such short pulses. These include fluorescence, ionic conductivity, electron spin resonance, nuclear magnetic resonance, resonance Raman scattering, and polarography.

Energy transfer. Energy is transmitted to irradiated material by momentum transfer and by excitation and ionization. The latter two always accompany the former. Momentum transfer is

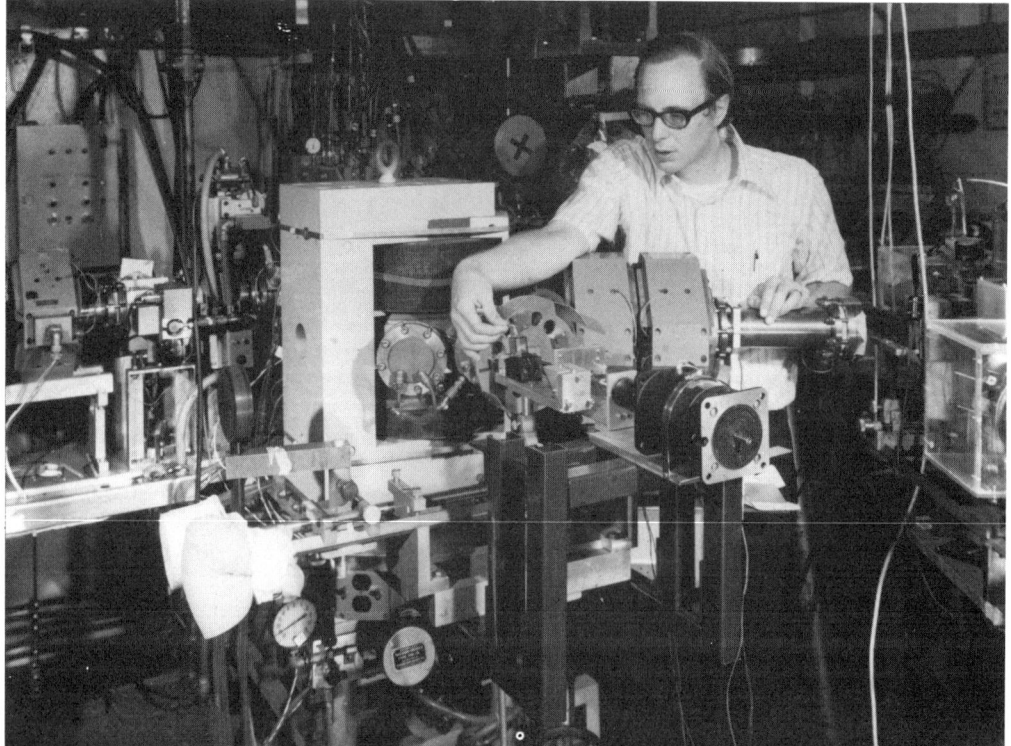

Fig. 1. Linear accelerator delivering single 25-picosecond pulses of 13-MeV electrons, together with associated equipment designed to identify and measure the reactivity of species produced by the pulse in chemical systems by using fast optical detection techniques. (*Argonne National Laboratory*)

characteristic of processes involving neutrons; it is always involved to some extent in particle effects and is an important contribution to heavy-particle (for example, proton) effects.

In a momentum transfer interaction, the usual effect is the ejection of a nucleus from its molecular or crystalline structure. In solids, this is known as the Wigner effect. When crystalline material is involved, the process is called discomposition. Discomposition results not only from neutron impact, but also secondarily from impacts involving high-energy displaced nuclei. Momentum transfer effects in solids are detected by changes in electronic and thermal conductivity, elastic moduli, and dimensions. In crystalline compounds, effects include chemical change, electron trapping, and color production.

Chemical yields are expressed in the older literature as ion-pair yields M/N—a yield being the number of molecules converted or produced per ion pair initially produced by the radiation. The modern literature uses the 100-eV yield G, which is the number of molecules converted or produced per 100 eV of energy absorbed. The term M/N is now used only in cases in which N, the number of ion pairs, is actually determinable from experimental data. A convenient rule of thumb for reading the older literature is $G - 3M/N$. Yields range from values such as $G - 0.01$ (copper phthalocyanin decomposition) to values such as those in notation (1).

$$G(H_2 + Cl_2 \rightarrow 2HCl) \sim 10^5 \tag{1}$$

Energy input may be determined directly from charge measurements by using Faraday cups when using machine sources or indirectly by chemical dosimetry. The Fricke dosimeter (aerated acidic ferrous sulfate solution) is such a secondary standard for high-energy electrons and for gamma rays and x-rays, where $G(Fe^{2+} \rightarrow Fe^{3+}) = 15.6$. These secondary standards have been calibrated against an absolute measurement of the energy input by using calorimetric methods. The units rad, roentgen, and rep are also used. One roentgen is equivalent to 6.08×10^{13} eV/g of water for x-rays and gamma rays. One gray (Gy) is equal to 1 joule/kg and is equivalent to 100 rads.

Theoretical considerations regarding primary physical processes indicate that for other than momentum transfer, the principal effects are due to fast-moving charged particles. A 1-MeV charged particle, unlike a parent photon which produces only one ionization (as in Compton process), may produce a total of 10^5 ions (and electrons) and excited molecules. The distribution of such primarily produced entities is inhomogeneous, and is affected greatly by the nature of the radiation and by the state of aggregation of the material irradiated. It is this inhomogeneity of the distribution of activated species which is an important factor differentiating radiation chemistry from photochemistry and thermally induced chemistry. In the latter two, the active species are produced with a homogeneous distribution. In radiation chemistry, it is only when the active species produced by the initial ionizing particle have diffused throughout the reaction volume that homogeneous reaction kinetics can be applied.

In condensed systems, ions and excited species tend to be formed in three quantitatively different kinds of groups or clusters: (1) spurs involving energy losses up to about 100 eV; (2) blobs of energy loss of approximately 100–500 eV; and (3) short tracks of relatively high energy of approximtely 500 eV to 5 keV. For a 1-MeV electron, the energy is deposited among the entities as follows: 67% spurs, 11% blobs, and 22% short tracks. In liquid water, the average spur contains about three ions and six excited molecules, and is thought to have a diameter of about 2 nanometers. A spacing of thousands of molecular diameters between spurs is typical for fast electrons such as those produced by ^{60}Co gamma rays. For heavy particles such as polonium alpha rays, the spurs overlap to form short tracks representing roughly finite cylindrical regions of high linear energy transfer (LET) to the medium. The existence and distribution of spurs, blobs, and short tracks affect the chemistry in liquid water. The approximate value of G for products at an early stage of the chemical effects in liquid water is shown in **Fig. 2**. In water vapor, the primary yield of molecules decomposed to radicals is $G \sim 8$.

Liquid water and other polar liquids display particularly interesting phenomena. The existence of solvated electrons with lifetime up to 40 ms in water (depending on the concentration and nature of the solutes present) was first established in the case of aqueous systems; the chemical characteristics of such species have been measured repeatedly by different techniques. The specific rate of the presumed reaction $2e_{aq}^- \rightarrow 2OH^- + H_2$ in water has been shown to have the surprisingly high value of 4.5×10^9 $M^{-1} s^{-1}$ at pH 10.3. Studies of high-pressure effects in the

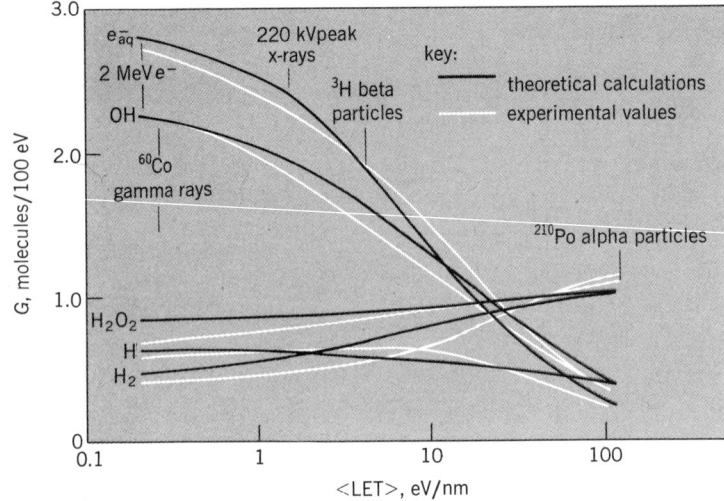

Fig. 2. Variation of G with linear energy transfer (LET) for the products: solvated electron (e_{aq}^-), OH, H_2O_2, H, and H_2. (*After R. C. Cooper and R. W. Wood, eds.,* Physical Mechanisms in Radiation Biology, *pp. 155–176, National Technical Information Service, 1974*)

radiolysis of aqueous systems, as well as theoretical calculations, indicate that the solvated electron in water (contrasted with ammonia) displaces very small volume.

Ionic processes also play a very significant role in the radiation chemistry of low dielectric constants such as cyclohexane. Electrons having a wide range of initial energies (from essentially zero to approximately 1 MeV in most cases) are generated in the liquid by high-energy radiation. Such electrons lose their energy in the medium, and are thermalized at various distances (related to their initial energies) from sibling cations. A certain small fraction corresponding to approximately 3% of the electrons can escape the coulombic field of the sibling cation. Such free ion pairs are produced with $G \sim 0.1$, as shown by electrical conductivity measurements in irradiated alkanes. The remaining 97% (corresponding to $G \sim 4$) of such ion pairs undergo recombination under the influence of their mutual coulombic fields, and are referred to as coupled or geminate ion pairs. For such coupled pairs, thermalization distances may range from about zero to approximately 30 nm and, according to theoretical estimates, have lifetimes for recombination (related to the thermalization distance) ranging from 10^{-7} s to less than 10^{-11} s. Either or both of the ions of such a coupled pair may enter into reactions with solutes present in the system, and thereby determine the nature and yields of certain chemical products.

Representative processes. Processes particularly characteristic of radiation chemistry are represented by reactions (2)–(7) in liquid water (the asterisk denotes a molecule in an excited

$$H_2O \rightsquigarrow H_2O^+ + e^- \quad (2) \qquad H_2O^+ + H_2O \rightarrow H_3O^+ + OH \quad (3) \qquad H_3O_{aq}^+ + e_{aq}^- \rightarrow H_2O + H \quad (4)$$

$$H_2O \rightarrow H + OH \quad (5) \qquad H_2O \rightsquigarrow H_2O^* \quad (6) \qquad H_2O^* \rightarrow H + OH \quad (7)$$

state). The H_2O undergoes a very rapid proton transfer reaction with a neighboring water molecule [ion-molecule reaction; reaction (3)]. The H_3O^+ becomes solvated into $H_3O_{aq}^+$ while the electron ejected in reaction (2) becomes thermalized and solvated into e_{aq}^- and then reacts through (4). The net result of reactions (2), (3), and (4) is reaction (5). These processes are the major ones in liquid water, but some minor ones occur, such as the dissociation of excited or superexcited molecules, reactions (6) and (7). Some of these reactive species react with each other before they escape the initial spurs and tracks forming the so-called molecular products (H_2 and H_2O_2), the yields of which are indicated in Fig. 2. When solutes are present, they can react with these H, OH, and e_{aq}^- species.

Other processes characteristic of radiation chemistry are reactions (8)–(12):

Ion molecule reactions: $CH_3^+ + CH_4 \rightarrow C_2H_5^+ + H_2$ \hfill (8)

Dissociative capture of an electron: $CH_3I + e^- \rightarrow CH_3 + I^-$ \hfill (9)

Nondissociative capture of an electron: $SF_6 + e^- \rightarrow SF_6^-$ (10)

Charge transfer: $C_6H_{12}^+ + C_6H_6 \rightarrow C_6H_{12} + C_6H_6^+$ (11)

Excitation transfer: $C_6H_6^* + p(C_6H_5)_2C_6H_4 \rightarrow C_6H_6 + p(C_6H_5)_2C_6H_4^*$ (12)

A net effect of certain of these reactions may be protection of solvent molecules by energy (or reactivity) transfer to a chemically stable receiver.

Among other processes are some that are observable also in photochemistry, such as radiosensitization, free-radical reactions, and induced internal conversions. Diffusion-controlled reactions of free radicals differ from those of photochemistry, in which radicals are formed initially only in pairs. In radiation chemistry, the existence of spurs can result in primary production of four or more free radicals in close proximity. SEE PHOTOCHEMISTRY.

Radiation effects. Chemical effects of high-energy radiation must be guarded against in nuclear reactors (where effects include radiation corrosion, water decomposition, and the Wigner effect) and in living systems (because of mutations, cancer production, and so on). Such effects may be deliberately employed to induce polymerization of special kinds: to cross-link, to graft, and to thermally stabilize polymers; to sterilize foods, medicinals, and surgical materials; to change the properties of catalysts; and to induce reactions not possible by other means or to induce them under unusual environmental conditions, such as under extremely low temperature, in very thick layers, and in heavy-walled (pressure) vessels. Applications include the curing of printing inks, coating of fabrics, wood-plastic combinations for musical instruments, plastic pipes for hot-water lines, and conversion of residual monomers in polymers.

Bibliography. G. Foldiak, *Radiation Chemistry of Hydrocarbons*, 1982; M. A. Rodgers, *Radiation Chemistry: Principles and Applications*, 1986.

SOLID-STATE CHEMISTRY
R. J. THORN

The science of the elementary, atomic compositions of solids and the transformations that occur in and between solids and between solids and other phases to produce solids. Solid-state chemistry deals primarily with those microscopic features which are uniquely characteristic of solids and which are the causes for the macroscopic chemical properties and the chemical reactions of solids. As with other branches of the physical sciences, solid-state chemistry also includes related areas that furnish concepts and explanations of those phenomena which are more characteristic of the subject itself.

The overlap of solid-state chemistry and solid-state physics is extensive. However, the perspectives of the two are sufficiently different that they can easily be identified. In general, solid-state physics treats properties, such as energy and entropy, which are continuously variable in the solid, whereas solid-state chemistry concerns those properties which are discontinuous because of chemical reactions; in a sense, by definition the properties are chemical because they are discontinuous. In another perspective, solid-state chemistry tends to be based on structure in configuration space, whereas solid-state physics tends to be based on momentum space.

STRUCTURES

A study of structures provides a basis for understanding the composition of solids and the reactions in which they participate.

Atomic geometries (configuration). The solid-state chemist is interested in the idealized structures described by space groups as a basis to derive information about bonding. However, the principal interest in structure is in the deviations from these space groups, because they reveal additional information about bonding or because they are the causes (mechanisms) or chemical processes in solids. These idealized structures exist, strictly stated, only at 0 K. At any finite temperatures, these structures are imperfect or distorted in at least two general respects: the atoms or ions vibrate with amplitudes which increase with increasing temperature; and the lattice sites of the space groups become vacant or interstitial positions become occupied, or both occur, with the valencies of the cations or anions becoming altered.

The concentration and spatial distribution of the defects in the solid produce various kinds of distortions of the idealized structures. When the defects interact only weakly and do not aggregate, the distortions are mostly localized near the point defects, which may be either randomly distributed or ordered on a long range. For this distribution, the x-ray or neutron diffraction pattern is essentially that of the idealized structures with a charged internuclear distance or lattice parameter and with additional diffraction maxima caused by the long-range order, that is, superstructure. This configuration of point defects occurs especially in several nonstoichiometric phases. In some cases the defects appear to be aggregated, with the result that a structured distortion occurs. Such a situation exists in cases known as shear structures that occur in WO_3 and MoO_3, where extensive regions appear as an idealized structure connected by a plane in the lattice in which a displacement equivalent to a shear translation exists.

Layered structures. Numerous binary compounds or phases have layered structures which in themselves or in combination with other elements can be considered as extensions of the descriptions given above. The layers are displaced relative to each other in such a way that the structures can be classified in the space groups, but when other elements or compounds are incorporated between the layers, the structure is highly distorted. The most widely recognized material to possess the layered structure and the associated solid-state chemistry is graphite (**Fig. 1**). Carbon atoms in this structure are arranged in hexagons connected along the edges, and thus form a plane throughout the crystal. These planes are stacked in layers, with each layer displaced with respect to its adjacent layers by a distance such that the structure still belongs to one of the space groups. The bonding within a layer is covalent; between the layers it is metallic. Consequently, graphite is highly anisotropic. When it is exposed to bromine, alkali metals, and other chemical substances, these substances, at high temperatures, enter the spaces between the layers to form intercalation compounds; the idealized structure thus becomes highly distorted. A large number of compounds with layered structures are known to form a variety of intercalation compounds. Among them are the disulfides TiS_2 and MoS_2. Two others, which are similar in that additional elements or compounds are located between lattice planes, are the tungsten bronzes and the β-aluminas.

Multicomponent phases. For three or more components, the structural situation is generally complex. The number of possible compounds and phases is extremely large, and so is the variety of structures and distortions. Classification of the distorted space groups is difficult and is still being developed. The actual structures are generally described as distortions of some well-recognized structures, which in themselves are complex. Because many of these complex structures were found in minerals, the distorted structures are frequently classified as being derived from these minerals. Examples of such are spinels, perovskites, bixbyites, feldspars, and garnets.

Glasses. All of the structures described above are based on a lattice with defects but with sufficient long-range order to produce well-defined diffraction patterns. There exists another group or class of solids in which the long-range order is insufficient to yield such patterns. These are the glasses or amorphous solids which generally exist only in metastable conditions. The most readily known among these are the silicate glasses, but there are many others, including some even in metallic systems.

Molecular solids. In all the cases described, there occurs no decisive tendencies toward molecular formation in the solids. However, when the intramolecular forces are great enough, as

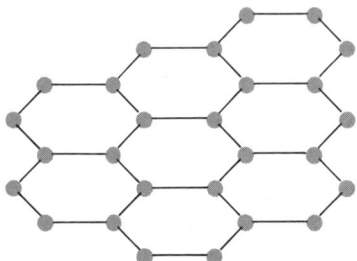

Fig. 1. Structure of graphite. Only a single layer of carbon atoms is shown.

in organic molecules, the solids are composed of molecular crystals. These solids in general have high volatilities and insufficient defects to cause reactions associated with the transfer of mass in the solids, but they have electronic structures which participate in photochemical reactions in organic solids.

Electronic structures. Aside from combinations near the center of the periodic system where covalency is dominant, the elements exist in solids as ions. For the regular elements, the outer, valence orbitals of the ions are closed, with the inert gas shells having s-orbital configurations on the cations and hybridized p-orbitals on the anion. In both cases, the ions have filled orbitals and tend to be spherical; they are not highly directional. These ions have no low-lying states, and the crystal fields are insufficient to decouple the spins. Consequently, the electronic structures of solids formed with these ions are simple, and they contribute a basis for solid-state chemistry in terms of ionicity and normal polarization. These solids, which are essentially those formed from groups I, II, VI, and VII in the periodic system, are transparent in the visible and near infrared. In addition, some of the cations of the transition, lanthanide, and actinide groups have closed p-orbitals with no low-lying electronic states. These are the scandium and titanium subgroups (thorium is included in the latter). All the other transition, lanthanide, and actinide ions in solids display electronic transitions in their infrared and visible spectra. A large number of the observed transitions are attributable to the fact that the d^n or f^n degenerate states are separated by the electrostatic field of the crystal. In the cases of the lanthanide and actinide ions, some of the observed maxima are also attributable to f-to-d transitions. The presence of all of these states contributes to anomalous polarization or ionicity and a complexity of the chemistry. SEE CRYSTAL FIELD THEORY.

Photoacoustic spectroscopy with powdered materials can be used to study the spectra of such solids. In this technique, a beam of light from a monochromator is chopped with an audio frequency, and then it is absorbed by the particles in the powder. The energy is absorbed in the electronic transitions and then transferred to vibrational states. The energy stored as heat in these states is transferred to an ambient gas in the form of sound waves of the same frequency as the chopper on the incident light. The sound waves are detected by a microphone. The study of the energies and mechanisms associated with these electron-phonon interactions and with the phonon-electron transfer furnishes microscopic information for the solid-state chemistry associated with charge transfer, polarization, and valence states.

A feature of the electronic structure of solids, which is particularly essential to solid-state chemistry, is the valence state of the ion and the energy required to change the valency in the solid. For the transition, lanthanide, and actinide ions with open d- or f-orbitals, these energies are sufficiently small that transitions from one valency to another can be promoted by thermal (phonon) activation or the oxidation potentials of anions. The relative concentrations of two valencies in solids can be determined chemically through the stoichiometry or through chemical reactions which presumably do not change the relative concentrations. The known stoichiometries in such solids as Fe_3O_4, Fe_2O_3, and U_3O_8 imply mixed valence states. If these phases are dissolved under nonoxidizing conditions, then the initial ratio of the two valence states can be determined. The question of the existence of two distinct valence states in the solid has been debated for decades. Some have argued that a resonance condition exists so that two well-defined oxidation states cannot be identified. Most of these arguments are based on the results of x-ray diffraction observations, which give no evidence of two different cation sites. Two techniques have been developed which enable direct observation of the valence states in the solid. These are photoelectron spectroscopy and Mössbauer spectroscopy.

CHEMICAL BONDING

The structure of a solid is the result of the operation of interatomic or interionic forces and the size and shape of the atoms or ions. Hence, logically, bonding should be described first and structure second. However, the detailed role of the electrons in interionic forces is so complex, and the quantitative aspects of the problem of minimizing the potential energy with respect to all the possible configurations is so difficult, that structures cannot be derived. Rather, it is necessary to derive some information about bonding from structures, cohesive energies, refractive indices, electron binding energies, polarizabilities, and other properties through the use of models. Because of the number of ions and electrons involved, the wave-mechanical formulation of bonding in solids is extremely difficult to solve. Fortunately, some rather simple, classically based models modified by quantum-mechanical concepts have been and continue to be quite useful. These can be classified generally as ionic, covalent, and metallic bonding and combinations of the three. SEE CHEMICAL BONDING.

Ionic bonding. In the ionic model, the solid is composed of positive (cations) and negative (anions) charges located on the respective lattice sites. These imagined point charges create through their collective coulombic interactions a net attractive force. The repulsive forces to balance the attractive forces at equilibrium consist essentially of the overlap of the valence electron orbitals. Additional attractive forces are frequently included in the model. These are occasionally referred to as the dispersive forces because they are the ones related to the dipole and multipole moments in ionic crystals and their indices of refraction. Solids having the most ionic character (ionicity) are those formed from combinations of the most electropositive with the most electronegative elements; thus CsF is highly ionic. In general, the ionicity increases with increasing cationic radius and with decreasing anionic radius.

Covalent bonding. In the covalent model, the electrons in the bond are shared equally between the atomic cores, and the electrons involved are in bound orbitals, frequently referred to as molecular orbitals. In essence, the sources of the attractive and repulsive forces are electrostatic, but they are formulated wave-mechanically. In purely covalent bonding, the atomic cores are identical, as in solids of the inert gases, halogens, oxygen, and sulfur.

Metallic bonding. In the metallic model, the solid is composed of cations on lattice sites surrounded by a uniform negative charge of the conduction or free electrons.

Combination bonding. In all real solids composed of different elements, the bonding consists of admixtures of these models. The bonding in ioniclike crystals composed of the regular metallic elements and the more electronegative elements can be rationalized as admixtures of ionic and covalent bonding. The degree of ionicity or covalency can be evaluated through optical dispersion theories of crystals, wherein the ionicity is evaluated from the index of refraction, the density of valence electrons, and some measure of the separation of the bonding and antibonding orbitals. In these formalisms, the group IV elements carbon, silicon, and so on, are used as the reference for covalency. Thus covalency increases as the pairs of elements in the solid become more nearly the same as group IV elements. In the case of the transition, lanthanide, and actinide elements in valence states having open shells, rationalization of the bonding on the basis of this description is inadequate. The crystal field of the solid removes the degeneracy of the orbitals, and the electrons in the bond can occupy states which increase the strength of the chemical bond. Thus there occurs a crystal field stabilization and the attendant anomalous dispersion accompanying the absorption in the infrared and visible spectra. Low-lying f-to-d transitions may also contribute to the increased bonding. In a sense, the role of these crystal field splittings and d states is to increase the polarizabilities of the cations.

In metals formed by higher-valent cations, starting at least with $3+$, the anionic-forming elements can react to produce compounds and phases which still contain free electrons. Thus solids such as subhalides of scandium, the monosulfides and carbides of several elements, and some oxides such as TiO or VO contain cations, anions, and conduction electrons or a bandgap sufficiently small to be n-type semiconductors. The roles of the valence states of the cations and the oxidation of the free electrons are illustrated through the variation of the conductivities and lattice parameters of the lanthanide monosulfides and monoselenides. All of these are conductors except those of europium, ytterbium, and samarium. The rationalization of this behavior is that in the metals the cations are in the $3+$ valency in all cases except europium, ytterbium, and samarium, in which the valency is $2+$. Thus the anionic elements, sulfur and selenium, are reduced by the free electrons to the $2-$ state so that the three cases with $2+$ cations have no free electrons, whereas each of the others has approximately one free electron. The cationic radii and the oxidation-reduction potentials are consistent with this description.

Measurement. One of the more direct ways of measuring the bonding in a solid, and particularly one which measures it at the cationic and anionic sites separately, is by measuring the electron binding energies by photoelectron spectroscopy. Thus, through the ejection of electrons from valence orbitals with x-rays or ultraviolet radiation $h\nu$ and the measurement of the electrons' kinetic energy, their binding energies are determined by reaction (1). In this reaction

$$M^{q+}\text{cs} \xrightarrow{h\nu} M^{(q+1)+} \text{cs}^* + e(g) \qquad (1)$$

cs represents a cation site; the asterisk indicates that the cation site is unrelaxed after the ionization; and the electron is in the free gaseous (g) state. Solid-state physicists generally refer binding energies to electrons in the Fermi level. In reaction (1) the reference is the free electron; the two differ by the instrumental work function. This technique permits determination of the valence state, $q+$, and the lattice self-potential from which measures of polarization, ionicity, and so on are derived. The lattice self-potential is derived through a comparison of the binding energy with the ionization potential of the gaseous cation.

CHEMICAL COMPOSITION

At finite temperatures, and particularly at high temperatures, the partial vapor pressures of the components in a solid are different. Consequently, in general there occurs a preferential loss of one component so that any solid at equilibrium tends to contain lattice defects and to become nonstoichiometric. However, in a large number of cases the deviations from stoichiometry are not detectable, so that the number of nonstoichiometric phases that can be studied in solid-state chemistry is not extremely large. Among those which have been studied extensively are $Fe_{1-x}O$, CeO_{2-x}, and various metal hydrides, as well as the oxides, carbides, and hydrides of uranium. Because of its use in reactor technology, UO_{2+x} may be the most extensively studied nonstoichiometric phase.

Whenever the structure of a solid is such that an interstitial position of one of the sublattices can be occupied, both interstitials and vacancies occur. However, over the compositional range of nonstoichiometry, one or the other of the defects is usually in the higher concentration. Thus UO_2 has a fluorite structure such that the position at the center of the unit cell can accommodate an oxygen ion with an attendant shift in the oxidation state of the uranium ions. In UO_{2+x} the defects are predominantly oxygen ions on interstitial sites and U^{5+} ions on some of the cation sites. In $UO_{2-x'}$ which exists at temperatures near 2000°C (3600°F), vacancies on the oxygen sublattice are at higher concentration. In a phase such as $Th_yU_{1-y}O_{2+x'}$ thorium ions are substitutional, and in $Pu_yU_{1-x}O_{2\pm x}$ plutonium is substitutional. In the last case, because plutonium has the 3+ and 4+ oxidation states accessible and uranium has the 4+ and 5+ states, the anion composition has an extensive range on both sides of stoichiometry. In metallic carbides, the carbon is often interstitial, and in UC_{1+x} the carbon tends to be incorporated in the UC lattice as a C_2 unit.

The role of the compositional variable in oxidation at the electronic level in the solid is illustrated in the x-ray photoelectron spectra of the valence band region of uranium at four stages of oxidation from metal to UO_3. For uranium metal and dioxide, the intensities of the 5f-orbital electrons are the same; no change occurs in the number of 5f electrons. Hence, the oxidation of the metal to UO_2 involves only the conversion of the free electrons to bound electrons in the hybridized 2p-orbital on the oxygen ion. When UO_2 is oxidized to U_4O_9, however, the intensity of the 5f part decreases and that of the 2p increases. With further oxidation to U_3O_8, the 5f intensities decrease further and the 2p intensity increases. Finally, in UO_3, the 5f electrons are completely oxidized to 2p electrons on the oxygen ion.

One of the measurements, which furnishes information needed to understand the compositional variable in solid-state chemistry, is the observation of partial vapor pressures, and their dependence on temperatures is determined. The chemical potentials and the partial molar enthalpies and entropies and their variation with composition are derived from these measurements. For a binary system, the chemical potential increases monotonically in the single-phase region, and is constant in the diphasic region. The construction and mathematical evaluation of statistical models to describe the chemical potential throughout both regions from the interionic forces, valence states, and defect energies contain conceptual problems associated with discontinuities at the diphasic region. Models can be constructed and evaluated to describe the homogeneous regions. One model is described by Eq. (2), in which θ_i is the fraction of interstitial sites occupied,

$$\mu = \mu_0 + RT\left[\ln\left(\frac{\theta_i}{1-\theta_i}\right) - E_i - 2\theta_i E_{ii}\right] \quad (2)$$

E_i is the energy required to remove an ion from the interstitial site, and E_{ii} is the energy of interaction between two occupied interstitial sites. Above some critical temperature ($T_c = E_{ii}/2R$), this function has a sigmoidal shape with no maximum or minimum. Thus for $T > T_c$, the variation of μ with θ_i (composition) as shown in **Fig. 2** represents a monophasic region. At the critical temperature, the isotherm has a zero slope. Below this critical temperature, the function has a maximum, a minimum, and an inflection between them as shown by the isotherm for $T < T_c$ in Fig. 2. Because the slope of μ versus composition in a real system cannot change sign, this behavior is unrealistic. The artifice which is introduced to excuse these van der Waal loops is to construct a horizontal line through the inflection point and use this to represent the diphasic region. A realistic model which describes the evolution to a diphasic region would be one in which the horizontal line is contained in the mathematics. In one technique which has been suggested, the chemical potential is described through the complex (mathematical) variable. In some cases (praesodymium oxides), the compositional isotherm displays hysteresis loops because of the defect complexes which form differently in the two directions of μ versus composition.

When a substantial nonstoichiometric phase is formed between an insulator and a metallic conductor, the electrical conductivity and concentration of free electrons of course change continuously. However, in some cases and perhaps in all cases, the change is so rapid over a small compositional range that the change can be viewed as an insulator-to-metal transition. One system which displays this behavior is $Sm_{1-x}Nd_xSe$. SmSe is an insulator with Sm^{2+} and S^{2-}, and NdSe is a conductor with Nd^{3+}, S^{2-}, and one free electron per unit. As the composition is varied, the conductivity and concentration of free electrons change significantly, but at $x \cong 0.1$ the two change by five orders of magnitude.

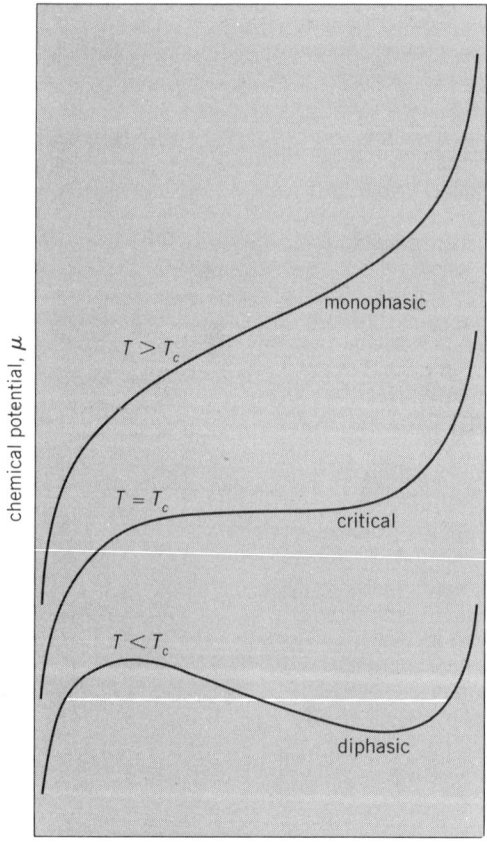

Fig. 2. Three isotherms for the variations of chemical potential with composition of defects.

CHEMICAL REACTIONS

The mechanisms of chemical reactions within and between solids are through lattice vibrations, lattice defects, and changes in valence states. These are the structural features through which migration of mass, charge, and energy occur. Consequently, diffusion and conductivity are integral, basic parts of solid-state chemistry, even though their quantitative roles in the totality of chemical reactions have not been developed. So long as the solid phase produced during reaction has a density nearly the same as those for the reactants, the microscopic description of the reaction in terms of mass and charge transfer is feasible. However, in a number of reactions the molar volumes differ sufficiently to destroy the integrity of the product, so that much of the reaction then proceeds at interfaces, microcracks, and fissures. Thus the total net mechanism becomes a composite process so complex that a comprehensive theory is lacking. The theories that have been developed are generally discussed under topics such as corrosion. Thus, no formal classification of reactions in solids is universally recognized, or, in fact even cited, in the literature. A categorization which identifies the scope and suggests a basis is the following: condensation, internal processes, interphase reactions, electrochemical reactions, photochemical reactions, and sublimation. Another possible basis of classification is one which recognizes that all reactions occur because of gradients in the chemical potential. These gradients may be caused by gradients in mass, concentration, electrical charge, or energy (heat), or combinations of these.

Condensation. Although it is well recognized that condensation to form solids occurs when the translational energy of the gaseous reactants is dissipated into vibrational and electronic states, the processes involved are sufficiently complex that they are difficult to study and to describe in detail. In general, two kinds of condensation need to be described. In a supersaturated vapor, condensation is a stepwise process in which molecules form in successively larger clusters if the rate of growth of the clusters exceeds the rate of dissociation. At some critical size, the process results in condensation, because the translational energy can be dissipated into the cluster. This process occurs in shock tubes and jets which produce super-saturated vapors and in hydrodynamically flowing gaseous beams in which adiabatic expansion occurs. Another kind of condensation occurs if atoms or molecules in the gas are sufficiently accommodated thermally on a colder substrate. In this case, the growth occurs primarily through the flow of translational energy into the substrate. However, the atoms or molecules bound to the surface can be mobile on the surface so that growth can occur through a stepwise process therein. Usually the process is nonequilibrious, and an amorphous, metastable product is formed. For instance, gaseous silicon monoxide can be condensed on a cold substrate to produce solid, amorphous, metastable silicon monoxide. This is the only known way of producing it. Condensation to produce epitaxial layers on substrates is used to produce various kinds of solid-state devices.

An elementary condensation which can be used to illustrate the processes involved and which represents what could be called the first step in a solid-state chemistry is the formation of a metal. Because the process is one in which the neutral gaseous atom is converted to a collection of cations and free electrons, it involves internal oxidation and reduction, represented by reaction (3). Thus, in this reaction the gaseous atom with the electronic configuration of the core electrons

$$M(---np^6) \rightarrow M^{q+} (---np^{(6-q)})e_q \tag{3}$$

(---) and the outer six np electron condenses to form something resembling a cation of charge $q+$ associated with q free electrons. In the supersaturated vapor, the critical step probably is the one in which the cluster is large enough to have free or conduction electrons to serve to dissipate the translation energy. When gaseous metal-forming atoms impinge on a metal substrate, the free electrons in the latter serve to transfer the energy, so that the accommodation and condensation coefficients of metal on metal are generally unity. On nonmetallic insulators, however, the coefficients are not unity, and condensation occurs through nucleation on the surface. The significance of recognizing that the condensation to form a metal is a redox process is contained in what is involved upon further oxidation. When the solid metal undergoes further reaction with an anion-forming element, it is the free electrons which are used first, and the cation subsequently oxidized.

Internal processes. One of the simplest reactions that occurs in solids is a phase transition. These transitions may be either first- or second-order. In the first case, a change occurs in the crystal structure, sometimes with the higher-temperature form having an apparently higher symmetry. Second-order transitions are usually order-disorder transitions such as the λ-type transitions which occur on only one of the sublattices.

The primary internal reactions which involve a chemical reaction are those associated with site occupancy, changes in valency, and clustering. Because it has been extensively studied, uranium dioxide serves as a useful illustration. In this phase, the interstitial position (*is*) of the fluorite lattice can accommodate an oxygen ion. Thus, an equilibrium exists between the occupancy of this site and the regular anion sites (*as*), as shown in reaction (4). Whenever uranium

$$O^{2-}as + is \rightarrow O^{2-}is + as \qquad (4)$$

dioxide is oxidized, the valence of uranium is increased from 4+ to 5+ so that reaction (5) is the

$$2U^{4+}cs + O(g) + is \rightarrow 2U^{5+}cs + O^{2-}is \qquad (5)$$

one occurring primarily. Reactions similar to these are responsible for the existence of nonstoichiometric phases. The dual valency on the cation also is involved in hole conductivity in *p*-type semiconductors. Although the precise mechanism involved in hole conduction is not known, some kind of a charge exchange promoted by phonons and involving cations and perhaps anion or anion sites occurs. A plausible one in uranium dioxide is shown in reaction (6).

$$2U^{4+}cs \rightarrow U^{5+}cs + U^{3+}cs \qquad (6)$$

A comparison of the energies associated with these reactions in solids and in gases illustrates the differences in the two cases. In the gas phase, the O^{2-} ion is unstable; in the solids, it is stabilized by the lattice self-potential. The energy associated with reaction (6) in the gas phase equals the difference between the two ionization potentials. Usually, this difference is of the order of tens of electronvolts. However, in the solid, this difference is compensated by the difference in the lattice self-potentials. Consequently, the energy associated with reaction (6), in the solid phase, is only a few electronvolts.

Interphase reactions. Reactions between phases accompanied by product formation with integrity maintained occur via chemical diffusion. This situation occurs when the reactants and the product have the same structure and comparable ionic radii. But even in this case, there generally is a volume change, the concentration of defects changes, and the initial interface between the reactant moves. A classical example is the reaction (that is, diffusion) between copper and copper-zinc alloy. The zinc diffuses more rapidly out of the alloy than the copper diffuses in, and the original interface moves toward the alloy. This phenomenon is known as the Kirkendall effect; it is one of the evidences that diffusion occurs via vacancies and interstitials rather than through positional exchange. SEE PHASE EQUILIBRIUM.

Two well-studied solid-state chemical reactions are sintering and corrosion. In the first reaction, mass is transferred between particles in a solid so that densification and frequently plastic deformation or creep occurs. The details of the transfer between particles is not well understood, but self-diffusion is generally involved. For instance, sintering and creep in uranium dioxide occur at temperatures where self-diffusion of oxygen through the interstitial position becomes significant. At these temperatures, self-diffusion of uranium is insignificant. In corrosion, diffusion through the product layer generally controls the rate of reaction, although electromigration may also be significant.

In both cases, the mechanism can involve either the migration of a reacting component from the substrate through the layer or the migration of a corroding component from the external surface. When the layer becomes sufficiently thick and its density is sufficiently different, the layer develops fissures and cracks. Then the mechanism occurs along these, but the primary chemical driving force may still be the same as before.

Electrochemical reactions. There are scientific as well as technological interests in electrochemical reactions in solids. Electrolytic cells with solid electrodes and electrolytes have been used to determine chemical potentials, especially in nonstoichiometric phases. An example is a cell composed of nonstoichiometric oxides, UO_{2+x} and Fe–$Fe_{1-y}O$, as electrodes and stabilized zirconia as the electrolyte: Fe–$Fe_{1-y}O|ZrO_2|UO_{2+x}$. The electrolyte is an oxygen ion conductor, so that the voltage developed is produced by the difference in the chemical potential of oxygen in the two solid electrodes. Rechargeable batteries in which either the electrolyte or the electrodes are solids have been investigated for technological use. Among these are one in which β-alumina is used as an electrolyte and one in which FeS or FeS_2 is used as a cathode and aluminum-lithium alloy is used as an anode. In the first example, sodium and sulfur are used as electrodes; in the second one, molten lithium-potassium chloride is used as an electrolyte. SEE ELECTROCHEMISTRY.

Photochemical reactions. Many of the photo-induced processes which occur in solids can be imagined to be solid-state chemical reactions. The classical ones, of course, are those involved in photographic plates and films. SEE PHOTOCHEMISTRY.

Sublimation. At sufficiently high temperatures, the rate of sublimation of all solids can be measured. A technique which has been especially useful for inorganic solids (and a few organic solids) is the Knudsen effusion technique. Thus the molecular composition of the vapor and the partial vapor pressures are derived through the measurement of the rate at which the saturated vapor effuses through a small orifice in a cell into a vacuum and through the measurement of the mass spectrum of the effusate. The total vapor pressures of several solids have also been determined from measurements of the momentum of the effusate. The partial vapor pressures of most of the inorganic fluorides, chlorides, and oxides and several of the carbides and sulfides have been determined. In a number of cases, the vapor contains complex molecular species, even though the temperature required is high. The presence of these molecules is, in a sense, a consequence of the relatively high vapor pressure and not the high temperature, which of course at the constant pressure is a degradative factor. The chemical potentials and the partial molar enthalpies and entropies of sublimation are derived from the partial vapor pressures and their dependencies on temperature. These are direct measures of the bonding in the solid.

Although it might be argued that the species observed in the vapor are unrelated to the structure of the solid, it is apparent that there must be a relation because the same basic elements of bonding are involved. Thus ionic solids tend to sublime to ionic molecules; the bonding in gaseous alkali and alkaline earth fluorides and chlorides tends to be ionic. The more covalent oxides of tungsten and molybdenum, WO_3 and MoO_3, sublime to covalently bonded polymers, trimers, tetramers, and pentamers. SEE HIGH-TEMPERATURE CHEMISTRY.

Bibliography. D. M. Adams, *Inorganic Chemistry: An Introduction to Concepts in Solid State Structural Chemistry*, 1974; S. Holt et al. (eds.), *Solid State Chemistry: A Contemporary Overview*, 1980; P. Kofstad, *Nonstoichiometry, Diffusion, and Electrical Conductivity in Binary Metal Oxides*, 1972, reprint 1983; R. Metselaar et al., *Solid State Chemistry*, 1982; G. M. Rosenblatt and W. L. Worrell (eds.), *Progress in Solid State Chemistry*, vol. 13, 1982, vol. 15, 1985; H. Schmalzried, *Solid State Reactions*, 2d ed., 1981.

CONTRIBUTORS

CONTRIBUTORS

Allred, Dr. Albert L. Department of Chemistry, Northwestern University.

Atkins, Dr. P. W. Department of Chemistry, Oxford University, England.

Bagley, Dr. Brian G. Bell Laboratories, Murray Hill, New Jersey.

Bender, Prof. Paul J. Department of Chemistry, University of Wisconsin.

Bolton, Dr. James R. Department of Chemistry, University of Western Ontario, London, Canada.

Brooks, Prof. Philip R. Department of Chemistry, Rice University.

Brown, Prof. Frederick C. Department of Physics, University of Illinois.

Brown, Dr. Glenn H. Professor of Chemistry and Director of the Liquid Crystal Institute, Kent State University.

Browne, Prof. Cornelius P. Department of Physics, University of Notre Dame.

Buck, Dr. Richard P. Department of Chemistry, University of North Carolina.

Burdett, Prof. Jeremy K. Department of Chemistry, University of Chicago.

Burwell, Dr. Robert L., Jr. Department of Chemistry, Northwestern University.

Caldwell, Prof. C. Denise, Department of Physics, Yale University.

Chu, Prof. Benjamin. Department of Chemistry, State University of New York, Stony Brook.

Clark, Dr. Noel A. Department of Physics, University of Colorado.

Curtiss, Prof. Charles F. Department of Chemistry, University of Wisconsin.

Daniels, Dr. Farrington. Deceased; formerly, Professor Emeritus, Solar Energy Laboratory, University of Wisconsin.

Davis, Dr. Jack. Deceased; formerly, Research and Development, Bright Star Industries, Inc., Clifton, New Jersey.

Delahay, Prof. Paul. Department of Chemistry, New York University.

DiSalvo, Dr. Frank J. Bell Laboratories, Murray Hill, New Jersey.

Eliel, Prof. Ernest L. Department of Chemistry, University of North Carolina.

Endicott, Prof. John F. Department of Chemistry, Wayne State University.

Ewing, Prof. George E. Department of Chemistry, Indiana University.

Eyring, Dr. Edward M. Department of Chemistry, University of Utah.

Fano, Prof. U. James Franck Institute, University of Chicago.

Forster, Dr. Denis. Monsanto Company, St. Louis, Missouri.
Fowkes, Prof. Frederick M. Department of Chemistry, Lehigh University.
Franzese, Dr. Kenneth. Research and Development, Bright Star Industries, Inc., Clifton, New Jersey.

Gaines, Dr. George L., Jr. Research and Development Center, General Electric Company, Schenectady, New York.
Glicksman, Dr. Richard. Solid State Division, RCA Corp., Somerville, New Jersey.
Gordon, Dr. Sheffield. Chemistry Division, Argonne National Laboratory, Argonne, Illinois.
Gruen, Dr. Dieter M. Argonne National Laboratory, Argonne, Illinois.

Hamer, Dr. Walter J. Institute for Basic Standards, National Bureau of Standards.
Hartley, Dr. A. M. Department of Chemistry and Chemical Engineering, University of Illinois, Urbana-Champaign.
Hildebrand, Dr. Joel H. Deceased; formerly, Department of Chemistry, University of California, Berkeley.
Hirschfelder, Dr. J. O. Department of Chemistry, University of Wisconsin.

Jaep, William F. Central Research Department, Experimental Station, E. I. Du Pont de Nemours and Company, Wilmington, Delaware.
Johnston, Dr. Francis J. Department of Chemistry, University of Georgia.
Jones, Prof. R. Alan. School of Chemical Sciences, University of East Anglia, England.

Knotek, Dr. M. L. Sandia National Laboratories, Albuquerque, New Mexico.

Laitinen, Prof. Herbert A. Department of Chemistry, University of Florida.
Laurie, Prof. Victor W. Department of Chemistry, Princeton University.
Levine, Prof. Ira N. Department of Chemistry, Brooklyn College.
Lineberger, Dr. W. C. Department of Chemistry, University of Colorado.

Madison, Dr. Vincent. Department of Medicinal Chemistry, School of Pharmacy, University of Illinois.
Mahoney, Dr. Lee R. Department of Chemistry, Ford Motor Company, Dearborn, Michigan.
Mantell, Dr. Charles L. Consulting Chemical Engineer, Electrochemical Engineering, Process and Plant Design, Manhasset, New York.
Margrave, Prof. John L. Dean of Research, Rice University.
Matijevic, Prof. Egon. Chairperson, Department of Chemistry, Clarkson College of Technology.
Mill, Dr. George S. Research Chemist, Shell Oil Company, New York, New York.
Miller, Dr. Glenn H. Weapons Effects Division, Sandia Laboratories, Albuquerque, New Mexico.
Miller, Dr. William H. Department of Chemistry, University of California, Berkeley.
Milligan, Dr. W. O. Robert A. Welch Foundation, Houston, Texas.
Moore, Dr. Paul B. Department of the Geophysical Sciences, University of Chicago.
Mulliken, Dr. Robert S. Professor Emeritus, Departments of Physics and Chemistry, University of Chicago.

Nachtrieb, Prof. Norman H. Chairperson, Department of Chemistry, University of Chicago.
Nedelsky, Prof. Leo. Department of Physical Science, University of Chicago.
Nesbitt, Dr. David J. Joint Institute for Laboratory Astrophysics and Department of Chemistry, University of Colorado.
Noyes, Richard M. Department of Chemistry, College of Arts and Sciences, University of Oregon.

O'Dom, Dr. George W. Research Scientist, TRW, Inc., Redondo Beach, California.

Parr, Prof. Robert G. Department of Chemistry, Johns Hopkins University.
Pecora, Prof. Robert. Department of Chemistry, Stanford University.
Perone, Prof. Sam P. Department of Chemistry, Purdue University.

Pierotti, Prof. Robert A. Department of Chemistry, Georgia Institute of Technology.

Riggs, Harold C. Retired; formerly, Manager, Marketing New Product Development, Electric Battery Company, Philadelphia, Pennsylvania.

Rochow, Prof. Eugene G. Retired; formerly, Department of Chemistry, Harvard University.

Rozeanu, Prof. L. Department of Material Science, Technion–Israel Institute of Technology, Haifa.

Sandler, Prof. Stanley I. Department of Chemical Engineering, University of Delaware.

Schaefer, Dr. Henry F., III. Department of Chemistry, University of California, Berkeley.

Scott, Prof. Robert L. Department of Chemistry, University of California, Los Angeles.

Slabaugh, Dr. Wendell H. Department of Chemistry, Oregon State University.

Smith, W. W. Technical Adviser, Legal Department, ESB Inc., Philadelphia, Pennsylvania.

Steele, William A. Department of Chemistry, Pennsylvania State University.

Stevens, Dr. Brian. Department of Chemistry, University of South Florida.

Taube, Dr. Henry. Department of Chemistry, Stanford University.

Thoma, Roy E. Oak Ridge National Laboratory, Oak Ridge, Tennessee.

Thomas, Dr. J. Kerry. Department of Chemistry, University of Notre Dame.

Thorn, R. J. Chemistry Division, Argonne National Laboratory, Argonne, Illinois.

Vanderzee, Dr. Cecil E. Department of Chemistry, University of Nebraska.

Van Winkle, Prof. Quentin. Department of Chemistry, Ohio State University.

Waddington, Prof. Thomas C. Department of Chemistry, University of Durham, England.

Wakeham, Dr. W. A. Department of Chemical Engineering, Imperial College, London, England.

Ware, Dr. Bennie R. Department of Chemistry, Syracuse University.

Wasserman, Dr. Zelda R. Department of Central Research and Development, E. I. Du Pont de Nemours and Company, Wilmington, Delaware.

Weber, Harold C. Chemical Engineer, Boston, Massachusetts.

Weller, Prof. Robert A. A. W. Wright Nuclear Structure Laboratory, Yale University.

Wenkert, Ernest. Department of Chemistry, University of California, San Diego.

Wentorf, Robert H., Jr. General Electric Research Laboratory, Schenectady, New York.

Wentworth, Prof. R.A.D. Department of Chemistry, Indiana University.

Wilcox, Prof. William R. Department of Chemical Engineering, Clarkson College of Technology.

Wilen, Dr. Samuel H. Department of Chemistry, City College of the City University of New York.

Wilhoit, Dr. Randolph C. Thermodynamics Research Center, Texas A & M University.

Wilke, Prof. Charles R. Department of Chemical Engineering, University of California, Berkeley.

Williams, Dr. Jack M. Chemistry Division, Argonne National Laboratory, Argonne, Illinois.

Wilson, Prof. E. Bright, Jr. Department of Chemistry, Harvard University.

Wilson, Dr. Therese. Biological Laboratories, Harvard University.

Young, Prof. Thomas F. Department of Chemistry, University of Chicago.

Zuman, Dr. Peter. Department of Chemistry, Clarkson College of Technology.

INDEX

INDEX

Asterisks indicate page references to article titles.

Absolute entropy 44
Acid-base catalysis 65–66
Activity 47–49*
  coefficients 48–49
  standard states 47–48
  theory 47
Adsorption 103–107*
  applications 103–104
  chemisorption of vapors 106
  kinetics 106–107
  from nonaqueous solutions 106
  physical adsorption of vapor 104–106
  of solutes from aqueous solutions 106
Amorphous solid 204–205*
  preparation 204
  types 205
Antiferromagnetism 377
Atomic diamagnetism 376
Atomic paramagnetism 376–377
Atomic structure and spectra: valence and structure 188
Atomic weight 230–232*
  accurate mass values 231
  mass tables 231–232
  unified scale 231
Autoxidation 67

Avogadro's number 225–227*
  determination 226–227
  significance 225–226

Battery 242, 301–302*
  applications 301
  life 302
  primary *see* Primary battery
  ratings 302
  reliability 302
  reserve *see* Reserve battery
  size 302
  storage *see* Storage battery
  types 301
Becker, P. 227
Benesi, H.A. 218
Bertholet, M.P.E. 217
Bertholet, P.E.M. 199
Biot, J.B. 180
Bohr, N. 188
Boiling point 237*
Bond angle and distance 142–144*
  biological effects 143
  bonding information 143
  determination 143–144
  physical properties 142–143
Buffers 75–77*
  effectiveness 75–76
  water as solvent 76–77

Calorimetry: measurements 19
Capillarity 98–99
Carter, H.M. 217
Catalysis 64–65*
  acid-base 65–66
  categories 65
  heterogeneous *see* Heterogeneous catalysis
  homogeneous *see* Homogeneous catalysis
  monomolecular film 111–117*
  phase-transfer catalysis 72–75*
  selectivity 65
Chain isomers 163–164
Chain reaction 63–64*
  photochemical reactions 63
  thermal reactions 63–64
Chapman, L. 123
Chapman, S. 123
Chemical bond theory 137–140*
  complex molecules 139–140
  covalent bond 137
  hydrogen molecule 138–139
  molecular mechanics model 367
Chemical bonding 140–142*
  bond angle and distance 142–144*

Chemical bonding (cont.):
  bond theory see Chemical bond theory
  conjugation and hyperconjugation 148–150*
  covalent bonding 141–142
  crystal field theory 144–148*
  electron orbitals see Molecular orbital theory
  hydrogen bond see Hydrogen bond
  ionic bonding 140–141
  metallic bonding 142
  solids 383–385
  valence 187–191*
  van der Waals forces 142
Chemical dynamics 54–63*
  adsorption kinetics 106–107
  bimolecular processes 54
  bimolecular reactions 55–57
  electric-field jump experiment 61
  energy distribution in exothermic reactions 58
  high-temperature chemistry 374–375
  lasers as reaction probes 58
  molecular dynamics 55–57
  most effective energy for reactions 58
  pressure-jump experiment 62–63
  reaction kinetics 54
  relaxation methods 59–63
  temperature jump experiment 60–61
  theoretical models and methods 58–59
  ultrasonic absorption experiment 61–62
  unimolecular processes 54
  unimolecular reactions 57
Chemical equilibrium 21–27*
  activity and standard states 22–23
  chemical potential 21–22
  equilibrium constant 23–25
  heterogeneous equilibria 26–27

Chemical equilibrium (cont.):
  homogeneous equilibria 25–26
  pK 51*
  thermodynamics 12–13
Chemical thermodynamics 8–17*
  activity 47–49*
  affinity and chemical equilibrium 12–13
  basic concepts 8–9
  chemical potential 13–15
  entropy see Entropy
  first law of thermodynamics 9–10
  free energy see Free energy
  fugacity 49–50*
  Gibbs free energy see Gibbs function
  Helmholtz free energy see Helmholtz free energy
  internal energy see Internal energy
  irreversible processes 16–17
  phase equilibrium see Phase equilibrium
  second law of thermodynamics 10–12
  specific heat 240*
  thermodynamical relationships 15–16
  third law of thermodynamics 16
  see also Thermochemistry
Chemiluminescence 339–340*
  molecular dynamics studies 55
Circular dichroism 333
Colloid 208–211*
  coagulation and flocculation 210–211
  dispersion stability 210
  isoelectric point 240*
  preparation 209–210
  properties 209
Colloidal crystals 211–213*
  complex systems 212
  crystal structure and properties 211–212

Combining volumes, law of 137*
Computational chemistry 365–368*
  interactive computer graphics 367–368
  molecular dynamics 366–367
  molecular mechanics 367
  molecular orbital theory 366
Conjugation and hyperconjugation 148–150*
Constitutional isomers 163–165
  functional isomers 163
  positional and chain isomers 163–164
  properties 164–165
Coulometric analysis 286–288*
  controlled-current analysis 287
  controlled-potential analysis 286–287
  flow electrolysis 287
  reaction kinetics 288
  reaction mechanisms 287
  thin-layer methods 287
Covalent bonding 141–142
  bond theory 137
  solids 384
Critical opalescence 341
Crystal field theory 144–148*
  anomalous effects 147–148
  complete theory 148
  number of electrons 145–146
  tetrahedral array 146
Crystallization 223–225*
  melts 225
  solutions 224–225

Davy, H. 69
Density 230*
Deslattes, R. 227
Desorption 107–111*
  stimulated 107–110
  thermal 110–111
Dialysis 132–133*
  Donnan equilibrium 39

Dialysis (cont.):
  electrodialysis 268
Diamagnetism: atomic 376
  molecular 376
Diastereomers 166, 181
Diffusion in gases and liquids
    120, 122–128*
  convection or bulk flow
    124
  correlation of mass-transfer
    coefficients 126–127
  dialysis 132–133*
  eddy diffusion 123
  equivalent film thickness
    126
  forced diffusion 124
  general equation for mass-
    transfer flux 124–125
  interdiffusion of two fluids
    128
  mass-transfer coefficients
    125–126
  molecular diffusion 122–123
  thermal diffusion 123–124
  unsteady-state diffusion
    and convection 127–128
Digital computer: computa-
    tional chemistry
    365–368*
Dipole moments 169–170
Donnan, F.G. 38
Donnan equilibrium 38–39*
Dootsen, F.W. 124
Dry cell 310–314*
  chemistry 312–313
  construction 310–312
  mercury battery 317–318*
  modifications 314
  operating characteristics
    313
  recharging 314
  shelf life 313
  temperature effect 313
  zinc chloride cell 314

Eddy diffusion 123
Einstein, A. 226
Electric accumulator see Stor-
    age battery

Electroanalytical chemistry
    243
  coulometric analysis
    286–288*
  electrodeposition analysis
    292–295*
  ion-selective membranes
    and electrodes 282–286*
  polarographic analysis
    288–292*
Electrochemical process
    256–270*
  alkali-chlorine processes
    260–264
  electrodialysis 268
  electroendosmosis 268–269
  electrolysis in aqueous solu-
    tions 256
  electrolytic corrosion of
    metals 260
  electromagnetic separation
    268
  electrophoretic deposition
    268
  electrothermics 268
  fused-salt electrolysis
    265–268
  in gases 268
  inorganic 256–269
  ion-permeable membrane
    cells 265
  metallurgical applications
    256–260
  organic 269–270
  oxidations and reductions
    265
Electrochemical series
    247–249*
Electrochemical techniques
    252–256*
  alternating-current polaro-
    graphy 254–255
  alternating potential meth-
    ods 254–255
  chronopotentiometry 255
  constant potential with con-
    vection 252
  controlled current methods
    255

Electrochemical techniques
    (cont.):
  controlled potential meth-
    ods 252–253
  coulostatic analysis 255
  cyclic chronopotentiometry
    255
  linearly varying potential
    253–254
  potentiostatic chronoampe-
    rometry 252–253
  pulse polarography 255
  square-wave polarography
    255
  thin-layer electrochemistry
    256
  variable potential methods
    253–254
Electrochemistry 242–243*
  electroanalytical chemistry
    243
  electrode kinetics 243
  electrodeposition 242
  electrolytic processes 242
  electrothermics 242
  galvanic cells 242
  miscellaneous phenomena
    243
  solid-state chemical reac-
    tions 388–389
  techniques for
    high-temperature chemis-
    try 374
Electrode 249*
  kinetics 243
Electrode potential 249–251*
  electromotive force
    298–301*
  gas electrodes 250
  measurements 250–251
  metal-metal ion electrodes
    249–250
Electrodeposition analysis
    292–295*
  constant-current analysis
    294–295
  constant-potential electroly-
    sis 293–294
  internal electrolysis 294
Electrodialysis 268

Electrokinetic phenomena 271–274*
  displacement of charged layers 272–274
  electrically charged layers 272
Electrolysis 242, 270–271*
  alkali halides 260–264
  applications 271
  in aqueous solutions 256
  corrosion of metals 260
  fused-salt 265–268
  theory 270–271
Electrolyte 243*
Electrolytic conductance 244–247*
  equivalent conductance 245–246
  ion conductance 246–247
  measurement 244–245
  nonaqueous systems 247
Electrometallurgy 242, 256–260
Electromotive force 298–301*
  cell types 298–299
  energy relations 299
  measurements 299
  standard cells 299–301
Electron affinity 184–185*
Electronegativity 185–187*
Electrophoresis 274–278*
  electrophoretic deposition 268
  gel techniques 275
  isoelectric focusing 275
  isoelectric point 240*
  isotachophoresis 276
  laser applications 276–277
  Tiselius cell 274–275
Electrothermics 242, 268
Emulsion 215–216*
  classification 215–216
  properties 216
Enantiomers 165–166
  see also Stereochemistry
Enthalpy 17–18, 41–42*
  free energy 44–46*
  of reaction 18–19
Entropy 42–44*
  absolute entropy 44

Entropy (cont.):
  degradation of energy 43
  function 42
  heat flow 42–43
  increasing entropy and mixing 43–44
  nonconservation of 43
  reversible processes 42
Enzyme: catalysis 65

Fabry-Perot interferometry 337
Feibelman, P.J. 107
Ferromagnetism 377
First law of thermodynamics 9–10
  internal energy see Internal energy
Fletcher, H. 226
Forced diffusion 124
Franck-Condon principle 177–178
Free energy 44–46*
  Gibbs see Gibbs function
  Helmholtz see Helmholtz free energy
  heterogeneous systems 45–46
  partial molal quantities 45
  theory 44–45
Free radical 194–198*
  detection and estimation 196–197
  mechanisms 197–198
  production 195–196
Fuel cell 314–317*
Fugacity 49–50*
Functional isomers 163
Fused-salt electrolysis 265–268
Fused-salt phase equilibria 35–38*
  binary systems 36
  methods of investigation 36
  ternary systems 36–38
  types 36–38

Gas 198–202*
  combining volumes, law of 137*
  diffusion in see Diffusion in gases and liquids
  empirical equations of state 199–200
  liquid-gas equilibrium 28–30
  molecular basis of viscosity 129–130
  principle of corresponding states 200–201
  theoretical considerations 201–202
Gas constant 227*
Gel 214–215*
Gibbs, J.W. 35, 39
Gibbs free energy see Gibbs function
Gibbs function 11, 44, 46–47*
Goldschmidt, V.M. 35
Gram-equivalent weight 236*
Gram-molecular weight 236–237*
Guldberg, C.M. 24
Gurney-Mott theory 347–348

Hartree-Fock formalism 361–365
Heat: first law of thermodynamics 10
  thermochemistry see Thermochemistry
Heat capacity: first law of thermodynamics 10
Heating value 20–21
Heats of reaction 18–19
Helfand, E. 367
Helmholtz free energy 11, 44
Herzberg, G. 362
Heterogeneous catalysis 65, 69–72*
  catalysts 70–71
  history 69
  mechanism 71–72
  reactors 69–70

High-pressure chemistry 368–372*
  equipment 369
  high-pressure effects 370–372
High-temperature chemistry 372–375*
  calorimetry 374
  chemical kinetics 374–375
  condensed systems 373–374
  electrochemical techniques 374
  gaseous species 373
  sublimation 389
Hildebrand, J. 217, 218
Homogeneous catalysis 65–69*
  acid-base catalysis 65–66
  metal complexes 66–67
  oxidation 67–69
Hydrogen bond 152–153*
  bond energies 152
  neutron diffraction 152–153
  spectroscopy 152
  theory 153
Hydrogen ion: buffers 75–77*
  pH 50*
Hydrogen molecule: bond in 138–139
  molecular orbital theory 154–155
Hyperconjugation see Conjugation and hyperconjugation

Indirect oxidation 67–69
Inhibitor 77–79*
Inorganic photochemistry 349–353*
  bimolecular excited-state reactions 352–353
  excited states 350
  unimolecular reactions 350–352

Intensity fluctuation spectroscopy 336–337
Intercalation compounds 150–152*
  intercalation reaction 151
  physical properties 151–152
  types of hosts 151
Interface of phases 102–103*
  adsorption see Adsorption
  contact angle 102–103
  monomolecular film 111–117*
  surface-active agent 99–101*
  surface energy 102
  surface tension 98–99*
Intermolecular forces 191–194*
  description 191
  occurrence 192–193
  origin 192
  study methods 193–194
  types in solution 216–222
Internal energy 17, 41*
  enthalpy see Enthalpy
  first law of thermodynamics 9–10
  free energy 44–46*
Ion 198*
  Donnan equilibrium 38–39
Ion-selective membranes and electrodes 282–286*
  applications 285–286
  classification and responsive ions 283–285
  properties 282–283
Ionic bonding 140–141
  solids 384
Ionic equilibrium 79–81*
  pK 51*
  quantitative relationships 80–81
  types 79–80
Ionization 79*
Irreversible processes: thermodynamics of 16–17
Isoelectric point 240*
Isomer: molecular isomerism 162–168*
Isotachophoresis 276

Joule, J.P. 43
Kirkwood, J.G. 218
Knotek, M.L. 107
Kollman, P.A. 368

Laser: chemical dynamics studies 58
Laser photochemistry 353–358*
  advantages of lasers 353–354
  laser-initiated chemistry 354–355
  laser isotope separation 357–358
  photon dissociation 356–357
  picosecond photochemistry 355–356
Law of combining volumes 137
LeBel, J.A. 180
Lewis, G.N. 47, 49, 189, 218
Liquid 202–204*
  boiling point 237*
  diffusion in see Diffusion in gases and liquids
  solid-liquid equilibrium 31–33
  theoretical explanations 203–204
  thermodynamic relations 202–203
  transport properties 203
  viscosity 130–131
Liquid crystals 205–208*
  applications 208
  classification and structure 205–208
Liquid-gas equilibrium 28–30
Liquid-liquid equilibrium 30–31
Lithium primary cell 318–320*
London, F. 216
London forces: solutions 216–217
Loschmidt, J. 226
Luminescence: chemiluminescence 339–340*

Magnetochemistry 375–377*
    antiferromagnetism 377
    atomic diamagnetism 376
    atomic paramagnetism 376–377
    ferromagnetism 377
    molecular diamagnetism 376
    molecular paramagnetism 377
Mass 227–230*
    atomic weight 230–232*
    density 230*
    and energy 229–230
    gravitation and inertia 228–229
    properties 229
Melting point 238*
Mercury battery 317–318*
Metallic bonding 142
Micelle 213–214*
Millikan, R.A. 226
Mole: Avogadro's number 225–227*
Molecular adhesion 101–102*
Molecular beams: molecular dynamics studies 55–56
Molecular diamagnetism 376
Molecular diffusion 122–123
Molecular isomerism 162–168*
    classification of isomers 162–163
    constitutional isomers 163–165
    interconversion limitations 166–168
    stereoisomers 165–166
Molecular orbital theory 154–162*
    computational chemistry 366
    first-row diatomics 155–157
    frontier orbitals 160–161
    hydrogen molecule 154–155
    multicenter bonding 157–158
    orbital symmetry 160
    symmetry considerations 158–160

Molecular orbital theory (cont.):
    transition-metal complexes 158
Molecular paramagnetism 377
Molecular structure and spectra 168–179*
    dipole moments 169–170
    electronic band spectra 175–178
    electronic band structures 178–179
    magnetochemistry 375–377*
    molecular electronic states 178–179
    molecular energy levels 170–173
    molecular polarizability 170
    molecular sizes 168–169
    molecular spectra 173–178
    polyatomic electronic spectra 179
    pure rotational spectra 174
    vibration-rotation bands 174–175
Molecular weight 232–236*
    colligative properties of solutions 234
    gram-molecular weight 236–237*
    importance of determination 233
    light scattering 235–236
    mass spectrometry 232–233
    measurement of auxiliary parameters 236
    miscellaneous determination methods 236
    sample purity 236
    sedimentation 234–235
    weighing of gases 233
Molecule 198*
Monolayer see Monomolecular film
Monomolecular film 111–117*
    chemical reactions 115–117
    experimental techniques 112–114
    transfer of spread monolayers 114–115

Mullikan, R.S. 186, 218

Nucleation oscillators 83
Opalescence 340–341*
    critical 341
    time dependency 341
Optical activity 332–335*
    circular dichroism 333
    correlation with molecular structure 333–335
    methods of measurement 332–333
    optical rotation 332–333
Optical rotation 332–333
Oscillatory reaction 81–84*
    nucleation oscillators 83
    reactions on surfaces 84
    redox oscillators 82–83
    thermokinetic oscillators 83–84
Osmosis 128–129*
    Donnan equilibrium 39
Oxidation: autoxidation 67
    indirect 67–69
Oxidation number 84–86
    valence 189
Oxidation-reduction 84–89*
    inorganic electrochemical process 265
    mechanisms 87–89
    organic electrochemical process 269–270
    oxidation number 84–86
    reactions 86–87
    redox oscillators 82–83

Paramagnetism: atomic 376–377
    molecular 377
Pasteur, L. 180
Pauling, L. 185
Periodic table: electron affinities 185
Perrin, J.B. 226
pH 50*
    buffers 75–77*
    isoelectric point 240*
Phase equilibrium 27–35*

INDEX **405**

Phase equilibrium (*cont.*):
binary systems 28–33
fused-salt *see* Fused-salt phase equilibria
liquid-gas system 28–30
liquid-liquid system 30–31
multicomponent systems 33–35
phase rule 39–40*
solid-liquid equilibrium 31–33
solid-state condition 33
triple point 239–240*
Phase rule 39–40*
derivation 40
examples 40
Phase-transfer catalysis 72–75*
applications 74–75
solid-state 75
Phase transitions: boiling point 237*
solid-state chemical reactions 387–388
transition point 238–239*
Photochemistry 342–347*
chain reactions 63
inorganic *see* Inorganic photochemistry
laser *see* Laser photochemistry
mechanisms 343
photochemical reactions 343–346
quantum yield 342–343
solid-state chemical reactions 389
theoretical aspects 346–347
Photolysis 347–349*
Gurney-Mott theory 347–348
large crystals 348–349
Pitzer, K.S. 182
pK 51*
Planck, M. 44, 226
Polarographic analysis 288–292*
apparatus 288
applications 290–291
evaluations 288–289

Polarographic analysis (*cont.*):
miscellanous polarographic techniques 291
polarographic curves 288
polarographically active species 289–290
related methods 291–292
solutions 288
Positional isomers 163–164
Primary battery 302–304*
aqueous-electrolyte batteries 302
dry cell *see* Dry cell
fused-electrolyte battery 304
lithium primary cell 318–320*
mercury battery 317–318*
solid-electrolyte battery 303
waxy-electrolyte batteries 304
wet cell *see* Wet cell

Quantum chemistry 360–365*
basis sets 361–362
definitions 360–361
energetic predictions 364–365
structural predictions 362–364
valence theory 189–190
Quasielastic light scattering 335–338*
Fabry-Perot interferometry 337
intensity fluctuation spectroscopy 336–337
miscellaneous applications 338
rotational diffusion coefficients 338
static light scattering 335–336
translational diffusion coefficients 337–338
Quincke, G.H. 278

Radiation chemistry 377–381*
energy transfer 378–380

Radiation chemistry (*cont.*):
radiation effects 381
representative processes 380–381
Redox oscillators 82–83
Reduction *see* Oxidation-reduction
Relative atomic mass 230*
Relative molecular mass 232*
Reserve battery 304–307*
electrolyte-activated batteries 305
gas-activated batteries 305–306
thermal batteries 306–307
water-activated batteries 305
Reuss, F.F. 274
Reversible processes: entropy 42
Rule of eight (valence) 189

Salem, L. 365
Schleyer, P.v.R. 366
Scott, R. L. 218
Second law of thermodynamics 10–12
Secondary battery *see* Storage battery
Semipermeable membrane: osmosis 128–129*
Solid: melting point 238*
Solid-liquid equilibrium 31–33
Solid solution 222–223*
occurrence 223
omission type 223
ordered arrangements 222–223
Solid-state chemistry 381–389*
atomic geometries 381–383
chemical bonding 383–385
chemical composition 385–386
chemical reactions 387–389
combination bonding 384
condensation reactions 387
covalent bonding 384
electrochemical reactions 388–389

Solid-state chemistry (cont.):
  electronic structures 383
  internal reactions in phase transition 387–388
  interphase reactions 388
  ionic bonding 384
  measurement of bonding 384–385
  metallic bonding 384
  photochemical reactions 389
  sublimation 389
  structures 381–383
Solid-state equilibrium 33
Solution 216–222*
  crystallization 224–225
  dipole interaction 217–218
  electron donor-acceptor interaction 218
  ideal 218–219
  ion-dipole interaction 218
  ion-ion interaction 218
  London forces 216–217
  molecular weight measurement 219–220
  nonideal 220–221
  osmotic pressure 220
  regular 221–222
  solid see Solid solution
  solubility of crystalline solid 219
  types of intermolecular force 216–218
Soret, J.L. 124
Specific heat 240*
Stereochemistry 179–184*
  configuration of stereoisomers 180–182
  conformational analysis 182–183
  optical activity 332–335
  resolution of enantiomers 182
  significance 180
  stereocontrolled synthesis 183–184
  symmetry of stereoisomers 180
Stereoisomers: diastereomers 166
  enantiomers 165–166

Stereoisomers (cont.):
  stereochemistry 179–184*
Steric effect 89–96*
  addition reactions 94–95
  electrophilic substitution 96
  elimination reactions 95–96
  rearrangement 91–93
  saturated nucleophilic substitution 89–91
  unsaturated nucleophilic substitution 93
Stimulated desorption 107–110
Storage battery 320–329*
  cell construction 323–325
  charging 326
  freezing of alkaline electrolyte 328
  freezing of electrolyte 325
  nickel-cadmium alkaline cells 327–328
  nickel-iron alkaline cell 326–327
  principles of operation 321
  reactants 321–323
  silver oxide-zinc-alkaline cell 328
  venting 328–329
Streaming potential 278–280*
Surface-active agent 99–101*
  classification 100–101
  mechanism 99–100
Surface tension 98–99*
  capillarity 98–99
  surface energy 98

Thenard, L.J. 69
Thermal desorption 810–811
Thermal diffusion 123–124
Thermochemistry 17–21*
  calorimetric measurements 19
  equilibrium constant 24–25
  heating values 20–21
  sources of data 20
  thermal chain reactions 63–64
  thermodynamic principles 17–19

Thermochemistry (cont.):
  units and symbols 19–20
Thermodynamics see Chemical thermodynamics
Thermokinetic oscillators 83–84
Third law of thermodynamics 16
Tiselius electrophoresis cell 274–275
Transition point 238–239*
Transport processes 120–122*
  diffusion see Diffusion in gases and liquids
  osmosis 128–129*
  thermal conduction 120–121
  thermal processes 121
  transport coefficients 121–122
  viscosity see Viscosity
Triple point 239–240*
Tyndall effect 339*

Valence 187–191*
  combining power of an element 188–189
  electronegativity 185–187*
  quantum theory 189–190
van der Waals, J. 199
van der Waals force 142
van't Hoff, J.H. 180
Viscosity 121, 129–132*
  flow behavior of complex fluids 132
  of liquids 130–131
  measurement 131–132
  molecular basis in gases 129–130

Waage, P. 24
Werner, A. 180
Wet cell 307–310*
  air-depolarized alkaline cell 308–309
  Lalande cell 307–308
  organic electrolyte cell 310
  Weston standard cell 309–310
Wohler, F. 163